Advances in the Assessment of Dietary Intake

Advances in the Assessment of Dietary Intake

Edited by

Dale A. Schoeller

Margriet S. Westerterp-Plantenga

CRC Press
Taylor & Francis Group
Boca Raton London New York

CRC Press is an imprint of the
Taylor & Francis Group, an **informa** business

CRC Press
Taylor & Francis Group
6000 Broken Sound Parkway NW, Suite 300
Boca Raton, FL 33487-2742

Printed on acid-free paper

International Standard Book Number-13: 978-1-4987-4932-9 (Hardback)

Library of Congress Cataloging-in-Publication Data

Names: Schoeller, Dale A. | Westerterp-Plantenga, Margriet S.
Title: Advances in the assessment of dietary intake / Dale A. Schoeller & Margriet S. Westerterp-Plantenga.
Description: Boca Raton : CRC Press, 2017. | Includes bibliographical references.
Identifiers: LCCN 2016053172 | ISBN 9781498749329 (hardback : alk. paper)
Subjects: LCSH: Nutrition surveys. | Nutrition. | Diet. | Biochemical markers.
Classification: LCC TX551 .S36 2017 | DDC 613.2--dc23
LC record available at https://lccn.loc.gov/2016053172

Visit the Taylor & Francis Web site at
http://www.taylorandfrancis.com

and the CRC Press Web site at
http://www.crcpress.com

Dedication

This book is dedicated to Dr. Ira Wolinsky, PhD, who conceptualized the aims of this book and then initiated the process leading to its publication.

Dr. Wolinsky passed away on November 26, 2014, after a long career as an investigator and an editor. He earned his doctorate in biochemistry from the University of Kansas, Lawrence, Kansas, and went on to investigate the role of bone metabolism and human health at Pennsylvania State University, State College, Pennsylvania. He later moved to the University of Houston, Houston, Texas, from which he retired as Professor Emeritus in 2005. From the very beginning of his scientific career, Dr. Wolinsky worked to apply the advances in basic science to the improvement of clinical science with respect to both treatment and prevention of human disease. During his PhD training at the University of Kansas, he applied the state-of-the-art biochemistry to increasing the health of human teeth. Later at Pennsylvania State University, he studied the interaction of skeletal muscle and bone to improve bone health. At the University of Houston, he examined the impact of weightlessness during space flight in order to prevent the loss of bone density.

Dr. Wolinsky published more than 40 journal articles and selected to expand the impact of original scientific research on the human condition by authoring and editing of more than 148 books. This included editing the influential CRC Series in Modern Nutrition, which he aimed toward a "scholarly audience ... to explain, review, and explore present knowledge and recent trends, developments, and the advances in nutrition." Although Dr. Wolinsky welcomed investigator-initiated book proposals for this series, he went beyond and monitored novel developments in nutrition recruiting others to create state-of-the-art volumes that enhanced and expanded upon those key basic developments.

Although I only knew Dr. Wolinsky through our correspondence, it was he who first engaged my efforts as a coeditor along with Dr. Helen Lane on a book that encapsulated the breadth of knowledge on nutrition that was necessary for spaceflight. In 2014, it was again because of Dr. Wolinsky's foresight and creativity that I was recruited to coedit this book in an effort to change the field of diet assessment. He contacted and convinced me that recent breakthroughs in technology were such that, if focused on diet assessment in the form of a state-of-the-art book, could help revolutionize the methods used for dietary assessment. Unfortunately, Dr. Wolinsky passed away before Dr. Margriet S. Westerterp-Plantenga and I, along with the many authors of Advances in the Assessment of Dietary Intake, could complete this book; however, because of his efforts as a creative and engaged editor, we dedicate this book to the memory of Dr. Wolinsky's achievements.

Dale A. Schoeller
Margriet S. Westerterp-Plantenga
Coeditors

Contents

Editors

Dr. Dale A. Schoeller is a director of the Isotope Ratio Laboratory in the Biotechnology Center and is the professor emeritus in Nutritional Sciences and the University of Wisconsin-Madison, Madison, Wisconsin. He was the first investigator to use the doubly labeled water method to measure total energy expenditure in humans, which led to the observation that self-reported dietary assessment instruments systematically underestimated dietary energy intake compared to the amount of energy that individuals were actually expending. This work by him and many others resulted in paradigm shift in validating and modeling the error structure in dietary assessment instruments. More recently, he collaborated with leading nutrition epidemiologists to use the doubly labeled water method as a biomarker of habitual energy intake to quantify systematic errors in dietary energy intake instruments forming the basis of his work that resulted in this book.

He earned his undergraduate degree from the University of Wisconsin-Milwaukee, Milwaukee, Wisconsin and his PhD in Chemistry from Indiana University, Bloomington, Indiana. He was trained as a post-doctoral fellow in Biomedical Sciences at the Argonne National Laboratory, Lemont, Illinois and joined the faculty of Medicine at the University of Chicago (Illinois, Chicago), where he directed the Stable Isotope Core Laboratory as part of the Clinical Nutrition Research Unit. For the last 20 years, he continued his research on the energetics of obesity at the University of Wisconsin-Madison. His research has been recognized through multiple awards including the Herman Award for Clinical Research from the American Society for Clinical Nutrition, the O.W. Atwater Award for Research from Agricultural Research Service, and the Friends of Mickey Stunkard Lifetime Achievement Award from the Obesity Society.

Dr. Margriet S. Westerterp-Plantenga is a professor of Food Intake Regulation in Humans, at Maastricht University Faculty of Health, Medicine and Life Sciences, School of Nutrim, Department of Human Biology, Maastricht, the Netherlands. In her productive career in human research, she has focused on the mechanisms of energy and reward homeostasis. This has involved diverse research on satiety and the role of orexigenic and anorexigenic hormones and glucose; the physiology of diet-induced thermogenesis and body-temperature regulation; circadian rhythm and sleep; and energy balance, macronutrient balance, and substrate oxidation. These studies have made major contributions to the understanding of food reward and related brain-signaling pathways and have done so with a strong perspective in their roles in the regulation of body weight and body composition.

She earned her undergraduate degree in Biology from the University of Groningen, Groningen, the Netherlands and her PhD degree in Biomedical Sciences from Maastricht University. She was a member of the Faculty Board of the Faculty of Health Sciences at Maastricht University and was responsible for the Research portfolio. She is currently an associate editor of the International Journal of Obesity and a member of the Editorial Board of Physiology and Behavior. She has been a project leader at the Leading Technological Institute of Food and Nutrition in the Netherlands and one of the leaders of several European projects, such as Diogenes, Full4Health, and at present PREVIEW. She has published about 240 papers in internationally peer-reviewed journals and authored and edited several books on food intake and energy expenditure.

Contributors

David B. Allison
Department of Nutrition Sciences, Nutrition Obesity Research Center
University of Alabama at Birmingham
Birmingham, Alabama

Vincent W. Antonetti
Department of Mechanical Engineering
Manhattan College
Riverdale, New York

John W. Apolzan
Clinical Nutrition and Metabolism
Pennington Biomedical Research Center
Baton Rouge, Louisiana

Gary Beecher
Beltsville Human Nutrition Research Center
Agricultural Research Service
United States Department of Agriculture
Beltsville, Maryland

Carl Brunius
Department of Food Science
Swedish University of Agricultural Sciences
Uppsala, Sweden

and

Department of Biology and Biological Engineering
Chalmers University of Technology
Göteborg, Sweden

Brooke Colaiezzi
Friedman School of Nutrition Science and Policy
Tufts University
Medford, Massachusetts

Muhammad Farooq
Department of Electrical and Computer Engineering
The University of Alabama
Tuscaloosa, Alabama

Hamid R. Farshchi
School of Life Sciences
Queen's Medical Centre
University of Nottingham
Nottingham, United Kingdom

Pietro Ferrari
Nutritional Methodology and Biostatistics Group
International Agency for Research on Cancer (IARC)
Lyon, France

Marta Garaulet
Department of Physiology
Faculty of Biology
Biomedical Research Institute of Murcia (IMIB-Arrixaca)
University of Murcia
Murcia, Spain

Lisa J. Harnack
Division of Epidemiology and Community Health
School of Public Health
University of Minnesota
Minneapolis, Minnesota

Keely R. Hawkins
Ingestive Behavior and Metabolism
Pennington Biomedical Research Center
Baton Rouge, Louisiana

Adam Hoover
Department of Electrical and Computer Engineering
Clemson University
Clemson, South Carolina

Reed W. Hoyt
Biophysics and Biomedical Modeling Division
U.S. Army Research Institute of Environmental Medicine
Natick, Massachusetts

Joseph J. Kehayias
Biophysics and Biomedical Modeling Division
U.S. Army Research Institute of Environmental Medicine
Natick, Massachusetts

Pekka Keski-Rahkonen
Nutrition and Metabolism Section, Biomarkers Group
International Agency for Research on Cancer
Lyon, France

Sharon I. Kirkpatrick
School of Public Health and Health Systems
University of Waterloo
Ontario, Canada

Rikard Landberg
Department of Food Science
Swedish University of Agricultural Sciences
Uppsala, Sweden

and

Department of Biology and Biological Engineering
Chalmers University of Technology
Göteborg, Sweden

and

Department of Medical Epidemiology and Biostatistics
Karolinska Institutet
Stockholm, Sweden

Ian Macdonald
School of Life Sciences
Queen's Medical Centre
University of Nottingham
Nottingham, United Kingdom

Ameneh Madjd
School of Life Sciences
Queen's Medical Centre
University of Nottingham
Nottingham, United Kingdom

Corby K. Martin
Pennington Biomedical Research Center
Baton Rouge, Louisiana

Ryan Mattfeld
Department of Electrical and Computer Engineering
Clemson University
Clemson, South Carolina

Holly L. McClung
Biophysics and Biomedical Modeling Division
U.S. Army Research Institute of Environmental Medicine
Natick, Massachusetts

Suzanne McNutt
Health Studies Sector
Westat, Inc.
Rockville, Maryland

Edward Melanson
Division of Endocrinology, Metabolism, and Diabetes/Division of Geriatric Medicine
University of Colorado
Aurora, Colorado

Eric Muth
Department of Psychology
Clemson University
Clemson, South Carolina

Candice A. Myers
Pennington Biomedical Research Center
Baton Rouge, Louisiana

Sarah H. Nash
Alaska Native Tumor Registry
Alaska Native Epidemiology Center
Anchorage, Alaska

Vanessa Neveu
Nutrition and Metabolism Section, Biomarkers Group
International Agency for Research on Cancer
Lyon, France

Diane M. O'Brien
Institute of Arctic Biology
University of Alaska–Fairbanks
Fairbanks, Alaska

Mark A. Pereira
Division of Epidemiology and Community Health
School of Public Health
University of Minnesota
Minneapolis, Minnesota

Ross L. Prentice
Fred Hutchinson Cancer Research Center
University of Washington
Seattle, Washington

Joseph A. Rothwell
Nutrition and Metabolism Section, Biomarkers Group
International Agency for Research on Cancer
Lyon, France

Edward Sazonov
Department of Electrical and Computer Engineering
The University of Alabama
Tuscaloosa, Alabama

Augustin Scalbert
Nutrition and Metabolism Section, Biomarkers Group
International Agency for Research on Cancer
Lyon, France

Dale A. Schoeller
Nutritional Sciences and Biotechnology Center
University of Wisconsin–Madison
Madison, Wisconsin

Phyllis Stumbo
Institute for Clinical and Translational Science
University of Iowa
Iowa City, Iowa

Amy F. Subar
Division of Cancer Control and Population Sciences
National Cancer Institute
Bethesda, Maryland

Moira A. Taylor
School of Life Sciences
Queen's Medical Centre
University of Nottingham
Nottingham, United Kingdom

Diana Thomas
Department of Mathematical Sciences
United States Military Academy
West Point, New York

Janet A. Tooze
Department of Biostatistical Sciences
Wake Forest School of Medicine
Winston-Salem, North Carolina

Klaas R. Westerterp
Department of Human Biology
Maastricht University
Maastricht, the Netherlands

Margriet S. Westerterp-Plantenga
Department of Human Biology
NUTRIM, FHML, MUMC+, Maastricht University
Maastricht, the Netherlands

Huaxing Wu
Department of Food Science
Swedish University of Agricultural Sciences
Uppsala, Sweden

Gary P. Zientara
Biophysics and Biomedical Modeling Division
U.S. Army Research Institute of Environmental Medicine
Natick, Massachusetts

Thea Palmer Zimmerman
Health Studies Sector
Westat, Inc.
Rockville, Maryland

Prologue

Margriet S. Westerterp-Plantenga and Dale A. Schoeller

WHY IS DIETARY ASSESSMENT NECESSARY?

Every day individuals are faced with multiple choices about what and how much to eat. Some of these choices are known to be associated with undernutrition, overfeeding, or chronic disease. How much these choices increase the dietary risk factors or lead to the potential beneficial factors in an individual or in populations has been revealed through dietary assessment, often in large-scale epidemiologic studies. Examples are numerous, and nutritionists employing dietary assessment have repeatedly demonstrated the importance of these choices.

Dietary research has reveled the impact of saturated fatty acids on low-density lipoprotein (LDL) cholesterol (Groen et al. 1952; Kinsell et al. 1952; Keys et al. 1965); in addition, the LDL cholesterol raising the effect of *trans* fatty acids was identified (Mensink and Katan 1990). These insights resulted in the producers reducing *trans* fatty acids in margarine. For the role of fatty acid consumption in cardiovascular diseases, dietary analysis was accompanied by newly developed objective measurements of intake such as the incorporation of fatty acids into blood cholesteryl esters, red blood cells, and fat tissue (Beynen and Katan 1985; Van Staveren et al. 1986; Katan et al. 1986, 1997). For the dietary bioactive phytochemical flavonoid, the assessment of intake led to insights in their role in the risk for coronary heart disease (Hertog et al. 1992, 1993a, 1993b, 1995; Rimm et al. 1996). Going into detail regarding fiber intake, determination of dietary fiber was developed, and among others, well-controlled intervention studies have shown that four major water-soluble fiber types (beta-glucan, psyllium, pectin, and guar gum) effectively lower serum LDL cholesterol concentrations, without affecting high-density lipoprotein (HDL) cholesterol or triacylglycerol concentrations. These findings underlie the current dietary recommendations to increase water-soluble fiber intake (Theuwissen and Mensink 2008). Foods containing plant sterol or stanol esters were shown to be useful for mildly hypercholesterolemic subjects (de Jong et al. 2003). Dietary assessments including comparisons between countries revealed differences in total cholesterol and HDL cholesterol levels in populations differing in fat and carbohydrate intake (Knuiman et al. 1987). Quite recently, among 9- to 11-year-old children in 12 countries, the home environment appeared to be more significant than the school environment for a healthy dietary pattern (Vepsäläinen et al. 2015).

Further research focused on the growing concern over excess body weight. Dietary studies have been performed, which found that high-fat diets in humans induced short-term hyperphagia and were associated with a higher adiposity (Tremblay et al. 1989). The fat content of the diet has an effect on body fat as a function of the effect of dietary fat on energy intake (Westerterp et al. 1996a), whereas dietary fat oxidation is a function of body fat (Westerterp et al. 2008). Although a low-glycemic index (GI) diet seems to modestly improve long-term glycemic control in subjects with type 2 diabetes (Brand et al. 1991), at present there is no evidence that low-GI foods are superior to high-GI foods in regard to long-term body weight control (Raben 2002). The applicability of the GI in the context of mixed meals and diets is still debatable, because the outcomes depend on the method used (Flint et al. 2004). Overweight individuals consuming artificial sweeteners instead of fairly large amounts of sucrose mostly as beverages did not increase their energy intake, body weight, fat mass, and blood pressure over 10 weeks, as their counterparts did, who consumed fairly large amounts of sucrose (Raben et al. 2002). In addition to general population studies, research questions aimed at promoting the performance in elite athletes by fine-tuning energy balance. Under extremely high-energy expenditures, such as in the Tour de France or Whitbread race around the world, it is actually difficult to ingest enough energy to prevent a negative energy balance, which would be disastrous for performance. High-energy drinks, consumed during the race, contribute significantly to daily intake (Westerterp et al. 1986). Sailors, during offshore races of several days or weeks, often fail to maintain energy balance due to their work on the deck, changing and trimming the sails, the continuous need to counterbalance movements of the boat, the watch system not allowing for a full night's sleep or a normal frequency, and the order of meals and activities (Branth et al. 1996). At high-altitude expeditions, it seems to be impossible to maintain the energy balance above 5000–5500 m. Maintenance of energy balance determines the success of these expeditions. Experienced negative energy balances are largely due to insufficient energy intake (Westerterp et al. 1992, 1994, 1996b, 2000). Hypoxia induces a negative energy balance by reduction of energy intake due to a more rapid increase in satiety and a decrease of hunger, probably caused by higher hypoxia-induced ventilation and diet-induced thermogenesis (Westerterp-Plantenga et al. 1999).

As evidenced previously, many of the advances in nutritional science have come from studies in which dietary assessment was central to the study. It should be noted, however, that epidemiologic studies of diet provided only part of the scientific story. Sometimes, the first notion came from the population dietary assessment data, which could be the last step in generalizing the findings.

WHY ARE NEW METHODS FOR DIETARY ASSESSMENT NECESSARY?

Although early methods for dietary assessment have evolved and improved since the 1930s, they still suffer from a lack of accuracy in reporting due to memory failure, response bias, and the daily variability in diet. The large meal-to-meal, day-to-day, and seasonal variations in food intake make it difficult to capture usual intake. Dietary reporting is inconvenient, and individual biases such as body image, weight consciousness, dietary restraint, and social expectations, may influence dietary reporting. Do the investigators help or hinder? Have they unwittingly contributed to the problem of misreporting? Do techniques such as the use of guided imagery or reminiscing, coercion, expectant looks, suggestive and repeated questioning, and conviction of importance of the survey further influence reporting accuracy?

On the positive side, accurate reporting of dietary intake has not been impossible to attain. Dieticians gave accurate reports of food intake of the Swedish Olympic cross-country skiing team during a training camp (Sjödin et al. 1994). After the first failing, dieticians learned to report their own food intake with full accuracy (Goris and Westerterp 1999, 2000). On the negative side, even in the hands of the most experienced nutritional epidemiologists, inaccuracies in dietary assessment remain (Freedman et al. 2014). Furthermore, even if reports were accurate, some research questions have been too specific to be approached using traditional dietary epidemiological methods. For instance, when the range of intakes of fatty acid distributions is small, and large-scale studies probably cannot find diet–health associations.

In the beginning of the 1980s, the emergence of biomarkers has been a major development for efforts to understand and increase the reliability of dietary assessment. Examples include 24 h urinary nitrogen (Huse et al. 1974; Isaksson 1980; Bingham and Cummings 1985) to validate protein intake, the PABA technique (Bingham and Cummings 1983) for recovery of intervention-induced changes of specific food items, fat biopsies (Dayton et al. 1966; Plakké et al. 1983) for determining the incorporation of fatty acids, and doubly-labeled water (Schoeller and van Santen 1982; Westerterp et al. 1986) for validating the energy intake under conditions of energy balance. With regard to the latter, major insights came through doubly-labeled water energy expenditure studies that have revealed that underreporting was endemic in dietary surveys (Livingstone et al. 1990), not method specific (Livingstone and Black 2003), mainly related to dietary fat intake in Western societies (Goris et al. 2000), and getting worse over time (Goris et al. 2001). Conversely, overreporting was identified, when caretakers reported on the patient's food intake (Simmons and Reuben 2000).

In the face of these shortcomings, experts still diverge in their opinions on dietary assessment and its traditional tools. Some advocate that "Something is not better than nothing" (Dhurandhar et al. 2015). Specifically with respect to the energy intake, they argue that "it is time to move away from the use of self-reports of energy intake as a research tool. They base this on the evidence that self-reports of energy intake are so poor that they are wholly unacceptable for scientific research on energy balance and that researchers and sponsors should develop objective measures of energy intake."

Other researchers admit that accuracy of reported or recalled absolute intakes fails, yet they believe that dietary patterns can be recognized, and thus, traditional dietary assessment should still be employed. It is added, however, that even if dietary assessment is imprinted by some inherent weaknesses, it should not be discredited and rejected for research purposes. Indeed, the information should be used with caution, should not be overinterpreted, and would be advantageously used with other supporting techniques. (Tremblay pers. comm. 2016). Many conclude that a multistep process should be used to reinforce the outcome of a traditional dietary assessment. Biomarkers and sensors are projected to add useful information that would compensate for shortcomings of traditional dietary assessment.

HOW PROMISING ARE THE NEW METHODS FOR DIETARY ASSESSMENT?

Expectations are that these shortcomings can be partly or completely circumvented during the twenty-first century. Biomarkers, improved sensors, and portable electronic technology are all under investigation. Photography may help to add information about ingredients, food preparation, and food consumption environment, including sociological and psychological aspects of eating. Three-dimensional scanning of foods may be obtained in canteens, restaurants, and shops to improve estimates of serving sizes, and a chip to measure nutrients in the blood is on the technical wish list. At the same time, updates of food composition databases (Murphy et al. 2016) and harmonization across cultures for international studies receive attention.

Reduction of the burden of the subjects, increase of acceptability and compliance, reduction of costs for the researchers, applicability to diverse population groups, higher data quality, consistency, completeness, and as such higher reliability of the outcome of dietary assessment are expected to lead to greater success in the assessment of dietary intake.

HOW RELEVANT IS ADVANCED DIETARY ASSESSMENT FOR CURRENT RESEARCH QUESTIONS?

With increasing health awareness, and numerous health-related advices, the question remains as to how detailed dietary assessment still can become more relevant. Current research questions are asked in the perspective of the leading causes of death: risks for cardiovascular diseases, chronic metabolic disorders, and cancer. Attention regarding the assessment of single nutrients seems to be shifting to attention on food patterns (Jacobs and Steffen 2003), regularity of meals (Pot et al. 2016), and a Mediterranean diet (Waijers et al. 2007; Knoops et al. 2004; Estruch et al. 2013; Berendsen et al. 2013; Santoro et al. 2014). Personalized nutrition is also a focus, with dietary assessment methods that are swift and on the spot, in an accessible attractive looking format, and with appropriate feedback and advice that are tailored to the individual's situation (Zeevi et al. 2015; Livingstone et al. 2016). To these ends, investigators are beginning to combine dietary assessment with a person's genetic profile, so that individuals may reduce their risk of dietary diseases as part of personalized medicine.

Dietary assessment, as currently performed, however, remains imperfect. Risk reduction for early mortality due to the leading chronic diseases in mainly Westernized societies remains a major focus for dietary assessment studies. Some success has been made, but further progress will require a wider perspective. Investigations have shown that not only food patterns but also the complete lifestyle are related to health and well-being. Thus, habitual physical activity is promoted along with regular timing of meals, activity, and sleep, and all are in line with the individual's circadian rhythm (Westerterp-Plantenga 2016); but exactly what food and exercise and how much of each remain controversial. To answer this, we need to learn to improve the accuracy and reliability of dietary assessment, and then how to disentangle that from other confounding lifestyle factors. Novel methods of dietary assessment are needed to augment the traditional dietary assessment methods.

ACKNOWLEDGMENTS

We thank Drs. Edith Feskens, Wageningen University, Wageningen, the Netherlands; Barbara Livingstone, Ulster University, Ulster, Ireland; Martijn Katan, Free University, Amsterdam, the Netherlands; Ronald Mensink and Klaas Westerterp, Maastricht University, Maastricht, the Netherlands; Angelo Tremblay, University of Laval, Quebec, Canada; Mikael Fogelholm, University of Helsinki, Helsinki, Finland; Anne Raben, University of Copenhagen, Copenhagen, Denmark; Nikhil Dhurandhar, Texas Tech University, Lubbock, Texas; Sally Poppit, University of Auckland, Auckland, New Zealand; Jennie Brand-Miller, University of Sydney, Sydney, Australia; Margarita Santiago Torrez, University of New Mexico, Albuquerque, New Mexico; Michele Mendez, University of North Carolina, Chapel Hill, North Carolina; Jean Jacques, Tufts University, Medford, Massachusetts; and Natalie M. Racine, University of Wisconsin-Madison, Wisconsin, for their valuable contributions.

REFERENCES

Berendsen, A., A. Santoro, E. Pini et al. 2013. A parallel randomized trial on the effect of a healthful diet on inflammageing and its consequences in European elderly people: Design of the NU-AGE dietary intervention study. *Mech Ageing Dev* 134(11–12):523–530.

Beynen, A. C. and M. B. Katan. 1985. Rapid sampling and long-term storage of subcutaneous adipose tissue biopsies for determination of fatty acid composition. *Am J Clin Nutr* 42:317–322.

Bingham, S. and J. H. Cummings. 1983. The use of 4-aminobenzoic acid as a marker to validate the completeness of 24 h urine collections in man. *Clin Sci* 64(6):629–635.

Bingham, S. A. and J. H. Cummings. 1985. Urine nitrogen as an independent validatory measure of dietary intake: A study of nitrogen balance in individuals consuming their normal diet. *Am J Clin Nutr* 42:1276.

Brand, J. C., S. Colagiuri, S. Crossman et al. 1991. Low-glycemic index foods improve long-term glycemic control in NIDDM. *Diabetes Care* 14(2):95–101.

Branth, S., L. Hanbraeus, K. Westerterp et al. 1996. Energy turnover during offshore sailing racing as studied by food inventories, doubly labeled water and anthropometry. *Med Sci Sports Exerc* 28:1272–1276.

Dayton, S., S. Hashimoto, W. J. Dixon et al. 1966. Composition of lipids in human serum and adipose tissue during prolonged feeding of a diet high in unsaturated fat. *J Lipid Res* 7:103.

Dhurandhar, N. V., D. Schoeller, A. W. Brown et al. 2015. Energy balance measurement: When something is not better than nothing. *Int J Obes* 39(7):1109–1113.

Estruch, R., E. Ros, J. Salas-Salvado et al. 2013. Primary prevention of cardiovascular disease with a Mediterranean diet. *N Engl J Med* 368(14):1279–1290.

Flint, A., B. K. Møller, A. Raben et al. 2004. The use of glycaemic index tables to predict glycaemic index of composite breakfast meals. *Br J Nutr* 91(6):979–989.

Freedman, L. S., J. M. Commins, J. E. Moler et al. 2014. Pooled results from 5 validation studies of dietary self-report instruments using recovery biomarkers for energy and protein intake. *Am J Epidemiol* 15:172–188.

Goris, A. H. and K. R. Westerterp. 1999. Underreporting of habitual food intake is explained by undereating in highly motivated lean women. *J Nutr* 129(4):878–882.

Goris, A. H. and K. R. Westerterp. 2000. Improved reporting of habitual food intake after confrontation with earlier results on food reporting. *Br J Nutr* 83(4):363–369.

Goris, A. H., E. P. Meijer, K. R. Westerterp. 2001. Repeated measurement of habitual food intake increases under-reporting and induces selective under-reporting. *Br J Nutr* 85(5):629–634.

Goris, A. H., M. S. Westerterp-Plantenga, K. R. Westerterp. 2000. Undereating and underrecording of habitual food intake in obese men: Selective underreporting of fat intake. *Am J Clin Nutr* 71(1):130–134.

Groen, J. J., B. K. Tjiong, C. E. Kamminga et al. 1952. The influence of nutrition, individuality and some other factors including various forms of stress, on the serum cholesterol: An experiment of nine months duration in 60 normal human volunteers. *Voeding* 13:556.

Hertog, M. G. L., P. C. H. Hollman, M. B. Katan. 1992. The content of potentially anticarcinogenic flavonoids of 28 vegetables and 9 fruits commonly consumed in the Netherlands. *J Agr Food Chem* 40:2379–2383.

Hertog, M. G. L., E. J. M. Feskens, P. C. H. Hollman et al. 1993a. Dietary antioxidant flavonoids and risk of coronary heart disease. The Zutphen elderly study. *Lancet* 342:1007–1011.

Hertog, M. G. L., P. C. H. Hollman, M. B. Katan et al. 1993b. Intake of potentially anticarcinogenic flavonoids and their determinants in adults in the Netherlands. *Nutr Cancer* 20:21–29.

Hertog, M. G., D. Kromhout, C. Aravanis et al. 1995. Flavonoid intake and long-term risk of coronary heart disease and cancer in the seven countries study. *Arch Intern Med* 155(4):381–386.

Huse, D. M., R. A. Nelson, E. R. Briones et al. 1974. Urinary nitrogen excretion as objective measure of dietary intake. *Am J Clin Nutr* 27:771–773.

Isaksson, B. 1980. Urinary nitrogen output as a validity test in dietary surveys. *Am J Clin Nutr* 33:4–5.

Jacobs, D. R., Jr. and L. M. Steffen. 2003. Nutrients, foods, and dietary patterns as exposures in research: A framework for food synergy. *Am J Clin Nutr* 78:508S–513S.

de Jong, A., J. Plat, R. P. Mensink. 2003. Metabolic effects of plant sterols and stanols (Review). *J Nutr Biochem* 14(7):362–369.

Katan, M. B., J. P. Deslypere, A. P. J. M. Van Birgelen et al. 1997. Kinetics of the incorporation of dietary N-3 fatty acids into cholesteryl esters, erythrocyte membranes and fat tissue—an 18-month controlled trial in man. *J Lipid Res* 38:2012–2022.

Katan, M. B., W. A. Van Staveren, P. Deurenberg et al. 1986. Linoleic and trans-unsaturated fatty acid content of adipose tissue biopsies as objective indicators of the dietary habits of individuals. *Prog Lipid Res* 25:193–195.

Keys, A., J. T. Anderson, F. Grande. 1965. Serum cholesterol response to changes in the diet: IV. Particular saturated fatty acids in the diet. *Metabolism* 14(7):776–787.

Kinsell, L. W., J. Partridge, L. Boling et al. 1952. Dietary modification of serum cholesterol and phospholipid levels. *J Clin Endocrinol Metab* 12(7):909–913.

Knoops, K. T., L. C. de Groot, D. Kromhout et al. 2004. Mediterranean diet, lifestyle factors, and 10-year mortality in elderly European men and women: The HALE project. *JAMA* 292(12):1433–1439.

Knuiman, J. T., C. E. West, M. B. Katan et al. 1987. Total cholesterol and high density lipoprotein cholesterol levels in populations differing in fat and carbohydrate intake. *Arteriosclerosis* 7:612–619.

Livingstone, M. B. and A. E. Black. 2003. Markers of the validity of reported energy intake. *J Nutr* 133:895S–920S.

Livingstone, K. M., C. Celis-Morales, S. Navas-Carretero et al. 2016. Effect of an Internet-based, personalized nutrition randomized trial on dietary changes associated with the Mediterranean diet: The Food4Me Study. *Am J Clin Nutr* 104(2):288–297.

Livingstone, M. B., A. M. Prentice, J. J. Strain et al. 1990. Accuracy of weighed dietary records in studies of diet and health. *BMJ* 300(6726):708–712.

Mensink, R. P. and M. B. Katan. 1990. Effect of dietary trans fatty acids on high-density and low-density lipoprotein cholesterol levels in healthy subjects. *N Engl J Med* 323(7):439–445.

Murphy, S. P., U. R. Charrondiere, B. Burlingame. 2016. Thirty years of progress in harmonizing and compiling food data as a result of the establishment of INFOODS. *Food Chem* 193:2–5.

Plakké, T., J. Berkel, A. C. Beynen et al. 1983. Relationship between the fatty acid composition of the diet and that of the subcutaneous adipose tissue in individual human subjects. *Hum Nutr Appl Nutr* 37A:365–372.

Pot, G. K., R. Hardy, A. M. Stephen. 2016. Irregularity of energy intake at meals: Prospective associations with the metabolic syndrome in adults of the 1946 British birth cohort. *Br J Nutr* 115(2):315–323.

Raben, A. 2002. Should obese patients be counselled to follow a low-glycaemic index diet? No. *Obes Rev* 3(4):245–256.

Raben, A., T. H. Vasilaras, A. C. Møller et al. 2002. Sucrose compared with artificial sweeteners: Different effects on ad libitum food intake and body weight after 10 wk of supplementation in overweight subjects. *Am J Clin Nutr* 76(4):721–729.

Rimm, E. B., M. B. Katan, A. Ascherio et al. 1996. Relation between intake of flavonoids and risk for coronary heart disease in male health professionals. *Ann Intern Med* 125:384–389.

Santoro, A., E. Pini, M. Scurti et al. 2014. Combating inflammaging through a Mediterranean whole diet approach: The NU-AGE project's conceptual framework and design. *Mech Ageing Dev* 136–137:3–13.

Schoeller, D. A. and E. van Santen. 1982. Measurement of energy expenditure in humans by doubly labeled water method. *J Appl Physiol Respir Environ Exerc Physiol* 53(4):955–959.

Simmons, S. F. and D. Reuben. 2000. Nutritional intake monitoring for nursing home residents: A comparison of staff documentation, direct observation, and photography models. *J Am Geriatr Soc* 48(2):209–213.

Sjödin, A. M., A. B. Andersson, J. M. Högberg et al. 1994. Energy balance in cross-country skiers: A study using doubly labeled water. *Med Sci Sports Exerc* 26(6):720–724.

Theuwissen, E. and R. P. Mensink. 2008. Water-soluble dietary fibers and cardiovascular disease. *Physiol Behav* 94(2):285–292.

Tremblay, A., G. Plourde, J. P. Despres et al. 1989. Impact of dietary fat content and fat oxidation on energy intake in humans. *Am J Clin Nutr* 49(5):799–805.

Van Staveren, W. A., P. Deurenberg, M. B. Katan et al. 1986. Validity of the fatty acid composition of subcutaneous fat tissue microbiopsies as an estimate of the longterm average fatty acid composition of the diet of separate individuals. *Am J Epidemiol* 12:455–463.

Vepsäläinen, H., V. Mikkilä, M. Erkkola et al. 2015. Association between home and school food environments and dietary patterns among 9–11-year-old children in 12 countries. *Int J Obes* 5(Suppl2):S66–S73.

Waijers, P. M., E. J. Feskens, M. C. Ocke. 2007. A critical review of predefined diet quality scores. *Br J Nutr* 97(2):219–231.

Westerterp, K. R., A. Smeets, M. P. Lejeune et al. 2008. Dietary fat oxidation as a function of body fat. *Am J Clin Nutr* 87:132–135.

Westerterp, K. R., B. Kayser, F. Brouns et al. 1992. Energy expenditure climbing Mt. Everest. *J Appl Physiol* 73:1815–1819.

Westerterp, K. R., B. Kayser, L. Wouters et al. 1994. Energy balance at high altitude: 6542m. *J Appl Physiol* 77:862–866.

Westerterp, K. R., E. P. Meijer, M. Rubbens et al. 2000. Operation Everest III: Energy and water balance. *Eur J Physiol* 439:483–488.

Westerterp, K. R., P. Robach, L. Wouters et al. 1996b. Water balance and acute mountain sickness before and after arrival at high altitude: 4350m. *J Appl Physiol* 80:1968–1972.

Westerterp, K. R., W. H. M. Saris, M. Van Es et al. 1986. Use of doubly labeled water technique in man during heavy sustained exercise. *J Appl Physiol* 61:2162–2216.

Westerterp, K. R., W. P. H. G. Verboeket-van de Venne, M. S. Westerterp-Plantenga et al. 1996a. Dietary fat and body fat: An intervention study. *Int J Obes* 20:1022–1026.

Westerterp-Plantenga, M. S. 2016. Sleep, circadian rhythm and body weight: Parallel developments. *Proc Nutr Soc* 27:1–9.

Westerterp-Plantenga, M. S., K. R. Westerterp, M. Rubbens et al. 1999. Appetite at "high altitude" [Operation Everest III (Comex-'97)]: A simulated ascent of Mount Everest. *J Appl Physiol* 87:391–399.

Zeevi, D., T. Korem, N. Zmora et al. 2015. Personalized nutrition by prediction of glycemic responses. *Cell* 163(5):1079–1094.

1 Benefits and Limitations of Traditional Self-Report Instruments

Hamid R. Farshchi, Ian Macdonald, Ameneh Madjd, and Moira A. Taylor

CONTENTS

1.1 INTRODUCTION

Dietary intake is an important and potentially modifiable determinant of health. Associations, in some cases demonstrated to be causal, have been reported between dietary intake and morbidity and mortality, for example, relating to cardiovascular diseases (Mente et al. 2009), cancer (Schwingshackl and Hoffmann 2015), diabetes (Khazrai et al. 2014), and obesity. Accurately assessing dietary intake, including the pattern and quantity of food, drinks, and supplement consumption is a fundamentally important element of nutrition and health research, across the spectrum of surveillance, epidemiological, and interventional studies. It is equally important in clinical practice, to be able to accurately establish an individual's current dietary intake, whether prior to or after giving dietary advice. Having confidence in the techniques used to assess dietary intake is essential if existing research outcomes are to be considered valid, and in order to justify funding for further techniques.

However, dietary intake assessment is not without challenges (Thompson et al. 2015b). Some caution that traditional techniques may provide incorrect data that may result in misguided health care policies, future research, and clinical judgment (Dhurandhar et al. 2015). Comparisons between self-reported intake and estimated energy expenditure measured using doubly-labeled water (DLW) have shown individual differences ranging from +25% to −76% (Schoeller et al. 1990), which assuming stable body weight, raises questions about their accuracy. It is thus essential that traditional methods are critically reviewed in order to inform whether there is justification for their continued use, or to identify priority areas for investing in the development of alternative techniques.

Dietary intake is a chronic and multidimensional phenomenon that may vary over time in relation to age, life stage, environment, and many other factors (Thompson et al. 2015b). Assessment methods are potentially required to establish nutrient intake, food choices, or broader patterns of eating, hence are attempting to measure what, when, how much, and how often foods and drinks are consumed. Further steps may then need to be taken to establish, for example, an estimate of nutrients using a food composition database. Food composition tables have many inherent limitations, for example, some foods may not be included, or the composition of the foods analyzed may differ from the foods consumed (Chapter 19) (Cantwell et al. 2006).

Many different methods have been developed for assessing dietary intake that vary with respect to cost and the burden placed on the researcher and respondent. Nutritional biomarkers of intake have been proposed but are not without limitations, such as weak associations with intake, and only providing information about one dietary component. (Thompson et al. 2010; Freedman 2011) (Chapters 15, 16, and 18). Nutrition researchers and practitioners thus may choose to rely on more traditional, self-reported measures. This chapter describes traditional methods for self-reported dietary intake assessment, considers benefits and limitations, and focuses on key issues in evaluating traditional dietary intake assessment methods.

1.2 SELF-REPORTED METHODS FOR ASSESSING DIETARY INTAKE

These methods include food records, 24-hour dietary recall (24HR), dietary history, and food frequency questionnaires (FFQs). Data are collected with the assistance of a trained interviewer or by self-report (after some or no training).

1.2.1 Food Records

A food record consists of a diary, in which participants record the type and amount of food and beverages consumed during a particular number of days, usually four to seven days. Participants in order to minimize reliance on memory are advised to record at the time of consumption, and this is thus viewed as a prospective method. The amounts consumed may be measured using a scale (i.e., weighed food record), or recorded using household measures (i.e., estimated food record). The former involves either a simple scale, or more sophisticated recording device that automatically records the weight and the respondent's verbal description of the food, early examples including the PETRA balance (Boutelle and Kirschenbaum 1998). In the latter, measures such as cups, tablespoons, or photographs of specific food portions may be used in an attempt to improve the estimate for the weight of food or drink consumed (Nelson et al. 1997). Food and drink may be photographed prior to and after eating to support interpretation, subsequently of the food intake record. In both methods, left-over items should be recorded too so that the weight of food or drink actually consumed can be calculated.

Generally multiple days are recorded. These are usually consecutive but not necessarily. Recording periods of more than four sequential days are usually nonproductive, as reported intakes decrease (Thompson et al. 2015b) because of participant's exhaustion. Respondents may adapt their intake or accuracy of recording differently, while keeping records, introducing interindividual confounding. Successive days may not be independent of each other, for example, if left-over foods are used, or if a day of high consumption is followed by a day of lower consumption. Nonsuccessive days may provide a more representative view of typical consumption patterns.

To complete a dietary record, each respondent must be trained in the level of detail required in order to adequately describe the foods and amounts consumed, including the name of the food (brand name, if possible), preparation and cooking methods, recipes for home prepared foods, and portion sizes. This places a considerable burden on the respondent. Dietary records may be recorded by someone else, such as parents reporting for their children, or staff in a residential care home providing that they are able to witness when food or drink is consumed.

Dietary records are usually collected using an unstructured diary. For example, there should not be sections for different meals, as this potentially communicates to the respondent that they should have such a meal pattern, which may introduce bias. However, some diaries are more structured (Schoeller et al. 1990; Black et al. 1997; Cantwell et al. 2006; Thompson et al. 2010). These forms contain checklists consisting of food groups, which the respondent reports on whether he or she consumed foods from a particular food group. The design of these checklists is similar to FFQs, but the time of the food intake report consumed is different between these two methods. In FFQ (described in a later section), subjects are asked about intake over a definite time period like the past year or month, while checklists should be completed at the time of intake or at the end of a day. A checklist may be designed to evaluate specific *main foods* that contribute markedly to consumptions of some nutrients (Kolar et al. 2005), and also to track food contaminants (Rebro et al. 1998). A high level of motivation is needed in the respondents, or those who complete the record on their behalf (Willett 1998).

Some protocols insist that the food record should be reviewed by a dietitian or other trained interviewer after one day of recording, and again once the record has been completed. The purpose is to prompt the recall of foods that may have been omitted (e.g., confectionery), clarify the interpretation of portion sizes, and add details such as the type of fat used, and amount of salt and sugar added at the point of consumption (Cantwell et al. 2006). Other protocols may not require this step (Kolar et al. 2005). The foods and drinks in the food record may then be matched to foods in a food composition database that are usually accessed with the assistance of computer software, and the program is used to derive variables such as the mean daily intakes by weight of a micronutrient, or percentage of total energy for a particular macronutrient. Data entry is time consuming, and it may be necessary to add foods to the database, having obtained information about composition from packets or direct from the manufacturer.

1.2.1.1 Benefits and Limitations of Food Records

In practical terms, food records have a range of benefits. First, because the format tends to be unstructured, plentiful information can be obtained from groups with diverse eating habits and then it can be analyzed using a multitude of techniques to address different research questions. Concurrent information may be collected, for example, location and mood at the time of eating. Providing records are completed at the time, as requested, there is a minimal reliance on the memory of the respondent. Where the participant is unable or unwilling to complete the record, it can be done by a proxy (e.g., to establish the intake of a child or patient with dementia). Completing a food record requires a minimum of equipment, facilitating completion of the record away from home and with minimal expense (with the exception of more sophisticated recording devices).

However, there are also limitations. The awareness that the type and amount of food must be recorded at the point of consumption may influence both the food selected and the amounts consumed (Rebro et al. 1998; Willett 1998; Vuckovic et al. 2000). This is a disadvantage when the goal is to assess typical (usual) dietary behaviors. However, during interventional studies in which the goal is to improve the awareness of dietary behaviors and change them, this effect can be considered an advantage (Glanz et al. 2006). Recording food intake, *per se*, supports effective weight loss (Boutelle and Kirschenbaum 1998; Hollis et al. 2008).

Selecting an appropriate number of recording days can result in a dilemma when the purpose is to capture habitual intake. Day-to-day variability in food consumed means that several days of records are needed to establish habitual intake (Bingham 1987). The number of days required depends on frequency of food consumed, and will be more important in some cases than others. Liver, for example, while infrequently consumed in the United Kingdom, is a rich source of iron, a mineral that can be stored. Insufficient days of recording that do not capture the consumption of this particular food, might erroneously lead to the conclusion that an individual's iron intake was inadequate. However, incomplete records increase significantly as more days of records are kept, and the validity of the collected information declines in the later days of a seven-day recording period, in contrast to collected information in the earlier days (Gersovitz et al. 1978).

Food records are unsuitable for individuals who are unable to read and write in the language selected for the assessment. Although translation may be an option, where basic literacy levels are low, issues remain. Respondents must be cooperative and motivated for this method to have validity as there is a large respondent burden (see summary in Table 1.1).

TABLE 1.1
Summary of Benefits and Limitations of Traditional Self-Report Instruments for Dietary Intake

	Benefits	Limitations
Food records	• Can provide typical meal and food pattern information • Recorded at the time of eating with no reliance on memory providing reliable data intake • Intake is quantified • Recordings made away from home suit for people with irregular lifestyle habits • Allows self-monitoring that can influence behavior change	• Can alter eating behaviors • May influence on both the food type and eaten amount • Requires multiple records to capture habitual intake • High subject burden • High staff cost and burden • Requires literate and motivated population
24-hour dietary recall	• Intake is quantified • Less subject burden • Does not alter eating behaviors • Does not require literate population for interviewer provided version	• Less accurate measurements of portion size • Relies on subject recall • High staff cost and burden • Requires multiple recalls over several months to capture habitual intake
Diet history	• Evaluate meal patterns and details of food intake • Catch details of the methods of food preparation • Does not require literacy • Less subject burden	• Rely on subject memory • May overreport *good* foods and underreport intake of *bad* foods • Quantify the relative amount and not the absolute intake • Not suitable for people with irregular meal pattern • High staff cost and burden
Food frequency questionnaires	• Less subject burden • Less staff cost and time burden • Does not alter eating behaviors • Captures habitual intake • Can be shorter and emphasis on foods rich in a specific nutrient or a specific group of foods, for example, fruits and vegetables	• Relies on subject recall • Not as quantifiably precise • Requires literate population • Does not provide meal pattern information • Bias with overestimation consumption of *good* foods or underestimation of *bad* foods

1.2.2 24-Hour Dietary Recall

The 24HR is a retrospective assessment method in which an interviewer prompts a respondent to recall and describe all foods and beverages consumed during the previous 24 hours or in the past day. The interview may be conducted face-to-face or by telephone (Buzzard et al. 1996; Casey et al. 1999), and may be with paper and pencil, or computer assisted (Vereecken et al. 2008; Arab et al. 2011). There is evidence of no significant differences in the dietary data obtained by face-to-face 24-hour recalls compared to telephone 24-hour recalls (Brustad et al. 2003).

Well-trained interviewers are essential because much of the dietary information are collected by asking investigative questions. Ideally, interviewers would be nutritionists; however, nonnutritionists who have been trained in the use of a standardized instrument can be effective. All interviewers should be well informed about foods available in the marketplace and about preparation practices, including prevalent regional or traditional foods.

As a retrospective method, it relies on an accurate memory of intake, reliability of the respondent not to under/misreport, and an ability to evaluate portion size, which is a limitation (Chapter 10). The interview is often structured, usually with specific probes, to help the respondent remember all foods consumed throughout the day. During the interview, the respondent may be encouraged to remember eating and drinking events by time periods, for example, starting with *on awakening* or relating to day time activities such as *arriving at work*. Probing is particularly valuable in collecting essential details (e.g., how foods were prepared). It also enables exploration of items not initially reported, such as common additions to foods (e.g., butter on toast) and eating episodes not firstly reported (e.g., snacks and beverage breaks). In addition, the interviewer may use prompts to assist the respondent to estimate portion sizes. A previous study showed that with interviewer probing, respondents reported 25% higher dietary intakes than respondents did without interviewer probing (Campbell and Dodds 1967). However, interviewers should use standardized neutral-probing questions in order to avoid leading the respondent to specific answers when they really do not know or remember. 24-hour recalls are particularly challenging in young children (Fisher et al. 2008) and the elderly (Tooze et al. 2007). A child's ability to provide a dietary recall noticeably rises after the age of eight years (Livingstone et al. 2003).

Once the 24-hour recall has been completed, foods and drinks are coded and analysis undertaken similar to that described in a food intake record. Recording consumption for a single day is rarely representative of an individual's usual intake due to day-to-day variation. Repeat 24-hour recall, also known as multiple recalls, can be used to assess a typical diet especially at an individual level. A single 24-hour recall, however, is more suitable for measuring intake in a large group and estimating group mean intakes. The number of recalls needed varies depending on the focus of the research, variability of the population, and the nutrients of interest. For example, studies evaluating diet and disease outcomes recommend 7–14 days depending on the nutrients (Hartman et al. 1990), whereas four repeat 24-hour recalls were recommended as the most appropriate method of dietary assessment in the UK Low Income Diet and Nutrition Survey (LIDNS) (Holmes et al. 2008a, b). In an Australian study in adults, eight repeat 24-hour recalls were recommended to investigate the variation in macronutrient intake (Jackson et al. 2008). In the United States, an interviewer-administered automated multiple-pass recall (AMPR), is the dietary assessment method used in National Health and Nutrition Examination Survey (NHANES) (Raper et al. 2004; Subar et al. 2007) to assess the diets of children and adults. The diet is evaluated with this tool over a period of three to five days. The multiple-pass recall (MPR) is a staged approach to the dietary recall. The precise number of stages or passes may differ between protocols, but they all monitor the pattern of a free and continuous recall of intake, followed by detailed and probing questions about intake (including quantities consumed) and concluding with a review of everything that was previously recalled.

1.2.2.1 Benefits and Limitations of 24-hour Dietary Recall

A benefit of this method is that when an interviewer applies the tool and records the responses, literacy of the respondent is not required. However, for versions completed by the participant, poor literacy can be a limitation. The immediacy of the recall period facilitates recall of foods and drinks; however, the method does rely on memory that may challenge young children and the elderly. The burden on the respondent is minimal, so the 24-hour recall method is useful across a wide range of populations. The 24HR is frequently used in national surveys (Kweon et al. 2014), randomized clinical trials, and cohort studies (Dauchet et al. 2007; Luke et al. 2011).

Furthermore, in contrast to the food record method, dietary recalls occur after the food has been consumed, so there is likely to be less impact on dietary habits. Finally, 24HR can be conducted by telephone or a web-based automated self-administered method (Thompson et al. 2015a) with minimal differences in comparison with interviewer-administered recalls.

However, on the other hand, potential limitations of the 24-hour recall method include inaccuracies in the assessment of portion size and individual food intake for a wide range of reasons relating to knowledge, memory, and the interview conditions. In addition, as a single observation provides a poor measure of individual intake, so multiple days of recalls may be needed. Although a single 24-hour recall can be used to describe the average dietary intake of a population, multiple recalls can improve precision (Blanton et al. 2006). Multiple days of recall inevitably increase cost and the burden placed on the respondent (see summary in Table 1.1).

1.2.3 Diet History

The diet history is another common method used in epidemiological studies (Heitmann et al. 1995; Van Staveren et al. 2002; Eiben et al. 2004; Rothenberg 2009) and clinical practice. To evaluate individual long-term dietary intake, Burke established a dietary history method in 1947, which is a detailed retrospective dietary assessment used to estimate a person's typical food and/or nutrient intake over a period of typically six months to one year. This method is furthermore used to determine the frequency and amount of food intake. However, it may not capture the entire intake of a particular nutrient of interest as not all foods consumed are probed.

A dietary history is a structured interview method. The traditional Burke diet history included three features: a comprehensive interview about usual pattern of consumption, a food list probing for quantity and usual frequency of consumption, and a three-day dietary record (Burke 1947), which either may not be used or may be replaced by a 24-hour recall. The detailed interview (that occasionally contains a 24-hour recall) is the essential feature of the Burke dietary history, with the food frequency checklist and the three-day diet record used as cross-checks of the history. Many variations of the Burke method have attempted to determine the usual eating patterns for an extended period of time, including type, frequency, and amount of foods consumed; many include a cross-check feature (Bloemberg et al. 1989). Highly expert professionals are required to collect information on the participant's usual diet using a detailed interview (approximately 90 minutes to complete). So, this method is used more in clinical practice instead of research and epidemiological studies.

1.2.3.1 Benefits and Limitations of Diet History

The major benefit of the diet history method is its evaluation of meal patterns and details of food intake rather than intakes for a short period of time (as in food records or 24HR), or for identifying the frequency of food consumption. Details of the methods of preparation of foods can be useful in better describing nutrient intake, such as frying versus baking, as well as exposure to other factors in foods (e.g., salt and flavors).

When the information is collected as discrete meals, analyses of the combined effects of foods eaten together are possible, such as effects on iron absorption of concurrent intake of tea. Although a meal-based method often needs more time from the respondent than does a food-based method, it may provide more cognitive support for the recall process. For example, the respondent may precisely report the total bread consumed by reporting bread typically eaten at each meal. Literacy is not required for individuals providing a diet history. This method may cover usual diet in detail, so only one interview is required for the specific time period.

However, respondents are asked to make many judgments about both the usual foods and the amounts of those foods consumed. So, respondents may overreport *good* foods and underreport intake of *bad* foods according to their perceptions. In addition, these subjective tasks may be difficult for many respondents. Burke cautioned that nutrient intakes estimated from these data should be interpreted as relative rather than absolute. The meal-based approach is not useful for individuals who have no particular eating pattern and may be of limited use for individuals who *graze* (i.e., eat throughout the day rather than at defined mealtimes) or shift workers who have varied meal patterns.

A further limitation of this method is reliance on the memory of the respondent that may result in recall bias. This may be a particular issue for older individuals who may become fatigued and unable to complete the interview in one session; a typical session lasts 60–90 minutes. All of these limitations are also shared with the food frequency method.

Given the need for a trained interviewer, the approach tends to be expensive and vulnerable to inconsistency in the exact method used. Poor standardization has resulted in difficulties in reproducing studies, and in comparing between studies (see summary in Table 1.1).

1.2.4 Food Frequency Questionnaire

The food frequency method interrogates respondents about their usual frequency of consumption of each food from a list of foods for a definite period, and therefore aims to obtain habitual intake (Zulkifli and Yu 1992; Willett 1998).

FFQs, generally, are designed to collect dietary information from larger populations (100 people or more) and are normally self-conducted, though interviewer-conducted and telephone interview are alternative approaches (Haraldsdóttir et al. 2001) (Chapter 4). The size of the food list may vary according to the nutrients or foods under consideration. If a variety of different nutrients and energy values are needed, the list of foods may be up to 150, while for specific, limited nutrients, the FFQ may contain just a few food items.

Using an appropriate food list is an essential issue in this method (Block et al. 1986). The whole range of an individual's diet, which includes many different foods, brands, and preparation practices, cannot be fully captured with a finite, predetermined food list. Obtaining precise reports for foods consumed both as single items and in mixtures is challenging.

FFQs usually deliver data on frequency, with minimal additional information about other components, such as the methods of cooking, or how foods are combined in meals. To improve the validity of the FFQ, researchers have developed a range of refinements, such as markers of seasonality. FFQs designed to evaluate total diet usually list for more than 100 specific line items, which take 30–60 minutes to complete. The degree of details sought may be inversely reflected in the response rate, although this has not been noted by all (Johansson et al. 1997; Morris et al. 1998; Subar et al. 2001).

A further refinement relates to gathering more detailed information about portion size in addition to frequency of consumption via semiquantitative FFQs. In most circumstances, the purpose of an FFQ is to obtain a rough estimate of total intakes during a designated time period. However, there is a debate whether portion size inquiries should be included in FFQs. Frequency has been established to be a greater contributor than serving size to the alteration in intake of most foods (Heady 1961); hence, some researchers favor FFQs without the extra respondent burden of recording serving sizes (Willett 1998). On the other hand, others report that including usual serving size for each food may improve the performance of FFQs (Block et al. 1992).

The time period over which the inquiry relates is an important factor to consider. Some tools ask about usual intakes during the past year (Rimm et al. 1992), but it may be asked about the past week (Dunn et al. 2011) or months (Segovia-Siapco et al. 2007), subject to designed research conditions. When investigating the previous year, the season in which the FFQ is administered may impact on the overall results (Fowke et al. 2004).

Finally, it is important to consider how missing data will be treated, especially in self-administered questionnaires. Zero values may be entered; having made the assumption that failure to answer questions relates to foods that respondents rarely or never consume (Rimm et al. 1992). Imputation of frequency values for unanswered questions is the other solution for this issue (Parr et al. 2008). However, it is now unclear whether imputation improves FFQ analyses (Fraser et al. 2009).

As a tool designed to obtain the habitual dietary intake, FFQs have been used commonly in epidemiological studies evaluating relations between diet and disease. In this case the aim may be to rank the intake of respondents relative to others, or as quantiles, as opposed to determining absolute intake. Two well-known FFQs (Willett et al. 1985; Block et al. 1986) with some adaptations have been used in many national studies (Lamb et al. 2007; Barclay et al. 2008).

1.2.4.1 Benefits and Limitations of Food Frequency Questionnaire

FFQ is a common tool to estimate usual dietary intake in large epidemiological studies due to the lower cost and time burden of data collection and processing on respondents and investigators than other dietary assessment methods. This tool may be self-administered via e-mail or the Internet that makes it relatively easy to administer compared with other methods.

FFQ may be used to assess habitual consumption over an extended period of time without impacting on eating behaviors. Semiquantitative portion-size estimates can be used to obtain absolute nutrient intakes. This tool may also be used to collect data on a variety of foods, or designed to be shorter with an emphasis on foods rich in a specific nutrient or a specific group of foods (e.g., fruits and vegetables).

On the other hand, FFQs have limitations, principally in terms of the measurement error (Prentice et al. 2011). Many components of food intake are not measured, and the quantification of intake is considered less accurate than recalls (Schatzkin et al. 2003), or records (Yang et al. 2010).

As this method relies on the memory of respondents attempting to estimate usual portion size, the accuracy of the data may suffer due to highly variable portion sizes across eating events (Hunter et al. 1988). On account of the inevitable error in the food-frequency approach, it is generally considered inappropriate to use FFQ data to estimate quantitative parameters, such as the mean and variance, of a population's usual dietary intake (Rimm et al. 1992; Carroll et al. 1996). Accordingly, evidence has suggested that FFQ data might be combined with recall or record data to improve estimated intakes (Haubrock et al. 2011; Carroll et al. 2012).

Bias may also be presented with respondents reporting eating *good* foods more often (overestimation), or the consumption of *bad* foods less frequently (underestimation). In addition, a fairly high grade of literacy and numeracy skills are necessary if self-administered, but interviewers can help overcome this problem.

As mentioned, self-administered FFQs may not be completed fully, which may cause problems with interpreting questions.

Due to variability in dietary habits and food composition in different areas, FFQs developed in one region, country, or for a specific subpopulation are unlikely to be suitable for use in another population. Critical to the process of developing a FFQ is that it is compared against other methods such as serial 24-hour recalls, food records, or biomarkers. Clearly the method is thus only as accurate as alternative methods used in development, and against what it is compared (Shahar et al. 2003) (see summary in Table 1.1).

1.3 KEY ISSUES IN EVALUATING DIETARY INTAKE ASSESSMENT METHODS

A range of traditional methods of dietary assessment have thus been described and their respective strengths and limitations considered. It is clear that there is no perfect method of assessing dietary intake that can be viewed as a *gold standard* against which others can be compared. When comparisons are made, different outcomes are found (Bingham 1987; Freedman et al. 2014). However, it is not known which result is closer to the *true intake*. In some individuals with a very varied diet this may be particularly difficult to capture. The following section considers, in more depth, issues that contribute to the growing concerns about the ability of traditional methods to deliver what is required of them.

1.3.1 Misreporting of Foods and the Amounts Consumed

Foods or drinks may be omitted, or inserted either deliberately, or due to a lapse in memory. Where portion size is included, this may be under or overestimated, again either deliberately or due to an inability to estimate portion size. The resulting estimate of food intake may be either an under or overestimate, either of which will result in incorrect conclusions, for example, if energy or nutrient intake is calculated from the food intake assessment (Boutelle and Kirschenbaum 1998).

The estimation of portion size has been known as a cause of error in studies measuring dietary intake for decades (Young et al. 1953). There is evidence that the errors related to calculating the portion of food eaten may be the largest quantifying error in most dietary assessment tools (Gibson 2005). Literacy, but not numeracy, has been showed as a main factor in a person's skill to precisely identify portion sizes (Huizinga et al. 2009). In addition, foods that are commonly eaten in distinct portions such as bread by the slice and beverages in cans or bottles may be more accurately reported than foods such as meat, and rice, or poured liquids (De Keyzer et al. 2011). Small portion sizes are also more prone to be overestimated, while large portion sizes are underestimated (Harnack et al. 2004; Ovaskainen et al. 2007). Inconsistencies may vary with the type and size of food (Gibson 2005). Large errors may happen, for instance, when estimating foods high in volume but low in weight (Gittelsohn et al. 1994).

Tools are available to try and improve accurate estimation of portion size including food photographs and food models (Cameron and van Staveren 1988; Nelson et al. 1997). Food models have been used extensively in large national studies, but there is debate regarding the accuracy of those reports especially in specific groups like children. Food models resulted in a somewhat larger error than using digital images (Foster et al. 2008). Digital technology can help to substitute for respondent judgment of portion size by assessing food consumption via digital photography. Computer software can also be applied to recognize food items and also to estimate the amount consumed (Chae et al. 2011) (Chapter 5). With an increasing use of preprepared foods in some cultures, there may be a reduction in food skills including, portion-size awareness, which may potentially make this a particularly important area for development.

1.3.2 Attempts to Quantify the Degree of Error in Dietary Intake Assessment

Identifying an inaccurate assessment of foods/drinks consumed and amounts is difficult, and once suspected, is difficult to quantify. Most evidence that methods of dietary intake assessment are inaccurate and relate to energy intake (EI) (Poslusna et al. 2009). Only a few studies have been designed to evaluate misreporting of the intake of micronutrients.

Energy intake is frequently used as a proxy for dietary intake, and the assumption made that if energy intake is underestimated, the intakes of other nutrients may also be underestimated (Livingstone and Black 2003).

To validate dietary reports, researchers commonly focus on comparing dietary energy intake data estimates with energy expenditure (EE) (Chapters 10 and 11). Previous evidence that compared reported energy intake with energy expenditure measured by DLW, indicated the presence of underreporting (Schoeller et al. 1990; Livingstone and Black 2003). Mean differences between self-reported energy intakes and expenditure in lean, nonathletic groups living in industrialized countries have been noted that are between 0%–20%. While obese populations demonstrated the largest mean differences of –35% and –50%. However, weight loss, due to energy restriction during data collection, should not be discounted. Within, individual variation was also noted (Black and Cole 2001). However, DLW is expensive; thus, it cannot be used as a routine tool for validating EI data (Livingstone and Black 2003). Further measures are also used to estimate the plausibility of self-reported intake such as urinary markers (Kahn et al. 1995), and estimated EE. However, questions have been raised about the validity of EE estimate (Johnstone et al. 2005; Dhurandhar et al. 2015). Where height, weight, age, and sex are known or estimated they have been used to calculate a ratio of EI: BMR ratio. However, different equations have been used to estimate BMR and different cut off points are used to classify underreporters (Schofield 1985; Goldberg et al. 1991; Livingstone and Black 2003). It has been shown that very little difference was found in the sensitivity of the Goldberg cut offs using measured and calculated BMR (Black 2000b). These cut offs were developed to establish the likelihood of a food intake estimate reflecting real intake. Two cut offs for the agreement between physical activity level (Hartman et al. 1990) and reported energy intake/BMR were identified. The first, *CUT-OFF 1*, PAL (TEE/BMR) was set at 1.35, the minimum plausible value for most individuals who are weight stable (Goldberg et al. 1991). It has thus been suggested that there is no benefit to using measured BMR in large epidemiological studies (Black 2000a). On the other hand, using measured BMR can avoid some misclassifications that might be important in small studies where individual data have greater influence on outcomes. Black et al. (1991) suggested that it was acceptable to use a PAL value suitable for the study population according to physical activity or lifestyle information (Black et al. 1991). This information is most often gained via physical activity questionnaires that are commonly used in large-scale studies to rank people in a wide range of activity levels and identifying patterns for general lifestyle, occupational, and leisure activities (Black 2000a). However, in small studies where the objective is to measure EE, detailed activity diaries or accelerometers are potential tools (Black et al. 1991). Attempts to evaluate the potential disparity between self-reported energy intake and estimated energy expenditure in large surveys produce concerning results, disparity values as high as −800 kcal per day have been reported (Archer et al. 2013).

1.3.3 Determinants of Underreporting Dietary Intake

Several determinants of underreporting dietary intake have been addressed, although the evidence was not always consistent (Poslusna et al. 2009).

1.3.3.1 Body Mass Index

Many studies have attempted to evaluate the link between body mass index (BMI) and underreporting, although with contradictory results. In most studies there was evidence of a positive association between BMI and underreporting (Briefel et al. 1997; Pryer et al. 1997; Johansson et al. 2001; Livingstone and Black 2003; Subar et al. 2003; Poslusna et al. 2009). On the other hand, this association was not supported by all (de Vries et al. 1994; Poppitt et al. 1998). Given the increasing proportion of the population who are overweight and obese, an increasing proportion of respondents may return in more inaccurate food records in surveys representative of the general population.

1.3.3.2 Age and Sex

It was reported that women and older respondents were more prone to underreporting (de Vries et al. 1994; Briefel et al. 1997). In the latter group this may be associated with memory deterioration (Johansson et al. 2001). However, others could only confirm the link between underreporting and female gender and not with age (Johnson et al. 1994). Given the changing demography toward an aging population in some countries, identifying methods that overcome potential difficulties in older respondents will be important.

1.3.3.3 Socioeconomic Levels and Education

Given that lower socioeconomic class and lower level of education were stated as predictors of underreporting (Pryer et al. 1997), methods of overcoming difficulties with reporting in this group are key to ensure that research is ethical and inclusive.

1.3.3.4 Eating Behaviors

A higher habitual (usual) energy intake has been linked with underreporting (Subar et al. 2003), which might be due to difficulties in remembering more foods or larger portion sizes, or in response to societal pressure to consume less and may be associated with a currently higher BMI. A higher proportion of energy from fat and an irregular meal pattern were also predictors of underreporting in women, whereas meal frequency was a predictor of underreporting in men (Tooze et al. 2004). Dieters were also reported to be more prone to underreporting (Briefel et al. 1997), again raising issues where rates of obesity are increasing (Gibson 2005).

1.3.3.5 Diet Variation

Habitual intake is usually the main objective of many dietary assessments to estimate average long-term intake of specific foods or nutrients for the group or the individuals of interest (MacIntyre 2009). In theory, assessing more days increases the probability of capturing habitual diet. Dietary intakes may increasingly vary from day-to-day due to the wide variety of foods that are available, the availability of ready prepared foods, and the opportunity to eat outside home. These changes in food availability and consumption patterns potentially mean that the dietary intake assessment of habitual diet poses an even greater challenge than in the past. The number of days of measurement using a food diary, or repeated 24HR, is affected by the dietary component, that is, assessed for its intraindividual and interindividual variability (Nelson et al. 1989). On the other hand, food frequency questionnaires and to some extent diet histories, result in an accumulated estimate of intake during a period of time.

Seasonal variations in foods, especially fruits and vegetables, availability and consumption may influence dietary assessment. In interventional studies evaluating dietary effects, dietary assessments should be undertaken at similar times of the year. It is also important for cross-sectional, or cohort studies that dietary assessments are performed throughout the year to decrease the effect of seasonal variation (Suga et al. 2014), and in all cases careful attention should be paid to limitations associated with the subsequent use of food-composition tables.

Various attempts have been made to estimate the number of days of data collection that are appropriate taking into account between and within subject variation (Nelson et al. 1989; Beaton et al. 1997; Gibson 2005).

The number of days required to assess the dietary intake varies depending on whether it is an individual- or group-level assessment, sample size, the desired precision, and whether the main objective is energy, macronutrients, or micronutrients measurement (Bingham 1987). According to this conventional review, the number of days needed for assessments of energy and macronutrient assessments varied from five to ten days, while this number may be even more for micronutrient measurements.

In national studies, the number of days for which data are collected has been decreased to four (Whitton et al. 2011) to improve recruitment. Analysis of dietary intake revealed a considerable difference between weekdays and weekends, and even between the two weekend days (Whitton et al. 2011). In order to overcome day-to-day variations for estimating means of reported intake, the start day of a prospective food diary or the recall days should be randomized.

1.3.3.6 Appropriate Selection of Administration Techniques

Some methods can be administered by an interviewer or a respondent. Interviewer-administered questionnaires may be completed in a face-to-face setting or by telephone, and a self-administered tool may be done on paper or electronically. There is a concern about response rates and coverage in telephone surveys as many people use only mobile phones (Grande and Taylor 2010), hence will not be included in surveys only using landline numbers. Response rates achieved using random digit dialing techniques have been decreasing (Blumberg et al. 2006). Nevertheless, some current national surveys use telephone contact to collect the dietary data.

Careful thought must be given to obtaining portion sizes and picture booklets, or other portion-size estimation tools may need to be posted to respondents before the interview. Previous studies indicated comparable reported food intake between dietary data collected by telephone and other methods like face-to-face interviews (Brustad et al. 2003; Godwin et al. 2004). Self-administration of the dietary assessment tools is less costly than interviewer-administered methods. Furthermore, self-administered surveys tend to limit social desirability bias (the desire of respondents to present themselves in the best possible light) compared with the interviewer- administered methods (Bowling 2005) as respondents are less likely to be motivated to reply in a manner that they feel will result in the approval of the interviewer when they do not have personal contact. However, for people with low literacy level or inadequate motivation, dietary report data may tend to be potentially misreported.

1.4 CONCLUSIONS

Traditionally a variety of methods have been recognized as providing dietary intake data, and certain methods have been promoted as being better suited to different purposes. However, there is a little question that none of these methods can be acknowledged confidently as providing a *gold standard* against which other methods can be measured. Modern techniques, such as using doubly-labeled water to measure energy expenditure, provide valuable benchmarks against which dietary intake data can be compared, assuming the individuals are in energy balance at the time of measurement. A complex picture becomes apparent with respect to the variable accuracy of traditional methods. The gauntlet has been thrown down to those working in the field to develop new creative ways of capturing accurately dietary intake across all populations, and for a variety of purposes.

REFERENCES

Arab, L., C.-H. Tseng, A. Ang, P. Jardack. 2011. Validity of a multipass, web-based, 24-hour self-administered recall for assessment of total energy intake in Blacks and Whites. *American Journal of Epidemiology* 174(11): 1256–1265.

Archer, E., G. A. Hand, S. N. Blair. 2013. Validity of U.S. nutritional surveillance: National health and nutrition examination survey caloric energy intake data, 1971–2010. *PLoS One* 8(10): e76632.

Barclay, A. W., V. M. Flood, J. C. Brand-Miller, P. Mitchell. 2008. Validity of carbohydrate, glycaemic index and glycaemic load data obtained using a semi-quantitative food-frequency questionnaire. *Public Health Nutrition* 11(6): 573–580.

Beaton, G. H., J. Burema, C. Ritenbaugh. 1997. Errors in the interpretation of dietary assessments. *The American Journal of Clinical Nutrition* 65(4): 1100S–1107S.

Bingham, S. 1987. The dietary assessment of individuals; methods, accuracy, new techniques and recommendations. *Nutrition Abstracts and Reviews* 57: 705–742.

Black, A. 2000a. Critical evaluation of energy intake using the Goldberg cut-off for energy intake: Basal metabolic rate. A practical guide to its calculation, use and limitations. *International Journal of Obesity* 24(9): 1119–1130.

Black, A. 2000b. The sensitivity and specificity of the Goldberg cut-off for EI:BMR for identifying diet reports of poor validity. *European Journal of Clinical Nutrition* 54(5): 395–404.

Black, A. E., B. S. Bingham, G. Johansson, W. A. Coward. 1997. Validation of dietary intakes of protein and energy against 24 hour urinary N and DLW energy expenditure in middle-aged women, retired men and post-obese subjects: comparisons with validation against presumed energy requirements. *European Journal of Clinical Nutrition* 51(6): 405–413.

Black, A. E. and T. J. Cole. 2001. Biased over- or under-reporting is characteristic of individuals whether over time or by different assessment methods. *Journal of the Academy of Nutrition and Dietetics* 101(1): 70–80.

Black, A. E., G. R. Goldberg, S. A. Jebb, M. B. Livingstone, T. J. Cole, A. M. Prentice. 1991. Critical evaluation of energy intake data using fundamental principles of energy physiology: 2. Evaluating the results of published surveys. *European Journal of Clinical Nutrition* 45(12): 583–599.

Blanton, C. A., A. J. Moshfegh, D. J. Baer, M. J. Kretsch. 2006. The USDA automated multiple-pass method accurately estimates group total energy and nutrient intake. *The Journal of Nutrition* 136(10): 2594–2599.

Block, G., A. M. Hartman, C. M. Dresser, M. D. Carroll, J. Gannon, L. Gardner. 1986. A data-based approach to doet questionnaire design an testing. *American Journal of Epidemiology* 124(3): 453–469.

Block, G., F. E. Thompson, A. M. Hartman, F. A. Larkin, K. E. Guire. 1992. Comparison of two dietary questionnaires validated against multiple dietary records collected during a 1-year period. *Journal of the American Dietetic Association* 92(6): 686–693.

Bloemberg, B. P. M., D. Kromhout, G. L. Obreman-de Boer, M. van Kampen-Donker. 1989. The reproducibility of dietary intake data assessed with the cross-check dietary history method. *American Journal of Epidemiology* 130(5): 1047–1056.

Blumberg, S. J., J. V. Luke, M. L. Cynamon. 2006. Telephone coverage and health survey estimates: Evaluating the need for concern about wireless substitution. *American Journal of Public Health* 96(5): 926–931.

Boutelle, K. N. and D. S. Kirschenbaum. 1998. Further support for consistent self-monitoring as a vital component of successful weight control. *Obesity Research* 6(3): 219–224.

Bowling, A. 2005. Mode of questionnaire administration can have serious effects on data quality. *Journal of Public Health* 27(3): 281–291.

Briefel, R. R., C. T. Sempos, M. A. McDowell, S. Chien, K. Alaimo. 1997. Dietary methods research in the third National Health and Nutrition Examination Survey: Underreporting of energy intake. *The American Journal of Clinical Nutrition* 65(4): 1203S–1209S.

Brustad, M., G. Skeie, T. Braaten, N. Slimani, E. Lund. 2003. Comparison of telephone vs face-to-face interviews in the assessment of dietary intake by the 24h recall EPIC SOFT program—the Norwegian calibration study. *European Journal of Clinical Nutrition* 57(1): 107–113.

Burke, B. 1947. The dietary history as a tool in research. *Journal of the American Dietetic Association* 23: 1041–1046.

Buzzard, I. M., C. L. Faucett, R. W. Jeffery et al. 1996. Monitoring dietary change in a low-fat diet intervention study. *Journal of the Academy of Nutrition and Dietetics* 96(6): 574–579.

Cameron, M. E. and W. A. van Staveren. 1988. *Manual on Methodology for Food Consumption Studies.* Oxford: Oxford University Press.

Campbell, V. A. and M. L. Dodds. 1967. Collecting dietary information from groups of older people. *Journal of the American Dietetic Association* 51(1): 29–33.

Cantwell, M. M., A. E. Millen, R. Carroll et al. 2006. A debriefing session with a nutritionist can improve dietary assessment using food diaries. *The Journal of Nutrition* 136(2): 440–445.

Carroll, R. J., L. S. Freedman, A. M. Hartman. 1996. Use of semiquantitative food frequency questionnaires to estimate the distribution of usual intake. *American Journal of Epidemiology* 143(4): 392–404.

Carroll, R. J., D. Midthune, A. F. Subar et al. 2012. Taking advantage of the strengths of 2 different dietary assessment instruments to improve intake estimates for nutritional epidemiology. *American Journal of Epidemiology* 175(4): 340–347.

Casey, P. H., S. L. P. Goolsby, S. Y. Lensing, B. P. Perloff, M. L. Bogle. 1999. The use of telephone interview methodology to obtain 24-hour dietary recalls. *Journal of the Academy of Nutrition and Dietetics* 99(11): 1406–1411.

Chae, J., I. Woo, S. Kim et al. 2011. Volume estimation using food specific shape templates in mobile image-based dietary assessment. *Proceedings of SPIE* 7873: 78730K.

Dauchet, L., E. Kesse-Guyot, S. Czernichow et al. 2007. Dietary patterns and blood pressure change over 5-y follow-up in the SU.VI.MAX cohort. *The American Journal of Clinical Nutrition* 85(6): 1650–1656.

De Keyzer, W., I. Huybrechts, M. De Maeyer et al. 2011. Food photographs in nutritional surveillance: Errors in portion size estimation using drawings of bread and photographs of margarine and beverages consumption. *British Journal of Nutrition* 105(7): 1073–1083.

de Vries, J. H., P. L. Zock, R. P. Mensink, M. B. Katan. 1994. Underestimation of energy intake by 3-d records compared with energy intake to maintain body weight in 269 nonobese adults. *The American Journal of Clinical Nutrition* 60(6): 855–860.

Dhurandhar, N. V., D. Schoeller, A. W. Brown et al. 2015. Energy balance measurement: When something is not better than nothing. *International Journal of Obesity* 39(7): 1109–1113.

Dunn, S., A. Datta, S. Kallis, E. Law, C. E. Myers, K. Whelan. 2011. Validation of a food frequency questionnaire to measure intakes of inulin and oligofructose. *European Journal of Clinical Nutrition* 65(3): 402–408.

Eiben, G., C. Andersson, E. Rothenberg, V. Sundh, B. Steen, L. Lissner. 2004. Secular trends in diet among elderly Swedes—Cohort comparisons over three decades. *Public Health Nutrition* 7(5): 637–644.

Fisher, J. O., N. F. Butte, P. M. Mendoza et al. 2008. Overestimation of infant and toddler energy intake by 24-h recall compared with weighed food records. *The American Journal of Clinical Nutrition* 88(2): 407–415.

Foster, E., J. N. S. Matthews, J. Lloyd et al. 2008. Children's estimates of food portion size: the development and evaluation of three portion size assessment tools for use with children. *British Journal of Nutrition* 99(1): 175–184.

Fowke, J. H., D. Schlundt, Y. Gong et al. 2004. Impact of season of food frequency questionnaire administration on dietary reporting. *Annals of Epidemiology* 14(10): 778–785.

Fraser, G. E., R. Yan, T. L. Butler, K. Jaceldo-Siegl, W. L. Beeson, J. Chan. 2009. Missing data in a long food frequency questionnaire: Are imputed zeroes correct? *Epidemiology* (Cambridge, Mass.) 20(2): 289–294.

Freedman, L. F. 2011. Combining self-report dietary intake data and biomarker data to reduce the effects of measurement error. *Measurement Error Webinar Series* Webinar 11, http://appliedresearch.cancer.gov/measurementerror/ (accessed May 29, 2016).

Freedman, L. S., J. M. Commins, J. E. Moler et al. 2014. Pooled results from 5 validation studies of dietary self-report instruments using recovery biomarkers for energy and protein intake. *American Journal of Epidemiology* 180(2): 172–188.

Gersovitz, M. M. J. and H. Smiciklas-Wright. 1978. Validity of the 24-hr. dietary recall and seven-day record for group comparisons. *Journal of the American Dietetic Association* 73(1): 48–55.

Gibson, R. S. 2005. *Principles of Nutritional Assessment*. Oxford: Oxford University Press.

Gittelsohn, J., A. V. Shankar, R. P. Pokhrel, K. P. West, Jr. 1994. Accuracy of estimating food intake by observation. *Journal of the Academy of Nutrition and Dietetics* 94(11): 1273–1277.

Glanz, K., S. Murphy, J. Moylan, D. Evensen, J. D. Curb. 2006. Improving dietary self-monitoring and adherence with hand-held computers: a pilot study. *American Journal of Health Promotion* 20(3): 165–170.

Godwin, S. L., E. Chambers 4th, L. Cleveland. 2004. Accuracy of reporting dietary intake using various portion-size aids in-person and via telephone. *Journal of the American Dietetic Association* 104(4): 585–594.

Goldberg, G. R., A. E. Black, S. A. Jebb et al. 1991. Critical evaluation of energy intake data using fundamental principles of energy physiology: 1. Derivation of cut-off limits to identify under-recording. *European Journal of Clinical Nutrition* 45(12): 569–581.

Grande, E. D. and A. W. Taylor. 2010. Sampling and coverage issues of telephone surveys used for collecting health information in Australia: Results from a face-to-face survey from 1999 to 2008. *BMC Medical Research Methodology* 10(1): 1–11.

Haraldsdóttir, J., L. Holm, A. V. Astrup, J. Halkjaer, S. Stender. 2001. Monitoring of dietary changes by telephone interviews: results from Denmark. *Public Health Nutrition* 4(6): 1287–1295.

Harnack, L., L. Steffen, D. K. Arnett, S. Gao, R. V. Luepker. 2004. Accuracy of estimation of large food portions. *Journal of the American Dietetic Association* 104(5): 804–806.

Hartman, A. M., C. C. Brown, J. Palmgren et al. 1990. Variability in nutrient and food intakes among older middle-aged men: Implications for design of epidemiologic and validation studies using food recording. *American Journal of Epidemiology* 132(5): 999–1012.

Haubrock, J., U. Nöthlings, J.-L. Volatier et al. 2011. Estimating usual food intake distributions by using the multiple source method in the EPIC-potsdam calibration study. *The Journal of Nutrition* 141(5): 914–920.

Heady, J. A. 1961. Diets of bank clerks: Development of a method of classifying the diets of individuals for use in epidemiological studies. *Journal of the Royal Statistical Society: Series A* 124: 336–371.

Heitmann, B. L., L. Lissner, T. I. Sørensen, C. Bengtsson. 1995. Dietary fat intake and weight gain in women genetically predisposed for obesity. *The American Journal of Clinical Nutrition* 61(6): 1213–1217.

Hollis, J. F., C. M. Gullion, V. J. Stevens et al. 2008. Weight loss during the intensive intervention phase of the weight-loss maintenance trial. *American Journal of Preventive Medicine* 35(2): 118–126.

Holmes, B., K. Dick, M. Nelson. 2008a. A comparison of four dietary assessment methods in materially deprived households in England. *Public Health Nutrition* 11(5): 444–456.

Holmes, B. A., C. L. Roberts, M. Nelson. 2008b. How access, isolation and other factors may influence food consumption and nutrient intake in materially deprived older men in the UK. *Nutrition Bulletin* 33(3): 212–220.

Huizinga, M. M., A. J. Carlisle, K. L. Cavanaugh et al. 2009. Literacy, numeracy, and portion-size estimation skills. *American Journal of Preventive Medicine* 36(4): 324–328.

Hunter, D. J., L. Sampson, M. J. Stampfer, G. A. Colditz, B. Rosner, W. C. Willett. 1988. Variability in portion sizes of commonly consumed foods among a population of women in the United States. *American Journal of Epidemiology* 127(6): 1240–1249.

Jackson, K. A., N. M. Byrne, A. M. Magarey, A. P. Hills. 2008. Minimizing random error in dietary intakes assessed by 24-h recall, in overweight and obese adults. *European Journal of Clinical Nutrition* 62(4): 537–543.

Johansson, G., Å. Wikman, A.-M. Åhrén, G. Hallmans, I. Johansson. 2001. Underreporting of energy intake in repeated 24-hour recalls related to gender, age, weight status, day of interview, educational level, reported food intake, smoking habits and area of living. *Public Health Nutrition* 4(4): 919–927.

Johansson, L., K. Solvoll, S. Opdahl, G. E. Bjørneboe, C. A. Drevon. 1997. Response rates with different distribution methods and reward, and reproducibility of a quantitative food frequency questionnaire. *European Journal of Clinical Nutrition* 51(6): 346–353.

Johnson, R. K., M. I. Goran, E. T. Poehlman. 1994. Correlates of over- and underreporting of energy intake in healthy older men and women. *The American Journal of Clinical Nutrition* 59(6): 1286–1290.

Johnstone, A. M., S. D. Murison, J. S. Duncan, K. A. Rance, J. R. Speakman. 2005. Factors influencing variation in basal metabolic rate include fat-free mass, fat mass, age, and circulating thyroxine but not sex, circulating leptin, or triiodothyronine. *The American Journal of Clinical Nutrition* 82(5): 941–948.

Kahn, H. A., P. K. Whelton, L. J. Appel et al. 1995. Validity of 24-hour dietary recall interviews conducted among volunteers in an adult working community. *Annals of Epidemiology* 5(6): 484–489.

Khazrai, Y. M., G. Defeudis, P. Pozzilli. 2014. Effect of diet on type 2 diabetes mellitus: A review. *Diabetes/Metabolism Research and Reviews* 30(S1): 24–33.

Kolar, A. S., R. E. Patterson, E. White et al. 2005. A practical method for collecting 3-day food records in a large cohort. *Epidemiology* 16(4): 579–583.

Kweon, S., Y. Kim, M.-J. Jang et al. 2014. Data resource profile: The Korea National Health and Nutrition Examination Survey (KNHANES). *International Journal of Epidemiology* 43(1): 69–77.

Lamb, M. M., C. A. Ross, H. L. Brady, J. M. Norris. 2007. Comparison of children's diets as reported by the child via the Youth/Adolescent Questionnaire and the parent via the Willett food-frequency questionnaire. *Public Health Nutrition* 10(7): 663–670.

Livingstone, M. B. E. and A. E. Black. 2003. Markers of the validity of reported energy intake. *The Journal of Nutrition* 133(3): 895S–920S.

Livingstone, M. B. E., P. J. Robson, A. E. Black et al. 2003. An evaluation of the sensitivity and specificity of energy expenditure measured by heart rate and the Goldberg cut-off for energy intake: Basal metabolic rate for identifying mis-reporting of energy intake by adults and children: a retrospective analysis. *European Journal of Clinical Nutrition* 57(3): 455–463.

Luke, A., P. Bovet, T. E. Forrester et al. 2011. Protocol for the modeling the epidemiologic transition study: A longitudinal observational study of energy balance and change in body weight, diabetes and cardiovascular disease risk. *BMC Public Health* 11(1): 1–10.

MacIntyre, U. 2009. Measuring food intake. In *Introduction to Human Nutrition*, ed. M. J. Gibney, S. A. Lanham-New, A. Cassidy, H. H. Vorster, pp. 238–275. Chichester: Wiley-Blackwell.

Mente, A., L. de Koning, H. S. Shannon, S. S. Anand. 2009. A systematic review of the evidence supporting a causal link between dietary factors and coronary heart disease. *Archives of Internal Medicine* 169(7): 659–669.

Morris, M. C., G. A. Colditz, D. A. Evans. 1998. Response to a mail nutritional survey in an older bi-racial community population. *Annals of Epidemiology* 8(5): 342–346.

Nelson, M., M. Atkinson, J. Meyer. 1997. *Food Portion Sizes: A Photographic Atlas.* London: MAFF publications.

Nelson, M., A. E. Black, J. A. Morris, T. J. Cole. 1989. Between- and within-subject variation in nutrient intake from infancy to old age: estimating the number of days required to rank dietary intakes with desired precision. *The American Journal of Clinical Nutrition* 50(1): 155–167.

Ovaskainen, M. L., M. Paturi, H. Reinivuo et al. 2007. Accuracy in the estimation of food servings against the portions in food photographs. *European Journal of Clinical Nutrition* 62(5): 674–681.

Parr, C. L., A. Hjartåker, I. Scheel, E. Lund, P. Laake, M. B. Veierød. 2008. Comparing methods for handling missing values in food-frequency questionnaires and proposing k nearest neighbours imputation: Effects on dietary intake in the Norwegian Women and Cancer study (NOWAC). *Public Health Nutrition* 11(4): 361–370.

Poppitt, S. D., D. Swann, A. E. Black, A. M. Prentice. 1998. Assessment of selective under-reporting of food intake by both obese and non-obese women in a metabolic facility. *International Journal of Obesity and Related Metabolic Disorders* 22(4): 303–311.

Poslusna, K., J. Ruprich, J. H. M. de Vries, M. Jakubikova, P. van't Veer. 2009. Misreporting of energy and micronutrient intake estimated by food records and 24 hour recalls, control and adjustment methods in practice. *British Journal of Nutrition* 101(Suppl S2): S73–S85.

Prentice, R. L., Y. Mossavar-Rahmani, Y. Huang et al. 2011. Evaluation and comparison of food records, recalls, and frequencies for energy and protein assessment by using recovery biomarkers. *American Journal of Epidemiology* 174(5): 591–603.

Pryer, J. A., M. Vrijheid, R. Nichols, M. Kiggins, P. Elliott. 1997. Who are the 'low energy reporters' in the dietary and nutritional survey of British adults? *International Journal of Epidemiology* 26(1): 146–154.

Raper, N., B. Perloff, L. Ingwersen, L. Steinfeldt, J. Anand. 2004. An overview of USDA's dietary intake data system. *Journal of Food Composition Analysis* 17: 545–555.

Rebro, S. M., R. E. Patterson, A. R. Kristal, C. L. Cheney. 1998. The effect of keeping food records on eating patterns. *Journal of the Academy of Nutrition and Dietetics* 98(10): 1163–1165.

Rimm, E. B., E. L. Giovannucci, M. J. Stampfer, G. A. Colditz, L. B. Litin, W. C. Willett. 1992. Reproducibility and validity of an expanded self-administered semiquantitative food frequency questionnaire among male health professionals. *American Journal of Epidemiology* 135(10): 1114–1126.

Rothenberg, E. M. 2009. Experience of dietary assessment and validation from three Swedish studies in the elderly. *European Journal of Clinical Nutrition* 63(Suppl 1): S64–S68.

Schatzkin, A., V. Kipnis, R. J. Carroll et al. 2003. A comparison of a food frequency questionnaire with a 24-hour recall for use in an epidemiological cohort study: Results from the biomarker-based Observing Protein and Energy Nutrition (OPEN) study. *International Journal of Epidemiology* 32(6): 1054–1062.

Schoeller, D. A., L. G. Bandini, W. H. Dietz. 1990. Inaccuracies in self-reported intake identified by comparison with the doubly labelled water method. *Canadian Journal of Physiology and Pharmacology* 68(7): 941–949.

Schofield, W. 1985. Predicting basal metabolic rate, new standards and review of previous work. *Human Nutrition – Clinical Nutrition* 39(Suppl 1): 5–41.

Schwingshackl, L. and G. Hoffmann. 2015. Adherence to mediterranean diet and risk of cancer: an updated systematic review and meta-analysis of observational studies. *Cancer Medicine* 4(12): 1933–1947.

Segovia-Siapco, G., P. Singh, K. Jaceldo-Siegl, J. Sabaté. 2007. Validation of a food-frequency questionnaire for measurement of nutrient intake in a dietary intervention study. *Public Health Nutrition* 10(2): 177–184.

Shahar, D., D. Fraser, I. Shai, H. Vardi. 2003. Development of a food frequency questionnaire (FFQ) for an elderly population based on a population survey. *The Journal of Nutrition* 133(11): 3625–3629.

Subar, A. F., V. Kipnis, R. P. Troiano et al. 2003. Using intake biomarkers to evaluate the extent of dietary misreporting in a large sample of adults: The OPEN study. *American Journal of Epidemiology* 158(1): 1–13.

Subar, A. F., F. E. Thompson, N. Potischman et al. 2007. Formative research of a quick list for an automated self-administered 24-hour dietary recall. *Journal of the Academy of Nutrition and Dietetics* 107(6): 1002–1007.

Subar, A. F., R. G. Ziegler, F. E. Thompson et al. 2001. Is shorter always better? Relative importance of questionnaire length and cognitive ease on response rates and data quality for two dietary questionnaires. *American Journal of Epidemiology* 153(4): 404–409.

Suga, H., K. Asakura, S. Sasaki et al. 2014. Effect of seasonality on the estimated mean value of nutrients and ranking ability of a self-administered diet history questionnaire. *Nutrition Journal* 13(51): 2–10.

Thompson, F. E., S. Dixit-Joshi, N. Potischman et al. 2015a. Comparison of interviewer-administered and automated self-administered 24-hour dietary recalls in 3 diverse integrated health systems. *American Journal of Epidemiology* 181(12): 970–978.

Thompson, F. E., S. I. Kirkpatrick, A. F. Subar et al. 2015b. The National Cancer Institute's dietary assessment primer: A resource for diet research. *Journal of the Academy of Nutrition and Dietetics* 115(12): 1986–1995.

Thompson, F. E., A. F. Subar, C. M. Loria, J. L. Reedy, T. Baranowski. 2010. Need for technological innovation in dietary assessment. *Journal of the American Dietetic Association* 110(1): 48–51.

Tooze, J. A., A. F. Subar, F. E. Thompson, R. Troiano, A. Schatzkin, V. Kipnis. 2004. Psychosocial predictors of energy underreporting in a large doubly labeled water study. *The American Journal of Clinical Nutrition* 79(5): 795–804.

Tooze, J. A., M. Z. Vitolins, S. L. Smith et al. 2007. High levels of low energy reporting on 24-hour recalls and three questionnaires in an elderly low-socioeconomic status population. *The Journal of Nutrition* 137(5): 1286–1293.

Van Staveren, W. A., L. C. de Groot, A. Haveman-Nies. 2002. The SENECA study: Potentials and problems in relating diet to survival over 10 years. *Public Health Nutrition* 5(6a): 901–905.

Vereecken, C. A., M. Covents, W. Sichert-Hellert et al. 2008. Development and evaluation of a self-administered computerized 24-h dietary recall method for adolescents in Europe. *International Journal of Obesity* 32(S5): S26–S34.

Vuckovic, N., C. Ritenbaugh, D. L. Taren, M. Tobar. 2000. A qualitative study of participants' experiences with dietary assessment. *Journal of the Academy of Nutrition and Dietetics* 100(9): 1023–1028.

Whitton, C., S. K. Nicholson, C. Roberts et al. 2011. National diet and nutrition survey: UK food consumption and nutrient intakes from the first year of the rolling programme and comparisons with previous surveys. *The British Journal of Nutrition* 106(12): 1899–1914.

Willett, W. C. 1998. *Nutrition Epidemiology*. New York: Oxford University Press.

Willett, W. C., L. Sampson, M. J. Stampfer et al. 1985. Reproducibility and validity of a semiquantitative food frequency questionnaire. *American Journal of Epidemiology* 122(1): 51–65.

Yang, Y. J., M. K. Kim, S. H. Hwang, Y. Ahn, J. E. Shim, D. H. Kim. 2010. Relative validities of 3-day food records and the food frequency questionnaire. *Nutrition Research and Practice* 4(2): 142–148.

Young, C. M., F. W. Chalmers, H. N. Church, G. C. Murphy, R. E. Tucker. 1953. Subjects' estimation of food intake and calculated nutritive value of the diet. *Journal of the American Dietetic Association* 29: 1216–1220.

Zulkifli, S. N. and S. M. Yu. 1992. The food frequency method for dietary assessment. *Journal of the American Dietetic Association* 92(6): 681–685.

2 Statistical Approaches to Mitigate Measurement Error in Dietary Intake Data Collected Using 24-hour Recalls and Food Records/Diaries

Sharon I. Kirkpatrick, Amy F. Subar, and Janet A. Tooze

CONTENTS

2.1 INTRODUCTION

In various types of nutrition research, dietary intake is measured using self-report instruments (Thompson and Subar 2013). These include 24-hour recalls and food records/diaries, which are considered to be *short-term instruments* in that they capture a detailed measure of all foods and beverages consumed over a day or for a limited number of days (National Cancer Institute 2015a; Thompson and Subar 2013; Thompson et al. 2015). It has long been recognized that there are numerous challenges in accurately capturing dietary intake using such instruments (National Cancer Institute 2015a; Thompson and Subar 2013) and that the resulting data are affected by *measurement error* (Beaton 1994; Beaton et al. 1979). Such errors can be *random* or *systematic* (also known as *bias*), with implications in terms of interpreting findings (Beaton 1994; Beaton et al. 1997). Measurement error must be recognized and, if possible, reduced, so that study findings better reflect truth (Beaton 1994; Beaton et al. 1997; Dodd et al. 2006; Freedman et al. 2011a; Kipnis et al. 1997, 2002; Subar et al. 2015; Thiébaut et al. 2007).

Dr. George Beaton noted in the mid-1990s that measurement error will continue to be inherent in the assessment of dietary intake and that the challenge lies in understanding the structure of the error so that it can be mitigated (Beaton 1994). At that time, there was already work underway to improve analytic methods and in the years that followed, improvements continued to be made to data collection tools themselves (refer to Chapters 1 and 3) as well as statistical approaches. Indeed, there has been significant work to quantify measurement error and develop strategies to reduce its effects using statistical methods. For example, studies making use of *recovery biomarkers* (described in Section 2.2.5 and Chapter 13) have resulted in a better understanding of the structure of the error in dietary data collected using different tools and informed strategies for improving resulting estimates for a limited number of nutrients (Arab et al. 2011; Kipnis et al. 2003; Lissner et al. 2007; Moshfegh et al. 2008; Neuhouser et al. 2008; Prentice et al. 2011; Subar et al. 2003). Sufficient large recovery biomarker-based validation studies have now been conducted to allow for data pooling (Freedman et al. 2014, 2015a, 2015b), providing additional insights into error in data collected using different self-report instruments and informing strategies to mitigate such error.

In this chapter, key concepts relevant to the analysis of intake data from recalls and records/diaries are described, the types of error in data collected using these methods and their potential implications are briefly reviewed, and analytic strategies for reducing error are discussed. Applications include describing usual dietary intake among groups as well as examining relationships between dietary intake and other variables, such as health or disease outcomes. It should be noted that the approaches reviewed here relate to the analysis of data collected from *groups of individuals* for the purposes of making inferences about dietary intake or relationships between diet and health outcomes at the level of *groups or populations* versus the individual level. The assessment of dietary intake at the individual level has been reviewed elsewhere (Barr et al. 2002; Institute of Medicine 2000).

2.2 KEY CONCEPTS

2.2.1 Usual Intake

Recalls and records/diaries capture foods and beverages consumed on a particular day or days. However, in nutrition research, we are often interested in habitual intake over a long period of time; this is referred to as *usual intake* (Dodd et al. 2006; Freedman et al. 2004; National Cancer Institute 2011; Tooze et al. 2010). Usual intake is of interest in regard to assessing the diets of a population relative to recommendations because these are meant to be met over time, not on any given day. Similarly, from the perspective of diet and disease, we are typically interested in how habitual diet, or chronic dietary exposure, is related to an outcome. Depending on our research questions, it may be necessary to estimate mean usual intake or distributions of usual intake; these concepts are expanded upon in Section 2.4.

The time period to which usual intake is defined may be specified vaguely (Freedman et al. 2015b) and/or relate to operational definitions, such as the period over which data were collected for surveillance or epidemiological studies. It has long been recognized that intake may vary between week days and weekend days and across seasons (An 2016; Haines et al. 2003; Ma et al. 2006; Thompson et al. 1986; Yang et al. 2014; Ziegler et al. 1987); indeed, data collection and analytic strategies often account for these so-called *nuisance effects* (National Cancer Institute 2015a). However, there is increasing attention to the concept that dietary intake is dynamic in that it varies temporally and may differ across potentially important time periods (e.g., early childhood and puberty), with analytic techniques emerging to account for these time-varying exposures (Freedman et al. 2015c; Liao et al. 2011). Long-term studies making use of the few available markers of true intake (i.e., recovery biomarkers, described below) are needed to better understand variations in true intake over time and to inform analytic strategies that account for such variations.

There are also cases in which *acute intake* (i.e., intake on a given day or days) is of interest; for example, one might be interested in calories contributed by alcohol on a given day (Nielsen et al. 2012). The methods reviewed here, however, relate primarily to applications in which usual intake is the focus.

2.2.2 Total Nutrient Intake

To arrive at estimates of *total nutrient intake*, it is necessary to account for contributions from vitamin and mineral *supplements* as well as foods and beverages. Data from the 2003 to 2006 National Health and Nutrition Examination Survey (NHANES) suggest that approximately half the U.S. population uses some type of dietary supplement (Gahche et al. 2011). Supplementation can make a significant contribution to intake of some nutrients, and thus, failing to account for it can result in over and underestimation of prevalences of inadequate and excess nutrient intake, respectively (Gahche et al. 2011). Similar to intake from foods and beverages, supplement intake can be ascertained using short-term methods (e.g., a module included in a 24-hour recall) or long-term methods (e.g., frequency of consumption over 30 days), with a combination of methods potentially resulting in the most accurate data (Nicastro et al. 2015). Consideration of the consistency of the time frames for which intake from food and beverages and from supplements are estimated is critical to combining the two sources to arrive at total nutrient intake.

2.2.3 Episodically-Consumed and Nonepisodically-Consumed Dietary Components

In considering how best to capture and analyze dietary intake data, it is important to consider both nutrients and foods that are consumed nearly daily and those consumed *episodically* by most persons (Dodd et al. 2006; Kipnis et al. 2009; Tooze et al. 2006). Examples of nonepisodically-consumed components among the U.S. population include nutrients like vitamin C and food groups like total grains. Episodically-consumed components include nutrients concentrated in a few foods like vitamin A, and food groups not commonly consumed, such as whole grains and dark green vegetables. Many individuals will report zero intake of episodically-consumed components on a given day, which poses a challenge for statistical modeling of usual intake (Dodd et al. 2006; Tooze et al. 2006). The regularity with which a dietary component is consumed can vary among subgroups; for example, milk may be episodically-consumed among adults but not among children (Krebs-Smith et al. 2010). It is also important to note that there are individuals within groups who can be characterized as *never consumers* of a particular dietary component (Haubrock et al. 2011; Kipnis et al. 2009). For example, those with a nut allergy avoid consumption of nut products, a vegetarian never consumes animal products, and some individuals abstain from alcohol consumption. There are also many who never consume vitamin or mineral supplements (Gahche et al. 2011).

2.2.4 Skewness of Dietary Intake Data

Skewness of dietary data is another issue that requires attention in estimating distributions of usual intake. Distributions of intake tend to be skewed to the right, sometimes quite severely depending on the dietary component of interest (Tooze et al. 2006) and particularly for those that are more episodically-consumed. This means that some persons in a sample have very high intake levels relative to others (National Cancer Institute 2011). Often transformations are applied so that the data are more normally distributed and meet the assumptions for statistical methods used to estimate usual intake distributions (Dodd et al. 2006; Tooze et al. 2006). In addition to skewness, supplement data may have spikes at levels corresponding with common dosages (National Cancer Institute 2011).

2.2.5 Reference Measures in Dietary Assessment

Reference measures, which are assumed to provide estimates that are closer to the underlying truth than the self-report instrument (Kipnis et al. 2003), can be used to understand both the measurement error properties of dietary intake data collected using particular tools and to calibrate self-report data so that they better reflect truth (Freedman et al. 2014, 2015a; Kipnis et al. 2003; Neuhouser et al. 2008). Examples of reference measures are *recovery biomarkers*, specific biologic products that are directly related to intake and provide unbiased estimates of true intake (Kaaks et al. 2002; National Cancer Institute 2011) (see Chapters 10 and 13). Recovery biomarkers have been identified for energy (doubly-labeled water, DLW) and protein (24-hour urinary nitrogen), potassium (24-hour urinary potassium), and sodium (24-hour urinary sodium) (Freedman et al. 2014, 2015a; Kipnis et al. 2003; Neuhouser et al. 2008; Subar et al. 2003).

Given that recovery biomarkers have been identified for only a few dietary components, often data from recalls or records are used as a reference measure for the purpose of reducing measurement error in other types of dietary intake data (e.g., food frequency questionnaires and brief dietary screeners) given *validation* studies that have shown that the former contain less bias than the latter (Kaaks et al. 2002; Kipnis et al. 2003; Freedman et al. 2014, 2015a). This approach has limitations but has been shown to be preferable to ignoring bias (Freedman et al. 2011a).

There are few reference measures other than recovery biomarkers available to mitigate the effects of measurement error in data from recalls or records. It is possible to use *observation* as a reference measure for short-term intake (National Cancer Institute 2015a). However, observation is generally not feasible for providing reference data for reducing bias in studies aimed at monitoring intakes of populations or elucidating relationships between diet and disease.

In addition to recovery biomarkers, there are two other classes of biomarkers that relate to food or nutrient intakes, *concentration* and *predictive* biomarkers. Concentration biomarkers reflect the concentration of a chemical or compound in blood, urine, or tissues after metabolism and are indirect measures of intake (Kaaks et al. 2002). Unlike recovery biomarkers, they are subject to between-individual differences related to physiology and health behaviors (e.g., smoking); for this reason, they are related to but cannot be used as an estimate of absolute intakes. Examples of concentration biomarkers are carotenoids. Predictive biomarkers have a stronger direct relationship with intake than concentration biomarkers (Tasevska et al. 2011), but they do not provide unbiased estimates of true intake. An example of a predictive biomarker is urinary sugar (Tasevska 2015; Tasevska et al. 2011). Although these latter classes of biomarkers are not reference measures, approaches have been developed to use them in addition to self-report data to improve estimates of intake (Freedman et al. 2010a, 2011b).

2.3 THE NATURE OF ERROR IN DIETARY INTAKE DATA COLLECTED USING RECALLS AND RECORDS/DIARIES AND ITS IMPLICATIONS

Measurement error refers to the difference between the true value of a parameter, such as dietary intake, and the value obtained from a measure. Data affected by random error are not biased or inaccurate, but are not precise. As can be seen in Figure 2.1, the distribution of intake with random

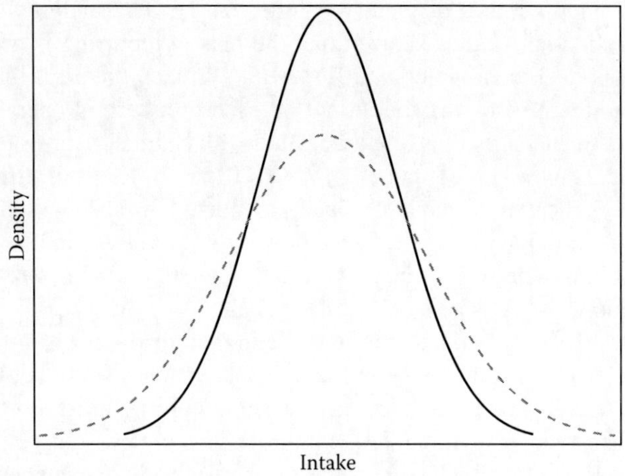

FIGURE 2.1 Hypothetical distributions of dietary intake without random error (solid line) and with random error (dashed line). The random error in data collected using 24-hour recalls and food records is mainly driven by day-to-day variation. The distribution with random error is wider and flatter than the distribution without this type of error.

error is wider than that without. With sufficient repeat measures, random error can be addressed using averaging (i.e., the random errors cancel each other out to arrive at truth). However, in practice in dietary assessment, the number of repeat measures needed to obtain an average not impacted by random error is often prohibitive. Random error is therefore addressed using a small number of repeated measures and statistical modeling (Dodd et al. 2006). Systematic error, or bias, results in measurements that consistently depart from the true value in the same direction. As can be seen in Figure 2.2, the distribution is shifted, in this case to the left, reflecting underestimation of true intake. To address systematic error, a reference instrument that allows us to estimate truth is needed.

Validation studies using recovery biomarkers indicate that data collected using recalls and records/ diaries are prone to substantial random error but less bias than food frequency data (Freedman et al. 2014, 2015a). Further, the extent of error differs by dietary component. Though recovery biomarkers

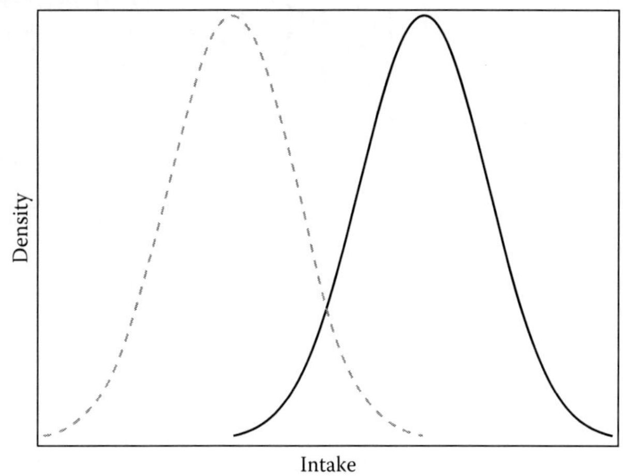

FIGURE 2.2 Hypothetical distributions of dietary intake without systematic error (solid line) and with systematic error (dashed line). Systematic error is also known as bias. The distribution with bias is shifted to the left, indicative of underreporting. Overreporting is also possible.

are available for a limited number of dietary components, recent analyses drawing upon pooled data from multiple validation studies showed that the bias in reporting is greater for energy than for protein and potassium (Freedman et al. 2014, 2015a). In fact, the error in energy reporting has been deemed so large that self-report should not be used to ascertain energy intake (Subar et al. 2015). Sources of bias include misreporting associated with characteristics such as age, educational status, and body weight status (Freedman et al. 2014, 2015a). In terms of differences in sources of bias by assessment tool, recall and frequency data are affected by biases associated with limitations in memory, whereas record/diary data are affected by *reactivity* biases in that individuals may alter their intake in response to monitoring (National Cancer Institute 2015a). See Chapter 4 for further discussion of this topic.

Random *within-person* error in data from recalls and records/diaries is mostly driven by *day-to-day variation* in intake. The fact that what we eat or drink changes from day-to-day and that this is reflected in intake data collected for a given day does not represent misreporting per se on the part of respondents. In fact, recovery biomarkers exhibit this type of within-person variation, reflecting both day-to-day variation in intake and biological variability. As this variation is random, on average, tools affected by this variation capture truth, that is, they are *unbiased*. We can therefore think of separating the variation in reported intake affected by random error into two components: the variation of the usual or long-term average intake and the day-to-day variation in that average. However, ignoring day-to-day variation and treating the sum of the two as the variance of usual intake will lead to error (either over or underestimation of intake) when the intent is to capture usual intake using data for one or a few days (Freedman et al. 2004). To illustrate, Figure 2.3, which assumes random error only and no bias, shows hypothetical distributions of intake based on one day and two days compared to the usual intake distribution, which is estimated using statistical modeling and replicate measures to adjust for day-to-day variation. The distribution based on a single day is wider and flatter than the usual intake distribution because of excessive within-person variation (Freedman et al. 2004). The distribution based on averaging over two days shows some improvement relative to the usual intake distribution but averaging does not adequately account for day-to-day variation. As a result, interpretations based on one or a few days without appropriate

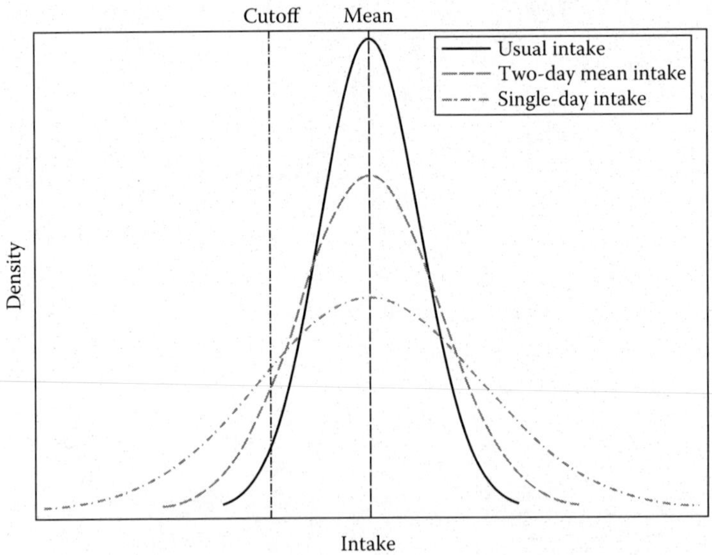

FIGURE 2.3 Simulated distributions of dietary intake based on a single day, two days, and usual intake. The distributions based on one and two days are wider and flatter than the usual distribution.

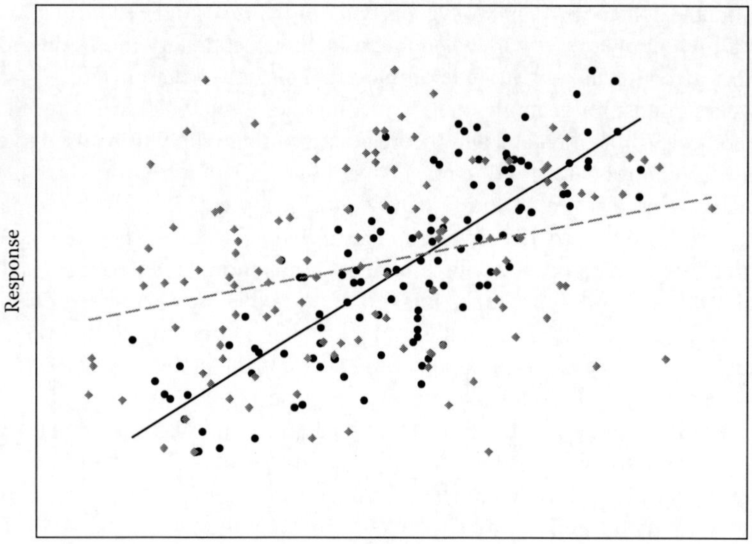

FIGURE 2.4 Hypothetical regression lines showing the estimated relationship between *true* dietary intake and a response such as a health outcome (solid line) and the estimated relationship between *reported* dietary intake and a health outcome (dashed line). The regression coefficient for the health outcome based on dietary intake measured with error is attenuated, or biased toward the null.

modeling to arrive at estimated usual intake are likely to be erroneous, particularly at the tails of the distribution. Given data that are also biased, the distribution can also be expected to be shifted, as shown in Figure 2.2.

Both random error and bias also have implications for assessing associations between diet and other factors, such as health or disease markers. Figure 2.4 illustrates hypothetical estimated associations between dietary intake and a health or disease outcome. As shown, using observed intake without taking measurement error into account can lead to erroneous results. Often, estimates of relationships are *attenuated* (i.e., biased toward the null) (Kipnis et al. 2003; Rosner et al. 1989). The degree of attenuation can be described using the *attenuation factor*, which indicates the extent to which the regression coefficient describing the relationship between diet and the health outcome is changed due to measurement error.

When only one exposure measured with error is included in a regression model, although the estimate of the regression coefficient is biased, the test of the null hypothesis (i.e., that there is no effect) remains valid (Freedman et al. 2011a). With additional covariates measured with error included in the model, the bias can work in different ways due to *contamination effects*, that is, the coefficients for two variables measured with error will each reflect part of the error of the other, and the test of the exposure effect may not be valid. However, the effect of contamination has been found to be much smaller than attenuation for energy, protein, and potassium; thus, the impact of contamination is not anticipated to be a large source of bias in estimated regression coefficients (Freedman et al. 2011a). In addition, measurement error causes loss of power to detect relationships that may actually exist (Kipnis et al. 2003), again leading to erroneous conclusions in that true effects are not observed. Unlike attenuation, which can be mitigated through the use of a recovery biomarker, power generally cannot be restored through calibration to a reference measure. However, including additional information through the use of multiple instruments may restore some power by improving the prediction of usual intake (Carroll et al. 2012; Freedman et al. 2011b; Kipnis et al. 2009; Prentice et al. 2009).

It is also salient to consider *differential* versus *nondifferential* error (Beaton et al. 1997; Freedman et al. 2011a). Differential error occurs when individuals report dietary intake differently in a manner that is associated with another variable of interest. This can occur when there are differences in the error between groups; for example, in a case-control study in which cases recall their diets differently than controls (though recalls and records/diaries are unlikely to be used to capture dietary intake data in case-controls given the retrospective design) or in a trial in which a group exposed to a dietary intervention might report differently than a group not exposed (Freedman et al. 2011a; Neuhouser et al. 2008; Willett 2013). Differential error may also arise when an outcome of interest, such as body mass index, is associated with differences in reporting or when participant characteristics are associated with error in both dietary intake and the outcome (Freedman et al. 2014, 2015a; Lissner et al. 2007) (see Chapter 10). Differential error is particularly tricky to address and is of concern because it can lead to spurious associations between dietary exposures and health outcomes or between an intervention or other factors (e.g., socioeconomic status) and dietary outcomes. Furthermore, even if the measurement error in a variable is nondifferential, if intake is categorized into quantiles or other groupings, the misclassification that occurs is often differential, even if data are collected prospectively (Flegal et al. 1991).

The best way to address differential error is by designing studies and statistical analyses to mitigate its effects. For example, prospectively collecting dietary intake data whenever possible is recommended to avoid differential error in the retrospective recalling of diet (however, this approach will not avoid differential error related to interventions). Further, information on personal characteristics known to be associated with differences in reporting and the outcome should be collected and included in models examining diet in relation to other factors (Tooze et al. 2016). Self-report dietary variables should be treated as continuous variables in statistical modeling of diet-outcome models rather than collapsing them into quantiles. Furthermore, studies that model health outcomes associated with differences in reporting, such as body mass index (Lissner et al. 2007), must be interpreted in light of possible differential error. This caution also applies to studies examining differences in dietary intake associated with an intervention in which groups may report differently due to differential exposure to the intervention itself.

2.4 STATISTICAL APPROACHES TO MITIGATE ERROR IN DIETARY INTAKE DATA COLLECTED USING RECALLS AND RECORDS/DIARIES

The remainder of the chapter addresses approaches to mitigate error in the analyses of data collected using recalls and records/diaries. We first address methods to account for error for the purposes of describing the diets of groups, for example, in surveillance or monitoring of the intake of populations. We then address methods for reducing error when examining relationships between diet and other variables, such as health or disease outcomes. For this latter discussion, we describe cases in which recalls or records/diaries are the main instrument used to collect dietary intake in a study, as well as those in which recalls or records/diaries are used as the reference instrument to reduce bias in data collected using food frequency questionnaires and screeners. In many cases, the approaches discussed have evolved from fruitful collaborations among statisticians, nutritionists, and epidemiologists. Given that dietary assessment is complex and the statistical procedures can be intensive and involve steep learning curves, collaborations of this nature in the analyses and interpretation of dietary data are beneficial and indeed, recommended.

As noted earlier, although random error can be addressed using repeated measures and statistical modeling, reference measures reflecting true intake are needed to reduce bias. Given the lack of such measures for most dietary components, the development and application of methods invoke a *working assumption that data from recalls are unbiased for true intake (i.e., are affected only by random error) for the given day* (Tooze et al. 2006). In other words, if we averaged many recalls, we would arrive at an estimate that approaches truth. This assumption is based on evidence noted above from biomarker-based studies that recall data contain less bias than do frequency data

(Freedman et al. 2014, 2015a; Kipnis et al. 2003; Subar et al. 2003). Although this assumption is flawed (i.e., recall data *do* contain some bias), until we are able to identify appropriate unbiased reference measures for a wider range of dietary components, proceeding with this assumption allows us to make the best use of available dietary intake data, with recognition of the limitations, as well as to continue to develop methods to address a wide range of research questions (National Cancer Institute 2011; Subar et al. 2015). The appropriateness of the assumption of unbiasedness depends on the dietary component of interest; for example, as noted above, energy is reported with greater error than are protein and potassium and it is recommended that self-report data not be used to arrive at estimates of energy intake (Subar et al. 2015). The assumption of unbiasedness may extend to records/diaries, but it has been made explicit by researchers developing and applying methods for recall data (Tooze et al. 2006, 2010).

A review of procedures for the processing of data collected using recalls and records is beyond the scope of this chapter. Briefly, prior to analysis, it is important to review data with attention to potential entry or coding errors. Although it is possible to examine outliers in nutrient and food group intakes or portion sizes, it is very difficult to differentiate reporting error from the wide range of intake values that are possible for a given day. Methods have also been developed for ascertaining energy misreporting; for example, the Goldberg equation uses the ratio of reported energy intake to basal metabolic rate (either from a prediction equation or measured by indirect calorimetry), and categorizes a person whose reported intake is much lower than expected based on the variability of the measures as an underreporter (Goldberg et al. 1991). However, excluding underreporters from analyses has been shown to be inadvisable. Recent work has explored the value of stratifying by underreporting status in statistical analyses under nondifferential error (Tooze et al. 2016). Stratifying has been shown to lead to less or more attenuation of diet-outcome relationships, dependent on the pairwise correlations between true exposure, observed exposure, and the ratio of energy intake to basal metabolic rate; the changes in the attenuation factors will occur in the same direction in each strata (Tooze et al. 2016). As the attenuation factors will either increase, decrease, or remain the same in both strata, including among plausible reporters, there is no benefit to excluding underreporters. Eliminating a large proportion of the sample also results in reduced power.

2.4.1 Describing Dietary Intake among Groups

Recalls and records have long been used to monitor the intake of populations, for example, in the NHANES (Centers for Disease Control and Prevention 2016), the Canadian Community Health Survey (Health Canada 2006), and in pan-European surveys using EPIC-Soft (De Keyzer et al. 2015). We may be interested in estimating mean intake among groups, or other aspects of the distribution such as percentiles or proportions above or below some cutoff or threshold (Institute of Medicine 2000). Statistical approaches for describing distributions of usual intake among groups are well developed and a number of specific methods to implement such approaches are available (Dekkers et al. 2014; Harttig et al. 2011; Haubrock et al. 2011; National Research Council 1986; Nusser et al. 1995; Tooze et al. 2006, 2010). These methods can be used with data from population-level surveys such as NHANES or other smaller-scale studies. The use of food frequency questionnaires for the purposes of estimating means and distributions of usual intake is not recommended because of the greater bias inherent in frequency data; however, some methods for estimating usual intake distributions using recalls and records/diaries allow for the use of covariates based on frequency data, as noted below.

2.4.1.1 Estimating Mean Usual Intake among a Group of Persons

If the mean usual intake for a group is the only desired statistic, no modeling is required and it is also not required to collect or make use of replicate recalls (Dodd et al. 2006; National Cancer Institute 2015a). As noted above, the extent of bias in misreporting differs across dietary components for which recovery biomarkers are available, with implications for the interpretation of estimates of mean intake. Furthermore, it is important that the data from the recalls account for potential factors that may affect intake, for example, by spreading data collection across the days of the week and across seasons. In studies such as NHANES, these factors are built into the sampling strategy (Centers for Disease Control and Prevention 2016), but this may not be the case for other sources of data.

Although there has been less study of the error in food records and diaries as compared to recalls, given the shared characteristics of the methods, a similar assumption of unbiasedness has been made in terms of a single administration providing the best estimate of usual mean intake for a group, without adjustment for within-person variation.

2.4.1.2 Estimating Distributions of Usual Intake among a Group of Persons

When we are interested in characterizing intake beyond the mean among a group or groups (e.g., for the purpose of estimating proportions of a population with inadequate or excess intake of given food groups or nutrients according to corresponding recommendations or reference values spelled out by the Dietary Guidelines for Americans [United States Department of Health and Human Services and United States Department of Agriculture 2015] or Dietary Reference Intakes [Institute of Medicine 2000]), it is necessary to estimate distributions of usual intake. Doing so based on recalls or records/diaries for a limited number of days requires accounting for random within-person variation, which is primarily driven by day-to-day variation, as noted above (Dodd et al. 2006). Without accounting for day-to-day variation, our estimates of characteristics of the distribution, such as the fraction of the population with usual intake below specific recommendations, may be erroneous, as shown in Figure 2.3; as can be seen, the proportion of the population with inadequate or excessive intake is overestimated on the basis of one or two recalls because the tails of the distribution are too wide. (In cases in which interest is in distributions of intake for a specific given day [i.e., acute intake], the observed distribution of intake can be used without adjustment for day-to-day variation.)

As noted above, random error can be addressed by averaging across repeated measures. However, in practice, we typically have too few repeats to sufficiently account for day-to-day variation. The number of days of intake data needed is dependent on the magnitude of the day-to-day variability as well as how episodically the dietary component of interest is consumed. A range of 31–433 days has been estimated to predict the usual nutrient intake of an individual, and 3–41 days for a group (Basiotis et al. 1987); theoretically, more days could be needed to predict intake of individual foods given greater day-to-day variability. To address this challenge, statistical modeling approaches are utilized (Dodd et al. 2006). After transforming the data to approximate normality, the basic steps in estimating the distribution of usual intake include: (1) estimating the mean and partitioning the variance into within-person and between-person and (2) estimating the percentiles of intake from the estimates of the mean and the variance components. In the second step, the within-person variation is removed so that the resulting distribution only reflects between-person variation. This may be referred to as adjusting for within-person variation or *shrinking* the distribution since accounting for the day-to-day variation pulls in the tails (Figure 2.3).

Application of this approach requires that replicate measures are available for at least a subsample; this subsample should be representative of the overall study sample. For dietary components that are not consumed frequently among most individuals within the population of interest, additional replicates (beyond two) may be needed so that there are a substantial number of persons who consumed the component on the recalled or recorded days. In general, studies should be designed so that at least 50 persons will have reports of the dietary component of interest on at least two days (sometimes referred to as double hits). For dietary components consumed on a daily or almost daily basis, this requires a subsample of this same size (i.e., 50); for episodically-consumed dietary components, the number of replicate recalls to be collected must take into account the probability of observing at least two days of intake for an individual (see Section 2.5). Given that the primary purpose of the replicate measures is to account for day-to-day variation, replicate measures should be nonconsecutive (as consecutive recalls may be highly correlated, for example, due to *leftover effects*, or autocorrelation) (Hartman et al. 1990; National Cancer Institute 2011). In practice, recalls are often separated by some number of days; for example, NHANES administers the second recall three to 10 days after the first (Centers for Disease Control and Prevention 2016). For records or diaries, a replicate measure refers to a subsequent administration, regardless of whether each record or diary is kept for multiple days (e.g., three or seven), since consecutive days of recording are unlikely to provide an adequate measure of day-to-day variation (National Cancer Institute 2015a).

A number of statistical methods have been developed to estimate usual intake distributions and corresponding proportions above or below a cutoff. These include the U.S. National Research Council or NRC method (National Research Council 1986), the Iowa State University or ISU method (Nusser et al. 1995), the National Cancer Institute or NCI method (Tooze et al. 2006, 2010), the Multiple Source Method (MSM) from the European Food Consumption Validation (EFCOVAL) Project (Harttig et al. 2011; Haubrock et al. 2011), and the Statistical Program to Assess Dietary Exposure (SPADE) (Dekkers et al. 2014) from the National Institute for Public Health and the Environment in the Netherlands. For the more recent methods, macros or software packages are available to execute the approach. For example, the ISU method is implemented using PC-SIDE software; the NCI method is implemented using a series of SAS macros that have been made publicly available online (National Cancer Institute 2015b) along with guidance on their use; MSM is executed using a web-based interface (MSM Development Team 2015); and SPADE is implemented in R using a package called SPADE.RIVM (National Institute for Public Health and the Environment 2016). Each of these methods addresses the basic steps of partitioning the variance and adjusting for within-person variation and includes transformations and back-transformations before and after modeling to address skewness in the intake data (Goedhart et al. 2012). The methods continue to evolve and differ somewhat in their features and approaches. They have been compared and contrasted elsewhere using simulation studies (Goedhart et al. 2012; Laureano et al. 2016; Souverein et al. 2011) and have been found to perform similarly. One such comparison yielded recommendations for caution in the case of dietary components that have high within-person variability or highly skewed distributions, as well as in cases in which distributions are generated using small sample sizes (Souverein et al. 2011).

Recent methods, including the NCI method, MSM, and SPADE, which build upon the two-part ISU method for foods, have been specifically designed to handle episodically-consumed dietary components (Souverein et al. 2011). For example, in the NCI method, usual intake of episodically-consumed dietary components is modeled in a two-part model as the product of the probability of consumption and the consumption day amount (Tooze et al. 2006). These two factors may be correlated in that those who consume a certain dietary component may tend to consume it in greater amounts; the two parts of the model in the NCI method are thus linked to accommodate this correlation. For nonepisodically-consumed components, the probability of consumption is close to 1 and so it is only necessary to model the consumption day amount (i.e., amount-only model) (Tooze et al. 2010). An approach to determining whether a component is episodically-consumed or not drawing from prior analyses is summarized in Box 2.1.

The MSM also accounts for components that are occasionally or rarely eaten and allows for inclusion of information on the proportion of the sample who are nonconsumers of a given dietary component (Harttig et al. 2011; Haubrock et al. 2011). With additional information about whether or not a food is ever consumed (such as from a food frequency questionnaire), MSM, SPADE, and the NCI method can be extended to incorporate never consumers. If interest is in the upper percentiles, this information is generally not informative; however, when interest is in the lower percentiles, it may be important to incorporate (Goedhart et al. 2012; Kipnis et al. 2009).

Some approaches, such as the NCI method and MSM, allow for the incorporation of covariates (Harttig et al. 2011; Haubrock et al. 2011; Tooze et al. 2006). Covariates may be used to account for time-dependent effects that affect true intake, such as the day of the week (e.g., week day versus weekend) or season (Tooze et al. 2010). They may also be used to address nuisance effects, which affect reporting of intake and may include recall sequence (e.g., whether the recall was the first or

BOX 2.1 DIFFERENTIATING EPISODICALLY-CONSUMED FROM NONEPISODICALLY-CONSUMED DIETARY COMPONENTS FOR MODELING PURPOSES

In prior applications of the NCI model for estimating usual intake distributions:

- Dietary components for which zero consumption was observed for less than 5% of each subgroup of interest were considered to be nonepisodically-consumed and an amount-only model (assuming that the probability of consumption was close to 1) was used.
- Dietary components for which more than 10% of each subgroup had zero intake of the dietary component were considered to be episodically-consumed and a two-part model considering both the probability of consumption and the consumption day amount was used.
- For dietary components for which the proportions with zero intakes fell between 5% and 10%, both models were run and the best-fitting was selected.

Sources: Kirkpatrick, S.I. et al., *J. Acad. Nutr. Diet.*, 112, 624–635.e6, 2012; Krebs-Smith, S.M. et al., *J. Nutr.*, 140, 1832–1838, 2010.

second completed by an individual) and the mode of administration in studies with multiple means of data collection (e.g., in-person versus telephone) (Tooze et al. 2010). In addition, covariates may be used to consider individual-level effects. For example, sex and age are often included to allow estimation of distributions for sex–age groups that correspond to nutrient requirements, such as the Dietary Reference Intakes. Including age and sex as covariates allows subgroup analyses (Tooze et al. 2010) and avoids several stratified models, which may suffer from small subsamples and thus, less precise estimates (National Cancer Institute 2011). Other potential individual-level covariates include indicators of socioeconomic status, for example, or behavioral characteristics associated with dietary intake, such as smoking status. Data from a food frequency questionnaire or screener can be included as covariates to augment that from the recalls or records/diaries; this may be most helpful in cases in which a dietary component is consumed episodically and few consumption days are available in the recalls or records/diaries due to a small number of replicates.

To estimate the distribution of usual intake in a population, there are two broad classes of approaches that are used, model-based and model-assisted (Goedhart et al. 2012). These are summarized in Box 2.2.

Due to the normality assumption underlying these methods and the skewness of dietary data, it is necessary to transform the data to approximate normality during the first step, and to back-transform the data to the original scale to obtain estimated percentiles of the distribution. As the mean of the transformed data is not equivalent to the transformation of the mean on the original scale, it is necessary to approximate the estimated percentiles when the data are back-transformed. The percentile can be approximated using a method such as Taylor linearization or numerical methods such as Gauss–Hermite quadrature. The latter method functions by breaking up the area under the curve into smaller pieces and summing them over a series of points using weighting coefficients for each point; it is used by most of the methods for estimating distributions described above.

Methods are also differentiated by their capacity to address single dietary components (i.e., univariate), ratios of two dietary components (i.e., bivariate), and multiple dietary components (i.e., multivariate). For example, within the NCI method, the univariate approach can be used to estimate distributions of usual intake for a single dietary component at a time (Kirkpatrick et al. 2012;

BOX 2.2 MODEL-BASED VERSUS MODEL-ASSISTED APPROACHES TO ESTIMATE THE DISTRIBUTION OF USUAL INTAKE IN A POPULATION

- In the *model-based approach*, distributions are derived from theoretically-derived quantiles of the distribution; the *model-assisted approach* rescales the observed individual mean distribution.
- In the model-based approach, it is assumed that the mean and between-person variance are normally distributed and the percentiles of interest are derived from tabulations of the standard normal distribution. One way this can be done is through a Monte Carlo simulation. In the simulation, values are drawn from the assumed normal distribution of between-person deviations. These values are then added to the estimated population mean to create values for pseudo-people, and the empirical distribution of these pseudo-people is used to estimate the usual intake distribution.
- In the model-assisted approach, each individual's mean intake is recentered by subtracting it from the population mean and scaled by multiplying by the square root of the ratio of the between-person variance to the variance of the within-person mean distribution; this value is then added back to the population mean. This results in rescaled means for each individual with a shrunken variance. The population distribution can be estimated from the empirical distribution of these shrunken means.
- **It is important to note that individual means predicted in these approaches should not be used as estimates of individuals' usual intake but only to estimate the distribution for the population.**
- The SPADE method uses a model-based approach, whereas the NRC method and MSM utilize a model-assisted approach. The ISU and NCI methods use a hybrid of the model-based and model-assisted approaches. For example, in the NCI method, the between-person deviations are assumed to have a normal (or bivariate normal, for the two-part model) distribution that is sampled, but these values are added to the individual mean predicted values; if no covariates are used, it is a model-based approach. In the modern application of the ISU method (C-SIDE or later software), a model-based approach is used, but earlier approaches (IML-SIDE) used a model-assisted shrinkage approach with a model-based transformation.

Krebs-Smith et al. 2010), whereas the bivariate approach can be used to describe the ratio of usual intake of two components simultaneously, such as the percentage of usual energy intake from total fat (Freedman et al. 2010b). A multivariate extension of the NCI method has been used to simultaneously model the components of the Healthy Eating Index (HEI)-2005 and HEI-2010 and to generate distributions of usual HEI scores (Guenther et al. 2014).

2.4.1.3 Estimating Total Usual Nutrient Intake among a Group of Persons

As noted above, to estimate total usual nutrient intake, it is important to consider nutrients contributed by supplements. In the case of group means, it is possible to simply add the nutrient intake from supplements to that from food and beverages, assuming that estimates of supplement usage can be arrived at for a consistent time period (e.g., over 24 hours or daily average). The assumption that reported supplement intake reflects usual intake may be more or less defensible, depending on how information about supplement use was collected. In terms of distributions of total usual nutrient intake, approaches have been developed to combine data on foods and beverages and supplements. These include *shrink, then add* or *add, then shrink* methods, referring to the order in which the usual intake distribution is modeled and the contribution of supplements is added. These approaches have been reviewed elsewhere (Garriguet 2010). The shrink, then add approach has been encouraged because it allows supplement users and nonusers to have different mean intake values (National Cancer Institute 2011). In methods for estimating distributions of usual intake that allow for covariates, a variable can be included to indicate whether a respondent is a supplement user or not. Inclusion of this covariate can allow for estimation of prevalences of inadequate and excessive intake among users and nonusers separately.

2.4.1.4 Additional Statistical Considerations

In applying approaches for describing dietary intake among groups to data collected using complex sampling frames, such as NHANES, it is necessary to account for sampling factors including stratification, clustering, and weighting (Korn and Graubard 1999) to avoid imprecise and biased estimates. Stratification refers to grouping individuals in the population by characteristics; generally demographic characteristics such as race are used. This is done to ensure that conclusions can be made about a particular stratum that would make up a relatively small proportion of a simple random sample. Clustering occurs when individuals are sampled from the same area, for example, counties, in which case they can no longer be assumed to be independent. Weighting is used to account for survey design and nonresponse. Survey-aware software can account for such strata, clusters, and weights. More complex approaches may be needed for estimating appropriate standard errors for distributions of usual intake, for example, Taylor linearization and resampling techniques such as bootstrapping, jackknife, or balanced repeated replication (BRR) (Korn and Graubard 1999), depending on the survey. The NCI method, for example, incorporates BRR weights for complex surveys such as NHANES (Tooze et al. 2010) but can also be used with bootstrapping (Kirkpatrick et al. 2015).

2.4.2 Examining Associations between Diet and Other Variables

In examining associations between diet and other variables, dietary intake can be treated as either an exposure (e.g., in relation to health or disease outcomes) or an outcome (e.g., in relation to sociodemographic predictors or an intervention).

2.4.2.1 Dietary Intake as the Exposure or Independent Variable

For the purposes of epidemiologic research, food frequency questionnaires have traditionally offered lower researcher and respondent burden and costs than recalls and records/diaries (Thompson and Subar 2013). Therefore, they have been the instruments of choice in large-scale cohort studies designed to assess diet–health relationships. Methods to mitigate error in dietary intake data in the context of understanding these relationships have thus largely focused on cases in which food frequency questionnaires (or screeners) are the main instrument (potentially with recalls or records/diaries as a reference), whereas approaches for situations in which recalls or records/diaries are the main instruments are newly emerging and have yet to be widely applied.

2.4.2.1.1 Recalls or Records/Diaries as the Reference Instrument in Examinations of Relationships between Diet and Health or Disease Outcomes

When we are interested in examining associations between dietary exposures and health, regression models are often used, typically with data from a prospective cohort study. For example, depending on the nature of the dependent variables, models include linear regression for continuous outcomes, logistic regression for binary outcomes, or Cox regression for time-to-event outcomes. Each of these approaches uses a linear function of the exposure and covariates to describe the strength of the relationship between the dietary exposure and the health outcome, which is quantified by the regression parameter or coefficient. For example, in logistic regression, the regression parameter is the log of the odds; in Cox regression, it is the log hazard ratio. Unbiased estimates of these coefficients can be obtained using statistical approaches (Kipnis et al. 2003; Neuhouser et al. 2008) if the requisite reference data are available. (For extensions to nonlinear models, see Carroll et al. 2006).

The most commonly used approach is *regression calibration*, which requires that data from a less biased or unbiased reference instrument are available for a random subsample (Neuhouser et al. 2008; Rosner et al. 1989). To obtain an unbiased estimate of the diet-outcome relationship, the reference measure should be unbiased and the error in the reference data should be independent of true intake and of error in the data collected using the main instrument for the study (Kipnis et al. 1999, 2003). Reference data may be collected using a calibration substudy (Kipnis et al. 2003), which may be internal or external to the main study. In cases in which data from an external calibration study are used, the population as well as the main instrument used to assess dietary intake should be similar (Freedman et al. 2011a).

As noted, ideally, the reference measure captures diet without bias. This is true for recovery biomarkers that provide a measure of true intake for only a few dietary components; other markers of true intake remain to be discovered. Thus, in practice, reference data for reducing bias in data from one self-report instrument, such as a food frequency questionnaire, are often obtained using another less biased self-report method, such as recalls or records/diaries. For example, 24-hour recalls are used as the reference instrument for the NIH–AARP cohort (Thompson et al. 2008). Further, food records have been used as a reference for evaluating a food frequency questionnaire in the Nurses' Health Study and the Health Professionals Follow-up Study (Feskanich et al. 1993; Rimm et al. 1992; Willett et al. 1985). It has been noted that these forms of reference data contain bias and likely do not meet the requirements stated above in terms of error that is independent of true intake and of the error in the data from the main instrument (Freedman et al. 2011a; Kipnis et al. 2003). However, the existing evidence suggests that even though data from recalls and records/diaries contain bias and violate assumptions, their use to mitigate the even-greater bias in frequency data is preferable to ignoring it, though this is based on comparisons with biomarker data available for only a few dietary components (Freedman et al. 2011a). Calibration can also be used to reduce bias in data from screeners.

In regression calibration, reported dietary intake values are replaced with expected values of true usual intake (Spiegelman et al. 1997). These expected values are the best predicted values based on measured intake as well as covariates (e.g., potential confounders that are theoretically related to intake and to the outcome) that will also be included in the diet–health model (Kipnis et al. 1999). Regression calibration includes two steps, both of which involve regression modeling. The predicted values are calculated using a prediction equation in which the values from the reference measure are regressed on reported intake from the main dietary instrument and the covariates using data from the calibration substudy (in which both the main instrument and the reference instrument were administered). The prediction equation can then be used to predict intake for all participants in the main study. These predicted or expected values are then used in the diet–health regression model. As a result, bias in the regression coefficients is reduced, or the coefficients are deattenuated. In *enhanced regression calibration*, the prediction equation includes additional covariates that are not related to the outcome (i.e., not confounders) but may help to predict true usual intake (Freedman et al. 2011b; Kipnis et al. 2009; National Cancer Institute 2011). For example, biomarkers of intake may be used in the prediction equation (Freedman et al. 2011b).

Regression calibration is appropriate in cases in which error in reported dietary intake is nondifferential. This is most likely to be the case in situations in which dietary intake data are collected prior to the occurrence of health or disease outcomes (i.e., prospective studies) (Freedman et al. 2011a). Further, it should be noted that regression calibration helps to reduce bias in estimation (i.e., by deattenuating estimates) but it does not recover the loss in power that arises due to measurement error (Freedman et al. 2011a).

In the Women's Health Initiative, calibration equations have been used to calibrate the estimates of dietary intake for participants, which are then associated with health outcomes using regression (Prentice et al. 2009, 2013; Tinker et al. 2011). The calibration equations are developed through linear regression of recovery biomarkers on estimates of intake from self-report instruments, including recalls and records/diaries, while also accounting for covariates. This method is further detailed in Chapter 13.

Another approach to deattenuate regression coefficients such as the log odds ratio is by dividing them by the *attenuation factor*, which is the slope from the regression of the true intake on self-report in a model also including the covariates (Rosner et al. 1989, 1990). The attenuation factor is estimated by the regression coefficient for the main instrument from the regression of the reference measure on self-report and covariates. In the second step, rather than regressing the outcome on the predicted value, the outcome is regressed on the main instrument, and this parameter estimate is deattenuated by dividing it by the attenuation factor. The standard error is then estimated by the delta method or bootstrap.

As stated above, absolute energy intake is poorly measured by any self-report dietary assessment instrument. If accurate energy intake estimates are necessary, the optimal means by which to capture them is to use DLW to allow for measurement error adjustment of self-reported energy intakes (Neuhouser et al. 2008). Such DLW data, however, should not be used to energy adjust other self-reported dietary constituents. Although flawed for making inferences about energy itself, self-reported energy intake estimates are useful in reducing measurement error in other self-reported dietary constituents (Kipnis et al. 2003), as the error in energy reporting is correlated with error in the reported intakes. In particular, nutrient densities based on food frequency questionnaires correlate with true intakes more closely than do absolute intakes. Generally, in multivariate models examining diet–disease associations, self-reported energy is controlled for by including it in the model even when nutrient density variables are included as it further reduces bias; however, the coefficients for energy should not be used to make inferences about energy intake and disease outcomes (Willett 2013).

To accommodate energy adjustment, the regression calibration approach can be extended to bivariate and multivariate applications. For example, to estimate relationships between energy-adjusted food group intakes and an outcome, a bivariate regression calibration approach that models the usual intake of the food group and energy simultaneously can be used (Kipnis et al. 2016; Liese et al. 2015).

2.4.2.1.2 Recalls or Records/Diaries as the Main Instrument in Examinations of Relationships between Diet and Health or Disease Outcomes

The use of recalls and records/diaries as the main dietary assessment instrument in epidemiologic research, such as cohort studies, has recently been made possible with technological advances (see Chapter 3) that have greatly reduced the cost of collecting and coding data. For example, web-based self-administered 24-hour recalls such as the Automated Self-Administered 24-hour (ASA24) Dietary Assessment Tool (Subar et al. 2012), which was developed for the United States and has been adapted for Canada and Australia, and myfood24 (Carter et al. 2015), which was developed in the United Kingdom, make it cost-effective to collect recalls in large-scale studies. Mobile food records or diaries have also been developed (see Chapter 3). As a result, statistical methods for situations in which recalls or records/diaries provide the main dietary assessment data for examining diet and health/disease relationships are emerging. These advances have the potential to reduce the power loss associated with measurement error because recalls and records capture dietary intake with less bias than the traditionally-used food frequency questionnaires (Freedman et al. 2011a). It should be noted that approaches to reduce error when recalls or records are the main instrument in diet–health studies require at least two nonconsecutive administrations of the instrument.

In cases in which recalls or records are the main instrument, the NCI method described earlier can be used to generate estimates of predicted usual intake that can be utilized to assess relationships between dietary exposures (both episodically and nonepisodically-consumed) and a health outcome. For example, the method has been applied to examine the relationship between fish intake and blood mercury using cross-sectional data from NHANES (Kipnis et al. 2009).

Further, regression calibration can be used to combine recalls/records and frequency data in cases in which both are collected, by utilizing the frequency data as a covariate in the enhanced regression calibration approach. This approach of combining instruments has been recommended for improving the precision of estimates and increasing the power to test diet–health relationships (Carroll et al. 2012). For nutrients that are not highly episodically-consumed, the use of two 24-hour recalls alone is comparable to that of the use of a single food frequency questionnaire, and using six 24-hour recalls can improve power from approximately 60% to approximately 80%; adding a food frequency questionnaire to the 24-hour recalls can produce even more dramatic gains in power, and is most beneficial for highly episodically-consumed foods, for which even a large number of 24-hour recalls (e.g., twelve) may be inadequate (Carroll et al. 2012). Therefore, in examining relationships between diet and health or disease outcomes, four to six administrations of 24-hour recalls and a food frequency questionnaire are recommended to enable examination of relationships between many nutrients and food groups and outcomes of interest. It is also possible to combine data collected using self-report measures with biomarker data using statistical approaches (Freedman et al. 2010a, 2011b). This is discussed in Chapter 13.

2.4.2.2 Dietary Intake as the Outcome or Dependent Variable

Relatively little work has been conducted to address error in cases in which diet is treated as the dependent variable, for example, for the purposes of elucidating associations between sociodemographic factors and dietary intake, or examining the influence of an intervention on diet. The statistical issues for diet as the outcome or dependent variable are different than discussed in the previous section. Random error alone in a predictor variable will lead to attenuation of diet–health relationships, and cannot be ignored. In contrast, for a *dependent* variable with random error only, the measurement error is essentially added to the variance of the true response, and thus may be considered *ignorable* (Carroll et al. 2006). However, when the dependent variable exhibits systematic error, as in the case of self-report dietary intake, the error can no longer be ignored. Furthermore, when the bias differs between two groups such as in an intervention study, then the error is differential between the groups, and the difference between the means of the self-reported intakes is different than the difference in true intakes between the two groups; the estimate of the intervention effect from self-report data is thus biased. When a biomarker is available for at least a subset of the sample, it is possible to obtain estimates of corrected group means by regressing the self-report instrument on the biomarker and using the intercept and coefficient for the biomarker to recenter and scale the group means. Next, a weighted average of the difference between the group biomarker means in the subsample with biomarkers and difference of the corrected group means is used to estimate the intervention effect; the weights are based on the variance and covariance of the two estimates of the difference (Buonaccorsi 1991; Keogh et al. 2016).

Addressing bias and differential error when diet is an outcome is an area for which additional methods development is needed. Biomarker-based studies to better understand intervention-related biases in reporting of intake would be desirable for informing such work.

2.5 IMPLICATIONS FOR DATA COLLECTION

The focus of this chapter is on analytic approaches rather than recommendations for the collection of intake data for specific research questions and designs, which are outlined elsewhere (National Cancer Institute 2015a). Nonetheless, it is important to note that the eventual application of recommended statistical approaches requires advance planning in terms of the types of data collected and the number of administrations of tools, as well as the execution of calibration substudies. It is thus critical that optimal analytic strategies are considered in advance to ensure that appropriate data are captured.

For surveillance purposes, estimating distributions of usual intake requires that replicate recalls be available at least for a subsample, with at least 50 *double hits* recommended. This subsample should be representative of the overall sample. Thus, NHANES collects multiple recalls from all participants (Centers for Disease Control and Prevention 2016), whereas the Canadian Community Health Survey administers a second recall to a subsample some days later (Health Canada 2006).

BOX 2.3 CALCULATION OF THE NUMBER OF RECALLS NEEDED TO ACHIEVE 50 (OR MORE) REPEAT REPORTS, OR *DOUBLE HITS*, OF A DIETARY COMPONENT

- To obtain at least 50 individuals who consume a dietary component on two recalls, divide the number of *double hits* desired by the probability of observing at least two reported consumption occasions for a person.
- The probability of having at least two recalls with consumption reports is calculated from the binomial formula (assuming independence of recalls):

$$\Pr(h \geq 2) = \sum_{h=2}^{n} \binom{n}{h} p^h (1-p)^{n-h}$$

where:

n is the number of recalls per person
h is the number of *hits* or consumption reports
p is the probability of consumption on a recall

- For example, if the probability of consuming the dietary component on any one recall is 20% and 50 *double hits* are desired:
  - If each person completes two recalls,

$$\Pr(h = 2) = \binom{2}{2} 0.20^2 0.80^0 = (1)(0.04)(1) = 0.04, \text{ and}$$

$$\text{the sample size needed is } \frac{50}{0.04} = 1250.$$

  - If each person completes four recalls, then

$$\Pr(h \geq 2) = \binom{4}{2} 0.20^2 0.80^2 + \binom{4}{3} 0.20^3 0.80^1 + \binom{4}{4} 0.20^4 0.80^0$$

$$= \frac{4!}{2!2!}(0.04)(0.64) + \frac{4!}{3!1!}(0.008)(0.80) + \frac{4!}{4!0!}(0.0016)(1)$$

$$= 0.1536 + 0.0256 + 0.0016$$

$$= 0.1808, \text{ and the sample size needed is } \frac{50}{0.1808} = 277.$$

- Note that for a complex survey, this estimated sample size should be multiplied by the design effect.

Source: Office of Science and Technology Standards and Health Protection Division Office of Water, United States Environmental Protection Agency, Guidance for conducting fish consumption surveys, Washington, DC, https://www.epa.gov/sites/production/files/2016-07/documents/guidance-conducting-fish-consumption-surveys-2016.pdf, accessed July 15, 2016.

BOX 2.4 SAMPLE SIZE CALCULATION FOR CALIBRATION STUDIES

A reasonable approach for calculating the sample size for a calibration study (N) is

$$N = 4\left(\frac{1}{f}\right)\left(\frac{\sigma_e^2}{\lambda_Q^2\sigma_Q^2}\right)$$

where:

f is the fraction of the uncertainty of the attenuated coefficient (such as 0.10)
σ_e^2 is the residual variance from regressing the reference instrument on the main instrument
σ_Q^2 is the variance of the main instrument
λ_Q is the attenuation factor

Sources: Carroll, R.J. et al., *Biometrics*, 53, 1440–1457, 1997a; Carroll, R.J. et al., *Am. J. Clin. Nutr.*, 65, 1187S–1189S, 1997b.

The sample size needed to obtain adequate multiple (at least two) reports of a given dietary component may be calculated by dividing the number of *double hits* needed by the probability of observing two or more reports (Box 2.3).

In cases in which replicate recalls are not available or feasible, it may be possible to obtain estimates of within-person variation from another study for use in estimating distributions of usual intake. However, such data for a similar population are not always available.

In epidemiological studies, the use of regression calibration to reduce bias in data collected using the main dietary assessment instrument requires that a reference method be administered in at least a subsample to allow for development of prediction equations for use in mitigating bias. This subsample may be internal or external to the main study, though external studies should be similar in terms of the population and the characteristics of the main dietary assessment instrument. The calibration study should be large enough to obtain precise estimates of the coefficient describing the relationship between the predictor of diet and the outcome in the model (Box 2.4). The uncertainty of this predictor is the sum of the uncertainty in estimating the attenuated coefficient and the uncertainty in estimating the attenuation factor in the calibration study of a certain size. To minimize increasing the variance from the calibration study, the second term is chosen to be a fraction of the first term, such as one-tenth.

Although regression calibration can alleviate bias, it does not restore power. (An exception is enhanced regression calibration, which restores some power by improving the prediction of intake in regression calibration.) It can be shown that the effective sample size of the study is equal to the sample size based on true intake multiplied by the correlation of self-report and true intake. (In other words, if N participants were required in a study in which intake was exactly measured, then the sample size would need to be inflated by a factor of $1/r^2$, where r is the correlation of dietary intake and true intake [Kaaks et al. 1995] in the study using the error-prone exposure.)

Therefore, when designing a diet–health study, estimated parameter estimates and variance from the self-report instrument and the regression of the outcome on the instrument should be used to estimate the sample size when this information is available, keeping in mind that the parameter estimates are attenuated. If estimates based on truth (e.g., derived from a feeding or biomarker study) are used to calculate sample size, the calculated sample size should be divided by r^2 to account for the substitution of the error-prone instrument for the true intake. Note that when coefficients are severely attenuated (for example, relative risks attenuated to observed values of 1.25 or less), this can lead to associations that are so close to zero that it can be difficult to separate the effect of diet from unmeasured confounders. As the impact of unmeasured confounders cannot be alleviated by increasing the sample size, even restoring power by increased sample size may not lead to the expected increased signal of the severely attenuated measure (Freedman et al. 2011a).

In intervention studies, it may be advantageous to collect multiple recalls from all participants to help reduce power loss, at least until further work is conducted to identify strategies for addressing intervention-related and other differential biases when diet is treated as an outcome. Whenever possible, objective data on dietary intake, such as a recovery biomarker, should be collected on at least a subset of participants in intervention studies. The self-report measure is most beneficial in terms of adding information about the intervention effect to the biomarker when its reliability is comparable to that of the biomarker (Keogh et al. 2016). For example, using four 24-hour recalls to measure sodium intake has been found to have similar reliability to a 24-hour urine measure of sodium (Keogh et al. 2016).

In cases of exploratory studies, such as prospective data collection initiatives that have the potential to examine multiple hypotheses, for example, driven by natural experiments (Craig et al. 2012), it may be strategic to adopt a data collection strategy that allows for maximum flexibility, such as the administration of multiple recalls and/or records, perhaps alongside a food frequency questionnaire to capture episodically-consumed dietary components (National Cancer Institute 2015a). As we continue to learn more about measurement error and strategies to alleviate it, it is likely that recommendations for assessment of dietary intake for different research questions will continue to evolve.

2.6 RESOURCES TO GUIDE COLLECTION, ANALYSIS, AND REPORTING OF DIETARY INTAKE DATA

Assessing dietary intake is central to much nutrition research. However, it is a complex endeavor. Resources continue to emerge to help researchers choose appropriate data collection and analysis strategies, as well as to clearly and fully report diet assessment methods used in studies. Tools that provide information to help guide method selection include the Diet Assessment Primer (National Cancer Institute 2015a) described in Thompson et al. (2015), the Diet and Physical Activity Measurement Toolkit (Medical Research Council 2016), and the Dietary Intake Assessment Method Selection Guide (Australasian Child & Adolescent Obesity Research Network 2011). The NHANES Dietary Web Tutorial (Centers for Disease Control and Prevention 2014) hosted by the Centers for Disease Control and Prevention provides guidance for both basic and advanced dietary analyses, including modeling of usual intake distributions and examining relationships between usual dietary intake and an outcome, as well as estimating total nutrient intake. In addition, Strengthening the Reporting of Observational Studies in Epidemiology-Nutritional Epidemiology (STROBE-nut), an extension of the STROBE Statement, has recently been released and provides reporting guidelines for nutritional epidemiology (Lachat et al. 2016). Guidelines within STROBE-nut, such as providing clear and detailed descriptions of the dietary assessment methods used as well as characteristics of the study that might have affected dietary intake, could be applied beyond epidemiology to improve the transparency of methods used to collect and analyze dietary data, informing appropriate interpretation.

ACKNOWLEDGMENTS

The authors of this chapter are part of a collaborative group based at the U.S. National Cancer Institute that planned and participated in a webinar series on measurement error in dietary intake data, which took place in the fall of 2011, as well as the National Cancer Institute's Dietary Assessment Primer, which was launched online in early 2015. The organization of the discussion in this chapter benefited from the preparation of materials for the Measurement Error Webinar Series and Dietary Assessment Primer, including the contributions of the collaborators on those projects. The authors are grateful to Dr. Susan M. Krebs-Smith for helpful comments on an earlier draft of this chapter. Sharon Kirkpatrick is funded by a Capacity Development Award (grant #702855) from the Canadian Cancer Society Research Institute. Janet Tooze is partially funded by a National Cancer Institute Cancer Center Support Grant P30 CA012197.

REFERENCES

An, R. 2016. Weekend-weekday differences in diet among U.S. adults, 2003–2012. *Ann Epidemiol* 26(1): 57–65.

Arab, L., C.H. Tseng, A. Ang, and P. Jardack. 2011. Validity of a multipass, web-based, 24-hour self-administered recall for assessment of total energy intake in blacks and whites. *Am J Epidemiol* 174(11): 1256–1265.

Australasian Child & Adolescent Obesity Research Network. 2011. Food and nutrition: Dietary intake assessment. http://www.acaorn.org.au/streams/nutrition/ (accessed July 15, 2016).

Barr, S.I., S.P. Murphy, and M.I. Poos. 2002. Interpreting and using the dietary references intakes in dietary assessment of individuals and groups. *J Am Diet Assoc* 102(6): 780–788.

Basiotis, P.P., S.O. Welsh, F.J. Cronin, J.L. Kelsay, and W. Mertz. 1987. Number of days of food intake records required to estimate individual and group nutrient intakes with defined confidence. *J Nutr* 117: 1638–1641.

Beaton, G.H., J. Milner, P. Corey et al. 1979. Sources of variance in 24-hour dietary recall data: Implications for nutrition study design and interpretation. *Am J Clin Nutr* 32(12): 2546–2559.

Beaton, G.H. 1994. Approaches to analysis of dietary data: Relationship between planned analyses and choice of methodology. *Am J Clin Nutr* 59(1): 253S–261S.

Beaton, G.H., J. Burema, and C. Ritenbaugh. 1997. Errors in the interpretation of dietary assessments. *Am J Clin Nutr* 65(4 Suppl): 1100S–1107S.

Buonaccorsi, J. 1991. Measurement errors, linear calibration and inferences for means. *Comput Stat Data Anal* 11(3): 239–257.

Carroll, R.J., L. Freedman, and D. Pee. 1997a. Design aspects of calibration studies in nutrition, with analysis of missing data in linear measurement error models. *Biometrics* 53(4): 1440–1457.

Carroll, R.J., D. Pee, L.S. Freedman, and C.C. Brown. 1997b. Statistical design of calibration studies. *Am J Clin Nutr* 65(4 Suppl): 1187S–1189S.

Carroll, R.J., D. Ruppert, L.A. Stefanski, and C.M. Crainiceanu. 2006. *Measurement Error in Nonlinear Models: A Modern Perspective*. Boca Raton, FL: Chapman & Hall/CRC.

Carroll, R.J., D. Midthune, A.F. Subar et al. 2012. Taking advantage of the strengths of 2 different dietary assessment instruments to improve intake estimates for nutritional epidemiology. *Am J Epidemiol* 175(4): 340–347.

Carter, M.C., S.A. Albar, M.A. Morris et al. 2015. Development of a UK online 24-h dietary assessment tool: Myfood24. *Nutrients* 7(6): 4016–4032.

Centers for Disease Control and Prevention. 2014. NHANES dietary web tutorial. http://www.cdc.gov/nchs/tutorials/dietary/ (accessed July 15, 2016).

Centers for Disease Control and Prevention. 2016. National Health and Nutrition Examination Survey. National Center for Health Statistics. http://www.cdc.gov/nchs/nhanes/ (accessed July 15, 2016).

Craig, P., C. Cooper, D. Gunnell et al. 2012. Using natural experiments to evaluate population health interventions: New medical research council guidance. *J Epidemiol Commun Health* 66(12): 1182–1186.

De Keyzer, W., T. Bracke, S. McNaughton et al. 2015. Cross-continental comparison of national food consumption survey methods—A narrative review. *Nutrients* 7(5): 3587–3620.

Dekkers, A.L., J. Verkaik-Kloosterman, C.T. van Rossum, and M.C. Ocke. 2014. SPADE, a new statistical program to estimate habitual dietary intake from multiple food sources and dietary supplements. *J Nutr* 144(12): 2083–2091.

Dodd, K.W., P.M. Guenther, L.S. Freedman et al. 2006. Statistical methods for estimating usual intake of nutrients and foods: A review of the theory. *J Am Diet Assoc* 106(10): 1640–1650.

Feskanich, D., E.B. Rimm, E.L. Giovannucci et al. 1993. Reproducibility and validity of food intake measurements from a semiquantitative food frequency questionnaire. *J Am Diet Assoc* 93(7): 790–796.

Flegal, K.M., P.M. Keyl, and F.J. Nieto. 1991. Differential misclassification arising from nondifferential errors in exposure measurement. *Am J Epidemiol* 134(10): 1233–1244.

Freedman, L.S., D. Midthune, R.J. Carroll et al. 2004. Adjustments to improve the estimation of usual dietary intake distributions in the population. *J Nutr* 134(7): 1836–1843.

Freedman, L.S., V. Kipnis, A. Schatzkin, N. Tasevska, and N. Potischman. 2010a. Can we use biomarkers in combination with self-reports to strengthen the analysis of nutritional epidemiologic studies? *Epidemiol Perspect Innov* 7(1): 2.

Freedman, L.S., P.M. Guenther, K.W. Dodd, S.M. Krebs-Smith, and D. Midthune. 2010b. The population distribution of ratios of usual intakes of dietary components that are consumed every day can be estimated from repeated 24-hour recalls. *J Nutr* 140(1): 111–116.

Freedman, L.S., A. Schatzkin, D. Midthune, and V. Kipnis. 2011. Dealing with dietary measurement error in nutritional cohort studies. *J Natl Cancer Inst* 103(14): 1086–1092.

Freedman, L.S., D. Midthune, R.J. Carroll et al. 2011b. Using regression calibration equations that combine self-reported intake and biomarker measures to obtain unbiased estimates and more powerful tests of dietary associations. *Am J Epidemiol* 174(11): 1238–1245.

Freedman, L.S., J.M. Commins, J.E. Moler et al. 2014. Pooled results from 5 validation studies of dietary self-report instruments using recovery biomarkers for energy and protein intake. *Am J Epidemiol* 180(2): 172–188.

Freedman, L.S., J.M. Commins, J.E. Moler et al. 2015a. Pooled results from 5 validation studies of dietary self-report instruments using recovery biomarkers for potassium and sodium intake. *Am J Epidemiol* 181(7): 473–487.

Freedman, L.S., D. Midthune, R.J. Carroll et al. 2015b. Application of a new statistical model for measurement error to the evaluation of dietary self-report instruments. *Epidemiology* 26(6): 925–933.

Freedman, L.S., D. Midthune, K.W. Dodd, R.J. Carroll, and V. Kipnis. 2015c. A statistical model for measurement error that incorporates variation over time in the target measure, with application to nutritional epidemiology. *Stat Med* 34(27): 3590–3605.

Gahche, J., R. Bailey, V. Burt et al. 2011. Dietary supplement use among U.S. adults has increased since NHANES III (1988–1994). *NCHS Data Brief* 61: 1–8.

Garriguet, D. 2010. Combining nutrient intake from food/beverages and vitamin/mineral supplements. *Health Rep* 21(4): 71–84.

Goedhart, P.W., H. Voet, S. Knüppel et al. 2012. A comparison by simulation of different methods to estimate the usual intake distribution for episodically consumed foods. *EFSA Supporting Publications* 9(6): 299E.

Goldberg, G.R., A.E. Black, and S.A. Jebb. 1991. Critical evaluation of energy intake data using fundamental principles of energy physiology: 1. Derivation of cut-off limits to identify under-recording. *Eur J Clin Nutr* 45: 569–581.

Guenther, P.M., S.I. Kirkpatrick, J. Reedy et al. 2014. The healthy eating index-2010 is a valid and reliable measure of diet quality according to the 2010 dietary guidelines for Americans. *J Nutr* 144(3): 399–407.

Haines, P.S., M.Y. Hama, D.K. Guilkey, and B.M. Popkin. 2003. Weekend eating in the United States is linked with greater energy, fat, and alcohol intake. *Obes Res* 11(8): 945–949.

Hartman, A.M., C.C. Brown, J. Palmgren et al. 1990. Variability in nutrient and food intakes among older middle-aged men. Implications for design of epidemiologic and validation studies using food recording. *Am J Epidemiol* 132(5): 999–1012.

Harttig, U., J. Haubrock, S. Knüppel, and H. Boeing. 2011. The MSM program: Web-based statistics package for estimating usual dietary intake using the multiple source method. *Eur J Clin Nutr* 65: 87–91.

Haubrock, J., U. Nothlings, J.L. Volatier et al. 2011. Estimating usual food intake distributions by using the multiple source method in the EPIC-Potsdam calibration study. *J Nutr* 141(5): 914–920.

Health Canada. 2006. Canadian Community Health Survey, cycle 2.2, nutrition (2004): A guide to accessing and interpreting the data. http://www.hc-sc.gc.ca/fn-an/surveill/nutrition/commun/cchs_guide_escc-eng.php (accessed July 15, 2016).

Huang, Y., L. Van Horn, L.F. Tinker et al. 2014. Measurement error corrected sodium and potassium intake estimation using 24-hour urinary excretion. *Hypertension* 63(2): 238–244.

Institute of Medicine Subcommittees on Interpretation and Uses of Dietary Reference and Upper Reference Levels of Nutrients and the Standing Committee on the Scientific Evaluation of Dietary Reference Intakes, Food and Nutrition Board. 2000. *Dietary Reference Intakes: Applications in Dietary Assessment*. Washington, DC: National Academies Press.

Kaaks, R., E. Riboli, and W. van Staveren. 1995. Calibration of dietary intake measurements in prospective cohort studies. *Am J Epidemiol* 142: 548–556.

Kaaks, R., P. Ferrari, A. Ciampi, M. Plummer, and E. Riboli. 2002. Uses and limitations of statistical accounting for random error correlations in the validation of dietary questionnaire assessments. *Public Health Nutr* 5(6A): 969–976.

Keogh, R.H., R.J. Carroll, J.A. Tooze, S.I. Kirkpatrick, and L.S. Freedman. 2016. Statistical issues related to dietary intake as the response variable in intervention trials. *Stat Med* 35(25): 4493–4508.

Kipnis, V., L.S. Freedman, C.C. Brown, A.M. Hartman, A. Schatzkin, and S. Wacholder. 1997. Effect of measurement error on energy-adjustment models in nutritional epidemiology. *Am J Epidemiol* 146(10): 842–855.

Kipnis, V., R.J. Carroll, L.S. Freedman, and L. Li. 1999. Implications of a new dietary measurement error model for estimation of relative risk: Application to four calibration studies. *Am J Epidemiol* 150(6): 642–651.

Kipnis, V., D. Midthune, L. Freedman et al. 2002. Bias in dietary-report instruments and its implications for nutritional epidemiology. *Public Health Nutr* 5(6A): 915–923.

Kipnis, V., A.F. Subar, D. Midthune et al. 2003. Structure of dietary measurement error: Results of the OPEN biomarker study. *Am J Epidemiol* 158(1): 14–16.

Kipnis, V., D. Midthune, D.W. Buckman et al. 2009. Modeling data with excess zeros and measurement error: Application to evaluating relationships between episodically consumed foods and health outcomes. *Biometrics* 65(4): 1003–1010.

Kipnis, V., L.S. Freedman, R.J. Carroll, and D. Midthune. 2016. A bivariate measurement error model for semicontinuous and continuous variables: Application to nutritional epidemiology. *Biometrics* 72(1): 106–115.

Kirkpatrick, S.I., K.W. Dodd, J. Reedy, and S.M. Krebs-Smith. 2012. Income and race/ethnicity are associated with adherence to food-based dietary guidance among US adults and children. *J Acad Nutr Diet* 112(5): 624–635.e6.

Kirkpatrick, S.I., K.W. Dodd, R. Parsons, C. Ng, D. Garriguet, and V. Tarasuk. 2015. Household food insecurity is a stronger marker of adequacy of nutrient intakes among Canadian compared to American youth and adults. *J Nutr* 145(7): 1596–1603.

Korn, E.L., and B.I. Graubard. 1999. *Analysis of Health Surveys*. Hoboken, NJ: John Wiley & Sons.

Krebs-Smith, S.M., P.M. Guenther, A.F. Subar, S.I. Kirkpatrick, and K.W. Dodd. 2010. Americans do not meet federal dietary recommendations. *J Nutr* 140(10): 1832–1838.

Lachat, C., D. Hawwash, M.C. Ocké et al. 2016. Strengthening the Reporting of Observational Studies in Epidemiology—Nutritional epidemiology (STROBE-Nut): An extension of the STROBE statement. *PLOS Medicine* 13(6): e1002036.

Laureano, G., V. Torman, S. Crispim, A. Dekkers, and S. Camey. 2016. Comparison of the ISU, NCI, MSM, and SPADE methods for estimating usual intake: A simulation study of nutrients consumed daily. *Nutrients* 8(3): 166.

Liao, X., D.M. Zucker, Y. Li, and D. Spiegelman. 2011. Survival analysis with error-prone time-varying covariates: A risk set calibration approach. *Biometrics* 67(1): 50–58.

Liese, A.D., J.L. Crandell, J.A. Tooze et al. 2015. Sugar-sweetened beverage intake and cardiovascular risk factor profile in youth with type 1 diabetes: Application of measurement error methodology in the SEARCH Nutrition Ancillary Study. *Brit J Nutr* 114(3): 430–438.

Lissner, L., R.P. Troiano, D. Midthune et al. 2007. OPEN about obesity: Recovery biomarkers, dietary reporting errors and BMI. *Int J Obesity* 31(6): 956–961.

Ma, Y., B.C. Olendzki, W. Li et al. 2006. Seasonal variation in food intake, physical activity, and body weight in a predominantly overweight population. *Eur J Clin Nutr* 60(4): 519–528.

Medical Research Council. 2016. Diet and physical activity measurements toolkit. http://dapa-toolkit.mrc.ac.uk/ (accessed July 15, 2016).

MSM Development Team. 2015. The Multiple Source Method (MSM). https://msm.dife.de/ (accessed July 15, 2016).

Moshfegh, A.J., D.G. Rhodes, D.J. Baer et al. 2008. The US department of agriculture automated multiple-pass method reduces bias in the collection of energy intakes. *Am J Clin Nutr* 88(2): 324–332.

National Cancer Institute. 2011. Measurement error webinar series. http://epi.grants.cancer.gov/events/measurement-error/ (accessed July 25, 2016).

National Cancer Institute. 2015a. Dietary assessment primer. https://dietassessmentprimer.cancer.gov/ (accessed July 15, 2016).

National Cancer Institute. 2015b. Usual dietary intakes: SAS macros for the NCI method. http://epi.grants.cancer.gov/diet/usualintakes/macros.html (accessed July 15, 2016).

National Institute for Public Health and the Environment. 2016. SPADE. http://rivm.nl/en/Topics/S/SPADE (accessed July 15, 2016).

National Research Council. 1986. *Nutrient Adequacy: Assessment Using Food Consumption Surveys*. Washington, DC: National Academies Press.

Neuhouser, M.L., L. Tinker, P.A. Shaw et al. 2008. Use of recovery biomarkers to calibrate nutrient consumption self-reports in the women's health initiative. *Am J Epidemiol* 167(10): 1247–1259.

Nicastro, H.L., R.L. Bailey, and K.W. Dodd. 2015. Using 2 assessment methods may better describe dietary supplement intakes in the United States. *J Nutr* 145(7): 1630–1634.

Nielsen, S.J., B.K. Kit, T. Fakhouri, and C.L. Ogden. 2012. Calories consumed from alcoholic beverages by U.S. adults, 2007–2010. *NCHS Data Brief*, 110.

Nusser, S.M., W.A. Fuller, and P.M. Guenther. 1995. Estimating usual dietary intake distributions: Adjusting for measurement error and nonnormality in 24-hour food intake data. *Dietary Assessment Research Series Report 6*. Staff Report 95-SR 80.

Office of Science and Technology Standards and Health Protection Division Office of Water, United States Environmental Protection Agency. 2016. Guidance for conducting fish consumption surveys. Washington, DC. https://www.epa.gov/sites/production/files/2016-07/documents/guidance-conducting-fish-consumption-surveys-2016.pdf (accessed July 15, 2016).

Prentice, R.L., P.A. Shaw, S.A. Bingham et al. 2009. Biomarker-calibrated energy and protein consumption and increased cancer risk among postmenopausal women. *Am J Epidemiol* 169(8): 977–989.

Prentice, R.L., Y. Mossavar-Rahmani, Y. Huang et al. 2011. Evaluation and comparison of food records, recalls, and frequencies for energy and protein assessment by using recovery biomarkers. *Am J Epidemiol* 174(5): 591–603.

Prentice, R.L., L.F. Tinker, Y. Huang, and M.L. Neuhouser. 2013. Calibration of self-reported dietary measures using biomarkers: An approach to enhancing nutritional epidemiology reliability. *Curr Atheroscler Rep* 15(9): 353.

Rimm, E.B., E.L. Giovannucci, M.J. Stampfer et al. 1992. Reproducibility and validity of an expanded self-administered semiquantitative food frequency questionnaire among male health professionals. *Am J Epidemiol* 135(10): 1114–1126.

Rosner, B., W.C. Willett, and D. Spiegelman. 1989. Correction of logistic regression relative risk estimates and confidence intervals for systematic within-person measurement error. *Stat Med* 8(9): 1051–1069; discussion 1071–1073.

Rosner, B., D. Spiegelman, and W.C. Willett. 1990. Correction of logistic regression relative risk estimates and confidence intervals for measurement error: The case of multiple covariates measured with error. *Amer J Epidemiol* 132(4): 734–745.

Souverein, O.W., A.L. Dekkers, A. Geelen et al. 2011. Comparing four methods to estimate usual intake distributions. *Eur J Clin Nutr* 65: S92–S101.

Spiegelman, D., A. McDermott, and B. Rosner. 1997. Regression calibration method for correcting measurement-error bias in nutritional epidemiology. *Am J Clin Nutr* 65(4 Suppl): 1179S–1186S.

Subar, A.F., V. Kipnis, R.P. Troiano et al. 2003. Using intake biomarkers to evaluate the extent of dietary misreporting in a large sample of adults: The OPEN study. *Am J Epidemiol* 158(1): 1–13.

Subar, A.F., S.I. Kirkpatrick, B. Mittl et al. 2012. The Automated Self-Administered 24-hour dietary recall (ASA24): A resource for researchers, clinicians, and educators from the national cancer institute. *J Acad Nutr Diet* 112(8): 1134–1137.

Subar, A.F., L.S. Freedman, J.A. Tooze et al. 2015. Addressing current criticism regarding the value of self-report dietary data. *J Nutr* 145(12): 2639–2645.

Tasevska, N., D. Midthune, N. Potischman et al. 2011. Use of the predictive sugars biomarker to evaluate self-reported total sugars intake in the Observing Protein and Energy Nutrition (OPEN) Study. *Cancer Epidemiol Biomarkers Prev* 20(3): 490–500.

Tasevska, N. 2015. Urinary sugars—A biomarker of total sugars intake. *Nutrients* 7(7): 5816–5833.

Thiébaut, A.C.M., L.S. Freedman, R.J. Carroll, and V. Kipnis. 2007. Is it necessary to correct for measurement error in nutritional epidemiology? *Ann Intern Med* 146(1): 65–67.

Thompson, F.E., and A.F. Subar. 2013. Dietary assessment methodology. In *Nutrition inthe Prevention and Treatment of Disease, 3rd edition*, ed. Coulston, A.M., Boushey, C.J., and Ferruzzi, M.G., pp. 5–46. New York: Academic Press.

Thompson, F.E., F.A. Larkin, and M.B. Brown. 1986. Weekend-weekday differences in reported dietary intake: The nationwide food consumption survey, 1977–78. *Nutr Res* 6(6): 647–662.

Thompson, F.E., V. Kipnis, D. Midthune et al. 2008. Performance of a food-frequency questionnaire in the US NIH-AARP (National Institutes of Health-American Association of Retired Persons) Diet and Health Study. *Public Health Nutr* 11(2): 183–195.

Thompson, F.E., S.I. Kirkpatrick, A.F. Subar et al. 2015. The national cancer institute's dietary assessment primer: A resource for diet research. *J Acad Nutr Diet* 115(12): 1986–1995.

Tinker, L.F., G.E. Sarto, B.V. Howard et al. 2011. Biomarker-calibrated dietary energy and protein intake associations with diabetes risk among postmenopausal women from the women's health initiative. *Am J Clin Nutr* 94(6): 1600–1606.

Tooze, J.A., D. Midthune, K.W. Dodd et al. 2006. A new statistical method for estimating the usual intake of episodically consumed foods with application to their distribution. *J Am Diet Assoc* 106(10): 1575–1587.

Tooze, J.A., V. Kipnis, D.W. Buckman et al. 2010. A mixed-effects model approach for estimating the distribution of usual intake of nutrients: The NCI method. *Stat Med* 29(27): 2857–2868.

Tooze, J.A., L.S. Freedman, R.J. Carroll, D. Midthune, and V. Kipnis. 2016. The impact of stratification by implausible energy reporting status on estimates of diet-health relationships. *Biom J* 58(6): 1538–1551.

U.S. Department of Health and Human Services and U.S. Department of Agriculture. 2015. 2015–2020 Dietary Guidelines for Americans. 8th edition. http://health.gov/dietaryguidelines/2015/ (accessed July 15, 2016).

Willett, W.C. 2013. *Nutritional Epidemiology*, 3rd ed. New York: Oxford University Press.

Willett, W.C., L. Sampson, M.J. Stampfer et al. 1985. Reproducibility and validity of a semiquantitative food frequency questionnaire. *Am J Epidemiol* 122(1): 51–65.

Yang, P.H., J.L. Black, S.I. Barr, and H. Vatanparast. 2014. Examining differences in nutrient intake and dietary quality on weekdays versus weekend days in Canada. *Appl Physiol Nutr Metab* 39(12): 1413–1417.

Ziegler, R.G., H.B. Wilcox, T.J. Mason, J.S. Bill, and P.W. Virgo. 1987. Seasonal variation in intake of carotenoids and vegetables and fruits among white men in New Jersey. *Am J Clin Nutr* 45(1): 107–114.

3 Computer-Assisted Dietary Assessment Methods

Suzanne McNutt, Thea Palmer Zimmerman, and Brooke Colaiezzi

CONTENTS

3.1 INTRODUCTION

Advances in technology are a constant of modern life, and are particularly evident in the field of dietary assessment. Today, thousands of computer-assisted dietary reporting tools (Bardus et al. 2016) are available to researchers, clinicians, and consumers to monitor the dietary intake of children, adults, and special populations for national surveillance, nutritional epidemiology, and behavior change. The vast majority of these tools are web accessible or smartphone applications designed for self-monitoring rather than research. This chapter focuses only on dietary assessment tools that are intended for dietary research and have published studies indicating the tool compared favorably to a recognized standard. Specifically, this chapter features tools used for two dietary assessment methods: Food Frequency Questionnaires (FFQs) and 24-hour dietary recalls (24HRs). We have also included a brief discussion of the use of 24HRs to collect food record (FR) data. The purpose and uses for these assessment methods are discussed in detail in Chapter 1. This chapter looks at the features, and strengths, and limitations of these tools.

3.2 HISTORY OF COMPUTERIZED DIETARY ASSESSMENT TOOLS

All dietary assessment methods began with paper-and-pencil data collection and manual data processing. Early applications of technology to tools used for research were driven by the desire to reduce burden, improve the quality of data collected, and decrease costs.

3.2.1 Computerized Food Frequency Questionnaires

FFQs are designed for assessing usual intake over a reference period (usually one year although shorter time frames are possible) and are useful for large-scale epidemiologic studies due to the relatively low cost of administration and analysis. However, FFQs have been shown to contain substantial measurement error (Kipnis et al. 2003) due to incomplete food lists, difficulties in estimating usual portion size over the reference period, and considerable respondent error when completing hard copy forms (Thompson and Subar 2008). Computer-assisted technology has ameliorated respondent error to some degree by automating skip patterns, forcing respondents to answer all questions, and using graphical images to reduce literacy issues. In addition, it has substantially reduced the cost of administering, processing, and analyzing FFQ data.

In the 1980s, two FFQ tools were converted to scannable paper documents that could be read using optical character recognition software. The Willet questionnaire developed by Walter Willett at Harvard University (Willett et al. 1985) and the Block FFQ developed at the United States National Cancer Institute (NCI) by Gladys Block (Block et al. 1986) were the first optically-scannable forms, which removed the need for manual data entry, allowing researchers to collect dietary information from large numbers of respondents and reducing the cost of administration. However, optical scanning did not address respondent burden and the potential for respondent error during data collection. As a result, there was great interest in an FFQ that could be completed by respondents via computer. This advancement was led by NutritionQuest's Block FFQ in 1998 (Boeckner et al. 2002); and was followed by the NCI DHQ*Web (Beasley et al. 2009), the University College Dublin's Food4Me (Forster et al. 2014), and the Cancer Council Victoria's (Australia) Dietary Questionnaire for Epidemiological Studies (DQES) (Bassett et al. 2016). These computerized tools improved FFQ accuracy by standardizing questions and their administration, simplifying the process of applying nutrient values, decreasing time between data collection and obtaining analyzed data, and further decreasing costs associated with dietary assessment. It also allowed for the adjustment of question paths based on responses, potentially reducing the time for respondents to complete the FFQ while maintaining accurate question skip patterns. Currently available FFQs automate the nutrient analysis and provide output instantly through a researcher site. Dietary screeners, which are short, targeted FFQs, have seen the same advances in technology as full FFQs.

3.2.2 Computerized 24-hour Dietary Recalls

The 24HR method collects all foods and beverages consumed by a respondent over the previous day or in the preceding 24 hours. The data collection is typically conducted by trained interviewers using a structured set of probes to elicit details about each food item. The interviewer-administered 24HR is a favored method of collecting dietary data for several reasons. Respondents can generally recall their intake for the past 24 hours more accurately than they can for the longer FFQ reference periods. In addition, the time it takes to complete the interview in comparison to a FFQ is minimal, and because respondents only respond to the interviewers' questions, they do not need literacy skills. Finally, unlike food records, respondents report foods that they have already consumed, making it less likely they would have changed their eating behaviors (Thompson and Subar 2008).

However, there are some limitations to 24HRs. Although the concept of collecting a 24HR is a straightforward sequential process of asking standard questions to elicit information about the foods the respondent consumed the previous day, the execution can be quite complicated and fraught with incomplete information should the interviewer forget to probe for critical food details. In addition, one 24HR does not generally represent an individual's diet because of day-to-day variations in individuals' intakes; multiple days of intake are required for individual diet assessments, which can be cost prohibitive. Further, traditional 24HR data collection relies on manual dietary coding to apply nutrients to the reported foods, which both slows and adds significant cost to the data processing step.

The desire to use the preferred diet assessment method to gain more accuracy and cost efficiency drove the development of computer-assisted 24HRs. In particular, computer-assisted systems reduce user error by requiring users (interviewer or respondent) to follow a sequence of steps to complete data collection, which include automatic skip patterns and range and logic checks. In addition, data analysis is streamlined by incorporating auto-coding to reduce or eliminate costs associated with manual dietary coding.

The first computer-assisted 24HR interview used for research was the Nutrient Data System (NDS), developed by the Nutrition Coordinating Center at the University of Minnesota, Minneapolis in 1988. The NDS software program allows dietary interviewers to conduct a dietary recall and enter the data directly into the NDS system (University of Minnesota Nutrition Coordinating Center 2016a) for real-time analysis. Interviewers must be highly trained, because they are required to formulate most of the questions to elicit enough details about the food to assign a food code. Several years later, EPIC-SOFT (now called GloboDiet) was developed by the International Agency for Research on Cancer (IARC) in collaboration with the European Prospective Investigation into Cancer and Nutrition (EPIC). The software is a standardized dietary assessment tool designed for global nutritional monitoring and surveillance. It was the first European interviewer-administered 24HR software for use in multiple countries, and is considered to be the most appropriate software for dietary data collection in a pan-European survey (IARC 2014; Voss et al. 1998). In 1999, the U.S. Department of Agriculture (USDA) computerized the paper-and-pencil interview used for the 1994–1996 Continuing Survey of Food Intakes by Individuals (Steinfeldt et al. 2000), and incorporated it into the National Health and Nutrition Examination Survey (NHANES) to collect dietary surveillance data. The automated interview, called the Automated Multiple-Pass Method (AMPM), guides the interviewer through specific question pathways in order to collect adequate information about the food for dietary coding. It is partially automated to simplify the dietary coding step.

Automated dietary coding reduces or, in the case of the NDS, eliminates the need for manual dietary coding to apply nutrients to the 24HR data. These three computer-assisted interview systems collect very high quality data and are regularly used by government and academic researchers to collect research data.

Nevertheless, because of the high cost to collect multiple days of 24HR data using interviewer-administered systems, researchers sought to design a self-administered version of the 24HR using internet-based technology and Web access. In 2009, DietDay (Arab 2010), a self-administered web-based pictorial 24HR, was launched by the University of California. DietDay presents food images to respondents, by meal, to report their day's food consumption. NCI released the first version of their self-administered system in 2011. The NCI Automated Self-Administered 24-hour (ASA24) dietary recall (NCI Division of Cancer Control and Population Science 2016) is based on the USDA AMPM and was designed to offer investigators the use of the 24HR method in a wider range of research. In 2012, researchers at Wageningen University in the Netherlands launched Compl-eat (Meijboom et al. 2016; Wardenaar et al. 2015), a web-based self-administered nutrition calculation system for the Dutch population; and most recently, Newcastle University in the United Kingdom released the Intake24 (Foster et al. 2014a) and Leeds University, UK, introduced the Measure your Food on One Day (myfood24) (Albar et al. 2016). These web-based 24HR systems provide analyzed data in real time at a very low cost.

3.2.3 Computerized Food Records

Researchers who want to collect computerized food records can use the DietDay, ASA24, and myfood24 computerized self-administered 24HRs by having respondents access the applications throughout the day to record foods and beverages. These systems provide analyzed data in real time using the same process as the 24HR. Other methods under development for FRs focus on the use of digital cameras to record foods before and after eating, to be used either as a memory aid or actual data collection tools. Chapter 5 presents a discussion of these image-based methods and their use.

3.3 FOOD FREQUENCY QUESTIONNAIRE FEATURES

Table 3.1 presents the features of five available computerized FFQs, and Table 3.2 summarizes their strengths and limitations. All are research tools that were used in some form for a number of years as hard copy instruments, and were found to be reliable and effective. Three were developed in the

TABLE 3.1
Features of Computer-Assisted Food Frequency Questionnaires

Features	Block FFQs	DHQ*Web	VioScreen	Food4Me	DQES
Country	U.S.	U.S./Canada	U.S.	Europe (7 countries)	Australia
Type	Various	Full FFQ	Full FFQ w/food images	Full FFQ	FFQ
Administration	Self; interviewer	Self	Self	Self	Self; interviewer
Delivery method	Web; phone app; hard copy scannable form	Web	Web	Web	Web; hard copy scannable form
Reference period	Past year; customizable	Past year; past month	Past three months; customizable	Past month	Past 12 months
Food items	Full: 127 Kids: 77–90 Screeners: 10–55	135	125	157	80
Time to complete	Full: 45 minutes Kids: 24–30 minutes Screeners: 5–12 minutes	45 minutes	<30 minutes	45 minutes	15 minutes
Literacy skills	Required	Required	Graphical images reduce literacy issue	Required	Required
Portions	Full: Yes, mapped to NHANES	With/without, mapped to NHANES	Yes, based on a variety of sources	Yes, mapped to National Adult Nutrition Survey	Full: Yes, mapped to 2010 National Nutrition Survey
Dietary supplements	Yes	Yes	Yes	Yes	Yes
Language	English Spanish some versions	English/Spanish for U.S. English/French for Canada	English/Spanish	English, German, Polish, Spanish, Dutch, and Greek	English
Analysis	Real time	Real time	Real time	Available on request	Separate processing step, two week delay
Data sources	SR, FNDDS, FPED, DSD Updated sporadically	FNDDS, FPED, DSD, NCC Food and Nutrient DB Updated sporadically	NCC Food and Nutrient DB Updated annually	Country-specific national food composition databases	Australian NUTTAB2010, AUSNUT2007
Output	>200 components Respondent report	176 nutrients, dietary constituents, and food group variables	>160 food components; respondent report	85 nutrients and food group variables	63 components
Images	Yes, portions	No	Yes, foods and portions	Yes, portions	Yes, portions

(Continued)

TABLE 3.1 (*Continued*)
Features of Computer-Assisted Food Frequency Questionnaires

Features	Block FFQs	DHQ*Web	VioScreen	Food4Me	DQES
Seasonal consumption	Full: Yes	Yes	No	No	No
Study management	Tracks and exports data; generates respondent reports	Tracks and exports data; generates researcher reports	Tracks and exports data; generates respondent reports	Tracks study progress; management company sends data when study is complete	Researcher site to monitor completion
Updated	Periodically	Periodically	Annually	No plans	Every 3–4 years
Fee	Yes, See website	No fee; public use	Yes, See website	Yes, hosting fees	Yes, See website
Website	http://nutritionquest.com/assessment/	http://epi.grants.cancer.gov/DHQ/webquest/	http://www.viocare.com/vioscreen.html	http://www.food4me.org/	http://www.cancervic.org.au/research/epidemiology/nutritional_assessment_services
Contact	sales@nutritionquest.com 510-704-8514	http://epi.grants.cancer.gov/DHQ/webquest/initiate.html	sales@viocare.com 609-497-4600	N/A	ffq@cancervic.org.au +61 3 9514 6264

U.S.—United States; FFQ—Food frequency questionnaire; FNDDS—USDA Food and Nutrient Database for Dietary Studies; FPED—USDA Food Pattern Equivalent Database; SR—USDA Food and Nutrient Database for Standard Reference.

TABLE 3.2
Strengths and Limitations of Computerized FFQs

Characteristics	Block FFQ	DHQ*Web	VioScreen	Food4Me	DQES
Delivery method	Web-based mobile system (computer, phone, tablet) and hard copy scannable form	Computer Web-based *but* must have access to the Web	Computer Web-based *but* must have access to the Web	Computer Web-based *but* must have access to the Web	Web-based system and hard copy scannable form
Literacy skills	Required	Required	Uses graphical images to reduce literacy issues	Required	Required
Reference period	Customizable	Past year and past month	Customizable	Past year	Past year
Analysis	Auto-code/real-time output	Auto-code/real-time output	Auto-code/real-time output	Mainly auto-coded/ processed on request	Mainly auto-coded/ processed on request
Researcher management system	Data-on-Demand allows for data transfer, download of data sets, and respondent reports	Admin interface allows tracking and export of data and researcher reports	Admin interface allows tracking and export of data and respondent reports	Researcher log-in allows monitoring study progress only	Researcher log-in allows monitoring study progress only
Cost	Set up, maintenance, analysis fees	Free public use	Set up, maintenance, user fees	Hosting fees	Set up, analysis

United States, one in the United Kingdom for use in Europe, and one in Australia. Most are full FFQs, but NutritionQuest also offers screeners that are targeted to certain nutrients and food components. Some of the FFQs are appropriate for and/or have been tested with children and teens; and some of the instruments are offered in Spanish or other languages. The reference period is generally the past year or month, but some tools can be customized for shorter periods of time. The FFQ tools take 30–45 minutes to complete, depending on the number of foods; and the screeners take 5–12 minutes. All FFQs include portion questions, but the DHQ*Web offers a *no-portion* version. In addition, they all include approximately the same number of questions, including supplements. All tools offer a study management system that provides study set up options for researchers and the ability to track completion. Only the NCI-sponsored DHQ*Web is a free public use tool; the other tools require a fee for set up, maintenance, and user/analysis. Additional features of particular importance are discussed below.

3.3.1 Block FFQ

The Block FFQ, distributed by NutritionQuest, is the oldest existing FFQ in the United States. It was originally developed as a paper-and-pencil questionnaire in 1982 and was called the Health Habits and History Questionnaire (HHHQ). It has been regularly revised over the years to reflect updated food composition tables, target populations of interest, and take advantage of technology. The Block FFQ has been extensively used in research, both to collect primary data and to serve as a comparison to other dietary assessment tools under development (Blalock et al. 1998; Taylor et al. 2003; Yamamoto et al. 2001). It was converted to a web-based computer-assisted instrument in 1998, and each release of the questionnaire since that time has been available as both a paper-and-pencil and web-based instrument. There are now a variety of Block computer-assisted FFQs for researchers, including full versions for adults and children as well as medium length FFQs and short screeners targeted to specific groups of nutrients and food components. Block uses the USDA Food and Nutrient Database for Standard Reference (SR), Food and Nutrient Database for Dietary Studies (FNDDS), Food Pattern Equivalents Database (FPED), and NHANES Dietary Supplements Database (DSD) as food composition sources for all questionnaire versions. Several validation studies have been conducted over the years demonstrating that the Block correlates with intakes assessed by four-day FRs and 24HRs, and the web version is comparable to paper-and-pencil (Block et al. 1990; Boeckner et al. 2002; Boucher et al. 2006).

The Block full-length FFQ and the shorter versions offer several strengths. Their reference period is generally the past year, but they can be customized as desired, and many of their tools are available in both English and Spanish. Standard portion images help respondents report usual amounts and seasonal consumption is accounted for. The Block full-length can be delivered in multiple modes for flexible data collection protocols. It is mobile responsive, available through the web on computer, tablet, and smartphone; can be downloaded for use as a computer-assisted interview; and administered as a hard copy scannable form for respondents without web access. Finally, NutritionQuest provides a very robust turn-key study management system that allows for tracking FFQ completion and real-time analysis.

The Block questionnaires are proprietary and must be licensed. The NutritionQuest website provides details on licensing the software.

3.3.2 Diet History Questionnaire

The NCI DHQ was developed more than 15 years after the Block with the goal of providing a better measure of frequency by making the FFQ easier for respondents to complete. A web-based computer-assisted version, the DHQ*Web, was developed in 2005 for the second release of the DHQ. In addition to the food composition sources used by the Block FFQ, the DHQ*Web also accesses data from the University of Minnesota's Nutrition Coordinating Center (NCC) Food and Nutrient Database. Validation studies have demonstrated that the paper-and-pencil DHQ correlates with intakes assessed by multiple 24HRs and biomarkers (doubly-labeled water and urinary nitrogen) and performs as well or better than the Block FFQ and Willett FFQs; and the web version is comparable to the paper-and-pencil one (Beasley et al. 2009; Kipnis et al. 2003; Subar et al. 2001, 2003;

Thompson et al. 2002). NCI offers versions of the DHQ*Web in English and Spanish and with reference periods of past year or past month. It also offers versions with and without portion questions; however, the portion questions do not include portion images.

The DHQ*Web has many of the same strengths as the Block FFQ. However, one particular advantage is that it streamlines the data collection process by allowing respondents to skip questions about foods in a food group they did not consume. For example, the Block FFQ sequentially asks the same set of questions (*how often?* and *how much?*) about each kind of fruit listed on the questionnaire. The DHQ*Web, on the other hand, displays all fruits on one page, asks respondents to choose the fruits they consumed during the reference period, and only asks the frequency and portion questions about the fruits the respondent reported consuming.

The DHQ*Web is publically available through NCI at no cost to researchers.

3.3.3 VioScreen

The VioScreen FFQ was developed specifically by Viocare® as a web-based system. VioScreen is an image-based FFQ that uses the NCC Food and Nutrient Database for its food composition data. Evaluation of an early version of the VioScreen FFQ, called the Graphical Food Frequency System (GraFFS), found that reliability was similar to or higher than those reported for paper-and-pencil FFQs (Kristal et al. 2014). The reference period for the VioScreen FFQ is three months, though it can be customized as desired. It is available in English and Spanish. VioScreen displays images of all foods and additional graphics and images to help users report frequency and portions. These features may help low literacy respondents identify foods and beverages consumed. There are no questions on seasonal consumption of foods or beverages.

The VioScreen FFQ is proprietary and must be licensed. The VioScreen website provides details on licensing the software.

3.3.4 Food4Me

The online Food4Me FFQ was based on the printed FFQ used in the EPIC, referred to as the EPIC-Norfolk FFQ. The Food4Me FFQ increased the number of foods queried from 130 in the EPIC-Norfolk FFQ to 157, and includes photographs of three portion sizes per food item. In a study comparing the Food4Me FFQ to the previously-validated EPIC-Norfolk FFQ, results showed good agreement between the two questionnaires for nutrient and food group intakes overall, though the Food4Me mean nutrient intakes were higher than those from the EPIC-Norfolk FFQ. Some of the differences observed may be due to the differing number of food items on each questionnaire (Forster et al. 2014). Note that there are no questions on seasonal consumption of foods or beverages. Food4Me is available for use in seven countries (UK, Ireland, Germany, Poland, Spain, The Netherlands, and Greece), translated into the country-specific language, and as the analysis has been harmonized across all the databases used, data collected across Europe is comparable. The Food4Me FFQ is proprietary and must be licensed. The Food4Me website provides details on licensing the software.

3.3.5 Dietary Questionnaire for Epidemiological Studies

The DQES was initially developed as a paper FFQ by the Cancer Council Victoria to assess dietary intake in the Melbourne Collaborative Cohort Study in the late 1990s. The DQES is now available as both a paper and electronic version. The DQES contains 80 food items with a limited number of portion photos per food item. It has been found to have good repeatability and to estimate intakes reasonably well for energy-adjusted intakes for Australian-born participants (Bassett et al. 2016).

A website allows researchers to monitor the completion of the DQES by participants. One of the limitations of the DQES is that no questions on seasonal consumption of foods or beverages are asked; additionally, intakes are centrally processed and it may take up to two weeks for researchers to receive analyzed data. The current version of the DQES was released in 2016.

Researchers using the DQES pay set up and processing fees.

3.4 24-HOUR DIETARY RECALL (24HR) FEATURES

3.4.1 Interviewer-Administered Systems

Table 3.3 presents the three computerized interviewer-administered 24HR systems available for use in research, and Table 3.4 summarizes their strengths and limitations. They all use a multiple-pass approach to help remind respondents of the food they consumed during the target day, and they take from 10 to 30 minutes to complete. As estimating portions for dietary assessment has been shown to be difficult for respondents, the systems use three-dimensional models and hard copy booklets with pictures of vessels and shapes that correspond to portion sizes in the system.

3.4.1.1 Automated Multiple-Pass Method

The USDA AMPM system produces the highest quality data of all the 24HR systems. This interviewer-administered windows-based system is used to collect U.S. national surveillance

TABLE 3.3
Features of Interviewer-Administered Computer-Assisted 24HR Dietary Tools

Features	AMPM	NDSR	GloboDiet
Country	U.S.	U.S.	14 countries
Administration	Interviewer	Interviewer	Interviewer
Assessment method	24HR	24HR	24HR
Delivery method	Windows-based	Windows-based	Windows-based
Approach	five-step multiple pass	five-step multiple pass	five-step multiple pass (no forgotten foods list)
Time to complete	20–30 minutes	20–30 minutes	20–34 minutes
Questions	Questions and sub-questions tailored for each food	General questions	Questions and subquestions tailored for each food
Literacy skills	N/A	N/A	N/A
Portions	Food Model Booklet and open-ended responses	Food Amounts Booklet and open-ended responses	Food portion packet and open-ended responses
Unfound foods	Yes	Yes	Yes
Dietary supplements	Yes	Yes	Yes
Language	English/Spanish	English	Each country-specific version is translated into local language
Analysis	PIPS and SurveyNet for dietary coding	Real time	For quality control purposes
Food composition database	FNDDS	NCC Food and Nutrient Database	One simplified nutrient database for each of 14 countries
Output	65 nutrients and food components; linked to FPED	165 nutrients, nutrient ratios, other food components; NDSR food group assignments	Energy and macronutrients only for QC purposes
Media	No	No	No
Study management	None provided	Software is part of license	None provided
Updated	Annually	Annually	
Fee	No fee; public use	Yes, see website	No fee
Website	www.ars.usda.gov/News/docs.htm?docid=7710	www.ncc.umn.edu/products/	None
Contact	alanna.moshfegh@ars.usda.gov 301-504-0170	ndsrhelp@umn.edu 612-626-9450	slimanin@iarc.fr

TABLE 3.4
Strengths and Limitations of Interviewer-Administered Computerized 24HRs

Characteristics	AMPM	NDSR	GloboDiet
Delivery method	Computer only	Computer only	Computer only
How portions reported	Exact amount consumed	Exact amount consumed	Exact amount consumed
Analysis	Requires two software programs and programming support	Auto-coding; generates real-time nutrients and other components	Coding and analysis are outside the system
Study management system	None	Select optional modules; track-study completion, download analysis files	None
Cost	No fee, public use; *but* labor costs for interviewing and coding	Licensing fee; and labor cost for interviewing	No fee; *but* labor costs for development of country-specific version

data in the *What We Eat in America survey*, the nutrition component of the NHANES. It has been used by the USDA since 2000. The AMPM has been validated in a number of studies using the doubly-labeled water technique, one-day feeding studies, and 14 days of food records (FR) as reference measures (Blanton et al. 2006; Conway et al. 2003, 2004; Moshfegh et al. 2008).

The AMPM is superior to other 24HR methods because of a combination of features designed to improve the quality of the data collection and processing. First, it is interviewer-administered, which allows interviewers and respondents to work together to report the details of the foods respondents consumed. Second, it incorporates the USDA five-step multiple-pass approach that has been shown to prompt respondents to remember additional foods at each of the five steps (Steinfeldt et al. 2013). Third, the questions are tailored to each food the respondent reports. That is, there is a question pathway for every food the interviewer chooses in the system. The interviewer moves through the question pathway capturing more detail with each question, ending up with the most precise food code possible based on the respondent's report. Fourth, the system can easily capture details about *unfound* foods. Unfound foods are foods the respondent reports for which the interviewer cannot find a match. The AMPM guides the interviewer through a series of questions that prompt the respondent to describe the food in general terms, which ensures complete data capture. Fifth, the AMPM is designed for use with measuring guides to help respondents report portions, including a custom-developed Food Model Booklet (McBride 2001). The models pictured in the Food Model Booklet are built into the AMPM response options to help increase the speed and accuracy of the portion reporting. Finally, the AMPM uses two software programs to process and code the data. The Post Interview Processing System (PIPS) automatically codes approximately 60% of all foods reported. These are foods that have been completely defined by the respondent such as 2% milk, McDonald's Big Mac, and Diet Coke. All other foods are coded by dietary coders using Survey Net, which also incorporates the codes from the Food Model Booklet and increases the speed and accuracy of coding the reported portions. Survey Net translates food intakes into component foods and nutrients by assigning food codes to the foods reported using the USDA FNNDS. FNDDS generates 65 nutrients and components and is linked to the FPED.

The advantages of collecting and coding data using the AMPM system also bring some limitations. An interviewer-administered dietary-coding system incurs relatively high labor costs for training staff and conducting and coding the data. Further, unlike other systems, the AMPM system does not offer a management system or user support, so investigators must provide their own programming to move the data into and out of each program. On account of the hybrid approach to data coding, the AMPM cannot provide data in real time. This limits a researcher's ability to monitor the data and requires all data to be coded before a nutrient report can be generated.

The AMPM system is publically available to researchers at no cost through the USDA Food Service Research Group.

3.4.1.2 Nutrition Data Systems for Research

The University of Minnesota NCC's Nutrition Data System for Research (NDSR) system is similar to the AMPM in that it is an interviewer-administered windows-based software program with many of the same features as the AMPM. The NDSR has been in use for more than 20 years in numerous dietary intake studies, but no specific validation studies of the NDSR system itself have been conducted (University of Minnesota Nutrition Coordinating Center 2016b). One distinct advantage for using the NDSR is that it calculates nutrients in real time immediately after the interview is completed. This eliminates the need for dietary coding, thus saving substantial labor costs. Another benefit is that NDSR uses the NCC Food and Nutrient Database as the source of food composition information. This proprietary database is drawn from many sources, including the USDA FNDDS. But the database is composed of many more nutrients, nutrient ratios, and other food components than FNDDS, and also contains approximately 8,000 name brand foods, which makes it easier for interviewers to find a correct match. The NCC Food and Nutrient Database is updated annually.

The one primary limitation to using the NDSR is that interviewers must be highly trained to both simultaneously interview and code the foods reported. Unlike the AMPM, the food questions are not specific to every food so an interviewer may need to form questions extemporaneously, based on the context of the interview. They must also be very familiar with the NCC Food and Nutrient Database as they must make decisions about how to code a food based on the unique characteristics and format of the database. This opens the door for interviewer bias should an interviewer not be well trained in all aspects of dietary data collection and coding. An additional consideration for researchers who want to compare their data to NHANES surveillance data is that there are differences in how nutrients are derived between the FNDDS and NCC Food and Nutrient Database. Further, food groups in the food composition database do not link directly to the Healthy Eating Index (HEI) food groups, though the NCC website provides information about how to create them using the NCC food groups.

Finally, NDSR is a proprietary system and must be licensed; the licensing option depends on the purpose of the research. The NDSR website provides details on licensing the software.

3.4.1.3 GloboDiet

GloboDiet is an interviewer-administered standardized dietary assessment tool. This tool offers country-specific versions for Belgium, Czech Republic, Denmark, France, Germany, Greece, Italy, Malta, the Netherlands, Norway, Republic of Korea, Spain, Sweden, and the United Kingdom. Versions for Brazil and Mexico are in development. It has been used since 1998, and has been validated in numerous studies (Crispim et al. 2011, 2012; Slimani et al. 2003).

One of GloboDiet's main strengths is that it is a highly standardized, international dietary assessment tool, which enables cross-country comparisons of reported dietary intakes. The program uses a stepwise approach to data collection, similar to the USDA's five-step multiple pass, and includes numerous data quality controls, which help to standardize the data collection process and ensure data are of high quality.

Though data collected using GloboDiet are detailed and of high quality, the number of personnel, expertise, and time needed to complete a survey are high (Ocke et al. 2011). GloboDiet provides limited real-time data processing for quality control purposes only. For more comprehensive information, researchers must *manually* link reported foods and mixed dishes with food composition tables after data collection has concluded. GloboDiet does not currently have a study management system, but there are plans to create one in the future (to be called *Interviewer Manager*) (Crispim et al. 2014).

3.4.2 Self-Administered Systems

Tables 3.5 and 3.6 present the features and strengths and limitations of the self-administered systems. These systems are all web-based. Two are used for research in the United States, two in the

TABLE 3.5
Features of Self-Administered Computer-Assisted 24HR Dietary Tools

Features	ASA24	DietDay	Myfood24	Intake24	Compl-Eat	Foodbook24
Country	U.S./Canada/Australia	U.S.	United Kingdom	United Kingdom	The Netherlands	Ireland
Administration	Self/interviewer	Self	Self/interviewer	Self	Self	Self
Assessment method	24HR, FR	24HR, FR	24HR, FR	24HR	24HR	24HR
Delivery method	Mobile system (computer, tablet, phone app)	Web	Web	Mobile system (computer, tablet, phone app)	Web	Web (laptop or desktop computer)
Approach	Five-step multiple pass	Meal-based navigation; separate tabs for food, beverages, additions	Meal-based navigation, optional quick pass	Five-step multiple pass	Five-step multiple pass	Five-step multiple pass
Time to complete	20–30 minutes	10–20 minutes	Unknown	12 minutes	Unknown	Unknown
Questions	Questions and subquestions tailored for each food	General questions	General questions	General questions	General questions	General questions
Literacy skills	Required	Required	Not required when using interviewer-administered format	Required	Required	Required
Portions	Discrete sizes (up to 8) using photos and sliding scale	Discrete sizes (up to 3) using photos and sliding scale	Discrete sizes (up to 7) using photos of portions served, allows entry by weight or volume measures	Discrete sizes (up to 7) using photos of portions served and leftovers	Entered using weight, volume, household measures, and standard portion sizes	Discrete sizes (up to 3) using photos and sliding scale
Unfound foods	Yes	No	No	Yes	Yes; must be manually coded	Yes, must be manually coded
Dietary supplements	Yes	Yes	Yes	No	Yes (reported but not calculated)	Yes
Language	English/Spanish	English	English	English	Dutch	English
Analysis	Real time	Real time	Real time	Real time	Real time	Real time (except text entries)

(Continued)

TABLE 3.5 (*Continued*)
Features of Self-Administered Computer-Assisted 24HR Dietary Tools

Features	ASA24	DietDay	Myfood24	Intake24	Compl-Eat	Foodbook24
Food composition database	FNDDS, FPED, NHANES DSD	Composite	Myfood24 database	McCance and Widdowson, manufacturer data	Dutch Food Composition Database, 2013	Irish National Adult Nutrition Survey database
Output	65 nutrients and food components; linked to FPED	Unknown	120 nutrients	Food groups and 20 nutrients	144 nutrients and food components	34 nutrients
Media	Portion photos	Food and portion photos	Portion photos	Portion photos	No	Portion photos
Study management	Researcher site tracks and exports data	Yes, contact for details	Researcher site tracks and exports data	Researcher site tracks and exports data	Researcher site tracks and exports data	Under development
Updated	Every two years	Periodically	Unknown	Unknown	Last updated in 2015	Unknown
Fee	No fee; public use	Yes, contact for details	Licensing fee	No fee, public use	Fees to host website	Fees to host website
Website	https://asa24.nci.nih.gov/researchersite/	http://www.ucladietday.com/	https://www.myfood24.org	https://intake24.co.uk/	http://www.compleat.nl	http://www.ucd.ie/foodbook24/
Contact	http://epi.grants.cancer.gov/asa24/	Centrax, Inc. eprentice@yepproject.com 312-946-2010	Myfood24@leeds.ac.uk	support@intake24.co.uk	Via website	Via website

TABLE 3.6
Strengths and Limitations of Self-Administered Computerized 24HRs

Characteristics	ASA24	DietDay	Myfood24	Intake24	Compl-Eat	Foodbook24
Delivery method	Flexible mobile system (computer, tablet, phone), but must have Web access	Computer Web-based system but must have Web access	Computer web-based only	Flexible mobile system (computer, tablet, phone), but must have Web access	Computer web-based only	Computer web-based only
How portions reported	Allows reporting of exact amount of beverages, but only displays eight discrete food portion-size images for each food	Only displays three discrete portion-size images for each food	Allows reporting of gram weight or milliliters for all foods, but only displays discrete portion sizes for top 100 foods consumed	Allows reporting of exact amount of beverages, but only displays up to seven discrete food portion-size images for each food	Exact amount consumed	Only displays three discrete portion-size images for each food
Analysis	Auto-coding; generates real-time nutrients, other components, and respondent report	Auto-coding; generates real-time nutrients, other components, and respondent report	Auto-coding; generates real-time nutrients and other components	Auto-coding; generates real-time nutrients, other components, and respondent report	Auto-coding; generates real time nutrients and other components	Auto-coding; generates real time nutrients
Researcher management system	Select optional modules, track study completion, and download analysis files	Yes, contact for details	Track study completion, download analysis files	Select optional modules, track study completion, and download analysis files	Track study completion and download analytic files	Under development
Cost	No fee *but* labor costs if interviewer-administered	Licensing fee	Licensing fee *but* labor costs if interviewer-administered	No fee	Hosting fee and optional fee for technical assistance	Hosting fee

United Kingdom, one in the Netherlands, and one in Ireland. They all use a multiple-pass approach to help remind respondents of the food they consumed during the target day, and they take from 10 to 30 minutes to complete. As estimating portions for dietary assessment has been shown to be difficult for respondents, the systems present a discrete number of images to help respondents estimate portion amounts.

3.4.2.1 Automated Self-Administered 24HR, United States

The NCI Automated Self-Administered 24HR recall is a self-administered web-based program that can be used to collect 24HR or food record data. The ASA24 is based on the methodology used in the AMPM, and has many of the same features. Evaluation and validation studies for the ASA24 indicate close agreement between data collected using the ASA24 and AMPM (Kirkpatrick et al. 2014; Thompson et al. 2015). The use of the ASA24 to obtain FR data has not been evaluated.

The primary strength of the ASA24 is that it replicates the AMPM methodology that results in high quality data without incurring dietary interviewer and coder costs. In particular, it links every AMPM scripted question pathway to an FNDDS food code and portion amount, resulting in real-time analysis. It also presents up to eight portion images of actual food items for users to report portions, and includes dietary supplements in the question pathways. An additional feature is that it is mobile-responsive, automatically resizing to a format suitable for display on a computer, tablet, or smartphone. ASA24 also incorporates some features developed for the ASA24-Kids program that may make it a useful tool for older children. (Diep et al. 2015) (ASA24-Kids is no longer a publically-supported tool). The main limitations of the ASA24 are related to self-administration. First, the data may not be as high quality as the AMPM or NDSR because respondents must make decisions about how to answer the food questions without interviewer guidance or coder understanding. In addition, self-administered tools require user literacy skills, and web-based self-administered tools require web access. To reduce the limitations of self-administration, the ASA24 has been used as an interviewer-administered tool.

ASA24 has been adapted for use in Canada and Australia; researchers from each country edited the foods included in their version of the ASA24 and changed the food composition data. The Canadian ASA24 uses the Canadian Nutrient File as the source of food composition data; the Australian ASA24 uses the Australian Food, Supplement and Nutrient Database.

All versions of the ASA24 are publically available through NCI at no cost to researchers.

3.4.2.2 DietDay, United States

DietDay is a self-administered web-based 24HR program, but is different from the ASA24 in that it uses a series of photos to help respondents report the foods they ate and drank during three periods of time during the day. The program was validated for energy using the doubly-labeled water technique and showed less underreporting than the DHQ (Arab et al. 2010, 2011).

Respondents are presented with 31 categories of food, and once they choose a category they are asked to select a specific food in that category. For example, the fast food category offers 16 types of fast foods (e.g., sandwiches, pizza, hamburgers, salads, wraps, breakfast, and fast foods). If a respondent chooses pizza, the program asks them to select one of the three types of pizza—with cheese, without cheese, and white pizza. Once they make their selection, they are asked a detail question (e.g., Did you eat deep-dish pizza?) and then asked to report the amount they consumed using a sliding scale that increases or decreases a picture of a slice of pizza. Foods are auto-coded and nutrient reports are generated in real time. A key advantage of DietDay is the graphic-based presentation that allows respondents to quickly move through the program and has been shown to produce high participation rates (Arab et al. 2010). However, there are several important limitations that suggest this tool does not capture an individual's diet at the same level of detail as the other systems. The food pictures restrict the food detail a user can report, there are a limited number of portion sizes per food, and the program does not guide the user through the interview. DietDay has shown to be easy to complete and may be a useful tool to provide an estimate of energy intake. DietDay is a proprietary system and may require a set up and user fee.

3.4.2.3 Myfood24, United Kingdom

Myfood24 is the first online 24HR designed for use with different age groups in the United Kingdom (Carter et al. 2015). A 2016 study in a sample of adolescents found that, in general, myfood24 recorded slightly lower energy intakes and macronutrient values compared to an interviewer-administered, multiple-pass recall (Albar et al. 2016). At press time, a large validation study is being conducted that includes biomarkers and an interviewer-administered multiple-pass recall as reference measures.

One of the strengths of myfood24 is the expanded database of approximately 45,000 UK branded and generic food items, developed specifically for myfood24 (Cade et al. 2014). The database combines the United Kingdom's McCance and Widdowson's Composition of Foods (Government of the United Kingdom 2015) with substantial manufacturer data. Myfood24 was based on extensive formative research conducted with multiple age groups (ranging in age from 11 to 65+ years) and incorporates a user-friendly design that also allows respondents to enter recipes. It presents up to seven images of actual food items for users to report portions, but also allows users to enter the exact weight or volume of the food or beverage consumed. Researchers have access to the myfood24's study management system, where they can specify the data collection method (FR or 24HR) and administration format (self-administered versus interviewer-administered), opt to add supplementary questions, prompt respondents to record their time of consumption, and decide whether to display a nutrient summary to respondents at the end of the survey.

One limitation of myfood24 is that not all foods within the database contain graduated portion images to assist respondents with quantifying intakes. Currently, the system includes images only for the 100 most commonly consumed food types, though respondents have the option to record the amount they ate using the weight or volume of the food, in addition to using photos of graduated portion amounts. In addition, while myfood24 contains features common to the five-step multiple-pass method (i.e., forgotten foods and final review), one of the most important passes, the quick list pass, is optional. In addition, myfood24 does not allow respondents to report unfound foods. Myfood24 is a proprietary system and may require a set up and user fee.

3.4.2.4 Intake24, United Kingdom

Intake24 is a web-based system developed for use in 11–24 year olds by researchers at Newcastle University in the United Kingdom, accessible on a standard laptop or desktop computer or mobile devices (smartphones and tablets). Intake24's design is based on the five-step multiple-pass method and is freely available to researchers by request. Intake24's development has been highly iterative and has capitalized on extensive research conducted on portion-size estimation methods in children, adolescents, and young adults in the United Kingdom. It was developed from a nonweb-based system called SCRAN24 (Self-Completed Recall and Analysis of Nutrition for use with children) (Foster et al. 2014b), which in turn, was based on IPSAS (Interactive Portion-Size Assessment Software) (Foster et al. 2014c), the United Kingdom's only validated computer-based, image-based tool for use in assessing the portion size of foods consumed by children (Foster et al. 2008). The relative validity of Intake24 has been assessed using an interviewer-administered 24HR as the reference method. Intakes of energy and macronutrients were within 1% on average, and 82% of foods reported in Intake24 were matched to those reported in the interviewer-administered 24HR (Foster et al. 2014a).

Currently, Intake24 contains more than 2500 food portion images and approximately 1560 foods (Newcastle University and United Kingdom Food Standards Agency 2013). It presents up to seven portion images of food items for users to report portions and allows users to also report the amount of food that was leftover. Two other key strengths of Intake24 are its database tool and study management system. The database tool allows Intake24 administrators to easily modify portion-size estimation methods within Intake24 or add or delete foods. Intake24's study management system enables researchers to start, suspend, or end a survey; monitor survey progress; update the survey schedule; upload usernames and passwords for participants; download survey data; and download an activity report. Survey data can be downloaded immediately after the respondent has submitted their 24HR data. Output information includes the reported food's portion size, energy content, macro and micronutrient values, and food group. The activity report informs the researcher of the number of recall submissions; the mean, minimum, and maximum completion times; and the submission dates for each participant (Newcastle University and United Kingdom Food Standards Agency 2013). It is important to note that Intake24 does not provide the total nutrient intake per intake day, only the nutrients per food item; researchers must analyze the data in Excel or a statistical program to derive total nutrient intake.

Intake24 does not offer respondents the option to report dietary supplements. Also, although Intake24 currently enables reporting unfound foods, plans are underway to remove this option and instead prompt respondents to select the most closely matched food within the system. Intake24 is freely available at no cost to researchers.

3.4.2.5 Compl-Eat, Netherlands

Compl-eat is a dietary assessment system developed in 2012 by researchers at Wageningen University in The Netherlands. It can be used as a web-based, self-administered 24HR, designed to collect and process dietary intake data in the Dutch population or as a standalone program to process 24HR data collected in any country. The 24HR program is intended for dietary research conducted within the Netherlands, as it is not available in any language besides Dutch and the foods, mixed dishes, and nutrient values are drawn from the 2013 Dutch food composition database (the Dutch food composition database is managed by the publically-funded Netherlands Nutrition Centre). The standalone processing program allows researchers to import their own food composition table and use Compl-eat as a dietary coding tool.

The Compl-eat 24HR system is based on the USDA's five-step multiple-pass approach (Trijsburg et al. 2015). It has been validated in a study in athletes that compared protein intakes assessed via repeat 24HR with urinary nitrogen as a reference measure (Wardenaar et al. 2015). One advantage of Compl-eat is that respondents can modify the type and quantity of ingredients in standardized recipes and may use a note field to include clarifying details on their reported intake. Compl-eat applies yield and retention factors, when appropriate, and automatically codes reported intakes. Portions are reported using household measures, weight, volume, or by selecting a standard portion amount. Although intake information is processed and available in real time, researchers must make adjustments to the data based on any notes made by respondents or any unusual portions reported. Researchers may request assistance with this from the Compl-eat research dietitians. One of the greatest advantages of Compl-eat is the range of output options that it offers to researchers, including analyses of total nutrient intake by respondent; nutrient ratios; nutrient intakes by food, food group, or meal; and calculation of dietary diversity scores. Researchers can also use Compl-eat to perform quality checks on the data and descriptive analyses of the data set.

One potential limitation of the Compl-eat is that it does not currently contain visual portion-size estimation aids for respondents, such as portion images. At present, researchers must provide respondents with a photobook in advance of the study if they would like to use visual aids during the quantification phase of the recall. In addition, although dietary supplements may be reported during the interview, analysis does not include nutrients from the supplements reported.

Compl-eat is currently available to researchers for a small fee to cover maintenance; assistance in resolving information in respondent notes is also available for a fee.

3.4.2.6 Foodbook24, Ireland

Foodbook24 is a web-based 24HR system intended for collecting dietary intake data from adults in Ireland. Development began in 2015 and was completed in 2016 (Timon et al. 2015). Foodbook24 uses a multiple-pass method and is fully auto-coded with real-time analysis for all but items entered using text entry. A comparison study investigating the comparability of Foodbook24 to an interviewer-led recall has completed analysis and is currently underway. A separate validation study where the agreement and differences in nutrient and food group intake recorded by Foodbook24 were compared to that of a four day semiweighed food diary and biological markers of nutrient intake from blood and urine samples has also been completed; analysis was still ongoing as this book went to press. (Communication with Eileen Gibney, University College Dublin, project lead for development of foodbook24.)

Strengths of foodbook24 include use of a multiple-pass method, ability to report unfound foods, and auto-coded, real-time analysis. Limitations include the need for literacy, as with all self-administered systems, and the lack of a researcher website for managing studies (though at press time there were plans to begin development). Foodbook24 is available to researchers for a nominal management fee.

3.5 CONCLUSIONS

Computer-assisted dietary assessment systems have significantly improved the ability of researchers to collect dietary data. Perhaps the most important aspect of these systems is the built-in skip patterns that guide the user through the interview based on their answer to a previous question. This feature alone reduces break-offs, increases accuracy, and reduces the time it takes to collect the data. Other attributes that make these systems powerful tools include auto-coding and real-time analysis, researcher management systems that allow tracking of study participants, and web-based delivery systems. Several of the self-administered 24HRs can also be used to collect real-time FRs. The appropriate choice is dependent on the study's purpose, size, and targeted sample. Table 3.7 summarizes the significant characteristics of each of the computer-assisted dietary assessment systems discussed in this chapter.

The field of computer-assisted dietary assessment is rapidly changing. New technologies will continually lead to more efficient and accurate data collection tools that can be used in a variety of settings, for a range of populations. Readers are advised to search not only for updated versions of the systems described here but also for newly released systems that may be better suited to the needs of their studies.

TABLE 3.7
Summary of Characteristics of Computer-Assisted Dietary Assessment Systems

	FFQ					Interviewer-Administered 24HR			Self-Administered 24HR					
									24HR/FR			24HR		
Characteristics	**Block**	**VioScreen**	**DHQ**	**Food4ME**	**DQES**	**AMPM**	**NDSR**	**GloboDiet**	**ASA24**	**DietDay**	**Myfood24**	**Intake24**	**Compl-eat**	**Foodbook24**
Country														
United States	X	X	X			X			X	X				
United Kingdom											X	X		
Ireland														X
The Netherlands													X	
Australia					X									
Multiple countries			X	X				X	X					
Food details[a]														
Groups	X	X	X	X	X									
General										X	X			X
Specific						X	X	X	X			X	X	
Respondent burden[b]														
Lowest						X	X	X						
Medium									X	X	X	X	X	X
Suitable for low literacy[c]														
Very						X	X	X						
Somewhat		X								X				

(Continued)

TABLE 3.7 (*Continued*)
Summary of Characteristics of Computer-Assisted Dietary Assessment Systems

	FFQ					Interviewer-Administered 24HR			Self-Administered 24HR					
									24HR/FR			24HR		
Characteristics	**Block**	**VioScreen**	**DHQ**	**Food4ME**	**DQES**	**AMPM**	**NDSR**	**GloboDiet**	**ASA24**	**DietDay**	**Myfood24**	**Intake24**	**Compl-eat**	**Foodbook24**
Population														
Adults	X	X	X	X	X	X	X	X	X	X	X	X	X	X
Children/teens	X	X				X	X	X	X		X	X		
Spanish-speaking	X	X	X	X		X	X	X	X					
Cost[d]														
Lower									X			X		
Higher						X	X	X						

[a] The level of detailed information collected. The *Groups* level represents groups of food such as meat dishes; *General* level represents a type of food such as beef stew or a limited list of variations; and *Specific* level represents precise foods such as beef stew with peas and carrots or unlimited variations.

[b] A combination of the time it takes an individual to complete the interview and the burden placed on the individual. *Lowest* represents low burden but more time; *Medium* represents less time but more respondent responsibility.

[c] The ability of a system to accommodate individuals with low literacy skills. *Very* suitable describes systems that place no burden on individuals to read and respond to questions; *Somewhat* describes systems that assist individuals with low literacy skills using food images.

[d] The relative cost of 24HR systems. *Lower* cost systems do not have user fees and are self-administered; *Higher* cost systems are interviewer-administered and may have user fees.

REFERENCES

Albar, S. A., N. A. Alwan, C. E. Evans, D. C. Greenwood, J. E. Cade. 2016. Agreement between an online dietary assessment tool (myfood24) and an interviewer-administered 24-h dietary recall in British adolescents aged 11–18 years. *Br J Nutr* 115 (9): 1678–1686.

Arab, L. 2010. Welcome to DietDay! http://www.ucladietday.com/ (accessed June 10, 2016).

Arab, L., C. H. Tseng, A. Ang, P. Jardack. 2011. Validity of a multipass, web-based, 24-hour self-administered recall for assessment of total energy intake in blacks and whites. *Am J Epidemiol* 174 (11): 1256–1265.

Arab, L., K. Wesseling-Perry, P. Jardack, J. Henry, A. Winter. 2010. Eight self-administered 24-hour dietary recalls using the internet are feasible in African Americans and Whites: The energetics study. *J Am Diet Assoc* 110 (6): 857–864.

Bardus, M., S. B. van Beurden, J. R. Smith, C. Abraham. 2016. A review and content analysis of engagement, functionality, aesthetics, information quality, and change techniques in the most popular commercial apps for weight management. *Int J Behav Nutr Phys Act* 13: 35.

Bassett, J. K., D. R. English, M. T. Fahey et al. 2016. Validity and calibration of the FFQ used in the Melbourne collaborative cohort study. *Public Health Nutr* 19: 1–12.

Beasley, J. M., A. Davis, W. T. Riley. 2009. Evaluation of a web-based, pictorial diet history questionnaire. *Public Health Nutr* 12 (5): 651–659.

Blalock, S. J., S. S. Currey, R. F. DeVellis, J. J. Anderson, D. T. Gold, M. A. Dooley. 1998. Using a short food frequency questionnaire to estimate dietary calcium consumption: A tool for patient education. *Arthritis Care Res* 11 (6): 479–484.

Blanton, C. A., A. J. Moshfegh, D. J. Baer, M. J. Kretsch. 2006. The USDA automated multiple-pass method accurately estimates group total energy and nutrient intake. *J Nutr* 136 (10): 2594–2599.

Block, G., A. M. Hartman, C. M. Dresser, M. D. Carroll, J. Gannon, L. Gardner. 1986. A data-based approach to diet questionnaire design and testing. *Am J Epidemiol* 124 (3): 453–469.

Block, G., M. Woods, A. Potosky, C. Clifford. 1990. Validation of a self-administered diet history questionnaire using multiple diet records. *J Clin Epidemiol* 43 (12): 1327–1335.

Boeckner, L. S., C. H. Pullen, S. N. Walker, G. W. Abbott, T. Block. 2002. Use and reliability of the World Wide Web version of the Block Health Habits and History Questionnaire with older rural women. *J Nutr Educ Behav* 34 Suppl 1: S20–S24.

Boucher, B., M. Cotterchio, N. Kreiger, V. Nadalin, T. Block, G. Block. 2006. Validity and reliability of the Block98 food-frequency questionnaire in a sample of Canadian women. *Public Health Nutr* 9 (1): 84–93.

Cade J. E., N. Hancock, M. Carter et al. 2014. PP38 Development of a new UK food composition database. *J Epidemiol Community Health* 68: A62–A63.

Carter, M. C., S. A. Albar, M. A. Morris et al. 2015. Development of a UK online 24-h dietary assessment tool: Myfood24. *Nutrients* 7 (6): 4016–4032.

Conway, J. M., L. A. Ingwersen, A. J. Moshfegh. 2004. Accuracy of dietary recall using the USDA five-step multiple-pass method in men: An observational validation study. *J Am Diet Assoc* 104 (4): 595–603.

Conway, J. M., L. A. Ingwersen, B. T. Vinyard, A. J. Moshfegh. 2003. Effectiveness of the US department of agriculture 5-step multiple-pass method in assessing food intake in obese and nonobese women. *Am J Clin Nutr* 77 (5): 1171–1178.

Crispim, S. P., J. H. de Vries, A. Geelen et al. 2011. Two non-consecutive 24 h recalls using EPIC-Soft software are sufficiently valid for comparing protein and potassium intake between five European centres-results from the European Food Consumption Validation (EFCOVAL) study. *Br J Nutr* 105 (3): 447–458.

Crispim, S. P., A. Geelen, J. H. de Vries et al. 2012. Bias in protein and potassium intake collected with 24-h recalls (EPIC-Soft) is rather comparable across European populations. *Eur J Nutr* 51 (8): 997–1010.

Crispim, S. P., G. Nicolas, C. Casagrande et al. 2014. Quality assurance of the international computerised 24 h dietary recall method (EPIC-Soft). *Br J Nutr* 111 (3): 506–515.

Diep, C. S., M. Hingle, T. A. Chen et al. 2015. The automated self-administered 24-hour dietary recall for children, 2012 version, for youth aged 9 to 11 years: A validation study. *J Acad Nutr Diet* 115 (10): 1591–1598.

Forster, H., R. Fallaize, C. Gallagher et al. 2014. Online dietary intake estimation: The Food4Me food frequency questionnaire. *J Med Internet Res* 16 (6): e150.

Foster, E., J. Delve, E. Simpson, S. P. Breininger. 2014a. Comparison study: INTAKE24 vs. Interviewer led recall. Final Report, pp. 1–45. https://www.food.gov.uk/sites/default/files/INTAKE24%20Comparison%20study%20final%20report.pdf (accessed June 3, 2016).

Foster, E., A. Hawkins, J. Delve, A. J. Adamson. 2014b. Reducing the cost of dietary assessment: Self-completed recall and analysis of nutrition for use with children (SCRAN24). *J Hum Nutr Diet* 27 Suppl 1: 26–35.

Foster, E., A. Hawkins, E. Simpson, A. J. Adamson. 2014c. Developing an interactive portion size assessment system (IPSAS) for use with children. *J Hum Nutr Diet* 27 Suppl 1: 18–25.

Foster, E., J. N. Matthews, J. Lloyd et al. 2008. Children's estimates of food portion size: The development and evaluation of three portion size assessment tools for use with children. *Br J Nutr* 99 (1): 175–184.

Government of the United Kingdom. 2015. Composition of foods integrated dataset (CoFID). https://www.gov.uk/government/publications/composition-of-foods-integrated-dataset-cofid (accessed June 25, 2016).

IARC. 2014. *Formal Announcement of the Name Change from EPIC-Soft to GloboDiet Software.* Lyon, France: IARC Communications Group.

Kipnis, V., A. F. Subar, D. Midthune et al. 2003. Structure of dietary measurement error: Results of the OPEN biomarker study. *Am J Epidemiol* 158 (1): 14–21.

Kirkpatrick, S. I., A. F. Subar, D. Douglass et al. 2014. Performance of the automated self-administered 24-hour recall relative to a measure of true intakes and to an interviewer-administered 24-h recall. *Am J Clin Nutr* 100 (1): 233–240.

Kristal, A. R., A. S. Kolar, J. L. Fisher et al. 2014. Evaluation of web-based, self-administered, graphical food frequency questionnaire. *J Acad Nutr Diet* 114 (4): 613–621.

McBride, J. 2001. Was it a slab, a slice, or a sliver? *Agric Res* 49: 4–7.

Meijboom, S., M. T. Streppel, C. Perenboom et al. 2016. *Evaluation of Dietary Intake Assessed by the Dutch Self-Administered Web-Based 24-hour Recall Tool (Compl-eat™) against Telephone-Based 24-hour Recalls.* In press.

Moshfegh, A. J., D. G. Rhodes, D. J. Baer et al. 2008. The US department of agriculture automated multiple-pass method reduces bias in the collection of energy intakes. *Am J Clin Nutr* 88 (2): 324–332.

NCI Division of Cancer Control and Population Science. 3-29-2016.ASA24. http://epi.grants.cancer.gov/asa24/

Newcastle University, and United Kingdom Food Standards Agency. 2013. Development of a web-based 24-hour dietary recall tool for use by 11–24 year olds: INTAKE 24. Final report, pp. 1–31. http://www.foodstandards.gov.scot/sites/default/files/890-1-1642_INTAKE24_Appendix_2_Development_of_a_web-based_24-hour_dietary_recall_tool_for_use_by_11-24_year_olds_final.pdf (accessed June 3, 2016).

Ocke, M. C., N. Slimani, H. Brants et al. 2011. Potential and requirements for a standardized pan-European food consumption survey using the EPIC-Soft software. *Eur J Clin Nutr* 65 Suppl 1: S48–S57.

Slimani, N., S. Bingham, S. Runswick et al. 2003. Group level validation of protein intakes estimated by 24-hour diet recall and dietary questionnaires against 24-hour urinary nitrogen in the European Prospective Investigation into Cancer and Nutrition (EPIC) calibration study. *Cancer Epidemiol Biomarkers Prev* 12 (8): 784–795.

Steinfeldt, L. C., J. Anand, T. Murayi. 2013. Food reporting patterns in the USDA automated multiple-pass method. *Proc Food Sci* 2: 145–156.

Steinfeldt, L. C., E. Anderson, J. Anand, N. Raper. (May, 2000). Using Blaise in a nationwide food consumption survey. *6th International Blaise Users Conference Proceedings.* Kinsale, Ireland.

Subar, A. F., V. Kipnis, R. P. Troiano et al. 2003. Using intake biomarkers to evaluate the extent of dietary misreporting in a large sample of adults: the OPEN study. *Am J Epidemiol* 158 (1): 1–13.

Subar, A. F., F. E. Thompson, V. Kipnis et al. 2001. Comparative validation of the Block, Willett, and National Cancer Institute food frequency questionnaires: The eating at America's table study. *Am J Epidemiol* 154 (12): 1089–1099.

Taylor, A. J., H. Wong, K. Wish et al. 2003. Validation of the MEDFICTS dietary questionnaire: A clinical tool to assess adherence to American heart association dietary fat intake guidelines. *Nutr J* 2: 4.

Thompson, F. E., S. Dixit-Joshi, N. Potischman et al. 2015. Comparison of interviewer-administered and automated self-administered 24-hour dietary recalls in 3 diverse integrated health systems. *Am J Epidemiol.* 181: 970–978.

Thompson, F. E. and A. F. Subar. 2008. Dietary assessment methodology. In *Nutrition in the Prevention and Treatment of Disease*, A. M. Coulston, C. J. Boushey, and M. G. Ferruzzi, eds., pp. 3–39. Amsterdam, the Netherlands: Academic Press.

Thompson, F. E., A. F. Subar, C. C. Brown et al. 2002. Cognitive research enhances accuracy of food frequency questionnaire reports: Results of an experimental validation study. *J Am Diet Assoc* 102 (2): 212–225.

Timon, C. M., K. Evans, J. Walton, A. Flynn, E. R. Gibney. 2015. The development of an innovative web based dietary assessment tool for an Irish adult population: The diet Ireland tool. *Proc Nutr Soc* 74: E274.

Trijsburg, L., J. H. de Vries, H. C. Boshuizen et al. 2015. Comparison of duplicate portion and 24 h recall as reference methods for validating a FFQ using urinary markers as the estimate of true intake. *Br J Nutr* 114 (8): 1304–1312.

University of Minnesota Nutrition Coordinating Center. 2016a. About NCC. http://www.ncc.umn.edu/about-ncc/ (accessed June 10, 2016a).

University of Minnesota Nutrition Coordinating Center. 2016b. FAQs. http://www.ncc.umn.edu/products/ndsrfaqs/ (accessed June 15, 2016b).

Voss, S., U. R. Charrondiere, N. Slimani et al. 1998. [EPIC-SOFT a European computer program for 24-hour dietary protocols]. *Z Ernahrungswiss* 37 (3): 227–233.

Wardenaar, F. C., J. Steennis, I. J. Ceelen, M. Mensink, R. Witkamp, J. H. de Vries. 2015. Validation of web-based, multiple 24-h recalls combined with nutritional supplement intake questionnaires against nitrogen excretions to determine protein intake in Dutch elite athletes. *Br J Nutr* 114 (12): 2083–2092.

Willett, W. C., L. Sampson, M. J. Stampfer, B. Rosner, C. Bain, J. Witschi, C. H. Hennekens, F. E. Speizer. 1985. Reproducibility and validity of a semiquantitative food frequency questionnaire. *Am J Epidemiol* 122 (1): 51–65.

Yamamoto, S., T. Sobue, S. Sasaki et al. 2001. Validity and reproducibility of a self-administered food-frequency questionnaire to assess isoflavone intake in a japanese population in comparison with dietary records and blood and urine isoflavones. *J Nutr* 131 (10): 2741–2747.

4 Strategies for Improving the Validity of the 24-hour Dietary Recall and Food Record Methods

Lisa J. Harnack and Mark A. Pereira

CONTENTS

4.1 INTRODUCTION

The 24-hour dietary recall (24HR) and food record methods are approaches for assessing food and nutrient intake that are commonly used in nutrition research and surveillance.

The 24-hour dietary recall method involves asking a participant to report all of the foods and beverages consumed the previous day. The participant is also asked a series of questions to collect detailed information about each food and beverage consumed (e.g., *What type of coffee did you drink?*, *Did you add anything to your coffee?*, and *How much coffee did you drink?*). The conventional approach is to have a trained interviewer carry out the recall interview.

With the food record method the participant is asked to record all the foods and beverages consumed over a day, conventionally via paper and pencil. They are asked to include detailed information about each food and the portion consumed. Measuring tools (e.g., measuring cups, ruler, and food scale) may be provided for use in quantifying food portions. Written and verbal instructions are typically provided in advance of food-record keeping, and completed records are reviewed by a trained study staff member and entered into a dietary analysis program for nutrient calculation. The participant may be contacted to address missing or unclear information in the record, or data entry rules may be followed to deal with missing information.

With both the dietary recall and food-record methods, participants are generally asked to report the time and name of each eating occasion (e.g., 8 am breakfast). Additional information, such as where eating occurred, may also be queried.

In this chapter the potential causes of error in reporting with the dietary recall and food-record methods are reviewed, and strategies that may improve the accuracy of reporting by addressing these causes are described.

4.2 BACKGROUND

The 24-hour dietary recall and food-record methods have numerous strengths in comparison to other self-report methods of dietary assessment such as food frequency questionnaires, making them preferred tools for monitoring food and nutrient intake of populations and studying diet–disease associations (Kipnis et al. 2003; Schatzkin et al. 2003; Moshfegh et al. 2008). But, along with strengths, such as collection of detailed information about foods consumed and ascertainment of meal patterns, there are limitations. As discussed in Chapter 1 and 2 (Taylor et al.; Kirkpatrick et al., this book) there is error in reporting with both the methods. Most notably, underestimation of energy intake is a pervasive problem (Madden et al. 1976; Acheson et al. 1980; Carter et al. 1981; Lanksy and Brownell 1982; Mertz et al. 1991; Johnson et al. 1998; Heitmann et al. 2000; Jonnalagadda et al. 2000; Tran et al. 2000; Barnard et al. 2002; Subar et al. 2003; Moshfegh et al. 2008; Freedman et al. 2014), with the magnitude of underestimation greater among those who are overweight or obese (Bandini et al. 1990; Briefel et al. 1997; Pryer et al. 1997; Johnson et al. 1998; Fischer et al. 2000; Bailey et al. 2007; Lissner et al. 2007; Moshfegh et al. 2008; Freedman et al. 2014), female (Briefel et al. 1997; Hebert et al. 1997; Lafay et al. 1997), and less educated (Lafay et al. 1997; Pryer et al. 1997; Johnson et al. 1998; Bailey et al. 2007). Underestimation of energy intake is particularly pronounced among those who are overweight or obese. For example, in a study of the validity of interviewer-administered 24-hour dietary recalls in a sample of mostly well-educated adults, the percentage of normal, overweight, and obese women classified as low-energy reporters was 14.4%, 24.7%, and 35.3% respectively (Moshfegh et al. 2008). Intake estimates for other nutrients (e.g., vitamin D, calcium, and iron) are believed also to be underestimated with the food record and 24-hour dietary recall methods.

Error in estimating food and nutrient intake has a host of adverse implications for nutrition research and surveillance (Kipnis et al. 2002, 2003; Schatzkin et al. 2003; Willett 2013). Measures of association in studies examining diet–disease associations may be attenuated or otherwise spurious depending on the type of misclassification that results from error in dietary reporting. In nutrition surveillance studies, the percentage of the population consuming nutrients below recommended intake levels may be overestimated when nutrient intake is underestimated.

Due to the aforementioned implications of measurement error, there is great interest in understanding the causes of error in reporting with the dietary recall and food-records methods. In addition, strategies to improve reporting are of high interest.

4.3 CAUSES OF ERROR IN 24-HOUR DIETARY RECALL AND FOOD-RECORD REPORTING

Multiple factors are believed to contribute to error in reporting with the dietary recall and food-record methods. Below we describe five important factors contributing to this problem.

4.3.1 Memory Failure

Numerous studies have documented that participants, with the 24-hour dietary recall method, are more apt to fail to report a food they consumed than to report a food not consumed (Linusson et al. 1974; Greger and Etnyre 1978; Schnakenberg et al. 1981; Krantzler et al. 1982; Karvetti and Knuts 1985). Memory failure is hypothesized to be one of the reasons for failure to report foods consumed, and memory failure potentially increases with higher energy intake because there is more food consumed, and thus more memory demand for recalling all those eating occasions and also the amount eaten on each occasion.

In theory, memory failure should be of little concern with the food-record method because participants are asked to record all the foods and beverages they consume as they consume them throughout the day. But, memory failure remains a potential concern because participants may forget to record some eating occasions. Also, some participants may wait until the end of the day to record foods eaten rather than recording in real time throughout the day. In this situation the food-record method becomes more akin to a dietary recall, with memory required to recall foods consumed throughout the day.

4.3.2 Social Desirability Bias

Social desirability bias is the tendency of respondents to answer questions in a manner that is perceived as being a socially acceptable response. With respect to the dietary recall and food-record methods, social desirability bias may result in underreporting foods, or eating behaviors viewed as undesirable and/or overreporting foods, or eating behaviors viewed as desirable or socially acceptable.

A number of studies have compared the accuracy of dietary reporting by the extent to which one is prone to social desirability bias, and results indicate that underestimation of energy intake is highest among those who score higher on social desirability scales compared to those who score lower (Maurer et al. 2006; Bailey et al. 2007; Miller et al. 2008). Women tend to score higher on social desirability scales than men, which may in part explain why underestimation of energy intake is greater among women than men (Maurer et al. 2006). Social desirability bias in reporting may also explain why underestimation of energy intake is greater among those who are overweight or obese, because these individuals may engage more frequently in eating behaviors that are inconsistent with social norms. Also, these individuals, especially women, may be dieting to lose weight, and thus may be prone to exaggerating their adherence to diet goals, such as lower fat intake.

4.3.3 Measurement Reactivity

Measurement reactivity occurs when a person's behavior is altered due to the awareness being measured and/or in response to being measured.

With regard to the dietary recall and food-record methods, measurement reactivity may occur in anticipation of the measure (e.g., *I am being asked to record everything I eat and drink today, so I better eat healthy today*) and as a result of the measure (e.g., *Wow, in keeping this food record I am beginning to realize I eat a lot of junk food. I better start cutting down on the junk food*).

Research evaluating the extent to which measurement reactivity occurs with the dietary recall and food-record methods is scant. In two studies participants who had completed food records were asked if they had changed eating habits while keeping the records and 30%–50% reported they had (Forster et al. 1990; Macdiarmid and Blundell 1997).

Several studies have reported declines in food and/or nutrient reporting with increasing days of reporting (Gersovitz et al. 1978; Rizek and Pao 1990; Rebro et al. 1998). For example, Rebro et al. compared mean food and nutrient intake estimates from the first and fourth day of food-record keeping in a sample of 176 older women (Rebro et al. 1998). The number of food items, food components, and snacks reported per day were significantly lower on the fourth compared to the first day. Mean nutrient intake estimates were also consistently lower on the fourth compared to first day of recording, although none of the differences were statistically significant. The magnitude of the differences was generally modest. As an example, the mean number of food items reported on day one and day four were 10.7 items versus 10.0 items respectively ($P < 0.005$). These observed declines are potentially attributable to measurement reactivity. Fatigue in food-record keeping is another potential explanation, however, research is lacking regarding whether fatigue contributes to underreporting.

4.3.4 Motivation and Participant Fatigue

Participant burden is significant with both the dietary recall and food-record methods. For example, collection of an interviewer-administered 24-hour dietary recall using the Nutrition Data System for Research (NDSR) software requires an average of 20 minutes (Derr et al. 1992). With the food-record method the participant must record information about the foods and beverages consumed throughout the day, with detailed information about each food recorded (e.g., all ingredients and their amounts should be provided for a turkey sandwich). In addition, the participant may be asked to weigh or measure foods to determine amounts, depending on the food-record protocol.

As a result of the burden involved, it is possible that limited motivation and/or fatigue may contribute to error in reporting with the dietary recall and food-record methods. As mentioned, participant fatigue/loss of motivation is among the potential explanations for the phenomenon of decreasing food and nutrient intake estimates with increasing days of reporting. However, research specifically evaluating the effect of fatigue on reporting is not available.

4.3.5 Difficulty Estimating Food Amounts

When collecting 24-hour dietary recalls, visual aids are generally provided to assist participants in reporting the amounts of foods consumed. A variety of visual aids may be used including plastic three-dimensional food models; mugs, glasses, bowls, and plates in various sizes; photographs or drawings of foods in various portions; a ruler or two-dimensional grid for estimating dimensions; and concentric circles of various sizes for describing the size of circular foods. When collecting 24-hour dietary recall over the telephone, a two-dimensional food amount booklet (FAB) is typically given to participants for use during the recall.

Even with the provision of visual aids to assist in amount estimation, error in reporting food amounts is notable (Bolland et al. 1988; Yuhas et al. 1989; Wein et al. 1990; Faggiano et al. 1992; Weber et al. 1997; Chambers et al. 1999; Harnack et al. 2004). Large food portions tend to be underestimated, whereas small food portions tend to be overestimated (Faggiano et al. 1992; Chambers et al. 1999). Also, error in estimation of amounts is believed to be more pronounced for foods that are amorphous in shape (e.g., spaghetti, rice, and French fries) (Bolland et al. 1988; Yuhas et al. 1989;

Wein et al. 1990; Weber et al. 1997; Harnack et al. 2004). Amorphous foods such as pasta, rice, and potatoes are among the leading sources of energy in the American diet (Subar et al. 1998). As a result, error in amount estimation for amorphous foods has the potential to contribute significantly to error in energy intake estimation.

4.4 POTENTIAL STRATEGIES FOR IMPROVING REPORTING

In this section a variety of strategies that have the potential to minimize error in reporting (see Table 4.1) are described and discussed in the context of the source of error that the strategy may address. Some of the strategies are currently used routinely in dietary recall and food-record collection, whereas others are not yet fully developed or used routinely. For some strategies research has been carried out to evaluate gains in validity that may be expected from implementation of the strategy. But, for others scientific evidence is lacking or limited. These considerations are included in the discussion of each strategy.

4.4.1 Strategies Related to Memory Failure

4.4.1.1 Multiple-Pass Approach to Dietary Recall Collection

Dietary recalls are typically collected using the multiple-pass approach (Conway et al. 2003). The multiple-pass approach is a highly structured methodology for collecting dietary recall information (see Table 4.2) that is believed to result in more accurate reporting of food and nutrient intake than a less structured approach.

Using the multiple-pass approach, first a comprehensive list of every eating occasion and all the foods and beverages consumed at each occasion over a 24-hour period is garnered (first pass, known as the *quick list*).

Next, the foods and beverages reported in the first pass are reviewed, and the participant is asked if there are any additions or corrections (second pass, known as the *review quick list* pass).

TABLE 4.1
Strategies for Minimizing Error in Reporting with the 24-hour Dietary Recall and/or Food Record Methods

Dietary Strategy	Food Recalls	Records
Multiple-pass approach to recall collection	x	
Shorten the retention interval	x	
Memory cues	x	
Technologies such as digital food records	x	x
Ecological momentary assessment	x	
Objectivity/neutrality	x	x
Bogus pipeline	x	x
Telephone administration	x	
Establish purpose	x	x
Avoid feedback/dietary analysis results	x	x
Unscheduled/unannounced	x	
Provide financial incentive	x	x
Include visual aids for range of portions	x	
Amounts in food specific units	x	x
Minimum and maximum food portion flags	x	x
Require weighing and measuring foods		x

TABLE 4.2
Multiple-Pass Approach to 24-hour Dietary Recall Collection

Pass	Sample Script
Pass 1: Quick list	*After midnight, what was the first time you had something to eat or drink? What did you have at that time? Did you have anything else at that time?*
Pass 2: Quick list review	*Now let us review what we have so far. At [insert time] you had [insert foods]. Did you have anything else to eat at that time? Did you have anything to eat between [insert time] and [insert time]?*
Pass 3: Food detail	*Now let us fill in your list with more details. Did you add anything to [insert food]? What brand/type of [insert food name] was it? How much did you eat?*
Pass 4: Final review	*Now we will review the record. Tell me if I have missed anything.*
Commonly forgotten food pass	*There are some foods that people tend to forget they ate. Did you have any crackers, breads, rolls, or tortillas that you have forgotten about? How about…*

Gaps between meals may also be probed during this pass (e.g., *Did you eat or drink anything between 1 pm and 6:30 pm*?).

The third pass (known as the *food detail* pass) involves asking the participant if anything was added to each food/beverage (e.g., *Did you add anything to your cereal*?). The participant is also asked to provide needed details about each food (e.g., *What type of cereal was it*?) and to report the amount of food consumed (e.g., *How much cereal did you eat*?).

The final pass involves reviewing the foods and beverages reported to allow the participant one last opportunity to make additions or corrections.

A commonly forgotten food pass may also be included in the multiple-pass approach. In this pass the interviewer probes for commonly forgotten foods (e.g., *Did you have a beverage with that meal*?).

The multiple-pass approach to 24-hour dietary recall collection is believed to result in more accurate and complete reporting because multiple opportunities are provided for making additions and corrections to foods reported, and prompts are included that may trigger recollection of foods added to other foods (e.g., *sugar added to coffee*) and foods that may be easily forgotten (e.g., beverages and desserts eaten with meals). Numerous studies have evaluated the validity of 24-hour dietary recalls collected using the multiple-pass approach (Johnson et al. 1996; Jonnalagadda et al. 2000; Conway et al. 2003; Blanton et al. 2006; Moshfegh et al. 2008). But, studies directly comparing the validity of 24-hour dietary recall collected using the multiple-pass approach versus a less structured interview format are lacking. Consequently, the extent to which the multiple-pass approach may result in more valid food and nutrient intake estimates is unknown.

4.4.1.2 Shorten the Retention Interval

The standard approach to collecting 24-hour dietary recalls involves asking the participant to report all the foods and beverages he/she consumed from midnight to midnight the previous day. 24-hour dietary recalls are often carried out in the afternoon or evening. Consequently, the retention interval is long and intervening meals occur that are not to be reported. This long retention interval is hypothesized to contribute to error in reporting due to memory failure. To test this hypothesis and evaluate whether shortening the retention interval improves accuracy of recall, Baxter et al. carried out a series of studies in children (Baxter et al. 1997, 2004, 2006, 2009, 2013, 2014, 2015). Results from these studies show that the accuracy of recall is significantly improved when the retention interval is shortened. As an example, in one of their studies, fourth-grade children's recall of foods observed eaten at breakfast and lunch was markedly improved when the retention interval was short (children interviewed in the afternoon about the prior-24-hour's intake) in comparison to long (children interviewed in the morning about the previous-day's intake). For the short versus long retention interval, accuracy was better by 33%–50% for each of the four accuracy measures—food item omission rate and intrusion rate, energy correspondence rate, and inflation ratio (Baxter et al. 2015).

Findings from the work of Dr. Baxter et al. suggest that marked improvements in the accuracy of 24-hour dietary recall reporting may be achieved by simply shortening the retention interval. For example, instead of the standard prompt in which the participant is asked to *Tell me everything you ate and drank from midnight to midnight yesterday*, the retention interval could be shortened by asking *Tell me everything you ate and drank from yesterday at this time until now.* This approach has not been tested in adults or validated for an entire 24-hour period. Consequently, more research on this strategy is warranted.

4.4.1.3 Use of Memory Cues for Snacks and Eating That Occurs While Engaged in Other Activities

It is speculated that foods and beverages consumed between meals (*snacks*) may be more easily forgotten than regular or routine meals because little time may be invested in preparing the types of foods and beverages consumed as snacks, and snacks may be eaten in a short period of time relative to time spent eating meals. Snacking is a common occurrence, with about two-thirds of American adults consuming two or more snacks daily (Kant and Graubard 2015). Consequently, incomplete reporting of snacks could be a significant contributing factor to underestimation of food and nutrient intake, especially among population groups who snack frequently.

Meals and snacks eaten while engaged in other activities such as watching TV or driving are also hypothesized to be more apt to be forgotten in 24-hour dietary recalls due to distraction. Historical data from the American Time Use Survey indicates that the time American adults spend in eating as the primary activity has declined over the past 30 years, but time spent eating while engaged in another activity has increased precipitously (Zick and Stevens 2009). This trend toward eating while doing other activities has spurred research on the topic. Most of the studies have focused on the influence of distracted eating on food and nutrient intake, but a few studies have experimentally examined the influence on memory (recollection of eating occasion) (Higgs and Donohuse 2011; Mittal et al. 2011; Oldham Cooper et al. 2011). These studies found that accuracy or clarity of memory was lower for meals/snacks eaten while engaged in other activities such as watching TV or playing a video game. For example, in one of the experimental studies the recall of snacks eaten resulted in greater underestimation of energy consumed for those who ate a snack while watching TV (−620.5 kJ) compared to those who ate a snack while not engaged in another activity (–410.1 kJ) ($p < 0.05$) (Mittal et al. 2011).

To improve reporting of snacks and eating that occurs while engaged in other activities; special memory prompts related to these types of eating occasions may be warranted. For example, a global prompt could be included in the instructions given to participants before the first pass of the 24-hour dietary recall is initiated (e.g., before asking the participant to report everything he or she ate, the participant could be asked to take a minute to think about the major things he or she did and places he or she traveled during the recall period). Alternatively, specific prompts could be in one or more passes. For example, at the end of the second (review quick list) pass participants could be told that people often forget to report foods eaten while doing things like driving or watching TV. Then, a few questions could be asked such as *Did you eat anything while driving*? An affirmative response could be followed with a question regarding whether they had remembered to report the foods eaten while doing this activity, and forgotten that foods could be added to the food list.

To date no research has been carried out to develop or evaluate this strategy. Although it has the potential to reduce error in reporting due to memory failure, it could increase the time required to complete 24-hour dietary recalls. If the participant burden is increased, error in reporting due to fatigue could potentially increase and possibly negate any benefit. Thus, research is needed to fully develop and evaluate this strategy.

4.4.1.4 Technology-Assisted 24-hour Dietary Recalls and Food Records

The development of new or adaptation of existing or emerging technologies has the potential to improve both the feasibility and validity of the 24-hour dietary recall and food-record methods. As discussed in Chapter 3 (McNutt and Zimmerman, this book), past use of computer technology has resulted in the development of software applications such as Nutrition Data System for Research (NDSR) (Feskanich et al. 1988) and the Automated Multiple-Pass Method (AMPM) system (Blanton et al. 2006). In comparison to paper-and-pencil approaches, these software applications may elicit more complete reporting of food intake due to inclusion of standardized prompts. In addition, the cost for collection may be lowered because automated coding requires less time than manual food coding (Feskanich et al. 1988).

Some of the technology-assisted 24-hour dietary recall and food-record methods under development will now be briefly described and discussed.

4.4.1.4.1 Computer/Mobile Phone Programs for Self-Administered 24-hour Dietary Recall Collection

Collection of 24-hour dietary recalls via self-administration with an online application, such as the National Cancer Institute Automated Self-Administered 24-hour (ASA24) Dietary Recall System, has the potential to substantially lower the cost of 24-hour dietary recall collection because an interviewer is not required for collection. In addition, it is possible that social desirability bias in reporting may be lower with a self-administered compared to interviewer-administered recall due to the anonymous nature of this form of data collection. Research on use of computer survey technology to assess sexual behavior (Turner et al. 1998) strongly supports this notion. Thus, a self-administered approach to 24-hour dietary recall collection has the potential to be less prone to this source of bias in reporting, although research evaluating this speculation is not yet available.

Potential concerns with a self-administered approach to recall collection include participant burden, since the time required for self-completion could be greater than that of an interviewer-administered recall. Completeness of reporting may be hindered if instructions for reporting are not clear, or there are usability concerns such as inability to readily locate foods consumed in the *food search* function in the tool.

A variety of self-administered 24-hour dietary recall applications have been evaluated. In most of these studies results from the self-administered recalls were compared to results from interviewer-administered recalls. Findings from these studies do not support the notion that reporting is more accurate with self-administration. Rather, it appears that underestimation of energy may be somewhat greater with self-administered dietary recalls (Illner et al. 2012). More research is needed, however, with biomarkers used as the criterion (reference) measure.

4.4.1.4.2 Digital Entry Food Record Mobile Phone Applications

Digital food records have the potential to be less costly than paper-and-pencil records because the participant records (enters) their food intake information into the application, which carries out nutrient intake calculations automatically. In contrast, with paper-and-pencil food records, a study staff member must enter food record information into a dietary analysis software program for nutrient intake calculation.

Reporting of food and beverage intake could be more complete and accurate with an electronic food record if the application includes probes and prompts to ensure needed food detail is entered (written food-record forms includes instructions on the level of food detail to enter, but no automated probes or prompts as such are possible with an electronic food record application. Also, participant burden might be lower if the application has good usability. But, reporting with an electronic food record could be no better or less accurate than traditional paper-and-pencil food records if the application lacks important prompts and probes and/or has poor usability (e.g., difficult to locate foods consumed using the application's *food search* function).

Results from studies evaluating various digital food-record applications are mixed (Illner et al. 2012; Hutchesson et al. 2015; Rangan et al. 2015). In a review of innovative technologies for measuring diet in nutritional epidemiology by Illner et al. the authors noted that evaluation results for some digital food-record apps indicated high levels of acceptance by respondents, whereas acceptance was poor in other evaluations (Illner et al. 2012). The authors speculated that differences in usability of various food-record applications could explain differences in acceptance. For example, the food search feature in one app may work well, whereas this feature may work poorly in another, thus leading to disparate levels of acceptance for the two food-record applications. It was also reported in a review by Illner et al. that results from studies evaluating the validity of food-record applications were also variable. Some validation studies have found reasonably good concordance between intake estimates from the food-record application being evaluated and a criterion or reference measure of intake such as energy intake estimated from doubly-labeled water or a conventionally collected food record. But, in other validation study issues with validity, including both underestimation and overestimation of nutrient intake, were found. The authors of the review speculated that varying findings for validity may also reflect the varying designs and features of each digital food-record application being evaluated for validity. The variable findings with respect to validity suggest that validation study data for specific applications should be used as a guide in choosing between different food-record applications.

4.4.1.4.3 Spoken (Audio) Food Record

With a spoken (audio) food record the study participant is instructed to record a verbal description of the foods and beverages they consume. A device, such as mobile phone application, is used for recording (Van Horn et al. 1990). Participant burden is potentially lower with a spoken as compared to written food record, but data are lacking to confirm this speculation.

Only one study has evaluated the validity of a spoken food record. Van Horn et al. carried out a study in which children performed tape-recorded food records ($n = 33$), whereas their parents completed food records concomitantly as a reference (criterion) measure of intake. Correlations between nutrient intake estimates from the tape-recorded food records and the parental record of the child's intake were = 0.80 or above for all the nutrients examined except calories ($r = 0.68$) (Van Horn et al. 1990).

Coding is a key feasibility issue with spoken food records. A variety of approaches may be taken to code an audio record for calculation of food and nutrient intake. One approach is to have a trained coder listen to the recordings, and enter the foods and beverages reported into a dietary analysis software program for nutrient calculation. This approach has the potential to be reasonably reliable and accurate if the participant has provided the needed food detail in the audio recording. But, the approach is apt to be very time consuming and hence costly. Another approach is to develop and use voice recognition software, so that the spoken record is transcribed into written words for manual or automated entry into a dietary analysis software application. Some work has been carried out to address this challenging aspect of spoken food-record collection and analysis (Lacson and Long 2006), but more work is needed if this approach to food-record collection is to become viable.

4.4.1.4.4 Photographic Food Record

As described in Chapter 5, photographic food records involve taking photos of foods and beverages consumed throughout the day, with annotation of the photos required to obtain some needed food detail (e.g., percent fat milk and whether soda is diet or regular). This approach to food-record collection has the potential to address a number of factors that contribute to error in reporting. Most notably, estimation of food amounts is not required with this method. Also, if a wearable camera that automatically detects and records eating occasions is used, the issue of forgetting to record eating occasions may be addressed as well.

To date a number of major obstacles have been encountered that limit the applicability of photographic food records for assessing diet, especially in larger-scale studies where cost is a key consideration (Illner et al. 2012; Gemming et al. 2015). See Chapter 5 for further detail.

4.4.1.4.5 *Devices That Detect Eating*

As discussed in Chapters 7 and 8, wearable devices that detect eating (e.g., wrist-worn device that detects eating by wrist-motion tracking) have the potential to improve dietary assessment. More specifically, these devices could be paired with the 24-hour dietary recall and food-record methods to improve their accuracy.

As an example, participants could be asked to wear a wrist device that detects eating episodes and sends these data to dietary recall interviewers. The 24-hour dietary recall could then be carried out using these data, with the interviewer asking the participant to report what he or she ate at each detected eating occurrence (e.g., *I see you ate or drank something at 2:25 pm yesterday. What did you eat or drink at that time?*) Thus, participants would no longer need to remember every eating occasion. Instead, their focus could be placed on remembering what and how much they ate at each eating occasion.

With respect to food records, wearable devices could also be used in conjunction with this method. For example, a wrist device could be designed to send a message to the participant each time eating is detected, with the message designed to remind or prompt the participant to record what was eaten.

4.4.1.4.6 *Dietary Recall Collection Using Ecological Momentary Assessment*

Ecological momentary assessment (EMA) is a method of collecting behavioral information close in time to the behavior and in the context of normal daily life. With this approach study participants are asked to record a particular behavior (e.g., foods and beverages consumed) for a specific time period that immediately precedes the request for reporting (e.g., in the past 15 minutes). The requests for information are sent at random time points throughout a period of time (e.g., across two weeks). The number of time points required and the time window during which time points will be randomly selected must be carefully determined to ensure the behavioral data collected with this method and are representative of the behavior of interest (e.g., usual food and nutrient intake) in the unit of measure of interest (e.g., total daily food and nutrient intake). Quantifying food and nutrient intake for a 24-hour period (a day) is typically the unit of interest because dietary recommendations are issued in this unit. But, other units could potentially be useful.

In theory, food and nutrient intake reporting may be more complete and accurate with an EMA versus conventional dietary recall collection approach because the retention interval (time that has transpired between eating and reporting what was eaten) is shorter (e.g., past 15 minutes vs. past 24-hours). At this very limited development and evaluation work has been carried out (Hand and Perzynski 2016).

4.4.2 Strategies to Minimize Social Desirability Bias

4.4.2.1 Objectivity/Neutrality

So that study participants do not feel like they are being judged or evaluated, dietary recall interviewers and those who interact with study participants for food record collection are generally trained to maintain an objective non-judgmental demeanor when interacting with participants. This includes both verbal and non-verbal communication. All written materials provided to participants and interview scripts should likewise be objective/neutral in tone and convey the message that the participant is not being judged or evaluated.

Participants sometimes vocalize self-criticism (e.g., "I know I should not have eaten that") or seek nutrition advice (e.g., "I eat avocados because I heard they are good for my heart. Is that true?") during dietary recall and food record collection. To avoid appearing judgmental, interviewers are generally trained to not respond or react to self-criticism. Also, interviewers are usually trained to avoid answering nutrition-related questions (e.g., trained to politely respond to participant nutrition questions by saying "I am sorry, I can not answer nutrition questions").

Research has not been conducted to evaluate the extent to which procedures to preserve objectivity/neutrality improve the accuracy of 24-hour dietary recall or food record reporting.

4.4.2.2 Use of Bogus Pipeline

The bogus pipeline is a technique that may be used to reduce false answers when attempting to collect self-report data (Baranowski et al. 2002). In particular, it may prevent or minimize incorrect responses due to social desirability bias. The technique involves telling the respondent that a measure is being collected that will accurately measure the behavior for which self-reported information is sought. For example, prior to asking a participant to answer questions about tobacco use (e.g., whether they smoke tobacco, number of cigarettes smoked per day) the participant would be informed that a sample of their blood is being collected and analyzed for serum cotinine, a biomarker that is indicative of tobacco use. By telling the respondent that a true measure of their behavior is being collected, the impetus for providing a socially desirable response may be reduced since the truth will be known regardless of their response.

Two studies have evaluated the use of a bogus pipeline to improve dietary recall and food record reporting (Muhlheim et al. 1998; Baranowski et al. 2002). Baranowski et al. evaluated whether use of a hair sample as a bogus pipeline increased the accuracy of food reporting among 4th grade children (n = 138) (Baranowski et al. 2002). Those children randomly assigned to the bogus pipeline condition had statistically significant higher percent agreement (69% vs. 49%) on foods reported when comparing 24 hour dietary recalls with observations of foods consumed (criterion measure). But, there was lower concordance on portion-size matches for children in the bogus pipeline group compared to the nonbogus pipeline group (r = 0.71 vs. 0.80) and there was no difference between the groups in percent intrusions (foods reported but not observed) and omissions (foods observed but not reported). Muhlheim et al. evaluated use of a bogus pipeline in a sample of adults identified as prone to under-reporting dietary intake. The participants (n = 28) completed food records for seven days under standard conditions, and were then randomized to either experimental (bogus pipeline) or control conditions. Those in the experimental group were instructed to complete food records for two weeks and were told the researchers would verify their intake through the use of doubly-labeled water (DLW). Those in the control condition were instructed to complete food records for two weeks (no bogus pipeline used). During the experimental period, reported energy intake for the control group was 1,298 kcal/day (48% of estimated energy intake) and reported energy intake of the experimental group was 1,689 kcal/day (61% of predicted energy intake). The effect of the experimental assignment was borderline significant for the total sample ($p = 0.05$) and significant for women ($p = 0.03$) but not men.

There are ethical concerns to consider when using the bogus pipeline technique. If a biomarker that is reflective of dietary intake is being collected and analyzed as indicated to the participant, there is no ethical concern. However, if a biomarker that is indicative of dietary intake is not being collected and analyzed as stated to the participant (bogus pipeline); the technique may be considered deceptive. When a deceptive technique is used in research most institutional review boards require that use of deception be disclosed to the participant at some point. Disclosing this information may lessen the trust the study participant has with the researcher. Thus, the trade-off between a potential benefit and harm may need to be carefully considered in deciding whether to use this technique.

4.4.2.3 Collect 24-hour Dietary Recalls Over the Telephone instead of In-Person

24-hour dietary recalls have traditionally been collected via an in-person interview. As a cost-saving measure and in recognition of the widespread presence of telephones in the home, research was initiated in the 1980s to evaluate whether 24-hour dietary recalls could be collected over the telephone (Posner et al. 1982; Van Horn et al. 1990). Collecting recalls over the telephone was speculated to have numerous possible benefits over recalls collected in-person including lower participant burden, lower costs associated with collection, and higher completion rates. In addition, it could be speculated that more accurate reporting may occur because the participant can check foods in the home for needed detail (e.g., look at carton of milk to report percent fat and look at package to determine portion size) and social desirability bias in reporting may be less acute because the participant may feel more anonymous when talking over the phone versus meeting face-to-face with an interviewer.

A number of studies have compared food and nutrient intake estimates from 24-hour dietary recalls collected via telephone versus in-person (Posner et al. 1982; Van Horn et al. 1990; Derr et al. 1992; Casey et al. 1999; Tran et al. 2000; Yanek et al. 2000; Bogle et al. 2001; Baxter et al. 2003). Results from these studies are uniform in finding that food and nutrient intake estimates from 24-hour dietary recalls collected over the telephone are similar to intake estimates for dietary recalls collected in-person. Consequently, with respect to validity the two modes of administration appear to be comparable.

4.4.3 Strategies to Prevent Measurement Reactivity

4.4.3.1 Establish Purpose of the Measure

In written and verbal instructions provided for 24-hour dietary recalls and food records, it may be important to explain the purpose of the dietary assessment so the participant knows the information is not being used to judge or evaluate them or their diet. Furthermore, it may be important for the participant to understand the study is interested in assessing what people typically eat and drink, and consequently they should continue their usual eating habits.

An example of the instruction that may be provided to participants to clarify purpose: "For this study we need to know what people like you typically eat and drink. The information being collected will not be used to judge or evaluate you. Remember, there are no right or wrong answers, so whatever you ate is OK."

Research is lacking to test the common belief that these strategies minimize subject reactivity.

4.4.3.2 Avoiding Feedback/Dietary Analysis Results

Providing nutrient analysis results to study participants may be used as an incentive for participating, and could potentially motivate participants to report more accurately. However, it could be speculated that providing feedback contributes to measurement reactivity. The validity of 24-hour dietary recalls and food records collected with and without the provision of dietary analysis results have not been compared. Consequently, research is lacking to confirm or quantify the effects of providing dietary analysis results.

4.4.3.3 Unscheduled/Unannounced 24-hour Dietary Recall Collection

When a specific date and time is scheduled for the collection of a 24-hour dietary recall, the participant could potentially change his/her eating habits the day prior to the recall in anticipation of the recall (e.g., "I better eat healthy today because tomorrow I'm going to have to report what I ate today"). Collecting 24-hour dietary recalls unscheduled/unannounced could be a way of avoiding this scenario.

Unscheduled/unannounced 24-hour dietary recall collection is possible when the recalls are being collected via telephone interview. To carry out unscheduled/unannounced recalls, participants are asked to provide their general availability for a telephone administered dietary recall (times each day of the week they are generally free). Participants are then called during the time periods they are generally free. If a participant answers the phone when called and is available for the length of time it takes to complete a recall, the 24-hour dietary recall is carried out.

The validity of 24-hour dietary recalls collected on a scheduled versus unscheduled/unannounced basis has not been compared. Consequently, research is lacking to confirm or quantify gains in validity that may be achieved by carrying out 24-hour dietary recalls unscheduled/unannounced.

4.4.4 Strategy to Motivate Participants

The time required to complete 24-hour dietary recalls and food records is significant. Consequently, to encourage study participation a financial incentive is often provided. In addition to encouraging participation, the provision of an incentive could potentially motivate participants to more diligently report foods and beverages consumed. To date only one study has tested this premise. Hendrickson and Mattes carried out a trial to determine whether offering an incentive ($50 or $100 for completion of four 24-hour dietary recalls) resulted in less under-estimation of energy intake compared to receiving no incentive (Hendrickson and Mattes 2007). No difference in under-estimation of energy intake was found between the groups receiving an incentive and the group receiving no incentive. The sample size in the study was small (n = 56) and homogenous (overweight females, most of whom were college students).

4.4.5 Improve Food Amount Estimation

4.4.5.1 Provide Food Amount Estimation Aids That Reflect Typical Food Portions

It has been speculated that larger food portions are more prone to underestimation (Faggiano et al. 1992; Chambers et al. 1999) because the food amount estimation aids typically used depict portions that are smaller than the portions common in the food marketplace today (Young and Nestle 1995). Consequently, it may be important to use food amount estimation aids that reflect typical food portions, including large portions.

Just one study has evaluated whether providing visual aids that more closely represent portions consumed improves accuracy of reporting of foods consumed in large portions. Harnack et al. (2004) found that underestimation was reduced from 27% to 40% for three large portioned food items with use of larger sized food amount estimation aids in comparison to smaller (typical) sized food amount estimation aids.

4.4.5.2 Allow Food Portions to Be Reported in Food Specific Units

Food amount estimation is challenging for participants, especially when estimating the amounts of amorphous foods such as pasta, rice, and French fries (Bolland et al. 1988; Yuhas et al. 1989; Wein et al.1990; Weber et al. 1997; Harnack et al. 2004). Consequently, it may be wise to minimize the need for estimating food amounts by allowing participants to report food amounts in food specific units (e.g., *large* McDonald's French fry order, *snack size* Snickers, etc.). To maximize the extent to which this is possible, it is important to choose a dietary analysis software application that includes food specific units for foods that may be reported by unit.

This common sense approach to improving accuracy of food amount reporting has not been evaluated to determine whether and to what extent it improves the validity of food and nutrient intake estimates.

4.4.5.3 Confirm Unusually Small and Large Food Amounts

Some dietary analysis software applications include minimum and maximum food amount flags, which require the interviewer confirm unusually small or large amounts with the study participants. In addition to potentially catching estimation errors by the participant, these flags may detect egregious data entry errors (e.g., interviewer mistakenly entering "1 teaspoon" instead of "1 cup" of Cheerios).

To date no studies have evaluated whether confirming unusually small and large food amounts reduces measurement error in 24-hour dietary recalls or food records.

4.4.5.4 Require Weighing and Measuring Food Amounts for Food Records

Requiring participants to measure and weigh foods for food records may minimize food amount reporting error. But, participant burden is increased and it is speculated that measurement reactivity may be greater when weighing and measuring foods is required. In addition, study costs are increased because participants must be provided with a food scale and volumetric measuring tools (e.g., measuring cups and spoons) in order to carry out the requested measures.

No studies have compared the validity of food records that require weighing and measuring versus estimating food portions. Consequently, it is unknown whether and to what extent food record accuracy may be improved by requiring weighing and measuring food portions.

4.5 FUTURE STUDIES

More than 20 strategies that could potentially improve the validity of the 24-hour dietary recall and food record methods were discussed in this paper. For many of these potential strategies little or no research has been carried out to evaluate whether and to what extent validity may be improved. Interestingly, some commonly used techniques, such as the multiple-pass approach to dietary recall collection, lack research confirming the approach improves validity.

Strategies for which further research is warranted include:

- Multiple-pass approach to 24-hour dietary recall interview: Does this highly structured approach to recall collection result in more valid reporting than a less structured approach? Are some passes more important than others?
- Shortening the retention interval for 24-hour dietary recalls: Studies in adults are needed, and effects on accuracy of reporting for an entire 24-hour period should be evaluated.
- Use of memory cues: A set of memory cue questions and a procedure for asking these questions in the context of a dietary recall need to be developed, and then evaluated.
- Technologies such as web-based self-administered dietary recalls, electronic food records, and spoken (audio) food records: Further development and evaluation required. Determining validity and acceptability in comparison to validity and acceptance of conventional methods is particularly important to establish.
- Commonly used common sense strategies such as training interviewers to maintain objective/non-judgmental demeanor, establishing purpose of the measure, providing no feedback/dietary analysis results, etc.: Ideally these commonly used procedures would be evaluated to determine whether they are important to the validity of dietary recalls and food records. But, they may be of lower priority for evaluation than strategies not yet regularly used that have the potential to markedly improve validity.
- Strategies for improving food amount reporting, such as use of food amount estimation that reflect typical portions and allowing food portions to reported in food specific units: In consideration of the extent to which food amount estimation errors are believed to reduce validity and the limited research carried out to evaluate these common sense strategies, further research on these strategies is warranted.

4.6 SUMMARY

In summary, the 24-hour dietary recall and food record methods elicit relatively high-quality data, making them preferred tools for monitoring food and nutrient intake of populations and studying diet-disease associations. However, there is error in reporting with both methods. Most notably, under-estimation of food and nutrient intake is a well-documented shortcoming of these methods. There are a variety of factors believed to contribute to error in reporting with dietary recalls and food records. These include memory failure, social desirability bias in reporting, measurement reactivity, fatigue/low motivation, and difficulty reporting food amounts. Potential strategies for minimizing error in reporting are as varied and numerous as the causes. Some of the strategies have a strong evidence base supporting their use. Others have been evaluated and found to offer no benefit with respect to validity.

REFERENCES

Acheson, K. J., I. T. Campbell, O. G. Edholm, D. S. Miller, M. J. Stock. 1980. The measurement of food and energy intake in man-an evaluation of some techniques. *Am J Clin Nutr* 33 (5): 1147–1154.

Bailey, R. L., D. C. Mitchell, C. Miller, H. Smiciklas-Wright. 2007. Assessing the effect of underreporting energy intake on dietary patterns and weight status. *J Am Diet Assoc* 107 (1): 64–71. doi:10.1016/j.jada.2006.10.009.

Bandini, L., D. Schoeller, H. Cyr, W. Dietz. 1990. Validity of reported energy intake in obese and nonobese adolescents. *Am J Clin Nutr* 52: 421–425.

Baranowski, T., N. Islam, J. Baranowski et al. 2002. The food intake recording software system is valid among fourth-grade children. *J Am Diet Assoc* 102: 380–385.

Barnard, J. A., L. C. Tapsell, P. S. Davies, V. L. Brenninger, L. H. Storlien. 2002. Relationship of high energy expenditure and variation in dietary intake with reporting accuracy on 7 day food records and diet histories in a group of healthy adult volunteers. *Eur J Clin Nutr* 56 (4): 358–367. doi:10.1038/ej/ejcn/1601341.

Baxter, S., A. F. Smith, M. S. Litaker. 2004. Recency affects reporting accuracy of children's dietary recalls. *Ann Epidemiol* 14: 385–390.

Baxter, S., A. F. Smith, M. S. Litaker. 2006. Body mass index, sex, interview protocol, and children's accuracy for reporting kilocalories observed eaten at school meals. *J Am Diet Assoc* 106: 1656–1662.

Baxter, S., W. O. Thompson, M. S. Litaker et al. 2003. Accuracy of fourth-graders' dietary recalls of school breakfast and school lunch validated with observations: In-person versus telephone interviews. *J Nutr Educ Behav* 35 (3): 124–134.

Baxter, S. D., A. F. Smith, D. B. Hitchcock et al. 2015. Effectiveness of prompts on fourth-grade children's dietary recall accuracy depends on retention interval and varies by gender. *J Nutr* 145 (9): 2185–2192. doi:10.3945/jn.115.213298.

Baxter, S. D., J. W. Hardin, C. H. Guinn, J. A. Royer, A. J. Mackelprang, A. F. Smith. 2009. Fourth-grade children's dietary recall accuracy is influenced by retention interval (target period and interview time). *J Am Diet Assoc* 109 (5): 846–856.

Baxter, S. D., D. B. Hitchcock, C. H. Guinn et al. 2014. A validation study concerning the effects of interview content, retention interval and grade on children's recall accuracy for dietary and/or physical activity. *J Academ Nutr Diet* 114: 1902–1914.

Baxter, S. D., W. O. Thompson, H. C. Davis, M. H. Johnson. 1997. Impact of gender, ethnicity, meal components, and time interval between eating and reporting on accuracy of fourth-graders' self-reports of school lunch. *J Am Diet Assoc* 97: 1293–1298.

Baxter, S. D., D. B. Hitchcock, C. H. Guinn et al. 2013. A pilot study of the effects of interview content, retention interval, and grade on accuracy of dietary information from children. *J Nutr Educ Behav* 45 (4): 368–373. doi:10.1016/j.jneb.2013.01.016.

Blanton, C. A., A. J. Moshfegh, D. J. Baer, M. J. Kretsch. 2006. The USDA automated multiple-pass method accurately estimates group total energy and nutrient intake. *J Nutr* 136: 2594–2599.

Bogle, M., J. Stuff, L. Davis et al. 2001. Validity of a telephone-administered 24-hour dietary recall in telephone and non-telephone households in the rural Lower Mississippi Delta region. *J Am Diet Assoc* 101 (2): 215–222.

Bolland, J. E., J. A. Yuhas, T. W. Bolland. 1988. Estimation of food portion sizes: Effectiveness of training. *J Am Diet Assoc* 88 (7): 817–821.

Briefel, R. R., C. T. Sempos, M. A. McDoweel, S. Chien, K. Alaimo. 1997. Dietary methods research in the third national health and nutrition examination survey: Underreporting of energy intake. *Am J Clin Nutr* 65 (4 Suppl): 1203S–1209S.

Carter, R. L., C. O. Sharbaugh, C. A. Stapell. 1981. Reliability and validity of the 24-hour recall. *J Am Diet Assoc* 79: 542–547.

Casey, P. H., S. L. Goolsey, S. Y. Lensing, B. P. Perloff, M. L. Bogle. 1999. The use of telephone interview methodology to obtain 24-hour dietary recalls. *J Am Diet Assoc* 11: 1406–1411.

Chambers, E., B. McGuire, S. Godwin, E. Edwards. 1999. *NHANES 1999+ Methodology Research: Approaches for Estimating Quantities of Foods Reported in a Dietary Interview Setting*. Hyattsville, MD: National Center for Health Statistics.

Conway, J. M., L. A. Ingwersen, B. T. Vinyard, A. J. Moshfegh. 2003. Effectiveness of the US department of agriculture's 5-step multiple-pass method in assessing food intake in obese and nonobese women. *Am J Clin Nutr* 77: 1171–1178.

Derr, J. A., D. C. Mitchell, D. Brannon, H. Smiciklas Wright, L. B. Dixon, B. M. Shannon. 1992. Time and cost analysis of a computer-assisted telephone interview system to collect dietary recalls. *Am J Epidemiol* 136 (11): 1386–1392.

Faggiano, F., P. Vineis, D. Cravanzola et al. 1992. Validation of a method for the estimation of food portion size. *Epidemiology* 3 (4): 379–382.

Feskanich, D., I. M. Buzzard, B. Welch et al. 1988. Comparison of a computerized and a manual method of food coding for nutrient intake studies. *J Am Diet Assoc* 88: 1263–1267.

Fischer, J. O., R. K. Johnson, C. Lindquist, L. L. Birch, M. I. Goran. 2000. Influence of body composition on the accuracy of reported energy intake in children. *Obes Res* 8 (8): 597–603.

Forster, J., R. Jeffery, M. Van Natta, P. Pirie. 1990. Hypertension prevention trial: Do 24-h food records capture usual eating behavior in a dietary change study? *Am J Clin Nutr* 51: 253–257.

Freedman, L. S., J. M. Commins, J. E. Moler et al. 2014. Pooled results from 5 validation studies of dietary self-report instruments using recovery biomarkers for energy and protein intake. *Am J Epidemiol* 180 (2): 172–188.

Gemming, L., J. Utter, C. N. Mhurchu. 2015. Image-assisted dietary assessment: A systematic review of the evidence. *J Acad Nutr Diet* 115: 64–77.

Gersovitz, M., J. P. Madden, H. Smiciklas Wright. 1978. Validity of the 24-hour recall and seven-day record for group comparisons. *J Am Diet Assoc* 73: 48–55.

Greger, J. L. and G. M. Etnyre. 1978. Validity of 24-hour dietary recalls by adolescent females. *Am J Public Health* 68: 70–72.

Hand, R. K. and A. T. Perzynski. 2016. Ecologic momentary assessment: Perspectives on applications and opportunities in research and practice regarding nutrition behaviors. *J Nutr Educ Behav*. doi:10.1016/j.jneb.2016.05.004.

Harnack, L., L. Steffen, D. K. Arnett, S. Gao, R. V. Luepker. 2004. Accuracy of estimation of large food portions. *J Am Diet Assoc* 104 (5): 804–806.

Hebert, J. R., Y. Ma, L. Clemow et al. 1997. Gender differences in social desirability and social approval bias in dietary self-report. *Am J Epidemiol* 146: 1046–1055.

Heitmann, B. L., L. Lissner, M. Osler. 2000. Do we eat less fat, or just report so? *Int J Obes* 24: 435–442.

Hendrickson, S. and R. Mattes. 2007. Financial incentive for diet recall accuracy does not affect reported energy intake or number of underreporters in a sample of overweight females. *J Am Diet Assoc* 107 (1): 118–121. doi:10.1016/j.jada.2006.10.003.

Higgs, S. and J. E. Donohuse. 2011. Focusing on food during lunch enhances lunch memory and decreases later snack intake. *Appetite* 2011: 202–206.

Hutchesson, M. J., M. E. Rollo, R. Callister, C. E. Collins. 2015. Self-monitoring of dietary intake by young women: Online food records completed on computer or smartphone are as accurate as paper-based food records but more acceptable. *J Acad Nutr Diet* 115 (1): 87–94.

Illner, A. K., H. Freisling, H. Boeing, I. Huybrechts, S. P. Crispim, N. Slimani. 2012. Review and evaluation of innovative technologies for measuring diet in nutritional epidemiology. *Int J Epidemiol* 41: 1187–1203.

Johnson, R. K., P. Driscoll, M. I. Goran. 1996. Comparison of multiple-pass 24-hour recall estimates of energy intake with total energy expenditure determined by the doubly labeled water method in young children. *J Am Diet Assoc* 96 (11): 1140–1144.

Johnson, R. K., R. P. Soultanakis, D. E. Matthews. 1998. Literacy and body fatness are associated with underreporting of energy intake in US low-income women using the multiple-pass 24-hour recall: A doubly labeled water study. *J Am Diet Assoc* 98: 1136–1140.

Jonnalagadda, S., D. Mitchell, H. Smiciklas-Wright et al. 2000. Accuracy of energy intake data estimated by a multiple-pass, 24-hour dietary recall technique. *J Am Diet Assoc* 100: 303–309, 311.

Kant, A. K. and B. I. Graubard. 2015. 40-year trends in meal and snack eating behaviors of American adults. *J Acad Nutr Diet* 115: 50–63.

Karvetti, R. L. and L. R. Knuts. 1985. Validity of the 24-hour dietary recall. *J Am Diet Assoc* 85: 1437–1442.

Kipnis, V., D. Midthune, L. Freeman et al. 2002. Bias in dietary-report instruments and its implications for nutritional epidemiology. *Public Health Nutr* 5 (6A): 915–923.

Kipnis, V., A. F. Subar, D. Midthune et al. 2003. Structure of dietary measurement error: Results from the OPEN Biomarker Study. *Am J Epidemiol* 158: 14–21.

Krantzler, N. J., B. J. Mullen, H. G. Schutz, L. E. Grevetti, C. A. Holden, H. L. Meilselman. 1982. Validity of telephoned diet recalls and records for assessement of individual intake. *Am J Clin Nutr* 36: 1234–1242.

Lacson, R. and W. Long. 2006. Natural language processing of spoken diet records (SDRs). *AMIA Annu Symp Proc*, pp. 454–458. Bethesda, MD: AMIA.

Lafay, L., A. Basdevant, M. A. Charles et al. 1997. Determinants and nature of dietary underreporting in a free-living population: The Fleurbaix Laventie Ville Sante (FLVS) Study. *Int J Obes* 21 (7): 567–573.

Lanksy, D. and K. D. Brownell. 1982. Estimates of food quantity and calories: Errors in self-report among obese patients. *Am J Clin Nutr* 35: 727–732.

Linusson, E., D. Sanjur, E. C. Erickson. 1974. Validating the 24-hour recall methods as a dietary survey tool. *Arch Latino Am Nutr* 24: 277–281.

Lissner, L., R. P. Troiano, D. Midthune et al. 2007. OPEN about obesity: Recovery biomarkers, dietary reporting errors and BMI. *Int J Obes (Lond)* 31 (6): 956–961. doi:10.1038/sj.ijo.0803527.

Macdiarmid, J. I. and J. E. Blundell. 1997. Dietary underreporting: What people say about recording their food intake. *Eur J Clin Nutr* 51: 199–200.

Madden, J. P., S. J. Goodman, H. A. Guthrie. 1976. Validity of the 24-hour recall. Analysis of data obtained from elderly subjects. *J Am Diet Assoc* 68: 143–147.

Maurer, J., D. L. Taren, P. J. Teixeira et al. 2006. The psychosocial and behavioral characteristics related to energy misreporting. *Nutr Rev* 64 (2): 53–66. doi:10.1301/nr.2006.feb.53–66.

Mertz, W., J. C. Tsue, J. T. Judd et al. 1991. What are people really eating? The relation between energy intake derived from estimated diet records and intake determined to maintain body weight. *Am J Clin Nutr* 54: 291–295.

Miller, T. M., M. F. Abdel-Maksoud, L. A. Crane, A. C. Marcus, T. E. Byers. 2008. Effects of social approval bias on self-reported fruit and vegetable consumption: A randomized controlled trial. *Nutr J* 7 (18). doi:10.1186/1475-2891-7-18.

Mittal, D., R. J. Stevenson, M. H. Oaten, L. A. Miller. 2011. Snacking while watching TV impairs food recall and promotes food intake on a later TV free test meal. *Applied Cog Psychol* 25: 871–877.

Moshfegh, A. J., D. G. Rhodes, D. J. Baer et al. 2008. The US department of agriculture automated multiple-pass method reduces bias in the collection of energy intakes. *Am J Clin Nutr* 88 (2): 324–332.

Muhlheim, L. S., D. B. Allison, S. Heshka, S. B. Heymsfield. 1998. Do unsuccessful dieters intentionally underreport food intake? *Int J Eat Disord* 24: 259–266.

Oldham Cooper, R. E., C. A. Hardman, C. E. Nicoll, P. J. Rogers, J. M. Brunstrom. 2011. Playing a computer game during lunch affects fullness, memory for lunch, and later snack intake. *Am J Clin Nutr* 93: 308–313.

Posner, B. M., C. L. Borman, J. L. Morgan, W. S. Borden, J. C. Ohls. 1982. The validity of a telephone-administered 24-hour dietary recall methodology. *Am J Clin Nutr* 36: 546–553.

Pryer, J. A., M. Vrijheid, R. Nichols, M. Kiggins, P. Elliott. 1997. Who are the "low energy reporters" in the dietary and nutritional survey of British adults. *Int J Epidemiol* 26 (1): 146–154.

Rangan, A. M., S. O'Connor, V. Giannelli et al. 2015. Electronic dietary intake assessment (e-DIA): Comparison of a mobile phone digital entry app for dietary data collection with 24-hour dietary recalls. *JMIR mHealth uHealth* 3 (4). doi:10.2196/mhealth.4613.

Rebro, S., R. Patterson, A. R. Kristal, C. L. Cheney. 1998. The effect of keeping food records on eating patterns. *J Am Diet Assoc* 98 (10): 1163–1165.

Rizek, R. and E. Pao. 1990. Dietary intake methodology 1. USDA surveys and supporting research. *J Nutr* 120: 1525–1529.

Schatzkin, A., V. Kipnis, R. J. Carroll et al. 2003. A comparison of a food frequency questionnaire with a 24-hour recall for use in an epidemiological cohort study: Results from the biomarker-based Observing Protein and Energy Nutrition (OPEN) study. *Int J Epidemiol* 32 (6): 1054–1062.

Schnakenberg, D. D. T., T. M. Hill, M. J. Kretsch, B. S. Morris. 1981. Diary-interview technique to assess food consumption patterns of individual military personnel. In *Assessing Changing Food Consumption Patterns*, edited by National Research Council Committee on Food Consumption Patterns, pp. 187–197. Washington, DC: National Academy Press.

Subar, A. F., V. Kipnis, R. Troiano et al. 2003. Using intake biomarkers to evaluate the extent of dietary misreporting in a large sample of adults: The OPEN study. *Am J Epidemiol* 158 (1): 1–13.

Subar, A. F., S. M. Krebs-Smith, A. Cook. 1998. Dietary sources of nutrients among US adults, 1989 to 1991. *J Am Diet Assoc* 98: 537–547.

Tran, K., R. Johnson, R. Soultanakis, D. Matthews. 2000. In-person vs telephone-administered multiple-pass 24-hour recalls in women: Validation with doubly labeled water. *J Am Diet Assoc* 100: 777–780.

Turner, C. F., L. Ku, S. M. Rogers, L. D. Lindberg, J. H. Pleck, F. L. Sonenstein. 1998. Adolescent sexual behavior, drug use, and violence: Increased reporting with computer survey technology. *Science* 280 (5365): 867–873.

Van Horn, L. V., N. Gernhofer, A. Moag-Stahlberg et al. 1990. Dietary assessment in children using electronic methods: Telephones and tape recorders. *J Am Diet Assoc* 90 (3): 412–416.

Weber, J. L., A. M. Tinskley, L. B. Houtkooper, T. G. Lohman. 1997. Multimethod training increases portion-size estimation accuracy. *J Am Diet Assoc* 97 (2): 176–179.

Wein, E. B., J. H. Sabry, F. T. Evers. 1990. Recalled estimates of food portion size. *J Can Diet Assoc* 51: 400–403.

Willett, W. 2013. *Nutritional Epidemiology*. 3rd ed. Vol. 40. New York: Oxfored University Press.

Yanek, L., T. Moy, J. Raqueno, D. Becker. 2000. Comparison of the effectiveness of a telephone 24-hour dietary recall method vs an in-person method among urban African-American women. *J Am Diet Assoc* 100: 1172–1177.

Young, L. and M. Nestle. 1995. Portion sizes in dietary assessment: Issues and policy implications. *Nutr Rev* 53 (6): 149–158.

Yuhas, J. A., J. E. Bolland, T. W. Bolland. 1989. The impact of training, food type, gender, and container size on the estimation of food portion sizes. *J Am Diet Assoc* 89: 1473–1477.

Zick, C. D. and R. B. Stevens. 2009. Trends in Americans' food-related time use: 1975–2006. *Public Health Nutr* 13 (7): 1064–1072.

5 The Assessment of Food Intake with Digital Photography

Keely R. Hawkins, John W. Apolzan, Candice A. Myers, and Corby K. Martin

CONTENTS

5.1 INTRODUCTION

The prevalence of overweight and obesity is increasing worldwide (Stevens et al. 2012) and based on the first law of thermodynamics a positive energy balance, which occurs when more energy is ingested than expended, is the culprit of weight gain and obesity (Schoeller and Thomas 2014). To decrease the prevalence of overweight and obesity, strategies that promote weight loss and prevent weight gain are needed. A key focus during weight loss interventions is to decrease energy intake, yet few methods are available that accurately measure energy and nutrient (i.e., food) intake in a natural or free-living environment. Self-report methods for measuring food intake have been used extensively, but they are subject to a number of limitations that include high user and experimenter

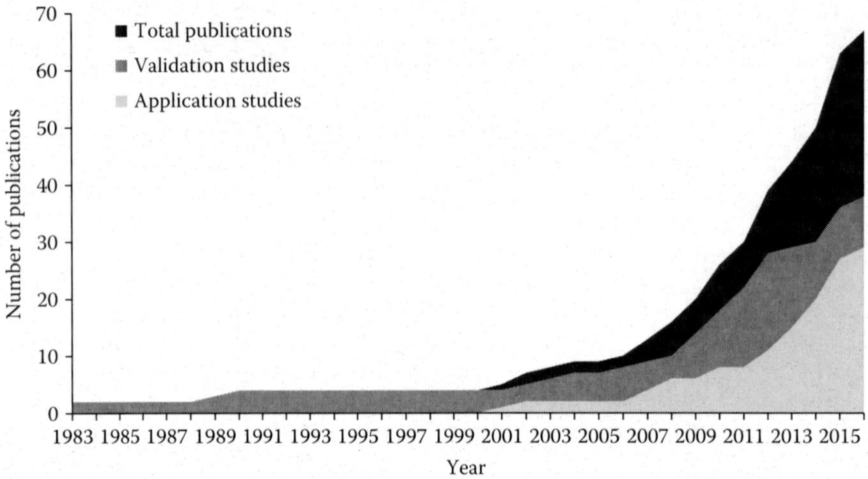

FIGURE 5.1 Number of publications over time that report studies where food photography approaches were examined for their validity (medium gray) or applied in an experimental or intervention setting (light gray). The total number of publications over time that report food photography studies is depicted in dark grey.

burden, interference with usual eating habits, decreased compliance over time, and underreporting bias. Consequently, this makes it difficult for researchers to investigate energy intake in relation to energy balance, weight gain, and obesity (Schoeller et al. 2013). Technology has provided alternatives to self-report methods; for example, digital photography has emerged as a viable method to assess food intake in individuals in cafeteria, laboratory, and free-living settings. Studies utilizing digital photography to assess food intake were first published in 1983 by Elwood and Bird (Bird and Elwood 1983; Elwood and Bird 1983) and beginning in 2002–2003, there has been a steady increase in the amount of publications that report measuring food intake with digital photography (Figure 5.1 and Table 5.1).

5.2 BACKGROUND

The ability to maintain energy balance is complicated by inaccurate measurements of energy intake, which makes it challenging for clinicians to effectively intervene and promote weight loss, as well as for individuals to manage their own weight (Hill 2006; Hill et al. 2012; Wing and Hill 2001). Self-report methods (e.g., food diaries/records, 24-hour dietary recall, and food frequency questionnaires [FFQs]) rely on individuals to report the types and amounts of foods they eat and are frequently used to estimate food intake. These methods however, have a number of limitations including interfering with their typical eating habits resulting in undereating, high burden on the experimenter and the user (Cordeiro et al. 2015), and decreased compliance over time (see Chapter 10). In addition, multiple studies have revealed that people tend to underreport energy intake by 10%–40% per day when using self-report methods (Dhurandhar et al. 2015), particularly when compared to the biomarker for measuring energy intake, which is the doubly-labeled water (DLW) method. DLW accurately measures total daily energy expenditure (TDEE), which is equal to energy intake when an individual is in energy balance (EB) (Chapter 10). When not in energy balance, short-term energy intake is estimated from DLW data by adjusting TDEE values for the energy content of the change in body mass, yet it is difficult to make these adjustments and obtain an accurate measure of energy intake when a large energy deficit is present during the DLW period (de Jonge et al. 2007). DLW has other limitations such as cost, it is not widely available, it requires specialized equipment and expertise, and it does not measure nutrient intake, making it impractical for most clinicians and researchers.

TABLE 5.1
Studies Assessing the Validity of Using Digital Photography to Assess Food Intake

Year	Authors	Title	Tech Used	Method Used	Reference Method	Setting	Population	Reference
1983	Elwood et al.	A photographic method of diet evaluation	Camera	Digital photography of foods	Weighed food records	Free-living	**Adults** N = 25	(Elwood and Bird 1983)
	Bird et al.	The dietary intakes of subjects estimated from photographs compared with a weighed record	Camera	Digital photography of foods	Weighed food records	Free-living	**Adults** N = 17	(Bird and Elwood 1983)
1989	Sevenhuysen et al.	Food image processing: A potential method for epidemiological surveys	Camera	Digital photography of foods	Weighed food records	Laboratory	**Pictures:** 23 commonly eaten food items	(Sevenhuysen and Wadsworth 1989)
	Sevenhuysen et al.	Comparison of food intake assessments obtained with recall interviews and photographic records	Camera	Digital photography of foods	Weighed food records, 24-hour recall	Free-living	**Adults** N = 13	(Sevenhuysen and Zacharia 1989)
1990	Sevenhuysen et al.	Estimates of daily individual food intakes obtained by food image processing	Camera	Digital photography of foods	Weighed food records, 24-hour recall	Free-living	**Adults** N = 10	(Sevenhuysen et al. 1990)
2002	Wang et al.	Validity and reliability of a dietary assessment method: The application of a digital camera with a mobile phone card attachment	PDA with digital camera and telephone card	Wellnavi	Weighed food records	Free-living	**Adults** female college students N = 20	(Wang et al. 2002)
2003	Williamson et al.	Comparison of digital photography to weighed and visual estimation of portion sizes	Digital video cameras	Digital photography of foods	Direct visual estimation + weighed foods	Cafeteria	**Adults** university cafeteria N = 60 meals	(Williamson et al. 2003)

(Continued)

TABLE 5.1 (*Continued*)
Studies Assessing the Validity of Using Digital Photography to Assess Food Intake

Year	Authors	Title	Tech Used	Method Used	Reference Method	Setting	Population	Reference
2004	Williamson et al.	Digital photography: A new method for estimating food intake in cafeteria settings	Digital video cameras	Digital photography of foods	Direct visual estimation	Cafeteria	**Adults** college students N = 130 meals	(Williamson et al. 2004)
2006	Wang et al.	Development of a new instrument for evaluating individuals' dietary intakes	PDA with digital camera and telephone card	Wellnavi	Weighed food records and 24-hour recall	Free-living	**Adults** college females N = 28	(Wang et al. 2006)
2007	Kikunaga et al.	The application of a handheld personal digital assistant with camera and mobile phone card (Wellnavi) to the general population in a dietary survey	PDA with digital camera and telephone card	Wellnavi	Weighed food records	Free-living	**Adults** N = 75	(Kikunaga et al. 2007)
2008	Swanson M.	Digital photography as a tool to measure school cafeteria consumption	Digital video cameras	Digital photography of foods Photos from child and photos from evaluator	Direct visual estimation	Cafeteria	**Children** 1st–5th grade N = 859 lunches	(Swanson 2008)
2009	Higgins et al.	Validation of photographic food records in children: Are pictures really worth a thousand words and quest	Disposable camera	Digital photography of foods	Weighed food records and written food records	Free-living	**Adolescents** N = 28 Age: 10–16 years	(Higgins et al. 2009)

(Continued)

TABLE 5.1 (*Continued*)
Studies Assessing the Validity of Using Digital Photography to Assess Food Intake

Year	Authors	Title	Tech Used	Method Used	Reference Method	Setting	Population	Reference
	Martin et al.	A novel method to remotely measure food intake of free-living individuals in real time: The remote food photography method	Mobile telephone	Remote food photography method (RFPM)	Weighed food records and laboratory lunches using RFPM and free-living conditions using RFPM	Laboratory and free-living	**Adults** N = 25	(Martin et al. 2009a)
	Boushey et al.	Use of technology in children's dietary assessment	PDA with camera and camera with notebook	Digital photography of foods	Multipass 24-hour recall, written food records, PDA with hierarchical menu, PDA with search menu	Laboratory and free-living	**Adolescents** N = 31	(Boushey et al. 2009)
2010	Chung et al.	Tele-dietetics with food images as dietary intake record in nutrition assessment	Digital camera	Compared 2D and 3D images	Weighed food records	Laboratory	N = 10 RDs evaluate 10 food items (10 2D, 10 3D)	(Chung and Chung 2010)
	Lassen et al.	Evaluation of a digital method to assess evening meal intake in a free-living adult population	Digital camera	Digital photography of foods	Weighed food records	Free-living (dinner only)	**Adults** N = 19	(Lassen et al. 2010)
	Six et al.	Evidence-based development of a mobile telephone food record	Mobile telephone	Mobile telephone food record (mpFR)	N/A	Study 1: Cafeteria Study 2: Free-living	**Adolescents** Age: 11–18 years Study 1: N = 63 Study 2: N = 15	(Six et al. 2010)

(*Continued*)

TABLE 5.1 (*Continued*)
Studies Assessing the Validity of Using Digital Photography to Assess Food Intake

Year	Authors	Title	Tech Used	Method Used	Reference Method	Setting	Population	Reference
2011	Matthiessen et al.	Convergent validity of a digital image-based food record to assess food group intake in youth	Digital camera	Digital imaging-based food record (DIFR)	24-hour recall	Free-living	**Adolescents** N = 26 Age: 9–12 years	(Matthiessen et al. 2011)
	Noronha et al.	Platemate: Crowdsourcing nutritional analysis from food photographs	Digital photography and crowdsourcing	Platemate	Weighed food records and 3 RD's and meal snap	Laboratory and free-living	Dietitians and meal snap provide expert estimates on 18 photographs with 36 distinct foods	(Noronha et al. 2011)
	Arab et al.	Feasibility testing of an automated image-capture method to aid dietary recall	Image diet day	Digital photography of foods	DLW and photos only and photos + 24-hour recall	Free-living	**Adults** N = 10	(Arab et al. 2011)
	Rollo et al.	Trial of a mobile phone method for recording dietary intake in adults with type 2 diabetes: Evaluation and implications for future applications	Mobile telephone	Nutricam	Estimated food record	Free-living	**Adults** with T2D N = 10	(Rollo et al. 2011)
2012	Daugherty et al.	Novel technologies for assessing dietary intake: Evaluating the usability of a mobile telephone food record among adults and adolescents	Mobile telephone	Mobile food record (take pics and give feedback only)	N/A	Study 1: cafeteria Study 2: Free-living	Study 1: **Adolescents** N = 78 Study 2: **Adults:** N = 57 **Adolescents:** N = 15	(Daugherty et al. 2012)
	Schap et al.	Advancing dietary assessment through technology and biomarkers	Mobile telephone	Digital food records	Weighed food records	Free-living	**Adults** Campus community N = 12	(Schap 2012)

(Continued)

TABLE 5.1 (*Continued*)
Studies Assessing the Validity of Using Digital Photography to Assess Food Intake

Year	Authors	Title	Tech Used	Method Used	Reference Method	Setting	Population	Reference
	Elinder et al.	Validation of personal digital photography to assess dietary quality among people with intellectual disabilities	Digital camera	Digital photography of foods	N/A	Cafeteria	**Adults** with intellectual disabilities N = 18 23–60 yrs	(Elinder et al. 2012)
	Martin et al.	Validity of the remote food photography method (RFPM) for estimating energy and nutrient intake in near real time	Mobile telephone	Remote food photography method (RFPM)	Study 1: DLW and standard EMA group and customized EMA group Study 2: Weighed food records and DLW	Free-living	**Adults** Study 1: N = 40 Study 2: N = 49	(Martin et al. 2012)
	Nicklas et al.	Validity and feasibility of a digital diet estimation method for use with preschool children: A pilot study	Digital camera	Digital diet estimation method	Weighed food records	Study 1: Laboratory Study 2: Cafeteria free-living	**Children** Preschool Study 1: N = 22 Study 2: N = 12 N = 12	(Nicklas et al. 2012a)
	Samaras et al.	Comparison of the interobserver variability of two different methods of dietary assessment in a geriatric ward: A pilot study	Digital camera	Digital photography of foods	Direct visual estimation	Cafeteria	**Adults** N = 6 meals	(Samaras et al. 2012)

(*Continued*)

TABLE 5.1 (*Continued*)
Studies Assessing the Validity of Using Digital Photography to Assess Food Intake

Year	Authors	Title	Tech Used	Method Used	Reference Method	Setting	Population	Reference
2013	Gauthier et al.	Evaluating the reliability of assessing home-packed food items using digital photographs and dietary log sheets	Digital camera	Digital photography of foods	Food logs and Rater A and Rater B	Cafeteria	**Children** 3rd–6th grade N = 60 meals	(Gauthier et al. 2013)
2014	Hanks et al.	Reliability and accuracy of real-time visualization techniques for measuring school cafeteria tray waste: Validating the quarter-waste method	Digital camera	Digital photography of foods	Weighed food records and direct estimation using ¼ waste and ½ waste methods	Cafeteria	**Children** N = 197 lunch trays	(Hanks et al. 2014)
2015	Casperson et al.	A mobile phone food record app to digitally capture dietary intake for adolescents in a free-living environment: Usability study	Mobile telephone w/ FRapp	FRapp	Determine *amendability*	Free-living	**Adolescents** N = 18	(Casperson et al. 2015)
	Henriksson et al.	A new mobile phone-based tool for assessing energy and certain food intakes in young children: A validation study	Mobile telephone	Tool for energy balance in children (TECH)	DLW and FFQ designed for children of ages 3–5 years	Free-living	**Children** N = 30 Age: 3 years	(Henriksson et al. 2015)
	Aflague et al.	Feasibility and use of the mobile food record for capturing eating occasions among children ages 3–10 years in Guam	Mobile telephone	mFR	N/A	Free-living (summer camp)	**Children** N = 126 Age: 3–10 yrs	(Aflague et al. 2015)

(*Continued*)

TABLE 5.1 (*Continued*)
Studies Assessing the Validity of Using Digital Photography to Assess Food Intake

Year	Authors	Title	Tech Used	Method Used	Reference Method	Setting	Population	Reference
	Boushey et al.	How willing are adolescents to record their dietary intake? The mobile food record	Mobile telephone	mFR	N/A	Free-living (summer camp)	**Adolescents** N = 41 Age: 11–15 yrs	
	Ptomey et al.	Digital photography improves estimates of dietary intake in adolescents with intellectual and developmental disabilities	Digital camera	Digital photography of foods	Proxy-assisted records	Free-living	**Adolescents** with intellectual disabilities N = 20	(Ptomey et al. 2015)
	Rollo et al.	Evaluation of a mobile phone image-based dietary assessment method in adults with type 2 diabetes	Mobile telephone	Nutricam dietary assessment method (NuDAM)	DLW and weighed food records	Free-living	**Adults** N = 10 BMI: 31	(Rollo et al. 2015)
	Doumit et al.	Effects of recording food intake using cell phone camera pictures on energy intake and food choice		Digital photography of foods		Free-living	**Adults** College students N = 76	(Doumit et al. 2015)
2016	Delisle Nystrom et al.	A mobile phone-based method to assess energy and food intake in young children: A validation study against the doubly-labeled water method and 24-hour dietary recalls	Digital camera	Tool for energy balance in children (TECH)	DLW and 24-hour recall	Free-living	**Children** N = 39 Age: 5.5 ± 0.5 years	(Delisle Nyström et al. 2016)
	Duhe et al.	The remote food photography method accurately estimates dry powdered foods—The source of calories for many infants	Mobile telephone	Remote food photography method (RFPM)	Weighed food records	Laboratory	**Infants** N = 28 prepared formula meals	(Duhe et al. 2016)

5.2.1 Advances in Technology

The emergence of technology brought about the development of software applications for mobile phones and personal digital assistants (PDAs) to record food intake with the intention of improving accuracy and easing the burden of keeping a food diary. For example, self-report methods were adapted for mobile phones (Adamson and Baranowski 2014; Rangan et al. 2015) and participants rated these methods as more acceptable than using pen-and-paper food records, yet recording food intake on these devices did not improve the accuracy of food intake estimates (Yon et al. 2006) and was no more accurate than traditional food records (Yon et al. 2007) and 24-hour recall (Beasley et al. 2005). Inaccurate estimates of portion size by the user are the greatest source of error when estimating food intake with self-report tools (Beasley et al. 2005), even when participants are trained to estimate portion size (Blake et al. 1989). Improving participants' portion size estimates is extremely challenging and resource intensive; for example, Martin and colleagues (2007a) found that extensive training increased the accuracy of portion-size estimates, yet portion-size estimates remained fairly inaccurate even after this improvement.

These findings demonstrated that alternative methods were needed that did not rely on users to accurately estimate portion size. Relying on trained observers to estimate portion size had been found to improve the accuracy of food intake estimates in cafeteria settings as early as the 1980s (Comstock et al. 1981; Wolper et al. 1995), yet this approach was not immediately applicable to free-living settings. The emergence and inclusion of digital cameras in portable handheld devices, such as PDAs, offered a method for users to capture images of foods in their natural environment and to transmit the images to trained *raters* to estimate food intake. These capabilities were accelerated by the advent, ubiquity, and technical capacity of smartphones, which contributed to the development, standardization, and validation of methods to remotely estimate food intake based on images of food selection and plate waste captured by users in their home environments (Martin et al. 2009b, 2012; Nicklas et al. 2012a; Six et al. 2010).

5.2.2 The Use of Digital Photography to Assess Food Intake

Using digital photography to assess food intake involves capturing images of foods before and after each eating episode. This method has shown to be highly valid and reliable when used in laboratory and cafeteria settings where images of foods are captured with a digital video camera or a digital camera (Daugherty et al. 2012; Dorman 2013; Duhe et al. 2016; Elinder et al. 2012; Samaras et al. 2012; Six et al. 2010; Swanson 2008; Williamson et al. 2002, 2003, 2004, 2013). The method has been adapted for free-living settings by having a research associate or the user capture images of food selection and plate waste with a digital camera or a camera-enabled mobile device. These methods have been found to produce reliable and valid estimates of food intake in free-living conditions (Bird and Elwood 1983; Higgins et al. 2009; Kikunaga et al. 2007; Lassen et al. 2010; Martin et al. 2009a, 2012; Nicklas et al. 2012a; Rollo et al. 2011, 2015; Sevenhuysen and Zacharias 1989; Wang et al. 2002, 2006).

When using digital photography to estimate food intake, the images are analyzed at a later date or in near real time, though analyzing the images in near real time requires additional resources. The process for analyzing the images to estimate food intake is similar among the groups who have developed such approaches, though some investigators have built computer software to streamline the analytic process (Martin et al. 2009a, 2012; Williamson et al. 2003, 2004). In brief, the images are processed by trained raters who first identify the foods in the images and find a match for each food item in a nutrient database. The raters then estimate the portion size of food selection and plate waste by comparing the user's food images to images of standard portions of foods. These images of standard portions of foods come from a number of sources. Specifically, in cafeteria and laboratory settings, investigators can capture images of measured portions of the foods served in the cafeteria on the day of data collection. In a free-living setting, where this is impractical, researchers rely on

FIGURE 5.2 When using digital photography, food intake is measured by trained raters who estimate portion size by comparing users' images of food selection (Panel A) and plate waste (Panel B) to an image of a standard portion of food (Panel C). Food intake is calculated by the difference (food intake = food selection minus plate waste) and energy and nutrient values are determined by identifying a match in a nutrient database for the foods in the images. (Reprinted with permission from Martin, C. K. et al., *J. Hum. Nutr. Diet.*, 27(Suppl. 1), 72–81, 2014.)

archives of standard portion food images. Once portion size is estimated for food selection and plate waste, food intake is calculated by difference (food intake = food selection minus plate waste), and the output for each of these three variables includes all of the data present in the reference nutrient database (e.g., energy, grams, vitamins, and minerals). Figure 5.2 illustrates the process of estimating portion size by comparing users' images of food selection (Panel A) and plate waste (Panel B) to the image of a standard portion of food (Panel C).

Importantly, agreement among raters who evaluate images has been consistently high (intraclass correlation coefficients around 0.90) across a number of studies (Martin et al. 2009a, 2012; Williamson et al. 2003, 2004, 2007, 2013). These findings demonstrate that multiple raters can estimate food intake using this method and estimates do not differ widely among raters, which facilitates use of this method with large samples that require multiple raters to analyze the images. Raters follow standard procedures when analyzing images and are trained prior to analyzing user's data, which contributes to the quality of the data and allows raters with different backgrounds to analyze food images. The analytic process is typically supervised by a Registered Dietitian Nutritionist or a doctoral level nutritionist, yet raters have included college students, nutrition professionals, and research associates with training outside of nutrition.

5.2.3 Development of Semi-Automated Approaches to Identify Foods and Estimate Food Intake from Food Images

Analyzing a large number of food images can be resource intensive, even if computer programs take human-generated portion-size estimates and calculate the energy and nutrient values of food selection, plate waste, and food intake based on a reference database housed in the software. In addition, collection of food images with smartphones allows the images to be transmitted instantly, which creates the opportunity to analyze the images in near real-time and quickly provide users with feedback about their food intake. These factors led investigators to examine the feasibility of creating computer programs to identify foods, estimate portion size, and estimate food intake (Dibiano et al. 2013; Martin et al. 2009b; Zhu et al. 2008, 2010). These methods have been referred to as *semi-automated* because human correction is needed at various points in the analytic process, as a mistake early in the process negates the ability to accurately estimate food intake (e.g., errors in food identification must be rectified before portion size is estimated). Figure 5.3 illustrates the semi-automated process of analyzing food images with computer software that is used to identify foods and estimate portion size (Dibiano et al. 2013), and the steps illustrated in the figure are described below.

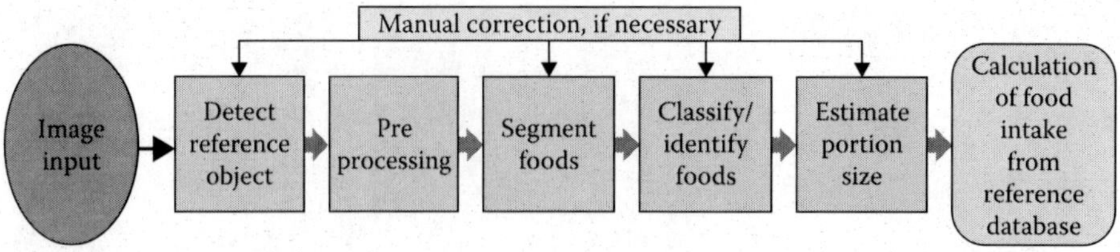

FIGURE 5.3 Illustration of a semiautomated process for identifying foods and estimating portion size and food intake with computer software. This semiautomated approach includes human or manual correction at certain stages of the analytic process to correct errors that the computer system makes. (Modified from Dibiano, R. et al., Food image analysis for measuring food intake in free living conditions. *Paper Presented at the SPIE Medical Imaging*, 2013.)

5.2.3.1 Training the Computer Program to Identify Foods and Estimate Portion Size

Computer programs identify foods by comparing the features (e.g., color, texture, hue, and shape) of the foods in the participants' images to a database that contains the features of food images that have been *trained* into the system. Similarly, computer programs estimate portion size by comparing the foods in the participants' images to images of foods of known portion sizes that have been trained into the program (Dibiano et al. 2013; Martin et al. 2009b). Hence, images of foods of known portion sizes must be entered/trained into the computer system prior to its deployment.

This highlights one limitation of using computer programs to identify foods and estimate portion size; they can only identify foods and estimate portion size/food intake for foods that have been trained into the program. To overcome this limitation, some programs include an online training feature, which allows the software to integrate data that it obtains as it analyzes participants' food images (Dibiano et al. 2013). In addition, the software can be programmed to estimate portion size and food intake from a similar food, which is analogous to what human raters do when they encounter a food for which they do not have a reference image (e.g., a standard portion image of white rice can be used to estimate portion size for a participant's image of fried rice, and the nutrient calculations remain accurate as they rely on the correct reference values from a nutrient database [fried rice, in this case]).

5.2.3.2 Detecting a Reference Object and Preprocessing Food Images

In order to estimate portion size, the computer software requires a reference object to scale the image. This reference object can be a plate or some other fiduciary marker, such as a reference card. For example, Figure 5.4 includes food selection and plate waste images captured with the SmartIntake® smartphone app, which participants use to collect food images for analysis using the Remote Food Photography Method© or RFPM (Martin et al. 2009, 2012; Williamson et al. 2002)*. The card (located to the left of the plate in Figure 5.4) is very similar to a credit card or driver's license, and two concentric rectangular bull's-eye patterns are printed in black on both sides of the white card. This feature allows the computer program to (1) correct the image for perspective/angle, (2) correct the image for the distance between the camera and food, and (3) color correct the image (Dibiano et al. 2013; Martin et al. 2014).

5.2.3.3 Segmenting and Classifying (Identifying) the Foods in the Images

In order to classify (identify) foods and estimate portion size, computer programs first must segment foods, or isolate the foods (foreground) from the background (the plate or tray) (Dibiano et al. 2013; Martin et al. 2009b). Once this is accomplished, the program extracts the features (e.g., color, texture, hue, and shape) of the participant's foods, and compares them to the features of foods that

* The technology surrounding the Remote Food Photography Method and SmartIntake app is owned by Louisiana State University/Pennington Biomedical Research Center and C. Martin is an inventor.

FIGURE 5.4 Images of food selection (Panel A) and plate waste (Panel D) captured with the SmartIntake® smartphone app, which participants use to collect food images for analysis using the RFPM. Foods in food selection images that are not easily identifiable are identified by users by either scanning barcodes or entering Price Look-Up (PLU) numbers (Panel B) or entering a brief description of the food via text message or voice recording (Panel C). The SmartIntake app® packages the food images and related data, and sends the information to a server for analysis.

have been trained into the database. The software then attempts to identify the food based on a comparison of these features.

5.2.3.4 Portion-Size Estimation

The computer program estimates portion size by using the fiduciary marker (the reference card) and geometric calibration to calculate the area of the segmented food region. Using the proportion between the area of the food region and the gram values learned during training, the amount of food in the image is estimated (Dibiano et al. 2013; Martin et al. 2014).

5.3 VALIDATION OF USING DIGITAL PHOTOGRAPHY TO ASSESS FOOD INTAKE

As illustrated by the red area in Figure 5.1, studies examining the validity of estimating food intake with photography have occurred steadily since the early 1980s, with an increase in publications from 2009 to 2014. Table 5.1 also includes a summary of validation studies.

5.3.1 Validity of Digital Photography in Cafeteria Settings

Digital photography to assess food intake shows great success and is accurate when used in a naturalistic, but confined settings such as cafeterias. Further, this method has proven effective at estimating the food intake of large numbers of diverse people sampled from a variety of geographical areas, including: (1) adults during basic combat training (Williamson et al. 2002) and in a college cafeteria setting (Williamson et al. 2003, 2004), (2) elementary school children (Swanson 2008; Williamson et al. 2007, 2013), and (3) preschool children (Nicklas et al. 2012a); and the validity of this method has been shown in adults, adolescents, and children.

Cafeteria settings provide an ideal venue for this method because information about each food provided during data collection, its method of preparation, and images of weighed portions can be captured (Figure 5.5, Panel C) which facilitates estimates of portion size and precise calculations of energy and nutrient intake. Further, images of food selection and plate waste can be captured efficiently and with very little user burden by using digital video cameras placed at the end of the food line (Figure 5.5, Panel A) and where people discard their plate waste (Figure 5.5, Panel B), respectively. The use of digital video cameras allows multiple frames (e.g., 30) to be captured per second, and these data are converted to still images for analysis in the laboratory. This procedure was used by Williamson and colleagues (2002) in a cafeteria setting to assess energy intake associated with basic combat training and in 2003 in a college cafeteria. Portion-size estimates for food intake correlated highly with weighed portion sizes ($r = 0.92$, $p < 0.0001$) and the mean difference between directly weighed total food intake and digital imaging estimates was 5.2 ± 0.95 g (mean ±

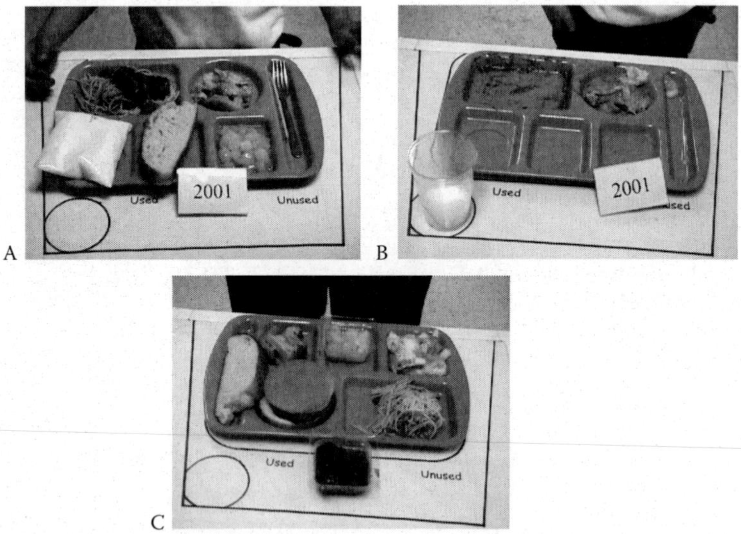

FIGURE 5.5 When using digital photography to assess food intake in cafeteria setting, digital video cameras can be set up at the end of the food line to capture images of food selection (Panel A) and where participants return their tray to capture images of plate waste (Panel B). In addition, images can be captured of carefully weighed standard portions of the foods provided during data collection (Panel C).

SEM), with no systematic bias over levels of food intake (Williamson et al. 2003). Among categories of foods, the mean error from digital imaging varies; for example, digital imaging overestimates beverage intake by 7.6 ± 3.07 g (mean ± SEM) or 4.3% and overestimates condiment intake by 4.9 ± 1.63 g (mean ± SEM) or 16.6% (Williamson et al. 2003).

Similar to previous studies by Williamson and colleagues (2004), Swanson (2008) used digital photography to measure children's food intake in a school cafeteria. They compared the use of digital video cameras versus direct visual estimation to measure food intake of school-aged children (1st–5th grade) from four lunches served in the cafeteria. They further investigated proper estimation techniques by arranging food items left on trays after the children ate and took photographs of the trays to assess differences caused by arrangement of foods on the tray. This helped to better estimate plate waste through consideration and items such as orange peels and apple cores. The results were promising; the analysts' consumption estimates compared to estimates from digital photography of the components and the meal were within 10% of each other in 92% of the cases. Nonetheless, the authors noted that getting proper consumption and plate waste data is difficult because it is impossible to accurately estimate some foods such as condiments served in individual packages and certain a la carte items such as a bag of chips without removing any leftovers.

Building on the concept of properly assessing plate waste, Hanks et al. (2014) compared direct visual estimation of three different methods versus weighing leftovers. They assessed the use of the photographic method to direct visual estimation using the quarter-waste method and the half-waste method. Their results showed low reliability of the photographic method due to the inability of photographs to determine the amount of food not consumed in packaged items. This research stresses the importance of proper planning of photographing food items prior to data collection, specifically when prepackaged foods are being served, or utilizing methods employed by others to overcome these limitations. For example, Williamson and colleagues (2002, 2003, 2004) and Martin and colleagues (2009a, 2012) instruct the people collecting data, whether they be staff members or participants, to turn empty beverage containers on their side if they are empty, and to empty any remaining beverages into a cup and empty packets of foods onto the plate prior to the plate waste image being captured.

In summary, digital photography can accurately assess food intake in cafeteria settings, particularly when standardized methods are used during data collection. Also, the method has been validated in diverse samples of people, including soldiers during basic combat training (Williamson et al. 2002), elementary (Swanson 2008; Williamson et al. 2007, 2013) and preschool children (Nicklas et al. 2012a), individuals with intellectual disabilities (Elinder et al. 2012), individuals in a geriatric ward (Samaras et al. 2012), and for assessing home packed food items (Dorman 2013).

5.3.2 Validity of Digital Photography in Laboratory and Free-Living Settings

5.3.2.1 Initial Validation Studies from Free-Living Conditions

One of the first validation studies conducted in free-living conditions was published by Bird and Elwood (1983) who compared dietary intake estimated from photographs to weighed food records. Their results showed that the difference between the two methods was trivial and that the photographic method of obtaining dietary intake is cost-effective, though challenges with the method were identified, including the quality of the photos and user compliance. Sevenhuysen and colleagues (1989b, 1990) subsequently assessed this method by comparing the photographic procedure to 24-hour recalls and weigh backs at lunchtime and found similar results. The photographic procedure predicted the actual weight of food better than the 24-hour recall, although further investigation of the methodology and procedures were warranted. In addition, missing data was common in these studies due to users forgetting to document or photograph all meals, and it was not until the 2000s when these issues could be better addressed by the real-time transfer of data, as detailed below.

Using similar data collection processes as previous studies, Wang et al. (2002, 2006) and Kikunaga et al. (2007) used a method of digital photography to assess food intake called Wellnavi.

This method of assessing food intake involves using a PDA with a built-in camera and telephone card to transmit photos through wireless technology. Individuals captured images at a 45° angle before and after eating, and used the stylus as the reference for portion-size estimation. After the images were captured, individuals were asked to record descriptions of foods and provide any ingredients that might be hard to judge on the screen using the stylus. Wang and colleagues (2002, 2006) used Wellnavi versus weighed food records in two pilot studies utilizing female college nutrition students to assess the reliability and validity of the method. They found no significant differences in energy or macronutrient intake between Wellnavi and weighed records. Kikunaga et al. (2007) subsequently used Wellnavi versus weighed food records in a larger validation study where 75 adults recruited from a general Japanese population simultaneously recorded dietary intake for seven days using both methods. Their results revealed underestimation of energy intake by Wellnavi by 13.1% (1977 ± 405 vs. 1718 ± 361; $p < 0.001$). Further it was shown to significantly underestimate all macronutrients. Participants reported difficulty in using the stylus on the screen to record notes; therefore, many images did not include written descriptions making complete analysis challenging using this method. Further limitations of using the PDA technology included poor image quality, battery life, and bulky devices.

Lassen and colleagues (2010) used digital photography to assess food intake in 2010 using a stand-alone camera with free-living adults ($n = 19$) versus weighed food records for five days during dinner only. Users were required to separate all food components on their plate and place a ruler beside foods as a reference for portion-size estimation. In addition, a notebook was provided to record additional information about the meal and ingredients such as recipes and measurements of ingredients used. Their results showed that the digital photographic method underestimated mean energy intake by 11.3% (526 ± 178 kcal vs. 471 ± 167 kcal; $p < 0.001$). Based on participant feedback, the authors concluded that compliance was responsible for most of the variation. Participants had trouble remembering to take photos and that it was awkward separating all foods on the plate before capturing the images, especially when eating out. Similarly, Helander and colleagues (2014) examined sustained use of a free mobile app (The Eatery) for dietary monitoring with photography and found that most people who used the app did not continue using the app actively.

5.3.2.2 Testing Validity in Both Free-Living and Laboratory Settings and Incorporation of Behavioral Assessment Principles to Improve Accuracy

The ability to transmit food images from free-living conditions to a researcher or clinician for analysis via a mobile device such as a PDA or smartphone in near real time allowed digital photography methods to be modified to more easily and accurately assess food intake while people live in their natural environment. However, obtaining a criterion measure for nutrient intake in free-living conditions is challenging. The doubly-labeled water method provides an excellent criterion measure for *energy* intake in free-living conditions, but it does not assess *nutrient* intake. This led some researchers to test their methods in free-living conditions, as well as laboratory or cafeteria settings to obtain a precise criterion measure (directly weighing foods) for nutrient intake. Testing these methods in laboratory conditions also provides important data to help identify why a method might not be as accurate in free-living conditions, and it serves as a benchmark for the accuracy of the method when used in controlled or semicontrolled conditions.

Near real-time transmission of food images provides an opportunity to deploy strategies to improve accuracy by improving data quality and reducing the amount of missing data. As noted in the prior section, when deployed in free-living conditions, digital photography suffers from not so dissimilar limitations as self-report methods, namely, users forget or fail to capture images of everything that they eat or drink. This led to the use of ecological momentary assessment (EMA) (Stone and Shiffman 1994) methods when collecting digital photography data. EMA is a method of using communication technology to sample and monitor phenomena (e.g., behaviors and mood) in real-time in natural settings and it can be used when collecting digital photography data in two important ways. First, EMA approaches can be used to prompt or remind participants to capture images of their foods. Second, near real-time data transfer allows the researchers or clinicians to review the completeness and quality of the transmitted images, and to alert the participant if data collection problems are apparent. This allows data collection challenges to be addressed in near real time before data for the entire monitoring period is compromised, and it increases accountability for the user to collect data as instructed. Again, the need for EMA or similar approaches was documented in the first free-living studies reviewed in Section 5.3.2.1, and latter research demonstrated that these approaches can be used successfully when collecting digital photography data (Martin et al. 2009a, 2012), as summarized below.

Martin and colleagues (2009a, 2012) developed the RFPM and later the SmartIntake® smartphone app that streamlines data collection and minimizes user burden. The RFPM and SmartIntake® app include EMA methods to promote data quality and completeness. When using the RFPM, participants use a cell or smartphone to capture images of their food selection and plate waste. Participants are asked to include a brief text or voicemail description for foods that are not immediately identifiable from the picture, and the food images and descriptions are transmitted wirelessly in near real time to a server for analysis (the SmartIntake® app collects and transmits RFPM data with little burden and includes a barcode scanner and Price Look-up number to automatically identify many foods and their nutrient information; see Figure 5.4 for screenshots from the SmartIntake® app). The analytic process follows those described in Section 5.2.2 and illustrated in Figure 5.2 (Martin et al. 2009a, 2012; Williamson et al. 2003, 2004), and relies on a computer program called the Food Photography Application to streamline the analysis and automate calculation of energy and nutrient values for food selection, plate waste, and food intake. Alternatively, the images can be analyzed using the semiautomated procedures outlined in Section 5.2.3.

The RFPM incorporates EMA in three ways. First, reminders or prompts, which participants respond to with a yes/no response, are scheduled for delivery to participants reminding them to capture images of their foods. Second, participants' responses to the reminders and the food images that they send are available for viewing in real time via a password protected website, allowing the research team to review participants' data in near real-time and identify and resolve data collection problems (a back-up method such as dietary recall is used if a food image is not captured). Third, the SmartIntake app includes a notification that is delivered 10 minutes after a food selection image is captured that reminds participants to capture images of their plate waste. The importance of these EMA methods was demonstrated during pilot testing and the validity studies of the RFPM (Martin et al. 2009a, 2012). One of the first tests of the validity of the RFPM found that, compared to directly weighed energy intake, the RFPM underestimated energy intake by 5% and 5.5% in two laboratory-based comparisons (Martin et al. 2009a). In free-living conditions, the RFPM (with EMA) underestimated energy intake by 7% when compared to directly weighed intake (participants were provided with a cooler containing their food and they returned the cooler to the center, allowing direct weighing of both food selection and plate waste as the criterion measure [Martin et al. 2009a]). These experiments demonstrated that the error from the RFPM in free-living conditions was similar to that in laboratory conditions, suggesting that the EMA methods resulted in fairly complete data. Furthermore, the error from the RFPM did not differ over energy intake levels, nor did error vary by body weight or body mass index (BMI) (Martin et al. 2009a), indicating that many of the shortcomings of self-report methods could be overcome with digital photography methods. Finally, participants also used pen-and-paper food records during these studies and, with few exceptions, they found the RFPM preferable compared to food records (93.6% of participants preferred the RFPM to food records) and they reported good satisfaction with the RFPM and found it easy to use.

Martin and colleagues (Martin et al. 2012) subsequently conducted two additional studies that tested the accuracy of the RFPM when used to measure energy intake over six days in adults in free-living conditions compared to DLW. One of these studies also tested if the rigor of EMA methods affected the RFPM's validity by randomly assigning participants to one of the two EMA groups. Group 1 received standard EMA prompts consisting of 2–3 prompts/day sent to participants via a smartphone around generic meal times (e.g., noon for lunch time). If data collection problems occurred, participants were contacted by study personnel within one to two days to correct the issue. Participants in group 2 received 3–4 customized EMA prompts/day scheduled around their personal meal times and they were contacted within 24 hours if data collection problems occurred. When standard EMA prompts were used, the RFPM, compared to DLW, significantly underestimated energy intake by 895 kcal/day or 34%. When customized EMA prompts were used, the RFPM did not differ significantly from DLW and underestimated energy intake by only 270 kcal/day or 9% (Martin et al. 2012). These results highlighted the importance of deploying EMA methods soundly, and resulted in modifications to RFPM procedures and a subsequent and more definitive validity study, during which energy intake over six days in free-living conditions was compared to energy intake measured by DLW. Energy intake from the RFPM did not differ significantly from DLW, and the RFPM underestimated energy intake by only 152 kcal/day or 4%. In addition, error did not differ over levels of energy intake and use of the RFPM was not associated with undereating or reactivity. Finally, laboratory-based test meals found that the RFPM also accurately measured macronutrient intake and intake of most nutrients and vitamins tested, and satisfaction remained high for the RFPM.

Using similar methods and technology as the RFPM, the mobile telephone food record (mpFR) was developed by Schap and colleagues (Schap 2012) and its validity was assessed with a sample of 12 adults recruited from a college campus. The study found that the mpFR produced energy intake estimates that did not differ significantly from estimated energy requirements.

5.3.3 The Application of Digital Photography for Assessing the Food Intake of Adolescents, Children, Infants, and Special Populations

Digital photography has been used and validated extensively in adults in cafeteria and free-living environments, and also in children in cafeteria settings. Recently, the validity of digital photography is being more extensively tested in special populations, such as young children. Using digital photography to estimate the food intake of young children is particularly challenging as food images must be captured by one or more adults throughout the day (e.g., parents or caretakers in the morning before school, school or daycare personnel during the day, and parents or other caretakers in the afternoon and evening).

5.3.3.1 Adolescents, Children, and Infants

Most of the adolescent studies that have been published to date evaluated the acceptability and reliability of the method rather than its validity/accuracy. Higgins and colleagues (Higgins et al. 2009) examined the use of digital photography to assess food intake in adolescents by asking them to use food record as well as digital photography. Their results showed no differences between the two methods for estimated energy and nutrient intake. They also reported that participants and their parents rated the photographic method as simpler, quicker, and preferred overwritten food records. Similarly, Matthiessen and colleagues (2011) used a digital image-based food record (DIFR) method to assess food group intake in adolescents from 5 p.m. to bedtime for seven days. To assess convergent validity, they assessed one-day and average weekly food group intakes using DIFR versus 24-hour dietary recalls. They found that one-day food group intakes assessed by DIFR were significantly correlated with 24-hour recall ($p < 0.001$) for all food groups. This shows promising results for DIFR to assess food group intakes among adolescents, although further investigation is warranted to assess overall dietary intake.

Boushey and colleagues (2009) also examined the feasibility and acceptability of using digital photography to assess food intake in adolescents by evaluating preferences for dietary assessment methods. Their results revealed that adolescents prefer to use technology, such as a PDA or disposable camera, compared to pen-and-paper records. In an additional study, they assessed the willingness of adolescents enrolled in a summer camp to use the mFR method to capture food and beverage intake for two consecutive days. Their results revealed that girls were more likely to capture images than boys and that participants were more likely to capture images of breakfast and lunch (90%) than afternoon (54%) and evening (40%) snacks (Boushey et al. 2015). Further, existing methods tend to be only useful for providing estimates of the energy intake of young children at the group level, and estimates for adolescents are particularly poor (Boushey et al. 2015).

In a larger study using the mpFR, Six and colleagues (Six et al. 2010) evaluated the ability of adolescents to identify foods and estimate portion size of foods consumed. Six and colleagues also evaluated the influence using this method in the laboratory setting on expected energy intake and 24-hour food intake. For this study, the mpFR incorporated a computer-automated system for volume estimation, segmentation, and classification of foods (see Section 5.2.3 for summary of such methods). This system labels images of foods and the results are sent back to the user to confirm and/or modify the description of foods or portion-size estimates calculated from the automated system. After confirmation of the foods by the user, the images are indexed with the Food and Nutrient Database for Dietary Studies and sent to trained analysts. An electronic back-up record is incorporated into the method for use when individuals forget to take pictures of foods during eating occasions. The results of this study revealed that adolescents are willing to adopt new technology using digital photography to measure food intake, although the design needs to be adapted to accommodate an adolescent lifestyle. Adolescents tend to have abnormal eating habits and behaviors that make data collection in this population difficult (Boushey et al. 2009), and more effective employment of EMA techniques are needed to establish the photographic method for dietary assessment in adolescents. Other studies investigating feasibility and acceptability among adolescents when using digital photography have found similar results (Boushey et al. 2015; Casperson et al. 2013, 2015; Daugherty et al. 2012).

Nicklas and colleagues (2012b) have extensively studied the ability to use digital photography methods with young children, namely, preschool children. In a study that recruited mothers in the Houston area, the research team trained the mothers to use a smartphone to capture images of foods consumed by their Head Start (HS) child in the home over a 24-hour period. Study personnel also used a smartphone to capture images of the foods consumed by the same children at HS on the same day. Another team of study personnel weighed food selection and leftovers during the same 24-hour period at HS and the children's homes. Total 24-hour food intake estimated using digital photography was compared to directly weighed food intake. There was no significant difference between study personnel and mothers in the mean amount of food consumed in the home based on the analysis of food images. In addition, no significant difference between the weighed intake of foods consumed in the home and the amount estimated from the mothers' food images were shown. Results revealed a 35 g difference in total food intake consumed in the home between weighed intake and intake generated from digital photography. Digital photography significantly overestimated food intake by a mean of 13%, though this mean difference was very small (10 g). Examination of the data identified foods that were difficult to analyze in this pilot study, and elimination of these foods decreased the percentage error by 45% to ~7%. The problematic foods included canned fruit and vegetables with liquid, condiments, mixed meals, layered foods, beverages, and foods with inedible portions. Procedures have more recently been established to better collect and analyze data from these problematic foods.

Nicklas and colleagues (2012b) have also examined acceptability ratings from mothers who have used digital photography to assess the food intake of their preschool children. The mothers reported that the smartphone is not difficult to use. Also all of the participants reported that it would not be difficult to capture food images with a smartphone when eating out. These data are consistent with results from Henriksson and colleagues (2015) and Delisle Nyström and colleagues (2016) who tested the tool for energy balance in children (TECH), which is a method to assess the food intake of children using digital photography. These studies compared energy intake using TECH versus DLW and assessed differences in intakes of food groups of interest in preschool children. In both studies, mean energy intake was not significantly different from total energy expenditure (TEE) assessed through DLW. These results demonstrate the potential for TECH to be a useful tool to assess dietary intake in preschool children.

Previous studies in children have utilized the caretaker in the data collection process to gather accurate and usable data. Aflague and colleagues (2015) used the mpFR method to examine the potential for children to independently capture eating occasions while at summer camp using this method versus written records. In study 1, children were asked to use the mpFR for one day and in study 2 to use this method for two consecutive days at two different time periods. Results from study 1 showed that 90% of the time children captured usable images, and all children returned the mpFR undamaged. In study 2, children used the mpFR at least one day 94% of the time during time period 1, 78% of the time during period 2, and 75% of the time at both time periods. These data showed promise for collecting usable data when assessing food intake in children using digital photography to capture eating occasions.

One study to date has examined the validity of using digital photography to estimate the energy intake in infants. Duhe and colleagues (2016) used the Remote Food Photography Method© and SmartIntake® app to estimate the weights of dry powdered infant formula before mixing the formula with water. They captured images of four different serving sizes (1 scoop, 2 scoops, 3 scoops, and 4 scoops) in 7 different bottles. In 1 bottle they used the recommended gram weight of powdered formula as suggested by the manufacturer and in 6 bottles they used 5%, 10%, and 15% more and less powdered formula than recommended for a total of 28 different bottles containing different amounts of formula. Their results showed a slight underestimation by the RFPM of 0.05 g (90% CI −0.49 to 0.40; $p < 0.001$), and the mean percentage error ranged between 0.3% and 1.6% among the four different serving sizes examined. These results support the use of digital photography as a valid measure to accurately estimate energy intake in infants through use of powdered formulas, and research is ongoing in this area.

5.3.3.2 Special Populations

Rollo and colleagues (2015) developed a method of assessing energy intake using digital photography called Nutricam. Their methods include a reference card to facilitate portion-size estimation, and an automatic prompt was incorporated to provide voice feedback about the name, brand, type, and preparation or cooking method for each food item. In a pilot study, adults diagnosed with type 2 diabetes simultaneously used Nutricam and food records for three days. The results show that Nutricam versus DLW underestimated mean daily energy intake by 9% ($p < 0.05$). As with earlier methods, poor compliance contributed to the observed underestimation of food intake. Only 71% of the entries had images of sufficient quality for analysis and only 66% of the entries included voice records making analysis difficult. In an additional study by Rollo and colleagues (2015), they assessed the performance of the Nutricam Dietary Assessment Method (NuDAM), a three-day mobile phone image-based dietary record, and weighed food records when compared against TEE measured using DLW in adults diagnosed with type 2 diabetes ($n = 10$). Energy intake was significantly underreported with both methods versus TEE with a mean ratio of 0.76 (EI:TEE) for both the NuDAM and weighed food records. These results demonstrate the need for modifications to the NuDAM method to improve validity when used with individual diagnosed with type 2 diabetes.

Digital photography has also been used to assess food intake in individuals with intellectual and developmental disabilities (IDD). Elinder and colleagues (2012) examined the feasibility of using digital photography to assess diet quality in adults with intellectual disabilities. They found that 85% of observed eating or drinking occasions were photographed when reminded by the staff. Interrater reliability was exceptional for all variables (≥0.88) except that meal quality (0.66) and correlations between items assessed from the photos and observations were robust (0.71–0.92). Ptomey et al. (2015) subsequently examined feasibility of using digital images to improve estimates if dietary intake versus a proxy assisted three-day food record in adolescents with IDD. The two methods were analyzed separately but results were reported for photo-assisted records using the photos to enhance the proxy-assisted food records. Their results revealed an overestimation of energy intake, grams of fat, carbohydrates, and proteins using the photo-assisted method versus proxy-assisted diet records alone.

These initial studies indicate that digital photography has promise for assessing the food intake of people diagnosed with conditions such as type 2 diabetes or IDD, but additional work is needed to enhance data completeness, facilitate compliance, and improve the accuracy of the methods.

5.4 THE APPLICATION OF DIGITAL PHOTOGRAPHY IN APPLIED SETTINGS AND CLINICAL RESEARCH

The number of publications over time that report using digital photography to assess food intake in an applied clinical or research setting is illustrated in Figure 5.1.

The utility of using digital photography to assess food intake in an applied setting was first demonstrated in a study assessing nutrient intake in association with food intake patterns in female shift workers in a computer factory in Japan (Ohtsuka 2001). This method was used for three working days and one day off. Results from this study provided valuable data about nutritional adequacy in shift workers and revealed that late-shift workers had more inadequacies in nutrient intake than morning or afternoon shift workers due to a lower meal frequency and poor meal quality. Williamson and colleagues (2002) subsequently utilized digital photography in a study examining changes in food intake and body weight related to basic training in Soldiers. Body weight was assessed throughout the eight-week period of basic training and digital photography was used to assess food selection and intake at baseline and follow-up. Their results revealed that basic training improved healthy eating and body weight. Healthy individuals tended to gain more body mass but not in the form of body fat and individuals who had higher amounts of body fat at the beginning of training tended to lose body mass, mostly in the form of body fat.

In addition, Crombie and colleagues (2013) demonstrated the effect of modifying foodservice practices in military dining facilities on food intake of soldiers by using digital photography during a test meal to assess dietary intake. They showed that modest changes in military dining facility serving practices can promote healthy eating habits and positive changes in the nutritional adequacy of soldiers. Finally, Belanger and colleagues (2016) used digital photography to assess the impact of implementing the initial military training menu standards in nontrainee military dining facilities in soldiers three weeks after menu changes were implemented. Substantial changes were observed in total energy and sodium intake after implementation of the menu standards.

Williamson and colleagues (2007, 2012) conducted two large intervention trials (WiseMind and LA Health) to prevent inappropriate weight gain in school children and used digital photography to evaluate the effect of the interventions on children's food intake in school cafeterias. In both studies, digital photography identified beneficial changes in food intake among the children, including decreased energy intake, specifically from fat and saturated fat (Williamson et al. 2013). Food selection and intake data from LA Health was also compared to the school meals initiative standards and the Institute of Medicine's recommendations, and it was found that children's food selection exceeded the recommendations for saturated fat and children were less likely to discard fat than carbohydrate, contributing to a higher proportion of fat being consumed (Martin et al. 2010). These results highlight the importance and benefit of obtaining estimates of both food selection and plate waste with digital photography. In a similar sample of school children, Martin et al. (2007b) also demonstrated utility of digital photography by quantifying the effects of second servings on food intake. Results from this study showed that when second servings were available, children who returned for second servings had higher energy intake than their peers. Other studies have similarly demonstrated the utility of using digital photography data to quantify changes in the food intake of children in school cafeterias after implementation of an intervention, including the Food Dudes program and the Smarter Lunchroom Intervention (Goldberg et al. 2015; Scherr et al. 2014; Smith and Cunningham-Sabo 2014; Upton et al. 2013, 2015).

Digital photography has also been deployed to assess the intake of preschool children and much of this work has been conducted by Nicklas and colleagues. Nicklas and colleagues (2013) examined the variability of portions of foods served and consumed by preschoolers attending sixteen Head Start centers and found that the amount of food served was highly associated with the amount of food consumed by the children. The same group of investigators found similar results in a sample of low income African American and Hispanic American preschool children (Nicklas et al. 2012b). Dinner meals were evaluated and it was found that substantial variation existed in the amount of food served and consumed at dinner, and the amount of food served was positively associated with the amount of food consumed.

Digital photography has been used in studies assessing the ability of photographic diaries to serve as intervention tools to change attitudes about nutrition and decisions involving food choice. One study found that photographic food diaries were more likely to alter attitudes and behaviors associated with food choice compared to written food records (Zepeda and Deal 2008). These data are important for understanding how digital photography can be utilized when employed in weight loss interventions in individuals with overweight or obesity. Donnelly and colleagues (2008) used digital photography to capture eating occasions of individuals with overweight and obesity in the Jayhawk Observed Eating Trial for the prevention of weight gain. They used this method to examine the effects of ad libitum diets on three levels of fat intake (<25% of energy from fat, 28%–32% of energy from fat, and >35% energy from fat). Results revealed that energy intake, not percentage of energy from fat, was responsible for the excess weight gain observed during the study.

The RFPM was used as part of a remotely delivered mobile health weight loss intervention called SmartLoss (Martin et al. 2015, 2016). Smartphone apps, such as the RFPM as data are now collected with the SmartIntake® app, offer promise at promoting behavior change and enhancing user satisfaction when paired with other health promotion interventions. One reason why these approaches facilitate behavior change is that objective data can be transferred in real or near real time. In the case of the RFPM, weight loss participants' food images are received in near real time, and clinicians can use these data to quickly send participants feedback about ways to modify their food intake to meet their energy and nutrient goals.

5.5 ADVANTAGES AND LIMITATIONS

Assessing food intake with digital photography has many advantages, including the accuracy of the methods when used in cafeteria settings. Although not all studies find that these methods are accurate when used in free-living settings, some have found them to be accurate, though those methods rely heavily on behavioral assessment principles and the near real-time transmission of data to facilitate data quality and completeness. An additional strength is the depth of the data generated, including energy and nutrient values for food selection, plate waste, and food intake. Digital photography also provides a unique and adaptable platform for assessing food intake that can integrate new technology as it is developed and validated. Also, digital photography has low user burden and eliminates the need for participants to estimate portion size, which likely contributes to the high user satisfaction observed with these methods, particularly when compared to traditional methods such as food records. Finally, the ubiquity of wireless devices such as smartphones allows the method to be more widely, easily, and affordably deployed. Indeed, when collecting digital photography data with a cellular or Internet connected device, data can be collected in near real time anywhere in the world with a cellular signal or an Internet connection.

However, digital photography methods are not without limitations. As noted earlier, people can forget to capture food images, lose the data collection device (e.g., smartphone), or experience technical problems that could impede collecting data. However, with the exception of forgetting to capture images of foods, these problems are infrequently reported in the literature. Users forgetting to capture images remains a challenge and, as outlined in this chapter, EMAs and other methods are needed to further refine the data collection process to reduce the frequency of missing images.

Most authors report that back-up methods, such as a food recall or food record, are used when a participant forgets or cannot capture an image of foods or beverages that they consumed, and Martin and colleagues (2012) found that these back-up data are very important. Specifically, they found that back-up methods were used on 9% of days that data were collected and constituted 10% of the total energy intake values. Further, the use of back-up methods can help promote adherence to using the digital photography method as capturing an image of one's food is considered easier than using a back-up method.

5.6 FUTURE RESEARCH

Wearable sensors and activity monitors have significantly advanced the objective assessment of activity and energy expenditure (Murakami et al. 2016), yet technology for assessing food intake has evolved at a much slower pace. The digital photography methods described in this chapter require active data collection, meaning the burden of capturing food images falls to personnel in cafeteria settings, or the participant in free-living settings. Hence, one area for future research is the development and validation of methods that rely on passive data capture.

The ability of sensors that detect sound muscle activity at the throat, ear, torso and limb movements to measure food intake have been tested in laboratory-based studies, yet some of these devices are cumbersome (O'Loughlin et al. 2013). However, recently, more passive and nonintrusive means of collecting food images have been developed. These methods rely on wearable cameras, such as the eButton (Sun et al. 2014) or Senscam (Hodges et al. 2006), which are pinned onto a user's shirt or attached to a lanyard that is worn around the neck. These devices capture multiple images per minute during eating occasions without the user physically capturing the images; hence, reducing user burden (see Chapters 6 through 9). Pettitt and colleagues (2016) recently developed and assessed the accuracy of using a lightweight wearable microcamera to collect food intake data. In this pilot study, six participants wore the devise for two days and data was compared to DLW measurements as well as a food diary. The wearable camera underreported energy intake by 34% when compared to energy expenditure. The authors concluded that while there is substantial underreporting, the wearable camera provides a richer data set because it offers information on macronutrient intake and eating rate. Although these methods of capturing images of eating occasions are promising, they also come with many limitations. The amount of images captured for each participant provides a very rich data set, but also presents data collection and analysis issues due to the massive amount of data that is collected. In addition, these can be considered obtrusive and they come with many privacy issues, especially when used in a free-living setting.

Aside from passive data capture, one final area for future research is the integration of new and modified EMA strategies into digital photography methods that rely on active data capture. Although EMA methods have been integrated into at least a few approaches, much work remains to better automate these procedures and fully exploit the ability to receive and send data and communications in real-time.

5.7 CONCLUSIONS

Digital photography has been found to accurately quantify food intake in cafeteria, laboratory, and free-living conditions. The method produces rich data on food selection, plate waste, and food intake and has been used successfully in diverse samples of participants, including infants, children, adults, soldiers, and individuals diagnosed with type 2 diabetes and intellectual and developmental disabilities. When used in free-living settings, data are transferred in near real-time, which provides researchers and clinicians the ability to more effectively survey data and intervene to help people change their eating habits and improve their health. Digital photography is not without its limitations, but its many strengths suggest that it is a useful method to quantify food intake.

REFERENCES

Adamson, A., and T. Baranowski. 2014. Developing technological solutions for dietary assessment in children and young people. *Journal of Human Nutrition and Dietetics* 27(s1): 1–4.

Aflague, T. F., C. J. Boushey, R. T. L. Guerrero, Z. Ahmad, D. A. Kerr, and E. J. Delp. 2015. Feasibility and use of the mobile food record for capturing eating occasions among children ages 3–10 years in Guam. *Nutrients* 7(6): 4403–4415.

Arab, L., D. Estrin, D. H. Kim, J. Burke, and J. Goldman. 2011. Feasibility testing of an automated image-capture method to aid dietary recall. *European Journal of Clinical Nutrition* 65(10): 1156–1162.

Beasley, J., W. T. Riley, and J. Jean-Mary. 2005. Accuracy of a PDA-based dietary assessment program. *Nutrition* 21(6): 672–677.

Belanger, M., L. Humbert, H. Vatanparast et al. 2016. A multilevel intervention to increase physical activity and improve healthy eating and physical literacy among young children (ages 3-5) attending early childcare centres: The Healthy Start-Depart Sante cluster randomised controlled trial study protocol. *BMC Public Health* 16(1): 313. doi:10.1186/s12889-016-2973-5.

Bird, G., and P. Elwood. 1983. The dietary intakes of subjects estimated from photographs compared with a weighed record. *Human Nutrition. Applied Nutrition* 37(6): 470–473.

Blake, A. J., H. A. Guthrie, and H. Smiciklas-Wright. 1989. Accuracy of food portion estimation by overweight and normal-weight subjects. *Journal of the American Dietetic Association* 89(7): 962–964.

Boushey, C. J., A. J. Harray, D. A. Kerr et al. 2015. How willing are adolescents to record their dietary intake? The mobile food record. *JMIR mHealth and uHealth* 3(2): e47.

Boushey, C. J., D. A. Kerr, J. Wright, K. D. Lutes, D. S. Ebert, and E. J. Delp. 2009. Use of technology in children's dietary assessment. *European Journal of Clinical Nutrition* 63: S50–S57.

Casperson, S. L., J. Reineke, J. Sieling, J. Moon, and J. Roemmich. 2013. Usability of mobile phone food records to assess dietary intake in adolescents. *The FASEB Journal* 27(1_MeetingAbstracts): 230.232.

Casperson, S. L., J. Sieling, J. Moon, L. Johnson, J. N. Roemmich, and L. Whigham. 2015. A mobile phone food record app to digitally capture dietary intake for adolescents in a free-living environment: Usability study. *JMIR mHealth and uHealth* 3(1): 30.

Chung, L. M. Y., and J. W. Y. Chung. 2010. Tele-dietetics with food images as dietary intake record in nutrition assessment. *Telemedicine and e-Health* 16(6): 691–698.

Comstock, E. M., R. G. St. Pierre, and Y. D. Mackiernan. 1981. Measuring individual plate waste in school lunches. Visual estimation and children's ratings vs. actual weighing of plate waste. *Journal of the American Dietetic Association* 79(3): 290–296.

Cordeiro, F., D. A. Epstein, E. Thomaz et al. 2015. Barriers and negative nudges: Exploring challenges in food journaling. *Paper Presented at the Proceedings of the 33rd Annual ACM Conference on Human Factors in Computing Systems*, April 18–23, Seoul, Republic of Korea.

Crombie, A. P., L. K. Funderburk, T. J. Smith et al. 2013. Effects of modified foodservice practices in military dining facilities on ad libitum nutritional intake of US army soldiers. *Journal of the Academy of Nutrition and Dietetics* 113(7): 920–927. doi:10.1016/j.jand.2013.01.005.

Daugherty, B. L., T. E. Schap, R. Ettienne-Gittens et al. 2012. Novel technologies for assessing dietary intake: Evaluating the usability of a mobile telephone food record among adults and adolescents. *Journal of medical Internet Research* 14(2): e58.

de Jonge, L., J. P. DeLany, T. Nguyen et al. 2007. Validation study of energy expenditure and intake during calorie restriction using doubly labeled water and changes in body composition. *The American Journal of Clinical Nutrition* 85(1): 73–79.

Delisle Nyström, C., E. Forsum, H. Henriksson et al. 2016. A mobile phone based method to assess energy and food intake in young children: A validation study against the doubly labelled water method and 24 h dietary recalls. *Nutrients* 8(1): 50.

Dhurandhar, N., D. Schoeller, A. Brown et al. 2015. Energy balance measurement: When something is not better than nothing. *International Journal of Obesity* 39(7): 1109–1113.

Dibiano, R., B. K. Gunturk, and C. K. Martin. 2013. Food image analysis for measuring food intake in free living conditions. *Paper Presented at the SPIE Medical Imaging*, February 09, Lake Buena Vista, FL.

Donnelly, J. E., D. K. Sullivan, B. K. Smith et al. 2008. Alteration of dietary fat intake to prevent weight gain: Jayhawk observed eating trial. *Obesity (Silver Spring)* 16(1): 107–112. doi:10.1038/oby.2007.33.

Dorman, S. 2013. Evaluating the Reliability of Assessing Home-Packed Food Items Using Digital Photographs and Dietary Log Sheets.

Doumit, R., J. Long, C. Kazandjian et al. 2015. Effects of recording food intake using cell phone camera pictures on energy intake and food choice. *Worldviews on Evidence-Based Nursing* 13(3): 216–223.

Duhe, A. F., L. A. Gilmore, J. H. Burton, C. K. Martin, and L. M. Redman. 2016. The remote food photography method accurately estimates dry powdered foods-The source of calories for many infants. *Journal of the Academy of Nutrition and Dietetics* 116(7): 1172–1177. doi:10.1016/j.jand.2016.01.011.

Elinder, L. S., A. Brunosson, H. Bergstrom, M. Hagstromer, and E. Patterson. 2012. Validation of personal digital photography to assess dietary quality among people with intellectual disabilities. *Journal of Intellectual Disability Research* 56(2): 221–226. doi:10.1111/j.1365-2788.2011.01459.x.

Elwood, P. and G. Bird. 1983. A photographic method of diet evaluation. *Human Nutrition. Applied Nutrition* 37(6): 474–477.

Gauthier, A. P., B. T. Jaunzarins, S. J. MacDougall et al. 2013. Evaluating the reliability of assessing home-packed food items using digital photographs and dietary log sheets. *Journal of Nutrition Education and Behavior* 45(6): 708–712.

Goldberg, J. P., S. C. Folta, M. Eliasziw et al. 2015. Great taste, less waste: A cluster-randomized trial using a communications campaign to improve the quality of foods brought from home to school by elementary school children. *Preventive Medicine* 74: 103–110. doi:10.1016/j.ypmed.2015.02.010.

Hanks, A. S., B. Wansink, and D. R. Just. 2014. Reliability and accuracy of real-time visualization techniques for measuring school cafeteria tray waste: Validating the quarter-waste method. *Journal of the Academy of Nutrition and Dietetics* 114(3): 470–474. doi:10.1016/j.jand.2013.08.013.

Helander, E., K. Kaipainen, I. Korhonen, and B. Wansink. 2014. Factors related to sustained use of a free mobile app for dietary self-monitoring with photography and peer feedback: Retrospective cohort study. *Journal of medical Internet Research* 16(4): e109.

Henriksson, H., S. E. Bonn, A. Bergström et al. 2015. A new mobile phone-based tool for assessing energy and certain food intakes in young children: A validation study. *JMIR mHealth and uHealth* 3(2): e38.

Higgins, J., A. LaSalle, P. Zhaoxing et al. 2009. Validation of photographic food records in children: Are pictures really worth a thousand wordsandquest. *European Journal of Clinical Nutrition* 63(8): 1025–1033.

Hill, J. O. 2006. Understanding and addressing the epidemic of obesity: An energy balance perspective. *Endocrine Reviews* 27(7): 750–761.

Hill, J. O., H. R. Wyatt, and J. C. Peters. 2012. Energy balance and obesity. *Circulation* 126(1): 126–132.

Hodges, S., L. Williams, E. Berry et al. 2006. SenseCam: A retrospective memory aid. In *UbiComp 2006: Ubiquitous Computing*, P. Dourish and A. Friday (Eds.), pp. 177–193. Berlin, Germany: Springer-Verlag.

Kikunaga, S., T. Tin, G. Ishibashi, D. H. Wang, and S. Kira. 2007. The application of a handheld personal digital assistant with camera and mobile phone card (Wellnavi) to the general population in a dietary survey. *Journal of Nutritional Science and Vitaminology* 53(2): 109–116.

Lassen, A. D., S. Poulsen, L. Ernst, K. K. Andersen, A. Biltoft-Jensen, and I. Tetens. 2010. Evaluation of a digital method to assess evening meal intake in a free-living adult population. *Food and Nutrition Research* 54: 1–9.

Martin, C. K., S. D. Anton, E. York-Crowe et al. 2007a. Empirical evaluation of the ability to learn a calorie counting system and estimate portion size and food intake. *British Journal of Nutrition* 98(2): 439–444.

Martin, C. K., J. B. Correa, H. Han et al. 2012. Validity of the remote food photography method (RFPM) for estimating energy and nutrient intake in near real-time. *Obesity* 20(4): 891–899.

Martin, C. K., L. A. Gilmore, J. W. Apolzan, C. A. Myers, D. M. Thomas, and L. M. Redman. 2016. Smartloss: A personalized mobile health intervention for weight management and health promotion. *JMIR mHealth and uHealth* 4: e18.

Martin, C. K., H. Han, S. M. Coulon, H. R. Allen, C. M. Champagne, and S. D. Anton. 2009a. A novel method to remotely measure food intake of free-living individuals in real time: The remote food photography method. *British Journal of Nutrition* 101(3): 446–456.

Martin, C. K., S. Kaya, and B. K. Gunturk. 2009b. Quantification of food intake using food image analysis. *Paper Presented at the Engineering in Medicine and Biology Society, 2009.* EMBC 2009. Annual International Conference of the IEEE, September 03–06, Minneapolis, MN.

Martin, C. K., A. C. Miller, D. M. Thomas, C. M. Champagne, H. Han, and T. Church. 2015. Efficacy of SmartLossSM, a smartphone-based weight loss intervention: Results from a randomized controlled trial. *Obesity* 23(5): 935–942.

Martin, C. K., R. L. Newton, S. D. Anton et al. 2007b. Measurement of children's food intake with digital photography and the effects of second servings upon food intake. *Eating Behaviors* 8(2): 148–156.

Martin, C. K., T. A. Nicklas, B. K. Gunturk, J. B. Correa, H. R. Allen, and C. Champagne. 2014. Measuring food intake with digital photography. *Journal of Human Nutrition and Dietetics* 27(Suppl. 1): 72–81. doi:10.1111/jhn.12014.

Martin, C. K., J. L. Thomson, M. M. LeBlanc et al. 2010. Children in school cafeterias select foods containing more saturated fat and energy than the Institute of Medicine recommendations. *The Journal of Nutrition* 140: 1653–1660.

Matthiessen, T. B., F. M. Steinberg, and L. L. Kaiser. 2011. Convergent validity of a digital image-based food record to assess food group intake in youth. *Journal of the American Dietetic Association* 111(5): 756–761.

Murakami, H., R. Kawakami, S. Nakae et al. 2016. Accuracy of wearable devices for estimating total energy expenditure: Comparison with metabolic chamber and doubly labeled water method. *JAMA Internal Medicine* 176(5): 702–703.

Nicklas, T. A., Y. Liu, J. E. Stuff, J. O. Fisher, J. A. Mendoza, and C. E. O'Neil. 2013. Characterizing lunch meals served and consumed by pre-school children in Head Start. *Public Health Nutrition* 16(12): 2169–2177.

Nicklas, T. A., C. E. O'Neil, J. Stuff, L. S. Goodell, Y. Liu, and C. K. Martin. 2012a. Validity and feasibility of a digital diet estimation method for use with preschool children: A pilot study. *Journal of Nutrition Education and Behavior* 44(6): 618–623.

Nicklas, T. A., C. E. O'Neil, J. E. Stuff, S. O. Hughes, and Y. Liu, 2012b. Characterizing dinner meals served and consumed by low-income preschool children. *Childhood Obesity* (*Formerly Obesity and Weight Management*) 8(6): 561–571.

Noronha, J., E. Hysen, H. Zhang, and K. Z. Gajos. 2011. Platemate: Crowdsourcing nutritional analysis from food photographs. *Paper Presented at the Proceedings of the 24th Annual ACM Symposium on User Interface Software and Technology*, October 16–19, Santa Barbara, CA.

O'Loughlin, G., S. J. Cullen, A. McGoldrick et al. 2013. Using a wearable camera to increase the accuracy of dietary analysis. *American Journal of Preventive Medicine* 44(3): 297–301.

Ohtsuka, N. S. R. 2001. Nutrient intake among female shift workers in a computer factory in Japan. *International Journal of Food Sciences and Nutrition* 52(4): 367–378.

Pettitt, C., J. Liu, R. M. Kwasnicki, G. Z. Yang, T. Preston, and G. Frost. 2016. A pilot study to determine whether using a lightweight, wearable micro-camera improves dietary assessment accuracy and offers information on macronutrients and eating rate. *British Journal of Nutrition* 115(1): 160–167.

Ptomey, L. T., E. A. Willis, J. R. Goetz, J. Lee, D. K. Sullivan, and J. E. Donnelly. 2015. Digital photography improves estimates of dietary intake in adolescents with intellectual and developmental disabilities. *Disability and Health Journal* 8(1): 146–150.

Rangan, A. M., S. O'Connor, V. Giannelli et al. 2015. Electronic dietary intake assessment (e-DIA): Comparison of a mobile phone digital entry app for dietary data collection with 24-hour dietary recalls. *JMIR mHealth and uHealth* 3(4): e98.

Rollo, M. E., S. Ash, P. Lyons-Wall, and A. Russell. 2011. Trial of a mobile phone method for recording dietary intake in adults with type 2 diabetes: Evaluation and implications for future applications. *Journal of Telemedicine and Telecare* 17(6): 318–323.

Rollo, M. E., S. Ash, P. Lyons-Wall, and A. W. Russell. 2015. Evaluation of a mobile phone image-based dietary assessment method in adults with type 2 diabetes. *Nutrients* 7(6): 4897–4910.

Samaras, D., N. Samaras, P. C. Bertrand et al. 2012. Comparison of the interobserver variability of 2 different methods of dietary assessment in a geriatric ward: A pilot study. *Journal of the American Medical Directors Association* 13(3): 309.e309–313. doi:10.1016/j.jamda.2011.06.006.

Schap, T. E. 2012. *Advancing Dietary Assessment through Technology and Biomarkers.* West Lafayette, IN: Purdue University.

Scherr, R. E., J. D. Linnell, M. H. Smith et al. 2014. The shaping healthy choices program: Design and implementation methodologies for a multicomponent, school-based nutrition education intervention. *Journal of Nutrition Education and Behavior* 46(6): e13–21. doi:10.1016/j.jneb.2014.08.010.

Schoeller, D. A., and D. Thomas. 2014. Energy balance and body composition. In *Nutrition for the Primary Care Provider*, Vol. 111, pp. 13–18, Karger Publishers, Basel, Switzerland.

Schoeller, D. A., D. Thomas, E. Archer et al. 2013. Self-report–based estimates of energy intake offer an inadequate basis for scientific conclusions. *The American Journal of Clinical Nutrition* 97(6): 1413–1415.

Sevenhuysen, G., W. van Staveren, K. Dekker, and E. Spronck. 1990. Estimates of daily individual food intakes obtained by food image processing. *Nutrition Research* 10(9): 965–974.

Sevenhuysen, G., and L. Wadsworth. 1989a. Food image processing: A potential method for epidemiological surveys. *Nutrition Reports International* (*USA*) 39: 439–450.

Sevenhuysen, G., and E. Zacharias. 1989b. Comparison of food intake assessments obtained with recall interviews and photographic records. *Nutrition Reports International* (*USA*) 40: 349–357.

Six, B. L., T. E. Schap, F. Zhu et al. 2010. Evidence-based development of a mobile telephone food record. *Journal of the American Dietetic Association* 110(1): 74–79.

Smith, S. L., and L. Cunningham-Sabo. 2014. Food choice, plate waste and nutrient intake of elementary- and middle-school students participating in the US National School Lunch Program. *Public Health Nutr* 17(6): 1255–1263. doi:10.1017/s1368980013001894.

Stevens, G. A., G. Singh, Y. Lu et al. 2012. National, regional, and global trends in adult overweight and obesity prevalences. *Population Health Metrics* 10(1): 1.

Stone, A. A., and S. Shiffman. 1994. Ecological momentary assessment (EMA) in behavioral medicine. *Annals of Behavioral Medicine* 16: 199–202.

Sun, M., L. E. Burke, Z.-H. Mao et al. 2014. eButton: A wearable computer for health monitoring and personal assistance. *Paper Presented at the Proceedings of the 51st Annual Design Automation Conference.*

Swanson, M. 2008. Digital photography as a tool to measure school cafeteria consumption. *Journal of School Health* 78(8): 432–437.

Upton, D., P. Upton, and C. Taylor. 2013. Increasing children's lunchtime consumption of fruit and vegetables: An evaluation of the Food Dudes programme. *Public Health Nutrition* 16(6): 1066–1072. doi:10.1017/s1368980012004612.

Upton, P., C. Taylor, and D. Upton. 2015. The effects of the food dudes programme on children's intake of unhealthy foods at lunchtime. *Perspectives in Public Health* 135(3): 152–159. doi:10.1177/1757913914526163.

Wang, D.-H., M. Kogashiwa, and S. Kira. 2006. Development of a new instrument for evaluating individuals' dietary intakes. *Journal of the American Dietetic Association* 106(10): 1588–1593.

Wang, D.-H., M. Kogashiwa, S. Ohta, and S. Kira. 2002. Validity and reliability of a dietary assessment method: The application of a digital camera with a mobile phone card attachment. *Journal of Nutritional Science and Vitaminology* 48(6): 498–504.

Williamson, D., H. Allen, P. D. Martin, A. Alfonso, B. Gerald, and A. Hunt. 2003. Comparison of digital photography to weighed and visual estimation of portion sizes. *Journal of the American Dietetic Association* 103(9): 1139–1145.

Williamson, D., H. Allen, P. D. Martin, A. Alfonso, B. Gerald, and A. Hunt. 2004. Digital photography: A new method for estimating food intake in cafeteria settings. *Eating and Weight Disorders-Studies on Anorexia, Bulimia and Obesity* 9(1): 24–28.

Williamson, D. A., C. M. Champagne, D. Harsha et al. 2012. Effect of an environmental school-based obesity prevention program on changes in body fat and body weight: A randomized trial. *Obesity* 20: 1653–1661.

Williamson, D. A., A. L. Copeland, S. D. Anton et al. 2007. Wise Mind project: A school-based environmental approach for preventing weight gain in children. *Obesity* 15(4): 906–917.

Williamson, D. A., H. Han, W. D. Johnson, C. K. Martin, and R. L. Newton. 2013. Modification of the school cafeteria environment can impact childhood nutrition. Results from the Wise Mind and LA Health studies. *Appetite* 61: 77–84.

Williamson, D. A., P. D. Martin, H. R. Allen et al. 2002. Changes in food intake and body weight associated with basic combat training. *Military Medicine* 167(3): 248–253.

Wing, R. R., and J. O. Hill. 2001. Successful weight loss maintenance. *Annual Review of Nutrition* 21(1): 323–341.

Wolper, C., S. Heshka, and S. B. Heymsfield. 1995. Measuring food intake: An overview. In D. Allison (Ed.), *Handbook of Assessment of Methods for Eating Behaviors and Weight-Related Problems*, pp. 215–240. Thousand Oaks, CA: Sage Publications.

Yon, B. A., R. K. Johnson, J. Harvey-Berino, and B. C. Gold. 2006. The use of a personal digital assistant for dietary self-monitoring does not improve the validity of self-reports of energy intake. *Journal of the American Dietetic Association* 106(8): 1256–1259.

Yon, B. A., R. K. Johnson, J. Harvey-Berino, B. C. Gold, and A. B. Howard. 2007. Personal digital assistants are comparable to traditional diaries for dietary self-monitoring during a weight loss program. *Journal of Behavioral Medicine* 30(2): 165–175.

Zepeda, L., and D. Deal. 2008. Think before you eat: Photographic food diaries as intervention tools to change dietary decision making and attitudes. *International Journal of Consumer Studies* 32(6): 692–698.

Zhu, F., M. Bosch, I. Woo et al. 2010. The use of mobile devices in aiding dietary assessment and evaluation. *EEE Journal of Selected Topics in Signal Processing* 4(4): 756–766.

Zhu, F., A. Mariappan, C. J. Boushey et al. 2008. Technology-assisted dietary assessment. *Paper Presented at the Electronic Imaging 2008*, January 27, San Jose, CA.

6 Meal Patterns, Physical Activity, Sleep, and Circadian Rhythm

Margriet S. Westerterp-Plantenga and Marta Garaulet

CONTENTS

6.1 INTRODUCTION

Assessment of food and energy intake takes place in the context of research on metabolic health and energy balance. However, as the metabolic processes of the ingested food depend on the time of ingestion on a day, timing of meals, and the relation to circadian rhythms should be taken into account. According to Froy (2007), Wolk and Somers (2007), Kohsaka et al. (2007), Mendoza (2007), Mendoza et al. (2008), Adamantidis and de Lecea (2008), Bechtold (2008), Laposky et al. (2008), Esquirol et al. (2009), Scheer et al. (2009), Arble et al. (2009), Garaulet et al. (2010), Cagampang and Bruce (2012), coordination of daily patterns of activity, feeding, energy utilization, and energy storage is supported by a synchronized pattern of release of the relevant endocrine components across the 24-hour cycle (see Figure 6.1). The master circadian clock, namely the suprachiasmatic nucleus (SCN) of the hypothalamus provides circadian alignment, and connected peripheral tissues of the body containing the molecular clock machinery and subsequently addresses the local circadian oscillation and rhythmic gene expression. However, metabolic processes are decoupled from the primarily light-driven SCN when food intake is desynchronized from normal diurnal patterns of activity resulting in changes in energy availability, substrate oxidation, storage, and metabolic status. If feeding becomes the dominant entraining stimulus, adaptation to the changed food intake patterns occurs, facilitated by an autonomous food entrainable oscillator (FEO) that governs behavioral rhythms. Consequently, energy metabolism can influence the clock mechanism of the SCN. This close interaction is critical for normal circadian regulation of metabolism, and desynchronization may underlie the disruption of proper metabolic rhythms observed in metabolic disorders,

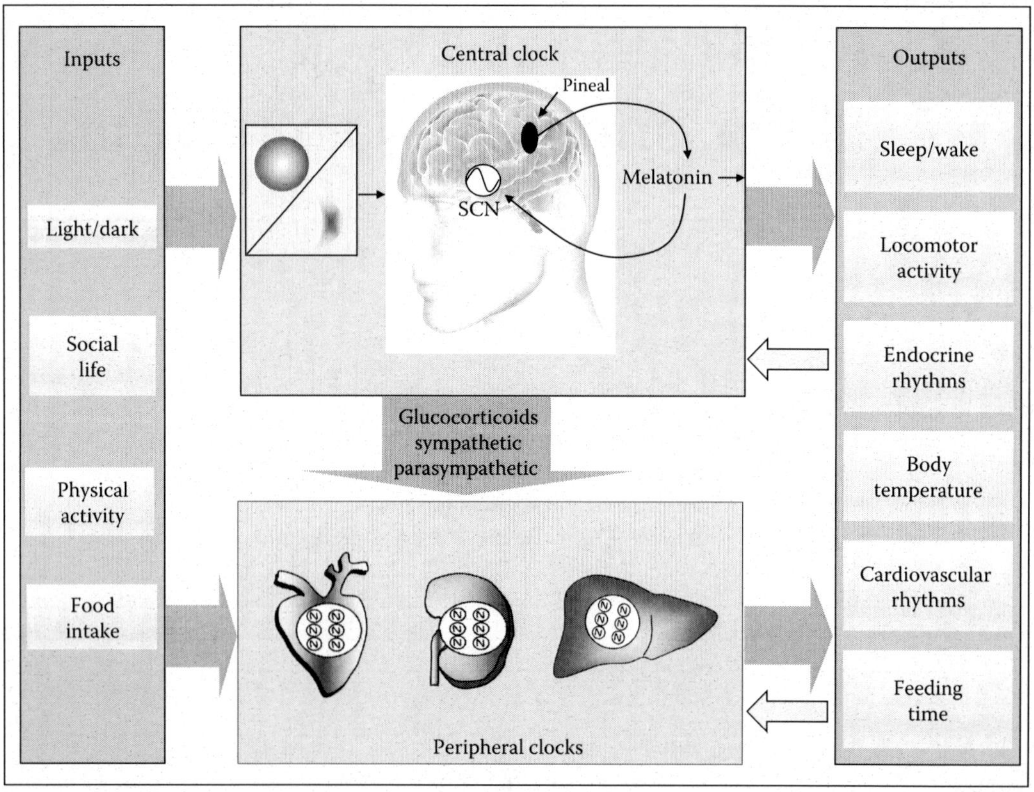

FIGURE 6.1 A general overview of the functional organization of the circadian system in mammals. Inputs: Environmental periodical cues can reset the phase of the central pacemaker, so that the period and phase of circadian rhythms became coincident with the timing of the external cues. Central pacemakers: The SCN is considered the major pacemaker of the circadian system, driving circadian rhythmicity in other brain areas and peripheral tissues by sending them neural and humoral signals. Peripheral oscillators: Most peripheral tissues and organs contain circadian oscillators. Usually they are under the control of the SCN; however, under some circumstances (i.e., restricted feeding, jet lag, and shift work), they can desynchronize from the SCN. Outputs: Central pacemakers and peripheral oscillators are responsible for the daily rhythmicity observed in most physiological and behavioral functions. Some of these overrhythms (i.e., physical exercise, core temperature, sleep–wake cycle, and feeding time), in turn, provide a feedback, which can modify the function of SCN and peripheral oscillators. SCN, suprachiasmatic nucleus. (From Garaulet, M. et al., *Int. J. Obes.*, 34, 1667–1683, 2010.)

such as obesity and type 2 diabetes (T2D) (Bechtold 2008; Esquirol et al. 2009; Arble et al. 2009; Hoogerwerf 2009; Schibler et al. 2003; Knutsson and Boggild 2010; Szosland 2010).

Other factors playing a role in circadian alignment are physical activity during the day, and the timing and duration of sleep. In this chapter, development of technical approaches for assessment of circadian rhythm, meal patterns, physical activity—and sleep patterns in outpatient settings are addressed.

6.2 CIRCADIAN RHYTHM

6.2.1 Background and Rationale

The daily patterns of feeding, energy utilization, and energy storage across the 24-hour cycle are based on a neuro-endocrinological system (Czeisler and Klerman 1999; Huang et al. 2011; Koren et al. 2011). Glucose and insulin levels peak during the late biological night (Kalsbeek and Strubbe

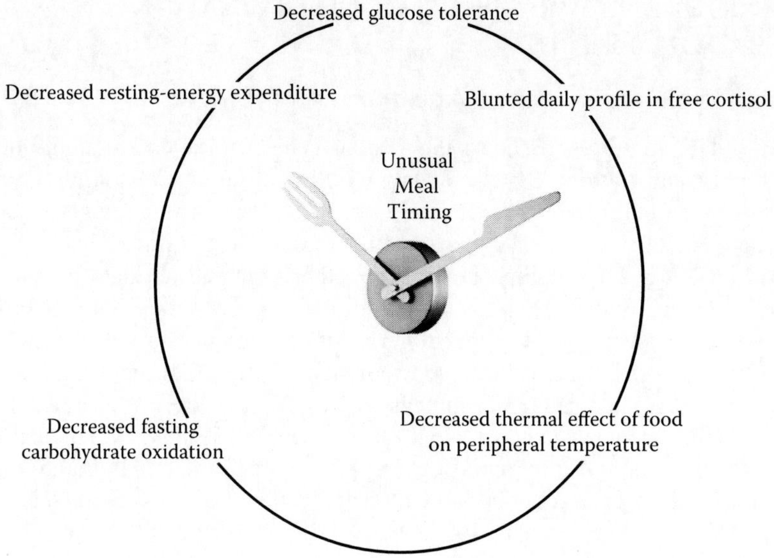

FIGURE 6.2 Effects of meal timing on metabolic health. (From Bandin, et al., *Int. J. Obes.*, 38, 8, 1083–1088, 2014.)

1999; Morgan et al. 2003). Leptin is secreted in a circadian manner (Schoeller et al. 1997; van Aggel-Leijssen et al. 1999; Wong et al. 2004; Lecoultre et al. 2011), with leptin levels being high at night but low during the day (Bado et al. 1998; Cinti et al. 2000, 2001). Ghrelin can alter clock function in the SCN in vitro (Yannielli et al. 2007; Yi et al. 2008). Ghrelin oscillates with feeding: elevated levels of ghrelin were found during the early part of the night in sleeping subjects, decreasing in the morning before awakening (Cummings et al. 2001), and may be a signal involved in the cross-talk between the peripheral and central circadian clock system. In relation to appetite, release of some of these endocrine products, such as glucose, insulin, GLP-1, ghrelin, and leptin concentrations, shift with meal patterns, whereas disruption of the circadian system is associated with a disturbed glucose and insulin metabolism and decreased GLP-1 concentrations indicating decreased satiety (Fletcher 1997; Ribeiro et al. 1998; Cinti et al. 2001; Gonnissen et al. 2012). Furthermore, large metabolic fluctuations in carbohydrate and fat oxidation were shown in a gorging food intake pattern, contrarily to the nibbling pattern (Verboeket-van de Venne and Westerterp 1991). Taken together, circadian alignment of meal patterns is crucial in energy- and nutrient-balance homeostasis, and in metabolic and cardiovascular health (Figure 6.2), (Bandín et al. 2015). Consequently, timing and regularity of meals is required and beneficial metabolic effects of regular timing and meal frequency have been shown (Garrow et al. 1981; Yki-Jarvinen et al. 1987; Jenkins et al. 1989, 1990; Wolever; 1990; Verboeket-van de Venne and Westerterp 1991; Arnold et al. 1993; Bertelsen et al. 1993; Jones et al. 1993; Verboeket-van de Venne et al. 1993; McGrath and Gibney 1994; Jenkins and Jenkins 1995; Thomsen et al. 1997; Ribeiro et al. 1998; Melanson et al. 1999a, 1999b, 1999c; Van Cauter et al. 2000; Westerterp-Plantenga et al. 2002; Westerterp-Plantenga et al. 2003; Farshchi et al. 2004a, 2004b, 2005; Chapelot et al. 2006; Smeets and Westerterp-Plantenga 2008; Esquirol et al. 2009; Nedeltcheva et al. 2009; Westerterp-Plantenga et al. 2009; Huang et al. 2011; Hursel et al. 2011; DeSantis et al. 2012; Gonnissen et al. 2013a, 2013b, 2013c).

Physical activity interferes with circadian rhythm parameters (Tranel et al. 2015). Circadian misalignment of sleep results in particular metabolic disturbances, including dysregulation in the hypothalamic–pituitary–adrenal (HPA) axis, increased cortisol secretion, higher fasting insulin concentrations, and a higher homeostasis model assessment of insulin resistance (HOMA-IR) index (Van Cauter et al. 2000; Knutson et al. 2011; Wu et al. 2011; Gonnissen et al. 2013a; Westerterp-Plantenga 2016). Therefore, sleep is the third factor to be included in the assessment of timing and patterns in energy- and nutrient-homeostasis.

6.3 TECHNIQUE AND PARAMETERS OF DETERMINATION OF CIRCADIAN RHYTHM

6.3.1 Theory of the Measurement, Validation, and Protocol

Measurement of circadianity is based on the responsiveness of the circadian pacemaker to light of the pineal melatonin rhythm. Synchronously with the onset of melatonin secretion, a fall in body temperature via cutaneous vasodilatation occurs in the evening, and the opposite occurs in the morning (Cagnacci 2007). The phases of the circadian rhythms of body temperature are used as a parameter for circadianity, possibly together with the physical activity pattern and sleep pattern. Circadian rhythms of body temperature are synchronized with the onset and offset of circadian melatonin secretion. The morning rise of core temperature is caused by the change from sleep to waking activity, the increase in ambient illumination, and the fall in melatonin secretion. In spite of a wide range of ambient temperatures or during exercise, the core body temperature is maintained at a rather stable level (Kräuchi 2007). The circadian changes in core temperature are due to a rhythmic input from the SCN acting on the hypothalamic thermoregulatory centers, modulating the set point, and altering the thresholds for cutaneous vasodilatation and sweating (Refinetti and Menaker 1992; Aizawa and Cabanac 2002). However, thus, when measuring the core temperature rhythm, a problem named *masking* of this rhythm by spontaneous and imposed activity, increasing core body temperature, and a masking effect of sleep, decreasing core body temperature due to unloading of the cardiovascular system occurs, and requires special attention. This masking effect needs to be corrected for, when the pure circadian core-body temperature rhythm is used as a parameter for circadian rhythm. Therefore, sleep and physical activity measurements need to be added to the temperature measurements, if possible. Sleep is typically initiated on the declining portion of the circadian rhythm of core body temperature (CBT) when its rate of change, and body heat loss, is maximal. Distal vasodilatation is strongly associated with sleepiness and sleep induction. In contrast, sleep has no or only a minor thermoregulatory function (Kräuchi 2007). The circadian modulation of sleepiness and sleep induction is clearly associated with thermoregulatory changes, but the thermoregulatory system seems to be independent of the sleepiness or sleep regulatory system (Kräuchi 2007). Thus, when discriminating exogenous and endogenous influences on the circadian rhythm, factors that influence deterioration of circadianity, for example, irregularities due to the individual's lifestyle causing metabolic disturbances can be unraveled. Absence of an overt rhythm may have several reasons, including the signal from the SCN becoming too weak or its transmission impaired (Weinert and Waterhouse 2007).

The rhythm of skin peripheral temperature has been proposed to measuring the central temperature. Core temperature rhythms are related to a cutaneous vasodilation in the evening with a distal increase in skin temperature, and a cutaneous vasoconstriction in the morning with a distal decrease in skin temperature. These diurnal changes in distal skin temperature are opposite to the changes in core body temperature. In real life distal skin temperature, that can be measured in the wrist (wrist skin temperature; WT) and is used as a representative parameter for measurement of reproducible and stable circadian rhythms of high amplitude, and with a characteristic phasing with respect to other biological processes and the external environment (Weinert and Waterhouse 2007), as shown in Figure 6.3. It has been shown that WT rhythm has a distinct endogenous component, even in the presence of multiple external influences, and also has a genetic component (Lopez-Minguez et al. 2015). Therefore, WT has been proposed as an informative and minimally invasive technique to measure circadian rhythm in free-living conditions (Martinez-Nicolas et al. 2013).

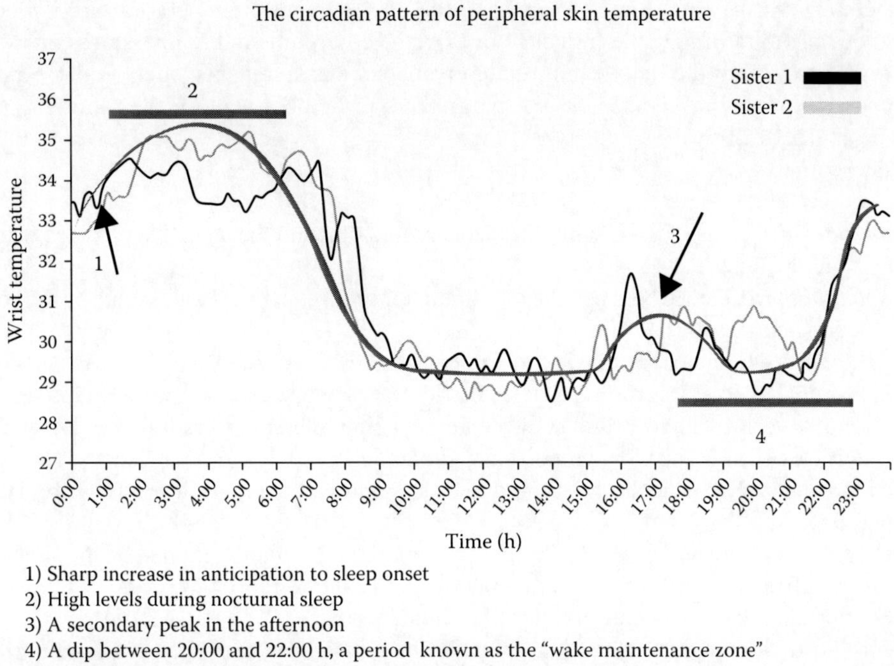

FIGURE 6.3 24-hour waveform of wrist skin temperature (WT) of two monozygotic sisters recorded continuously for seven days. The standard skin temperature rhythm exhibits a sharp increase in anticipation to sleep onset, it maintains high levels during nocturnal sleep and shows a secondary peak in the afternoon.

6.3.2 Wireless Determination of Skin Temperature Using iButtons

A wireless temperature system for human skin temperature measurements, that is, the Thermochron iButton DS1291H, has been validated, and its application in clinical and field measurements has been evaluated by van Marken Lichtenbelt et al. (2006). The authors show that iButtons have a mean accuracy of −0.09°C with a precision of 0.05°C; these properties can be improved by using calibration. With respect to circadian rhythm and sleep research, skin temperature assessment by iButtons is of significant value in clinical and field situations.

In addition, Hasselberg et al. (2013) conclude that the iButton can be used to obtain a valid measurement of human skin temperature. They support the ibutton as being a practical alternative to traditional measures of circadian rhythms in sleep or wake research.

6.3.3 Protocol

Wrist temperature rhythmicity is assessed continuously for seven days, using a temperature sensor (Thermochron iButton DS1921H, Dallas, Maxim, Dallas, TX) with a sensitivity of 0.125°C and programmed to collect information every 10 minute (Lopez-Minguez et al. 2015). It is attached to a bracelet or watch strap, and the sensor surface is placed over the inside of the wrist on the radial artery of the nondominant hand (Sarabia et al. 2008). The information stored in the iButton is transferred through an adapter (DS1402D-DR8; Dallas, Maxim) to a personal computer using iButton Viewer v.3.22 (Dallas Semiconductor Maxim software provided by the manufacturer). At the same time the data on the environmental temperature are collected, in order to be able to minimize the influence of extreme environmental temperatures on WT readings. After processing the data to eliminate erroneous measurements, such as those produced by temporarily removing the sensor, the parameters to be obtained are the following (Lopez-Minguez et al. 2015):

Cosinor's analysis:

- *Mesor*: Mean value of the rhythm fitted to a cosine function that usually coincides with the mean temperature.
- *Amplitude*: Difference between the maximum (or minimum or average) value of the cosine function and mesor.
- *Acrophase*: Timing of the maximum value of the cosine function.
- *Rayleigh test*: To assess the acrophase distribution within a 24-hour period. This test provides an r vector with its origin at the center of a circumference of radius one. The r vector length (between 0 and 1) is proportional to the degree of phase homogeneity during the period analyzed and can be considered to be a measure of the rhythm's phase stability during successive days.
- *Percentage of rhythmicity* (*PR*): Percentage of variance of data explained by the sinusoidal function. Higher values of this parameter mean to a more sinusoidal curve.
- Fourier analysis to calculate the first harmonic's power (*P*1) as a spectral power of the 24-hour rhythm and the second harmonic's power (*P*2) that gives the postprandial rise in temperature.

Nonparametric analysis (Van Someren et al. 1999):

- *Interdaily stability* (*IS*): The similarity of the 24-hour pattern over days. It varied between 0 for Gaussian noise and 1 for a perfect stability, where the rhythm repeated itself exactly day after day.
- *Intradaily variability* (*IV*): Fragmentation of the rhythm, its values oscillated between 0 when the wave was perfectly sinusoidal and 2 when the wave describes a Gaussian noise.
- *Five-hour of maximum temperature*: Average of made at 10-minute intervals for the five consecutive hours with the maximum temperature (*M*5).
- *Relative amplitude*: Difference between *M*5 and the average of measurements made for the 10 consecutive hours with the minimum temperature (*L*10) divided by the sum of both values (*M*5 + *L*10).
- *Circadian function index* (*CFI*) (Ullah et al. 2012): A numerical index that determines the circadian robustness and is based on three circadian parameters: IS, IV, and relative amplitude (Ortiz-Tudela et al. 2010). IV values are inverted and normalized between 0 and 1, with 0 being a noise signal and 1 a perfect sinusoid. Finally, CFI is calculated as the average of these three parameters performed by the program *Circadianware*. Consequently, CFI oscillates between 0 (absence of circadian rhythmicity) and 1 (a robust circadian rhythm). *Circadianware* is an integrated package for temporal series analysis *Circadianware* (Chronobiology Laboratory, University of Murcia, Spain, 2010).

6.4 DETERMINATION OF PHYSICAL ACTIVITY AND SLEEP PATTERNS

In order to relate physical activity to circadian rhythm, a validated accelerometer need to be used that measures patterns of physical activity, such as the GT3X++ (Actigraph, Pensacola, Florida, the United States) (Trost et al. 2005; Robusto and Trost 2012; Rowlands and Stiles 2012). Also see Chapter 11.

The GT3X actigraph accelerometer provides measurements of physical activities, physical activity patterns, and objective monitoring of sleep–wake rhythm [3] in free-living settings based on the recording of motion. Body movements are recorded by an accelerometer, which can be worn on the wrist, ankle, or hip. For sleep assessment, the device is typically worn on the nondominant wrist, and this technique has shown acceptable agreement with polysomnography (PSG), ranging between 85% and 95% for the identification of sleep–wake epochs (Sadeh 2008), and for a similar sleep–wake detection ability of systems (Tonetti et al. 2013; Cellini et al. 2015). Validity of detection of sleep–wake patterns against a concurrent PSG recording has been assessed (Cellini et al. 2013; Zinkhan et al. 2014; Slater et al. 2015). GT3X+ appeared to underestimate wakefulness (i.e., both sleep latency and wake after sleep onset [WASO]), as well as an effect of device placement. The accuracy, sensitivity, and specificity values were comparable to reports of other similar devices (Cellini et al. 2013; Slater et al. 2015), when the device was worn on the wrist, but not when worn on the hip (Zinkhan et al. 2014; Slater et al. 2015). The wrist–worn GT3X+ appeared to be valid for detecting sleep or wake patterns in a laboratory setting, yet the ability of this device to monitor sleep in a free-living environment remains to be studied. Furthermore, the Actiwatch-64 (AW-64; Phillips Respironics, Bend, Oregon, the United States) is a validated wearable monitor for sleep in adult populations (Rupp and Balkin 2011; Cellini et al. 2013; Natale et al. 2014), as well as in children (Hyde et al. 2007; Meltzer et al. 2012). Cellini et al. (2016) reported a good agreement of the GT3X with AW-64 for sleep assessment, especially when the GT3X is worn on the wrist. These accelerometers measure the time of going to sleep and waking up, as well as parameters of sleep architecture such as total sleep time (TST), sleep onset latency (SL), wake after sleep onset (WASO), and sleep efficiency (SE).

Recently another method, based on tilt sensing for the assessment of activity and body position has been proposed. In general, most actigraphic devices are placed on the wrist and their measures are based on acceleration detection. In this case, actigraphy is measured at the level of the arm for joint evaluation of activity and body position. This method analyzes the tilt of three axes, scoring activity as the cumulative change of degrees per minute with respect to the previous sampling, and measuring arm tilt for the body position inference. This novel method has been validated against (a) DLMO (dim light melatonin onset) (Bonmati-Carrion et al. 2014) and (b) actigraph device located on the wrist with acceleration (Bonmati-Carrion et al. 2015).

The study demonstrated that both the arm and the wrist activity devices were suitable for circadian phase prediction and for evaluating the sleep–wake cycle as assessed by comparison with the DLMO and sleep logs, respectively. Interestingly, all correlations between rhythmic parameters obtained from both actimetry methods were significant, and particularly strong for all circadian phase markers (Bonmati-Carrion et al. 2014).

To assess the position and the arm actigraphy, an actimeter (HOBO, Pendant G Acceleration Data Logger UA-004-64, Bourne, MA) was placed on the nondominant arm by means of a sports band, with its x-axis parallel to the humerus bone as already described (Ortiz-Tudela et al. 2010). The device was programmed to sample once every 30 s. The information stored in the actimeter was transferred through an optical USB Base Station (MANBASE-U-4, HOBO, Bourne, MA) to a personal computer, using the software provided by the manufacturer (HOBOware 2.2). The actimeter provided information on two variables: Arm actigraphy (AA) and position (P). AA was expressed as the accumulative changes in three-axis tilt with respect to the previous point and expressed as degrees per minute, and body position was calculated as the angle between the x-axis of the actimeter and the horizontal plane. Thus, P oscillated between 0 for maximum horizontality and 90 for maximum verticality (see Ortiz-Tudela et al. 2010 for details) (Bonmati-Carrion et al. 2015).

Toward analyzing physical activity and sleep together with the measurements of circadian wrist temperature, the software package *Circadianware* is used. Circadianware is a tool for the analysis of circadian rhythm of temperature, activity, and position of subjects.

6.4.1 Integrative Measures

In order to increase the reliability of circadian monitoring, integrated variables obtained from processing individual variables have been recently proposed. For example, the TAP (Thermometry Actimetry body Position) algorithm, proposed by Ortiz-Tudela et al. (2010), is based on integrating, after normalization, the following variables: skin temperature (T), motor activity (A), and body position (P). The first of these variables, skin temperature, is under endogenous control, whereas motor activity is modified voluntarily but it is also under endogenous control. Finally, of the three variables used for the integration, body position is most closely dependent on voluntary control. TAP is modular, thus, it can be amplified by incorporating new variables that complement the information even further. TAP variable permits us not only to determine how the individual's circadian system functions, but also to infer the sleep–wake rhythm with a precision higher than 90% according to polysomnography recording. This technique constitutes the base of ambulatory circadian monitoring procedure.

TAP presented highest values for agreement rate (86%) defined as an index that measures the coincidence between sleep or wake episodes measured by polysomnography and sleep or wake episodes estimated by TAP, compared with wrist temperature (80%), activity (61%), and body position. Moreover, it has been demonstrated that TAP is able to assess more homogeneous heritability estimates of the circadian system in a genetic study performed in 53 pairs of female twins (28 monozygotic [MZ] and 25 dizygotic [DZ]) than individual variables alone, suggesting that TAP can be more reliable and less subject to environmental artifacts than wrist temperature or actimetry separately (Lopez-Minguez et al. 2016).

6.5 CIRCADIAN RHYTHM AND MEAL PATTERNS

Once circadian rhythm has been established, meal patterns need to be included, in order to assess whether their regularity and frequency fit in the circadian system.

In relation to the well-known glucostatic theory, discovered in the animal model (Mayer 1953, 1955), and applied in research in humans (Campfield et al. 1996; Melanson et al. 1999a, 1999b, 1999c), it was observed that transient, spontaneously resolving declines in blood glucose precede meal initiation. Glucose concentrations do not necessarily act in a simple depletion-repletion model; the patterns serve as signals for meal initiation (Campfield and Smith 1990; Melanson et al. 1999a, 1999b, 1999c). In time-blinded humans, it was shown that in ~88% of the transient declines, a meal request followed, and that after a meal glucose concentrations rose dramatically, followed by a dynamic decline (Melanson et al. 1999c). Therefore, glucose monitoring can be used as a biomarker for physiological meal initiation indicated by the transient decline, and for actual feeding, indicated by the rise and fall of blood glucose concentrations. In practice, noninvasive glucose monitoring is needed. Noninvasive glucose monitoring is conducted using a glucose sensor that does not lead to blood collection and which does not need piercing of the skin with a solid object (Tura et al. 2007). The GlucoWatch (Cygnus Inc, Redwood City, CA) fulfills these requirements. It is a commercial device that monitors glucose noninvasively and in real time. Cygnus received FDA clearance for the GlucoWatch in 2001, updated in 2002 (Howsmon and Bequette 2015). The measurement is based on iontophoresis, a process by which an electric current brings interstitial glucose to the surface of the skin and then measures the amount of glucose via an electrochemical sensor (Tamada et al. 1999; Potts et al. 2002). The device is mainly used in diabetic patients in order to indicate hypoglycemic and hyperglycemic conditions (Dos Santos et al. 2006). Another possibility is the use of continuous subcutaneous glucose monitors (CSGMs). CGMs use small enzymatic sensors inserted beneath the skin to measure interstitial glucose. Subsequently, an oxidation–reduction reaction gives a measurable current that is calibrated with a blood glucose measurement (Gandrud et al. 2004). The commercial CGM named MiniMed (Medtronic, Northridge, CA) was approved by the FDA in 1999. Further development then refined these devices.

6.6 APPLICATIONS

Applications of measuring timing of food intake in the context of chronobiology lay in the newly developing field of chrono-nutrition (Figures 6.4 through 6.6) (Tahara and Shibata 2013). An increasing amount of chronobiological studies have started to raise awareness concerning the pivotal role of the circadian system in the development of, for instance, obesity and type 2 diabetes. For instance, the group of Research in Nutrition of the University of Murcia (Spain) has characterized biorhythms in a variety of population groups including babies (Zornoza-Moreno et al. 2011), young people (Sarabia et al. 2008; Martinez-Nicolas et al. 2011), and menopausal women (Gomez-Santos et al. 2016), and identified relationships with cardiovascular risk factors (Ortiz-Tudela et al. 2010), metabolic syndrome and obesity (Corbalan-Tutau et al. 2011, 2015; Bandin et al. 2014).

As described in the introduction, the internal timekeeping mechanism rhythmically regulates metabolic and physiological processes in order to meet the fluctuating demands in energy use and supply throughout the 24-hour day. The bidirectional interaction between the circadian system and metabolism, triggering the food entrainable oscillator (FEO) explains how disruption of body clocks

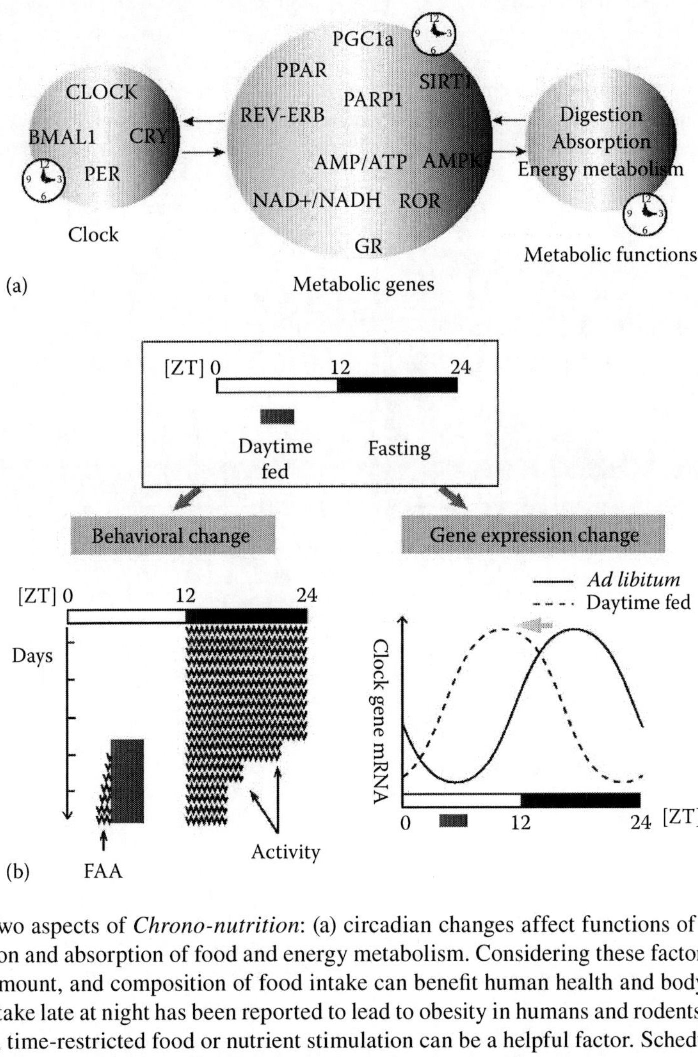

FIGURE 6.4 Two aspects of *Chrono-nutrition*: (a) circadian changes affect functions of food metabolism, including digestion and absorption of food and energy metabolism. Considering these factors when determining the timing, amount, and composition of food intake can benefit human health and body weight gain. For example, food intake late at night has been reported to lead to obesity in humans and rodents and (b) similar to light stimulation, time-restricted food or nutrient stimulation can be a helpful factor. Scheduled feedings during light periods in nocturnal mice cause food anticipatory activity (FAA) and phase-advance in clock gene expression rhythms. (From Tahara, Y. and S. Shibata, *Neuroscience*, 253, 77–88, 2013.)

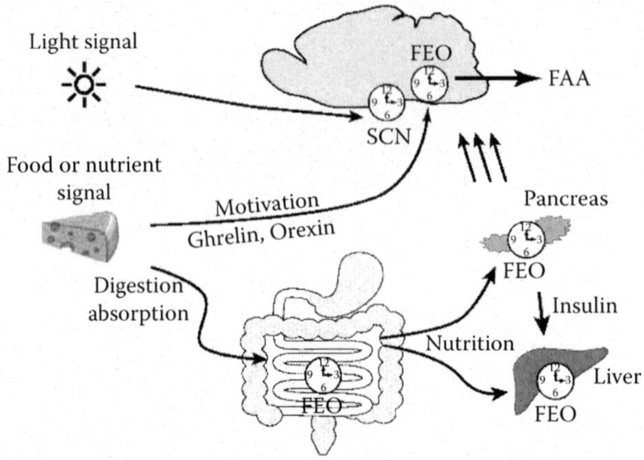

FIGURE 6.5 Food entrainable oscillator (FEO) in the brain and peripheral tissues. Light information goes to the SCN, the master oscillator in mammals, through the retina. Food information goes to the FEO, but the mechanism of FEO is not well understood. Food information goes to the hypothalamus, but not to the SCN, and then produces FAA before feeding time. Formation of FAA requires the circadian system and hormonal activity to induce hunger. Food information is directly relayed to the digestive and metabolic organs, and it alters the functions of food metabolism set to the feeding time. Nutrient signals (i.e., nutrient-induced hormonal secretions and activation of key proteins or nuclear receptors) are signaling pathways between the food and FEO in peripheral tissues. (From Tahara, Y. and S. Shibata, *Neuroscience*, 253, 77–88, 2013.)

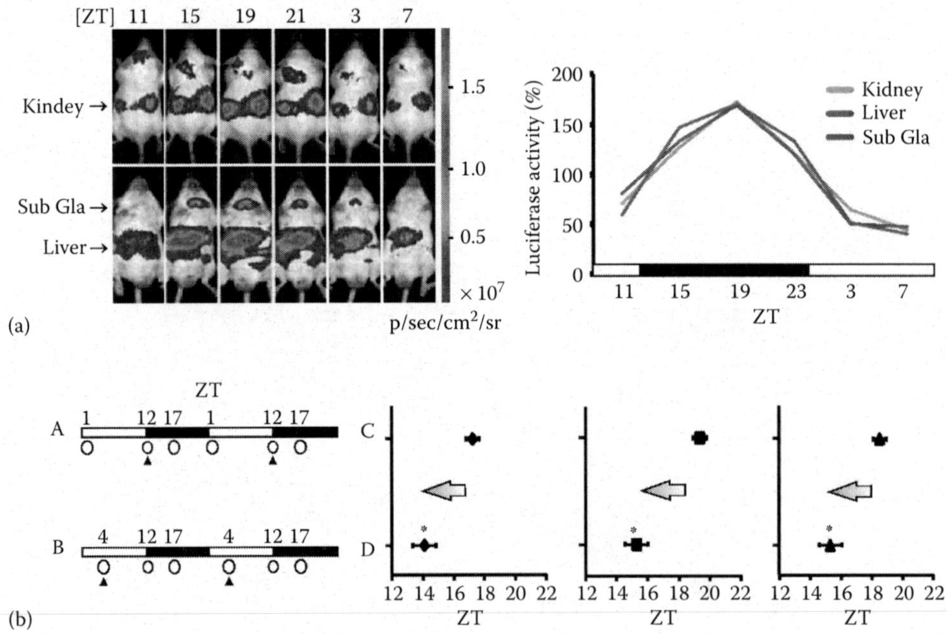

FIGURE 6.6 Food ingestion after long fasting periods increases entrainment during the phases of peripheral clocks. Phase differences during different feeding conditions by in vivo monitoring of peripheral PER2 LUCIFERASE bioluminescence rhythms: (a) representative imaging data of PER2 LUCIFERASE bioluminescence. Bioluminescence from the kidney, liver, and submandibular gland of PER2 LUCIFERASE mice was detected. Clear and stable oscillations of bioluminescence were detected in individual mice; (b) phase advances in peripheral clocks were induced by late night food intake (in vivo monitoring). The left panel (A) and (B) shows the experimental schedule of feeding times. The right panel (C) and (D) shows phases of the PER2 LUCIFERASE rhythm in each tissue and condition. $p < 0.05$ versus (a), by Student's t-test. *(Continued)*

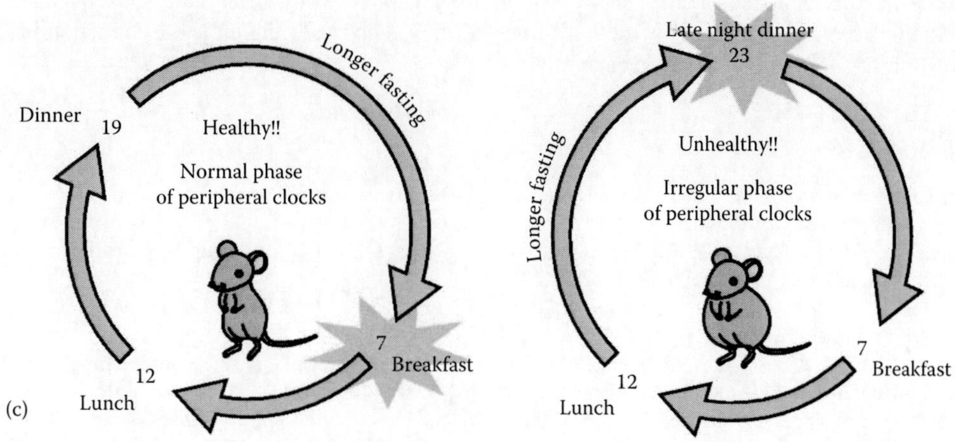

FIGURE 6.6 (Continued) Food ingestion after long fasting periods increases entrainment during the phases of peripheral clocks. Phase differences during different feeding conditions by in vivo monitoring of peripheral PER2 LUCIFERASE bioluminescence rhythms: (c) Late night food intake (e.g., at 23:00) had a greater impact on the changes in the peripheral clock phase, and promoted irregular feeding times. This could be because late night dinner comes after a longer fasting period compared to breakfast. (From Tahara, Y. and S. Shibata, *Neuroscience*, 253, 77–88, 2013.)

occur by means of shift work, frequent time zone traveling, nonstop consumption of calorie-dense foods, late dinners, and thus evokes detrimental metabolic alterations that contribute to obesity and T2D. Regulating the body's circadian rhythms by means of time-related dietary approaches (chrono-nutrition) may, therefore, represent a novel and interesting way to prevent or treat obesity and associated comorbidities (Tahara and Shibata 2013; Laermans and Depoortere 2016).

6.7 CONCLUSION

The actual approach of chrono-nutrition by determining meal patterns, physical activity patterns, sleep patterns, and circadian rhythm by continuous temperature measurement allows a complete picture of a lifestyle pattern. In the present context, meal patterns can be related to circadian rhythm, determined by continuous temperature measurement, and confirmed by the physical activity patterns and sleep patterns. The strength of the actual approach is the integrated way of determining lifestyle factors.

6.8 FUTURE RESEARCH

Future research is necessary with respect to the validations. Validations of the separate measurements differ from each other. The measurements of circadian rhythm and physical activity have been well validated. Although measurement of sleep using the Actigraph has been validated, more studies are needed for confirmation and building of experience. The use of integrative measures like TAP, allows for very accurate prediction of rest–activity periods, and contributes to the ambulatory evaluation of circadian system status in humans while providing a reliable evaluation of the sleep–wake rhythm validated with polysomnography (Ortiz-Tudela et al. 2014), but may need further development. In the present context, the validation of the continuous glucose monitoring system is a limitation. The noninvasive continuous glucose monitoring system has been validated for different purposes (Howsmon and Bequette 2015), whereas using transient and dynamic blood glucose declines as markers for meal initiation and meal termination mainly have been shown in the laboratory (Campfield et al. 1996; Melanson et al. 1999a, 1999b, 1999c). A sophisticated application of using this noninvasive continuous glucose monitoring system in the field requires future research.

Finally, although the integrative lifestyle approach for measurements, and subsequent recommendations seem promising, new methods may still be needed to disentangle the contributions from the different lifestyle factors from each other.

REFERENCES

Adamantidis, A. and L. de Lecea. 2008. Sleep and metabolism: Shared circuits, new connections. *Trends Endocrinol Metab* 19: 362–370.

Aizawa, S. and M. Cabanac. 2002. The influence of temporary semi-supine and supine postures on temperature regulation in humans. *J Therm Biol* 27: 109–114.

Arble, D. M., J. Bass, A. D. Laposky et al. 2009. Circadian timing of food intake contributes to weight gain. *Obesity* (*Silver Spring*) 17: 2100–2102.

Arnold, L. M., M. J. Ball, A. W. Duncan. 1993. Effect of isoenergetic intake of three or nine meals on plasma lipoproteins and glucose metabolism. *Am J Clin Nutr* 57: 446–451.

Bado, A., S. Levasseur, A. Attoub et al. 1998. The stomach is a source of leptin. *Nature* 394: 790–793.

Bandin, C., A. Martinez-Nicolas, J. M. Ordovas et al. 2014. Circadian rhythmicity as a predictor of weight-loss effectiveness. *Int J Obes* (*Lond*) 38 (8): 1083–1088.

Bandín, C., F. A. Scheer, A. J. Luque et al. 2015. Meal timing affects glucose tolerance, substrate oxidation and circadian-related variables: A randomized, crossover trial. *Int J Obes* (*Lond*) 39 (5): 828–833.

Bechtold, D. A. 2008. Energy-responsive timekeeping. *J Genet* 87: 447–458.

Bertelsen, J., C. Christiansen, C. Thomsen et al. 1993. Effect of meal frequency on blood glucose, insulin, and free fatty acids in NIDDM subjects. *Diabetes Care* 16: 4–7.

Bonmati-Carrion, M. A., B. Middleton, V. Revell et al. 2014. Circadian phase assessment by ambulatory monitoring in humans: Correlation with dim light melatonin onset. *Chronobiol Int* 31: 37–51.

Bonmati-Carrion, M. A., B. Middleton, V. Revell et al. 2015. Validation of an innovative method, based on tilt sensing, for the assessment of activity and body position. *Chronobiol Int* 32 (5): 701–710.

Cagampang, F. R. and K. D. Bruce. 2012. The role of the circadian clock system in nutrition and metabolism. *Br J Nutr* 108: 381–392.

Cagnacci, A. 2007. Influences of melatonin on human circadian rhythms. *Chronobiol Int* 14: 205–220.

Campfield, L. A. and F. J. Smith. 1990. Systemic factors in the control of food intake: Evidence for patterns as signals. In: Stricker, E. M. ed. *Handbook of Behavioral Neurobiology 10, Neurobiology of food and fluid intake*. New York: Plenum Press, pp. 183–206.

Campfield, L. A., F. J. Smith, M. Rosenbaum et al. 1996. Human eating: Evidence for a physiological basis using a modified paradigm. *Neurosci Biobehav Rev* 20: 133–137.

Cellini, N., M. P. Buman, E. A. McDevitt et al. 2013. Direct comparison of two actigraphy devices with polysomnographically-recorded naps in healthy young adults. *Chronobiol Int* 30: 691–698.

Cellini, N., E. A. McDevitt, S. Mednick et al. 2016. Free-living cross-comparison of two wearable monitors for sleep and physical activity in healthy young adults. *Physiol Behav* 157: 79–86.

Cellini, N., E. A. McDevitt, A. A. Ricker et al. 2015. Validation of an automated wireless system for sleep monitoring during daytime naps. *Behav Sleep Med* 13: 157–168.

Chapelot, D., C. Marmonier, R. Aubert et al. 2006. Consequence of omitting or adding a meal in man on body composition, food intake, and metabolism. *Obesity* 14: 215–227.

Cinti, S., R. de Matteis, E. Ceresi et al. 2001. Leptin in the human stomach. *Gut* 49: 155.

Cinti, S., R. D. Matteis, C. Pico et al. 2000. Secretory granules of endocrine and chief cells of human stomach mucosa contain leptin. *Int J Obes Relat Metab Disord* 24: 789–793.

Corbalán-Tutau, M. D., Gómez-Abellán, P., Madrid, J. A. et al. 2015. "https://www.ncbi.nlm.nih.gov/pubmed/24953771". Toward a chronobiological characterization of obesity and metabolic syndrome in clinical practice. *Clin Nutr.* 34(3): 477–83.

Corbalan-Tutau, M. D., J. A. Madrid, J. M. Ordovas et al. 2011. Differences in daily rhythms of wrist temperature between obese and normal-weight women: Associations with metabolic syndrome features. *Chronobiol Int* 28 (5): 425–433.

Cummings, D. E., J. Q. Purnell, R. S. Frayo et al. 2001. A preprandial rise in plasma ghrelin levels suggests a role in meal initiation in humans. *Diabetes* 50: 1714–1719.

Czeisler, C. A. and E. B. Klerman. 1999. Circadian and sleep-dependent regulation of hormone release in humans. *Recent Prog Horm Res* 54: 97–132.

DeSantis, A. S., A. V. DiezRoux, A. Hajat et al. 2012. Associations of salivary cortisol levels with inflammatory markers: The multi-ethnic study of atherosclerosis. *Psychoneuroendocrinology* 37: 1009–1018.

Dos Santos, M. L., F. F. Aragon, C. R. Padovani et al. 2006. Daytime variations in glucose tolerance in people with impaired glucose tolerance. *Diabetes Res Clin Pract* 74: 257–262.

Esquirol, Y., V. Bongard, L. Mabile et al. 2009. Shift work and metabolic syndrome: Respective impacts of job strain, physical activity, and dietary rhythms. *Chronobiol Int* 26: 544–559.

Farshchi, H. R., M. A. Taylor, I. A. Macdonald. 2004a. Regular meal frequency creates more appropriate insulin sensitivity and lipid profiles compared with irregular meal frequency in healthy lean women. *Eur J Clin Nutr* 58: 1071–1077.

Farshchi, H. R., M. A. Taylor, I. A. Macdonald. 2004b. Decreased thermic effect of food after an irregular compared with a regular meal pattern in healthy lean women. *Int J Obes Relat Metab Disord* 28: 653–660.

Farshchi, H. R., M. A. Taylor, I. A. Macdonald. 2005. Beneficial metabolic effects of regular meal frequency on dietary thermogenesis, insulin sensitivity, and fasting lipid profiles in healthy obese women. *Am J Clin Nutr* 81: 16–24.

Fletcher, E. C. 1997. Sympathetic activity and blood pressure in the sleep apnea syndrome. *Respiration* 64 Suppl 1: 22–28.

Froy, O. 2007. The relationship between nutrition and circadian rhythms in mammals. *Front Neuroendocrinol* 28: 61–71.

Garaulet, M., J. M. Ordovas, J. A. Madrid. 2010. The chronobiology, etiology and pathophysiology of obesity. *Int J Obes* 34: 1667–1683.

Garrow, J. S., M. Durrant, S. Blaza et al. 1981. The effect of meal frequency and protein concentration on the composition of the weight lost by obese subjects. *Br J Nutr* 45: 5–15.

Gandrud, L. M., H. U. Paguntalan, M. Van Wyhe et al. 2004. Use of the Cygnus GlucoWatch biographer at a diabetes camp. *Pediatrics* 113 (1): 108–111.

Gomez-Santos, C., C. B. Saura, J. A. Lucas et al. 2016. Menopause status is associated with circadian and sleep related alterations. *Menopause* 23: 682–690.

Gonnissen, H. K., T. Hulshof, M. S. Westerterp-Plantenga. 2013a. Chronobiology, endocrinology, and energy- and food-reward homeostasis. *Obes Rev* 14: 405–416.

Gonnissen, H. K., R. Hursel, F. Rutters et al. 2013b. Effects of sleep fragmentation on appetite and related hormone concentrations over 24h in healthy men. *Br J Nutr* 109: 748–756.

Gonnissen, H. K., C. Mazuy, F. Rutters et al. 2013c. Sleep architecture when sleeping at an unusual circadian time and associations with insulin sensitivity. *PLoS One* 8: e72877.

Gonnissen, H. K., F. Rutters, C. Mazuy et al. 2012. Effect of a phase advance and phase delay of the 24-h cycle on energy metabolism, appetite, and related hormones. *Am J Clin Nutr* 96: 689–697.

Hasselberg, M. J., J. McMahon, K. Parker. 2013. The validity, reliability, and utility of the iButton® for measurement of body temperature circadian rhythms in sleep/wake research. *Sleep Medicine* 14: 5–11.

Hoogerwerf, W. A. 2009. Role of biological rhythms in gastrointestinal health and disease. *Rev Endocr Metab Disord* 10: 293–300.

Howsmon, D. and B. W. Bequette. 2015. Hypo and hyperglycaemic alarms devices and algorithms. *J Diabetes Sci Technol* 9: 1126–1136.

Huang, W., K. M. Ramsey, B. Marcheva. 2011. Circadian rhythms, sleep, and metabolism. *J Clin Invest* 121: 2133–2141.

Hursel, R., F. Rutters, H. K. Gonnissen et al. 2011. Effects of sleep fragmentation in healthy men on energy expenditure, substrate oxidation, physical activity, and exhaustion measured over 48 h in a respiratory chamber. *Am J Clin Nutr* 94: 804–808.

Hyde, M., D. M. O'Driscoll, S. Binette et al. 2007. Validation of actigraphy for determining sleep and wake in children with sleep disordered breathing. *J Sleep Res* 16: 213–216.

Jenkins, D. J. and A. L. Jenkins. 1995. Nutrition principles and diabetes. A role for "lente carbohydrate?" *Diabetes Care* 18: 1491–1498.

Jenkins, D. J., T. M. Wolever, A. Ocana et al. 1990. Metabolic effects of reducing rate of glucose ingestion by single bolus versus continuous sipping. *Diabetes* 39: 775–781.

Jenkins, D. J., T. M. Wolever, V. Vuksan et al. 1989. Nibbling versus gorging: Metabolic advantages of increased meal frequency. *N Engl J Med* 321: 929–934.

Jones, P. J., C. A. Leitch, R. A. Pederson. 1993. Meal-frequency effects on plasma hormone concentrations and cholesterol synthesis in humans. *Am J Clin Nutr* 57: 868–874.

Kalsbeek, A. and J. A. Strubbe. 1999. Circadian control of insulin secretion is independent of the temporal distribution of feeding. *Physiol Behav* 63: 553–558.

Knutsson, A. and H. Boggild. 2010. Gastrointestinal disorders among shift workers. *Scand J Work Environ Health* 36: 85–95.

Kohsaka, A., A. D. Laposky, K. M. Ramsey et al. 2007. High-fat diet disrupts behavioral and molecular circadian rhythms in mice. *Cell Metab* 6: 414–421.

Koren, D., L. E. Levitt Katz, P. C. Brar et al. 2011. Sleep architecture and glucose and insulin homeostasis in obese adolescents. *Diabetes Care* 34: 2442–2447.

Knutson, K. L., E. Van Cauter, P. Zee et al. 2011. Cross-sectional associations between measures of sleep and markers of glucose metabolism among subjects with and without diabetes: the Coronary Artery Risk Development in Young Adults (CARDIA) Sleep Study. *Diabetes Care* 34: 1171–1176.

Kräuchi, K. 2007. The human sleep-wake cycle reconsidered from a thermoregulatory point of view. *Physiol Behav* 90: 236–245.

Laermans, J. and I. Depoortere. 2016. Chronobesity: Role of the circadian system in the obesity epidemic. *Obes Rev* 17: 108–125.

Laposky, A. D., J. Bass, A. Kohsaka et al. 2008. Sleep and circadian rhythms: Key components in the regulation of energy metabolism. *FEBS Lett* 582: 142–151.

Lecoultre, V., E. Ravussin, L. M. Redman. 2011. The fall in leptin concentration is a major determinant of the metabolic adaptation induced by caloric restriction independently of the changes in leptin circadian rhythms. *J Clin Endocrinol Metab* 96: E1512–E1516.

Lopez-Minguez, J., L. Colodro-Conde, C. Bandin et al. 2016. Application of multiparametric procedures for assessing the heritability of circadian health. *Chronobiol Int* 33 (2): 234–244.

Lopez-Minguez, J., J. R. Ordonana, J. F. Sanchez-Romera et al. 2015. Circadian system heritability as assessed by wrist temperature: A twin study. *Chronobiol Int* 32 (1): 71–80.

Martinez-Nicolas, A., E. Ortiz-Tudela, J. A. Madrid et al. 2011. Crosstalk between environmental light and internal time in humans. *Chronobiol Int* 28 (7): 617–629.

Martinez-Nicolas, A., E. Ortiz-Tudela, M. A. Rol et al. 2013. Uncovering different masking factors on wrist skin temperature rhythm in free-living subjects. *PLoS One* 8 (4): e61142.

Mayer, J. M. 1953. Glucostatic mechanisms in the regulation of food intake. *New Eng J Med* 249: 13–16.

Mayer, J. M. 1955. Regulation of energy intake and body weight, the glucostatic theory and the lipostatic hypothesis. *Ann N Y Acad Sci* 63: 15–43.

McGrath, S. A. and M. J. Gibney. 1994. The effects of altered frequency of eating on plasma lipids in free-living healthy males on normal self-selected diets. *Eur J Clin Nutr* 48: 402–407.

Melanson, K. J., M. S. Westerterp-Plantenga, L. A. Campfield et al. 1999a. Blood glucose and meal patterns in time-blinded males, after aspartame, carbohydrate, and fat consumption, in relation to sweetness perception. *Br J Nutr* 82: 437–446.

Melanson, K. J., M. S. Westerterp-Plantenga, L. A. Campfield et al. 1999b. Appetite and blood glucose profiles in humans after glycogen-depleting exercise. *J Appl Physiol* 87: 947–954.

Melanson, K. J., M. S. Westerterp-Plantenga, W. H. Saris et al. 1999c. Blood glucose patterns and appetite in time-blinded humans: carbohydrate versus fat. *Am J Physiol* 277: R337–R345.

Meltzer, L., H. Montgomery-Downs, S. Insana et al. 2012. Use of actigraphy for assessment in pediatric sleep research. *Sleep Med Rev* 16: 463–475.

Mendoza, J. 2007. Circadian clocks: Setting time by food. *J Neuroendocrinol* 19: 127–137.

Mendoza, J., P. Pevet, E. Challet. 2008. High-fat feeding alters the clock synchronization to light. *J Physiol* 586: 5901–5910.

Morgan, L., S. Hampton, M. Gibbs et al. 2003. Circadian aspects of postprandial metabolism. *Chronobiol Int* 20: 795–808.

Natale, V., D. Léger, M. Martoni et al. 2014. The role of actigraphy in the assessment of primary insomnia: A retrospective study. *Sleep Med* 15: 111–115.

Nedeltcheva, A. V., J. M. Kilkus, J. Imperial et al. 2009. Sleep curtailment is accompanied by increased intake of calories from snacks. *Am J Clin Nutr* 89: 126–133.

Ortiz-Tudela, E., A, Martinez-Nicolas, J. Albares et al. 2014. Ambulatory circadian monitoring (ACM) based on thermometry, motor activity and body position (TAP): A comparison with polysomnography. *Physiol Behav* 126: 30–38.

Ortiz-Tudela, E., A. Martinez-Nicolas, M. Campos et al. 2010. A new integrated variable based on thermometry, actimetry and body position (TAP) to evaluate circadian system status in humans. *PLoS Comput Biol* 6 (11): e1000996.

Potts, R. O., J. A. Tamada, M. J. Tierney. 2002. Glucose monitoring by reverse iontophoresis. *Diabetes Metab Res Rev* 18 (S1): S49–S53.

Refinetti, R. and M. Menaker. 1992. The circadian rhythm of body temperature. *Physiol Behav* 51: 613–617.

Ribeiro, D. C., S. M. Hampton, L. Morgan et al. 1998. Altered postprandial hormone and metabolic responses in a simulated shift work environment. *J Endocrinol* 158: 305–310.

Robusto, K. M. and S. G. Trost. 2012. Comparison of three generations of ActiGraph™ activity monitors in children and adolescents. *J Sports Sci* 30: 1429–1435.

Rowlands, A. and V. Stiles. 2012. Accelerometer counts and raw acceleration output in relation to mechanical loading. *J Biomech* 45: 448–454.

Rupp, T. L., and T. J. Balkin. 2011. Comparison of motionlogger watch and actiwatch actigraphs to polysomnography for sleep/wake estimation in healthy young adults. *Behav Res Methods* 43: 1152–1160.

Sadeh, A. 2008. Commentary: Comparing actigraphy and parental report as measures of children's sleep. *J Pediatr Psychol* 33: 406–407.

Sarabia, J. A., M. A. Rol, P. Mendiola et al. 2008. Circadian rhythm of wrist temperature in normal-living subjects A candidate of new index of the circadian system. *Physiol Behav* 95 (4): 570–580.

Scheer, F. A., M. F. Hilton, C. S. Mantzoros et al. 2009. Adverse metabolic and cardiovascular consequences of circadian misalignment. *Proc Natl Acad Sci USA* 106: 4453–4458.

Schibler, U., J. Ripperger, S. A. Brown. 2003. Peripheral circadian oscillators in mammals: Time and food. *J Biol Rhythms* 18: 250–260.

Schoeller, D. A., L. K. Cella, M. K. Sinha et al. 1997. Entrainment of the diurnal rhythm of plasma leptin to meal timing. *J Clin Invest* 100: 1882–1887.

Slater, J. A., T. Botsis, J. Walsh et al. 2015. Assessing sleep using hip and wrist actigraphy. *Sleep Biol Rhythms* 13: 172–180.

Smeets, A. J. and M. S. Westerterp-Plantenga. 2008. Acute effects on metabolism and appetite profile of one meal difference in the lower range of meal frequency. *Br J Nutr* 99: 1316–1321.

Szosland, D. 2010. Shift work and metabolic syndrome, diabetes mellitus and ischaemic heart disease. *Int J Occup Med Environ Health* 23: 287–291.

Tahara, Y. and S. Shibata. 2013. Chronobiology and nutrition. *Neuroscience* 253: 77–88.

Tamada, J. A., S. Garg, L. Jovanovic et al. 1999. Noninvasive glucose monitoring: Comprehensive clinical results. *JAMA* 282 (19): 1839–1844.

Thomsen, C., C. Christiansen, O. W. Rasmussen et al. 1997. Comparison of the effects of two weeks' intervention with different meal frequencies on glucose metabolism, insulin sensitivity and lipid levels in non-insulin-dependent diabetic patients. *Ann Nutr Metab* 41: 173–180.

Tonetti, L., N. Cellini, M. De Zambotti et al. 2013. Polysomnographic validation of a wireless dry headband technology for sleep monitoring. *Physiol Behav* 118: 185–188.

Tranel, H. R., E. A. Schroder, J. England et al. 2015. Physical activity, and not fat mass is a primary predictor of circadian parameters in young men. *Chronobiol Int* 32: 832–841.

Trost, S. G., K. L. McIver, R. R. Pate. 2005. Conducting accelerometer-based activity assessments in field-based research. *Med Sci Sports Exerc* 37: S531–S543.

Tura, A., A. Maran, G. Pacini. 2007. Non-invasive glucose monitoring: Assessment of technologies and devices according to quantitative criteria. *Diabetes Res Clin Pract* 77: 16–40.

Ullah, S., R. Arsalani-Zadeh, J. MacFie. 2012. Accuracy of prediction equations for calculating resting energy expenditure in morbidly obese patients. *Ann R Coll Surg Engl* 94 (2): 129–132. doi:10.1308/003588412x13171221501988.

van Aggel-Leijssen, D. P., M. A. van Baak, R. Tenenbaum et al. 1999. Regulation of average 24h human plasma leptin level; the influence of exercise and physiological changes in energy balance. *Int J Obes Relat Metab Disord* 23: 151–158.

Van Cauter, E., R. Leproult, L. Plat. 2000. Age-related changes in slow wave sleep and REM sleep and relationship with growth hormone and cortisol levels in healthy men. *JAMA* 284: 861–868.

van Marken Lichtenbelt, W. D., H. A. M. Daanen, L. Wouters et al. 2006. Evaluation of wireless determination of skin temperature using iButtons. *Physiol Behav* 88: 489–497.

Van Someren, E. J., D. F. Swaab, C. C. Colenda et al. 1999. Bright light therapy: Improved sensitivity to its effects on rest-activity rhythms in Alzheimer patients by application of nonparametric methods. *Chronobiol Int* 18: 809–822.

Verboeket-van de Venne, W. P. and K. R. Westerterp. 1991. Influence of the feeding frequency on nutrient utilization in man: consequences for energy metabolism. *Eur J Clin Nutr* 45: 161–169.

Verboeket-van de Venne, W. P., K. R. Westerterp, A. D. Kester. 1993. Effect of the pattern of food intake on human energy metabolism. *Br J Nutr* 70: 103–115.

Weinert, D. and J. Waterhouse. 2007. The circadian rhythm of core temperature: Effects of physical activity and aging. *Physiol Behav* 90: 246–256.

Westerterp-Plantenga, M. S. 2016. Sleep, circadian rhythm and body weightL: Parallel developments. *Proc Nutr Soc* 75: 431–439.

Westerterp-Plantenga, M. S., A. H. Goris, E. P. Meijer et al. 2003. Habitual meal frequency in relation to resting and activity-induced energy expenditure in human subjects: The role of fat-free mass. *Br J Nutr* 90: 643–649.

Westerterp-Plantenga, M. S., E. M. Kovacs, K. J. Melanson. 2002. Habitual meal frequency and energy intake regulation in partially temporally isolated men. *Int J Obes Relat Metab Disord* 26: 102–110.

Westerterp-Plantenga, M. S., A. Nieuwenhuizen, D. Tome et al. 2009. Dietary protein, weight loss, and weight maintenance. *Annu Rev Nutr* 29: 21–41.

Wolever, T. M. 1990. Metabolic effects of continuous feeding. *Metabolism* 39: 947–951.

Wolk, R. and V. K. Somers. 2007. Sleep and the metabolic syndrome. *Exp Physiol* 92: 67–78.

Wong, M. L., J. Licinio, B. O. Yildiz et al. 2004. Simultaneous and continuous 24-hour plasma and cerebrospinal fluid leptin measurements: dissociation of concentrations in central and peripheral compartments. *J Clin Endocrinol Metab* 89: 258–265.

Wu, H., W. S. Stone, X. His et al. 2011. Effects of different sleep restriction protocols on sleep architecture and daytime vigilance in healthy men. *Physiol Res* 59: 821–829.

Yannielli, P. C., P. C. Molyneux, M. E. Harrington et al. 2007. Ghrelin effects on the circadian system of mice. *J Neurosci* 27: 2890–2895.

Yi, C. X., E. Challet, P. Pevet et al. 2008. A circulating ghrelin mimetic attenuates light-induced phase delay of mice and light-induced Fos expression in the suprachiasmatic nucleus of rats. *Eur J Neurosci* 27: 1965–1972.

Yki-Jarvinen, H., C. Bogardus, B. V. Howard. 1987. Hyperglycemia stimulates carbohydrate oxidation in humans. *Am J Physiol* 253: E376–E382.

Zinkhan, M., K. Berger, S. Hense et al. 2014. Agreement of different methods for assessing sleep characteristics: A comparison of two actigraphs, wrist and hip placement, and self-report with polysomnography. *Sleep Med* 15:1107–1114.

Zornoza-Moreno, M., S. Fuentes-Hernandez, M. Sanchez-Solis et al. 2011. Assessment of circadian rhythms of both skin temperature and motor activity in infants during the first 6 months of life. *Chronobiol Int* 28 (4): 330–337.

7 Assessment of Ingestion by Chewing and Swallowing Sensors

Edward Sazonov, Muhammad Farooq, and Edward Melanson

CONTENTS

7.1 INTRODUCTION

Bites, chews, and swallows are naturally associated with the ingestion of food and liquid. On account of this, chewing and swallowing sensors provide an objective method to assess food intake. The physiological fundamentals of the ingestion process differ with age and social customs, but there are similarities in the function of the various organs, muscle groups, and the neural processes by which the ingestion is coordinated. A review of the factors that are similar across ages and social customs indicates several that may be markers of chewing and swallowing, and thus potential candidates for automatic monitoring by sensors. The evolution of specific sensor solutions for the detection of food intake are presented in brief along with the description of various sensors that utilize chewing sound, jaw motion, or electromyography for detection of chewing and food intake. The chewing sensors have been monitored using a variety of sensor modalities for the detection of swallowing, including acoustical, electromechanical, magnetic, bioelectric, and other sensors. There has been a surge of research on the use of chewing and swallowing data that indicates that they can be used to gain a deeper insight into the characteristics of ingested food (mass, food type, and caloric content).

7.2 BACKGROUND: PHYSIOLOGICAL FOUNDATIONS OF THE INGESTION PROCESS

The ingestion process varies with age and with societal norms, but shares commonality in the physiological and neurological foundations that can be utilized for detection and characterization of ingestion.

At birth, humans rely on the intake of liquids (milk or formula) to provide necessary nutrients. The ingestion process in infants relies on the tight coordination of sucking, swallowing, and breathing during the feeding cycle (Grassi et al. 2005). The coordination skills start to develop during the gestation period and mature after birth (Lau and Kusnierczyk 2001). Swallowing in fetuses is observable as early as 11 weeks of gestation and progressively develops into a coordinated and rhythmic activity (Grassi et al. 2005). The sucking starts to develop at 18–24 weeks of gestational age and matures at 26–29 weeks (Lau and Kusnierczyk 2001; da Costa et al. 2008). The majority of term infants are capable of oral feeding at birth, which is accomplished through nutritive sucking (as opposed to nonnutritive sucking, is also prevalent in infants). In the process of nutritive sucking, the jaw and the tongue depress while maintaining the seal around the nipple, creating negative intraoral pressure and drawing liquid into the mouth. Approximately 100 ms later, the tongue compresses the nipple to express the liquid and to propel it toward the pharynx for swallowing (da Costa et al. 2008). The pharyngeal and esophageal phases of swallowing are mediated by an involuntary swallowing reflex that relies on the signals from pressure, stretch, temperature, and chemoreceptors in the pharynx and the esophagus. The act of swallowing overrides the respiratory control mechanism to prevent aspiration of the fluid into the airways, creating a swallowing apnea. This seemingly simple process involves coordination among more than 30 nerves and muscles (Matsuo and Palmer 2008), including activity in trigeminal, facial, glossopharyngeal, vagus, hypoglossal, and the cervical and thoracic spinal cord segments (Barlow 2009). The rhythmic activity and coordination of the nutritive sucking, swallowing, and respiration are controlled by central pattern generator networks, which are adaptive networks of interneurons that modulate the duration and intensity of motor neuron bursts to generate task-specific motor patterns (Guandalini et al. 2015). The central pattern generators respond both to the central nervous system and sensory inputs, accommodating for volitional- and reflexive-control mechanisms. The coordinated patterns of sucks, swallows, and breathing during nutritive sucking may potentially be captured by sensors and used to detect and characterize feeding in infants. Similarly, the patterns of intraoral pressure and force profile of the tongue motion may be used to estimate the efficiency and maturity of nutritive sucking. Limited variability of the food type (milk or formula) and a consistently uniform feeding mechanism potentially simplify the task of detecting and assessing the food intake in infants. However, the prevalence of nonnutritive sucking potentially complicates the detection of ingestion based on the sucking alone.

As the infant matures, sucking and exclusively liquid feeding are replaced by mastication and increasing consumption of solid foods. The transition from sucking to chewing occurs gradually, over several months after birth. A key difference is that the mastication process is primarily driven by the jaw elevator muscles, whereas the sucking is driven by the jaw depressors. At this time, it is not clear if the process of mastication develops through the evolution of the central pattern generator for sucking or the emergence of a dedicated central pattern generator (da Costa et al. 2008). The mastication continues to evolve of the lifetime following the changes in dentition. Another distinct feature in transitioning from infant feeding is development of fine motor skills that result in emergence of self-feeding behaviors (Carruth and Skinner 2002), including using hands and silverware to bring the food into the mouth. The hand-to-mouth gestures of the self-feeding behaviors also make an attractive target for automatic detection of food intake and are described in Chapter 8 of this book.

In children, adolescents, and adults the process of normal eating is described by the four-stage model for drinking and swallowing liquid and the process model for eating and swallowing solid food (Matsuo and Palmer 2008). These models describe the biomechanics and movement of the food bolus during food intake. Similar to infants, adult ingestion requires tight coordination of the processes of mastication (chewing), swallowing, and breathing, which make them possible targets for objective monitoring of ingestive behavior. During chewing, food particles are reduced in size, and saliva is produced to moisten and lubricate the food for swallowing (Engelen et al. 2005). Swallowing is a physiological process of transporting solids and liquids from mouth to stomach (Engelen et al. 2005). After food is chewed, tongue movements initiate a swallowing reflex that activates the throat muscles and push the chewed food bolus through the throat and into the esophagus (Amft and Troster 2009). Deglutition is under central control to prevent the aspiration of ingested materials (Shaker et al. 2013). This *cortical swallowing network* encompasses the cerebral cortex, brain stem, medulla, motor neurons, and the swallowing musculature. The brain stem believed to be the central regulator of deglutitive neuromuscular activity, which involves a considerable number of muscle groups that act in a definite sequential pattern (Lear et al. 1965). However, the timing and relative contribution of muscles involved in swallowing process are not completely understood (Imtiaz et al. 2014). The cortical swallowing network receives sensory input from pharynx, larynx, and esophagus to modulate swallowing function (Shaker et al. 2013).

Deglutition occurs in three phases (Imtiaz et al. 2014). During the oral phase, chewed food is transported toward the pharynx. This involves a complex set of motor movements to coordinate the tongue, mandible, and hyoid bone. During the pharyngeal phase, the chewed bolus is transported through the pharynx and into the esophagus. Normal respiration stops during this phase by activating muscles that prevent aspiration into lungs and nasal passages (McFarland and Lund 1995). Finally, during the esophageal phase, the closing of upper esophageal sphincter prevents the bolus from going back into the pharynx, and at the same time, the opening of the lower esophageal sphincter to allow bolus to enter the stomach. During this final phase of deglutition, peristaltic contractions are initiated in the pharynx and transmitted to the esophagus, and the strength of contractions related to the volume of the bolus (Shaker et al. 2013).

As chewing (sucking in infants) and swallowing are tightly associated with the process of food ingestion, they make good targets as a foundation for detection and characterization of food intake. Following a hand-to-mouth gesture, chewing is the first step in ingestion of the solid foods. The association between chewing activity and ingestive behavior is complex, and is influenced by bite size, and the qualities of the consumed food (e.g., hardness and moisture content) (Hiiemae et al. 1996). The frequency of swallowing increases significantly during intake of solid and liquid foods, making the deglutition a good candidate for monitoring ingestive behavior. For example, Lear et al. (1965) reported that the mean number of swallows during consumption of a meal (180/hour) was substantially higher than that observed during a period of inactivity (reading, 23.5/hour) and during sleep (5.3/hour). They also reported a large interindividual variance in swallows during consumption of a fixed meal during 10 minutes (80–510 swallows). Initiation of swallowing is dependent on the volume of the bite and the characteristics of the food (e.g., water content and hardness). Dry and hard foods (e.g., carrot) require more chewing cycles before swallowing. Adding butter to foods reduces the number of chews before swallowing (Engelen et al. 2005). Thus, it is possible that monitoring both chewing and swallowing could also provide some information about the types of food consumed, as no single device can capture all dimensions of eating behavior (Amft and Troster 2009).

Chewing and swallowing can be measured by a variety of different sensors that rely on capturing various manifestations of the ingestion process such as sounds, muscle activity, motion, and others. To be usable in free living, most of the sensor systems reported in the literature are implemented as wearable devices. Due to unrestricted conditions of wear, the sensor information is often contaminated by the artifacts resulting from other daily activities. For example, chewing and talking both require motion of the jaw and may look similar on the sensor signals. The methods of signal processing and machine learning or pattern recognition are frequently employed to differentiate the activities, and to recognize chewing and swallowing on the background of all of the activities possible in daily life. Quite often, information from several sensors has to be combined through the sensor fusion to reliably assess food intake. The following sections review various sensor modalities for monitoring of chewing and swallowing reported in the recent literature, with the focus on the methods that demonstrate the best performance in the conditions of free living.

7.3 DETECTION OF CHEWING

The first and most fundamental task that is accomplished by the chewing sensors is measurement of the physical and/or physiological phenomena that accompany the chewing process. Several sensor modalities have been proposed for this task, including microphones to capture chewing sounds, electromyographic sensors to measure muscle activation during chewing, strain, and acceleration sensors to capture the chewing motion, and sensors that measure the deformation of the ear canal. The captured sensor signals can then be used to recognize the chewing episodes and characterize the metrics such as the number of chews, chewing rate, and others. This section introduces the various sensor modalities for chewing detection.

7.3.1 Chewing Sounds

The process of chewing mechanically reduces the size of food particles, generating the sound of food being crushed (Drake 1965; Lee Hi et al. 1988). Historically, the analysis of the chewing sound originated as a study of sensory perception in the food science and evolved into a method of ingestion monitoring. The chewing sound propagates well through the hard tissues of the teeth and the bones and may be detected at various locations, such as the ear canal or bones of the skull. The intensity of the chewing sound depends on the rheological properties (texture and composition) of the food, for example, dry crispy foods generate more intense sounds than soft moist foods. The initial cutting of the ingested food is followed by the gradual crushing of the food structure that results in the reduction of the intensity and changes in the spectral content of the chewing sound. The sensor approaches that utilize the chewing sound aim to capture and recognize the chewing sound using one or several microphones that may be placed at various locations on the head or neck. Several of such sensor systems also explored the possibility to recognize the food type through the analysis of the spectral content of the chewing sound.

One of the earliest works used a conduction microphone placed in the outer ear canal to capture sounds produced during chewing food and used those signals to classify four different food types using decision trees (Amft et al. 2005). In an extended work by the same authors, the conduction microphone was incorporated into an ear pad (Amft 2010). Spectral analysis was used to differentiate 19 food types. Researchers in Nishimura and Kuroda (2008) presented the prototype of a Bluetooth in-ear conduction microphone to capture chewing sounds. In the proposed method a two-stage recognition algorithm was used for chewing recognition; the first stage used log energy of 20 ms frames of the signal combined with a threshold, to separate chew-like signals. The second stage was the chewing sound verification stage where Linde–Buzo–Gray (LBG) codebook (Kinnunen et al. 2006) training algorithm with marginalized corrupted features (MCF) was used to classify the chewing sounds. One potential limitation of the use of microphones is that these sensors respond both to the chewing sounds and to environmental noise. A potential solution is the use of two microphones of which one is used for capturing environment noise and the second one is used for monitoring of chewing. The two microphone system then can utilize noise cancellation techniques to reduce the interference in the swallowing sound. Researchers in Passler et al. (2012); Päßler et al. (2012); Päßler and Fischer (2014) used a two-microphone system with adaptive sound modeling of chewing using the maximum posterior estimation algorithm. The study of Päßler et al. (2012) reported the accuracy of 83% in food intake detection and 79% in food classification of seven types of food. A substantially larger study by the same authors (Päßler and Fischer 2014) reported 80% accuracy of food intake detection on the data set of 68094 chewing events. The study of Masaki Shuzo (2010) proposed a portable and wearable sensor system using a conduction microphone placed in the user's ear for capturing internal body sounds (chewing sounds during food intake) and a condenser microphone for monitoring of surrounding sounds. A portable recorder was used to sample both sensors at 48 kHz. A KNN (k-Nearest Neighbor) classifier was used to recognize activities such as eating, drinking, and speech.

The reported accuracy of detecting food intake through recognition of the chewing sound is typically in the range of 80%–90% (Amft et al. 2005; Päßler and Fischer 2014). The interpretation of the accuracy numbers should pay close attention to the conditions of the experiment. All the reported works on the chewing sound rely on the experiments conducted in the controlled laboratory conditions, where the background noise may be artificially eliminated, participants may not be allowed to speak, and so on. Therefore, the sound recognition results may not represent the true performance expected in the conditions of free living, where the sensor may be exposed to a variety of sources of interference and generate false positive detections. Similarly, the classification of various food types (e.g., Amft et al. 2005; Nishimura and Kuroda 2008; Bi et al. 2015) cannot be directly extended to the wide variety of foods observed in daily life. Chewing sound primarily depends on the physical properties of food that have a limited number of dimensions, and therefore are not capable of differentiating the huge variability of the foods. A more realistic approach (Masaki Shuzo 2010) differentiates the food groups based on their physical properties, rather than the food type.

7.3.2 Electromyography

During chewing, the mandible movement is controlled by the temporalis, masseter, and pterygoid muscles. The force needed for mastication, and correspondingly, the strength of the muscle activity are dependent on the physical properties of the food and decreases as the food is broken down. For example, greater muscle activity is observed with increased hardness of the food. The strength of the muscle activation can be measured by observing the muscle contraction activity through electromyography (EMG) that measures the electrical activity of the motor neurons. Chewing can be monitored by measuring masseter and temporalis muscle activation (Mioche et al. 1999). These muscles are superficial and easy to measure (Amft and Troster 2009). Earlier studies used subcutaneous placement of electrodes to avoid interference from muscles in the neck, but more recent studies have utilized surface EMG (sEMG) that can be measured with small wireless sensors that are relatively noninvasive (Imtiaz et al. 2014).

The muscle activity during chewing is dependent on the texture of the food, and EMG has been used for studying the effect of food texture on muscle activity (Effect of Food Texture on EMG Profiles Detected from Masticatory Muscle 2016). The maximum bite force has been used as a determinant of the chewing and has been shown that individuals use a certain percentage of the bite force depending on the characteristics of the food (van den Braber et al. 2004). A number of studies have suggested the use of sEMG for measuring the bite force (Ottenhoff et al. 1996). Others have suggested that there is a relationship between the mastication performance and EMG activity of jaw-closing muscles (Tate et al. 1994; Wilding and Shaikh 1997). Some other studies have suggested that there is a potential relationship between the median particle size of the bolus of food and the mean voltage of sEMG measured at the masseter and anterior temporalis muscle (Tate et al. 1994; Wilding and Shaikh 1997). Mastication patterns change throughout the chewing sequences as the texture of the food changes. EMGs of the masseter and temporalis muscles have been used to study the relationship between the mastication and the hardness of the food (Foster et al. 2006).

To date, EMG sensors have been primarily used to estimate parameters of mastication rather than to detect and characterize food intake. A potential reason is that EMG requires good contact between the skin and the electrode that is typically achieved through the use of wet adhesive AgCl electrodes that may be bulky and not convenient to wear in the facial area. With the development of dry electrode technology, future may see EMG devices applied for food intake detection.

7.3.3 Jaw Motion

During food intake, mastication produces rhythmic up-and-down and side-to-side jaw movements in the frequency range of 1.25–2.5 Hz. The mechanical motion of the jaw during chewing can be measured at several locations. First, the motion of the mandible relative to the temporal bones can be captured immediately below the pinna of the outer ear (Sazonov et al. 2008; Fontana et al. 2011). Second, the deformation in the skin curvature created by the activation of the jaw muscles can be measured over the surface of the muscle. The sensors measuring the motion of the jaw are not sensitive to the environmental acoustic noise and can potentially provide more accurate and simpler solution to detection of chewing compared to monitoring of chewing sounds. However, jaw motion can also be observed during talking, yawning, and other activities, thus requiring a method to differentiate chewing from other activities.

Use of jaw motion for detection of food intake was reported in Sazonov and Fontana 2012; Fontana et al. (2014). The laboratory study of Sazonov and Fontana (2012) used a piezoelectric strain sensor placed below the outer ear to differentiate between episodes of chewing and non-chewing with an accuracy of 81% with a time resolution of 30 seconds. This work was extended to monitor food intake in free-living conditions for 24 hours on a population of 12 subjects (Fontana et al. 2014). The participants in the study wore a wearable device (automatic ingestion monitor) while performing usual daily activities, including sleep. Data were collected from three different sensors; a hand gesture sensor to detect hand-to-mouth gestures related to bites; piezoelectric film sensor for chewing detection (placed on the jaw below the ear as shown in Figure 7.1), and an accelerometer to monitor physical activity. Subject independent artificial neural network (ANN) models were able to differentiate between eating and other activities such as walking, talking, inactivity, and other daily activities with an average accuracy of 89%. The proposed system has been used in other studies to evaluate the feasibility of the system in multiday experiments as well as for selection of optimal feature set (Sazonov and Fontana 2012; Farooq et al. 2013; Fontana et al. 2013). Piezoelectric strain sensors are inexpensive and easy to

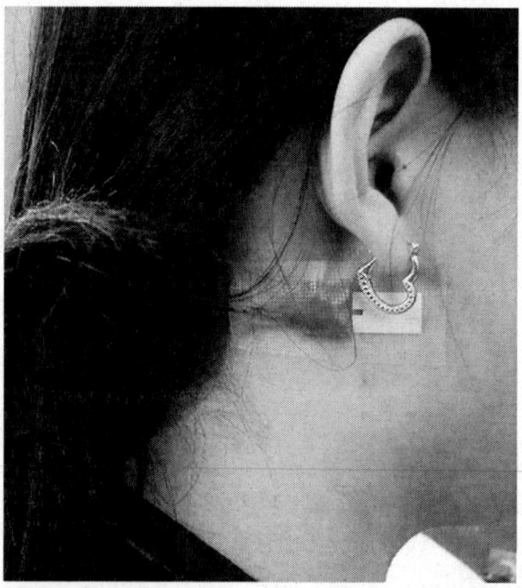

FIGURE 7.1 (See color insert.) A piezoelectric strain sensor attached to the skin below the ear to capture jaw motion associated with chewing.

use; however, they need to be attached using medical adhesive or medical tape and require exact positioning over the measurement location. Although not critical, these factors should be taken into consideration when using the piezoelectric sensor.

Jaw motion during chewing can also be monitored by placing accelerometer on the temporalis muscle or accelerometer attached to the temple of the glasses. For example, a single-axis accelerometer placed on the temporalis muscle was proposed in Wang et al. (2015) to monitor chewing during eating episodes in a laboratory experiment and was shown to be able to detect chewing with an accuracy of 96% in a small-scale laboratory study. Other researchers have proposed using eyeglasses instrumented with accelerometers for monitoring of temporalis muscle movements during chewing (Rahman et al. 2015; Ye et al. 2015).

Jaw motion during infant feeding is very similar to that during mastication in adults and was recently used to estimate sucking counts and sucking rate (Farooq et al. 2015a, 2015b). Although most of the modern sensor systems rely on video observation or measurement of intraoral pressure to measure the feeding behavior (da Costa et al. 2008; Taki et al. 2010; Taffoni et al. 2013; Chen et al. 2015), instrumentation of the infant's jaw is less intrusive and allows for monitoring of breast-fed and bottle-fed infants (Figures 7.2 and 7.3). In the study of Farooq et al. (2015b) the jaw motion sensor achieved a mean absolute error rate of about 7% compared to human annotated sucking count with an average intraclass correlation of 0.92 between the sensor and human raters.

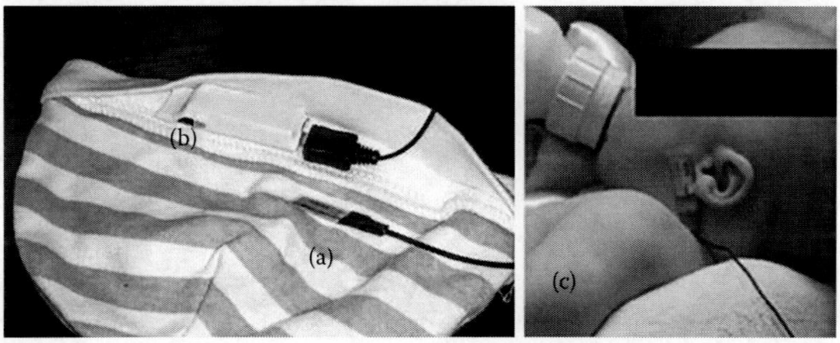

FIGURE 7.2 (See color insert.) Jaw motion sensor in the infant's hat: (a) jaw motion sensor, (b) wireless data acquisition module, and (c) a bottle-fed infant wearing the sensor.

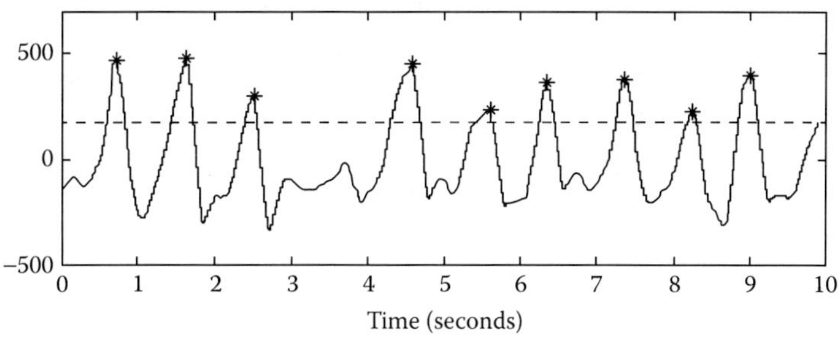

FIGURE 7.3 Jaw motion sensor signal. Detected sucks are represented by *. Horizontal dashed line shows the detection threshold. Amplitudes are in the ADC units.

7.3.4 Other Chewing Sensors

Chewing activity causes deformation in the ear canal wall that can be used as an indicator for monitoring of chewing and detection of food intake. In Bedri et al. (2015a), a wearable device in the form of an earpiece was proposed. The device utilized three infrared proximity sensors placed orthogonally with respect to each other. The proximity sensors measured the magnitude of deformations in the ear canal, and differentiated speech and mastication activities with an accuracy of 95.3% using subject-dependent HMM models. These experiments were performed in laboratory settings. In an extended work (Bedri et al. 2015b) by the same authors, subjects used the system in free-living conditions with six subjects for six hours performing normal daily activities. The device classified five-minute segments of time as eating or noneating with 93% accuracy using subject-dependent models.

Chewing can also be detected by intraoral devices that may rely on strain (Stellar and Shrager 1985) or inertial sensors (Li et al. 2013). One of the earliest devices reported that intraoral devices (Stellar and Shrager 1985) used a wired strain sensor embedded into a dental arch fastened to the back molars. The sensor demonstrated sensitivity both to chews and swallows, with the chewing resulting in *smaller and quicker* deformations of the strain gauge. The sensor has also shown sensitivity to talking, coughing, and other oral activities. No attempt was made to differentiate the activities registered by the sensor through automatic recognition. Recently, use of a wired triaxial accelerometer embedded in an artificial tooth was reported by Li et al. (2013). The data from the sensor were used to differentiate oral activities such as chewing, talking, drinking, and coughing. The accuracy of activity recognition was reported as 93.8% for a person-dependent classification and 59.8% for person-independent classification. Intraoral sensors present a unique opportunity because they come into the contact with the ingested food, and thus, may support direct sensing of the food being consumed, potentially allowing to measure the nutritional content. However, the challenges facing this approach are significant due to the limited space available inside of the oral cavity, need for custom fitting and/or installation of dental sensors and difficulty in wireless communications from the inside of the oral cavity. Future technology may resolve these issues and stimulate the development of intraoral sensors and devices.

Chewing was also explored in the framework of the sensors distributed in the environment, such as that of surveillance cameras used to detect faces and chewing activity (Schmalz et al. 2009). The proposal of a multistage vision recognition algorithm first detected the mouth region, which was followed by the computation of spatiotemporal frequency spectrum of the perioral region. The spectral data were then analyzed to detect chewing. Algorithm was tested on a group of seven individuals who performed activities like chewing with mouth open, chewing with mouth closed, talking, static-face presentation (control case), and moving-face presentation. A similar system proposed in Cadavid et al. (2011) used active appearance model to track subject's face across the video sequence, and the variations observed in the active appearance model parameters were used to differentiate between chewing and other facial movements such as speaking. For the data set of 37 subjects, the proposed system was able to get an agreement of 93% for automated chewing detection. The use of computer vision techniques combined with surveillance videos presents a system that poses the least amount of burden to the user. Users are not required to wear any sensors and are not required to record their eating patterns. The main limitation of such a system is that users are restricted to specially equipped spaces where they need to be constantly in the view of the camera. Such a system will fail if the user is out of the view of the camera. Another limitation is that these systems cannot be used in free-living conditions. Also, computer vision and image processing algorithms are highly sensitive to the quality of images captured, for example, such a system may fail in low lighting conditions.

7.4 DETECTION OF SWALLOWING

During food intake, a chewing sequence is usually followed by one or more swallows. The swallowing process involves the passage of a bolus of food or liquid from the mouth to the stomach and involves contraction and relaxation of muscles of the tongue (oral preparation), pharynx (the pharyngeal), and esophagus (esophageal phase) (Dodds 1989). A number of sensor modalities such as microphones to capture swallowing sounds, or surface electromyography to capture variations in the muscle activation patterns have been proposed for detecting swallowing. Swallowing is probably one of the most reliable indicators of food intake as all foods, even those that do not have to be chewed, need to be swallowed. However, swallowing is not always associated with food intake. Spontaneous saliva swallowing occurs approximately twice per minute and does not carry nutrition. The nutritive swallowing can be differentiated from spontaneous swallowing based on the measure of swallowing frequency (Lear et al. 1965; Sazonov et al. 2009; Lopez-Meyer et al. 2010).

7.4.1 Swallowing Sounds

The pharynx consists of a number of pumps and valves where reverberations are produced within the pharynx when a bolus of food passes through the pharynx and, hence, producing swallowing sound. Swallowing sounds have a distinct frequency spectrum and can potentially be distinguished from other noises such as speech, breathing, coughing, and other environmental noises (Matsuo and Palmer 2009). Sound-recognition approaches can be used to detect the passing of a bolus of food (both solid and liquid) through the pharyngeal phase.

A number of different locations can be used to capture the swallowing sound such as a miniature microphone placed in the ear, on the mastoid bone, or over the laryngopharynx. The earliest works used throat microphones placed over the laryngeal area to detect the swallowing events (Sazonov et al. 2008, 2009, 2010). The typical steps of automatic swallowing sound recognition included time-frequency decomposition of the microphone signal (e.g., short-time Fourier transform or wavelet decomposition [Makeyev et al. 2007, 2008]) followed by a classification step. The study of Sazonov et al. (2010) used a laboratory meal consumed under observation to train and validate a method for swallowing sound recognition. The data set contained more than 64 hours of annotated food intake including 9,996 swallows. The reported average accuracy of detecting swallowing events was 85%. The same study was later extended into a method for detection of food intake based on the recognized swallowing instances (Makeyev et al. 2012). The initial step of recognizing the swallowing instances was followed by the algorithm to detect food intake from the observed increase in the swallowing frequency. The reported average accuracies of >80% and >75% were obtained for intrasubject and intersubject models, correspondingly with a temporal resolution of the 30 s. The study of Lopez-Meyer et al. (2010) developed and compared several methods to classify each swallow either as nutritive or spontaneous. The classification was performed based on the swallowing frequency. The best subject-independent model was able to recognize food intake correctly with the average accuracy of 94% (range 78%–98% depending on the food type).

Some of the studies utilized multiple microphones. The study of Päßler et al. (2012) used an in-ear microphone that captured the swallowing and chewing sounds in combination with a reference microphone that captured environmental noise and applied noise cancellation methods to improve the detection of the swallowing (and chewing) sound in the noisy environment. The method was able to differentiate between seven different solid foods and two different beverages using a finite-state grammar decoder based on the Viterbi algorithm. Reported accuracy of food intake detection was 83% in the laboratory conditions. In a similar study (Bi et al. 2015), subjects wore a high-fidelity microphone around the neck. Hidden Markov models were used to detect chewing and swallowing events, which are then processed to extract time, frequency domain, and nonlinear features. The study reported the accuracy of 86.6% in food intake detection in a quiet laboratory setting.

In general, the methods relying on the acoustical detection of the swallowing sound were limited to the tests in the laboratory environment, either quiet or with simulated noise. Although the acoustical detection relies on the simple and reliable sensor (microphone), the method has several challenges that need to be overcome for practical utilization. First, even the best throat microphones are sensitive to the environmental noise and artifacts resulting from the head motion and other activities such as walking. The noise and artifacts complicate sound recognition and may potentially result in false positive detections of swallowing. Second, the sound recognition algorithms are computationally intensive as they rely on sophisticated time-frequency decomposition to represent the signal, thus severely limiting the battery life. Finally, a shared trait among the methods for swallowing detection is the use of collar-like sensor holders in the neck area. Such *collars* are not well tolerated and accepted by the potential users (Fontana and Sazonov 2013).

7.4.2 Electrical Swallowing Sensors

During swallowing, the electrical impedance across the larynx changes with the passing of the bolus of food as well as expansion and contraction of muscle, which can provide an indication of the swallowing.

The use of an electroglottography (EGG) device was proposed for automatic and objective monitoring of ingestive behavior using swallowing by monitoring the impedance changes across the larynx caused by the motion of cricoid cartilage and the passage of food bolus (Farooq et al. 2014). The assumption is that the passage of bolus will cause significant variation in the transverse impedance compared to the baseline, and those variations can be captured and used as an indicator of food intake. For a sample of 30 subjects, the performance of the EGG-based approach was compared to the acoustical approach that used a microphone placed at the larynx level with a collar (Figure 7.4). The EGG-based ANN classification models achieved higher accuracy (average accuracy 90%) compared to the acoustic-based ANN classification models (average accuracy 83%). The performance of EGG-based models was not influenced by the gender or the level of adiposity of the individuals. Further, the EGG-based methodology was insensitive to background noise that makes it potentially superior to the acoustical swallowing detection. However, the study was performed in controlled laboratory conditions compared to the unrestricted free-living environment. Further, there is a possibility that the sensor performance might be impacted by the presence of motion artifacts such as walking.

Another approach that can be used for measuring the muscle contraction and relaxation during swallowing is EMG (Ertekin et al. 2000). A combination of EMG and bioimpedance (EMBI) measurement was proposed for swallowing detection (Schultheiss et al. 2014). The system performed simultaneous measurements of bioimpedance and EMG by placing current electrodes bilaterally at on the sternocleidomastoid muscle, and the voltage measurement electrodes were placed bilaterally

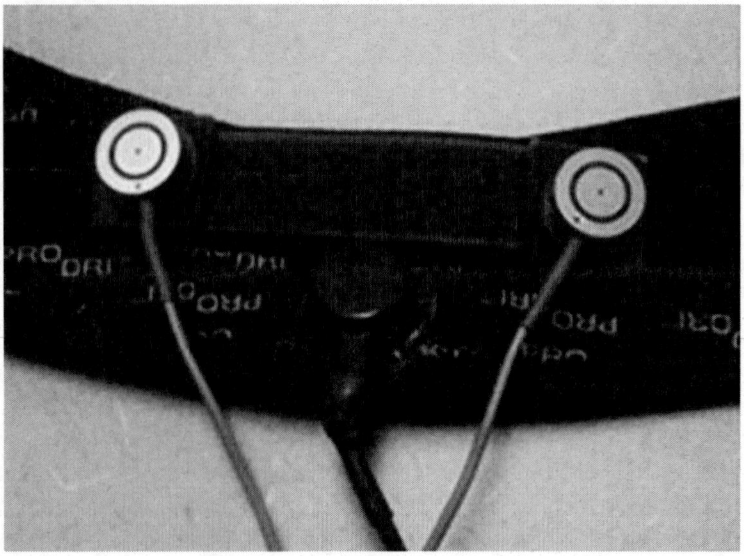

FIGURE 7.4 **(See color insert.)** A collar with a microphone and electrodes for EGG device (Farooq et al. 2014).

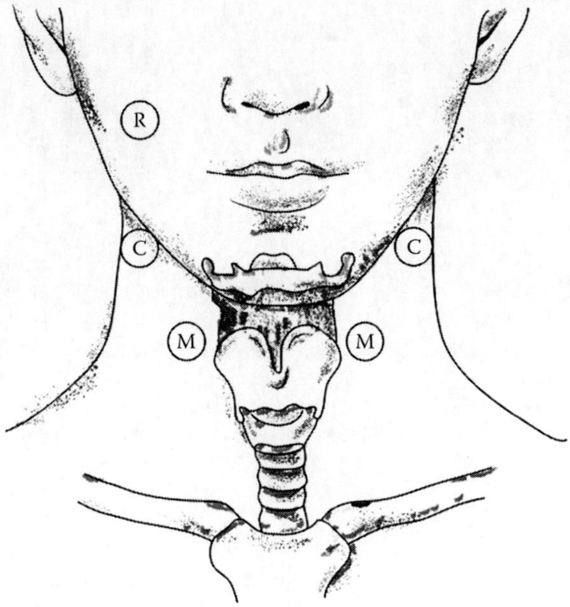

FIGURE 7.5 Electrode positioning: R = reference electrode, M = voltage measurement electrodes, and C = current source electrodes.

between the larynx and hyoid. The reference electrode was placed on the right cheekbone. Figure 7.5 shows the placement of electrodes. A total of 72 subjects were recruited for this study where subjects had to consume liquids, semiliquids, and solid foods of different volumes in one visit. A two-stage classification approach using support vector machines resulted in a sensitivity of 96% and specificity of 97%, for swallowing detection. The same approach when tested with patients with swallowing disorder resulted in a sensitivity of 84% and specificity of 85%, for swallowing detection. The experiments were performed in controlled laboratory settings where the subjects were in a sedentary posture. Differences in the sensitivity and specificity of swallowing detection between healthy subjects and patients with swallowing disorders were observed because of the characteristic functional changes in the pharynx between the two groups.

EMG and bioimpedance techniques for monitoring of swallowing are insensitive to the external acoustic noise that makes them a good candidate for use in free living. However, these measurement methods may be prone to motion artifacts such as head motions while walking. Another limitation common to both EMG and bioimpedance measurement technique is their reliance on the electrodes that need to have good contact with the skin and might cause discomfort to the user, which can impact user compliance to the system.

7.4.3 Laryngeal Movements in Swallowing

During swallowing of the food bolus, the contraction of the muscles results in the motion of the skin that can be used for monitoring of swallowing. The laryngeal movements during swallowing of food bolus can be recognized using a number of sensor modalities such as accelerometers or strain sensors.

Inertial sensors have been proposed for studying the swallowing disorders, for example, researchers used a piezoelectric pulse transducer (PPT) for studying the passage of food bolus through the esophagus (Toyosato et al. 2007). A PPT was placed on the right auricular at the level of the thyroid cartilage. Swallowing was also monitored using EMG and videofluorography. Their results suggest that PPT results in a characteristic waveform during swallowing that can be used to detect the

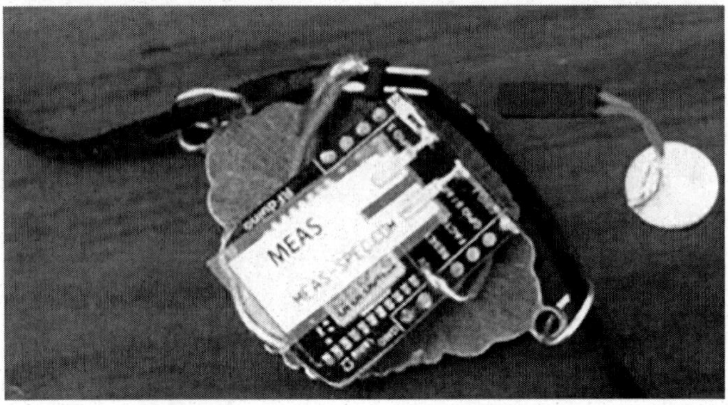

FIGURE 7.6 **(See color insert.)** Necklace with piezoelectric strain sensor for detection of swallowing. (From Kalantarian, H. et al., *Comput. Biol. Med.*, 58 [March], 46–55, 2015. doi:10.1016/j.compbiomed.2015.01.005.)

passage of bolus through the esophagus. A similar approach is the use of the accelerometer to track the laryngeal movements and have been proposed for automatic detection of swallowing in both healthy subjects and subjects with swallowing disorders (Damouras et al. 2010).

Researchers have studied the relationship between the laryngeal movements during pharyngeal swallowing and the bolus volume in the swallowing of tea (Miyaoka et al. 2011). Laryngeal movements were recorded by placing a piezoelectric sensor to the front of the neck in 11 subjects where they were asked to swallow six volumes of tea (10–32 mL). Four characteristic intervals were identified on the piezoelectric sensor signal for each swallow. General linear model-analysis of variance (GLM-ANOVA) results suggested that there was a relationship between volume swallowed and the intervals between the characteristic points on the sensor signal suggesting that this approach can potentially be used for estimating the bolus volume. Piezoelectric sensors have also been used for studying the swallowing disorders (Ertekin et al. 1996).

In more recent studies, piezoelectric strain sensors have been utilized for automatic monitoring of food intake by monitoring the laryngeal movements during swallows (Kalantarian et al. 2015). In a controlled laboratory study, a necklace with an embedded piezoelectric strain sensor was placed on the lower trachea to detect swallowing events. Figure 7.6 shows the necklace with a piezoelectric sensor. Muscle contractions during swallowing causes skin motion, which results in a mechanical stress in the sensor, by pushing it away from the skin toward the necklace. Authors presented two experiments where during the first experiment, 20 subjects were asked to consume sandwich, chips, and water. In the second experiment, they were asked to consume nuts, chocolate, and a meat patty. A classification algorithm achieved F-measures of 75% and 79% for first and second experiment respectively, where F-measure is used as an accuracy metric for algorithm performance. However, acoustic-based swallowing detection method (using a microphone) achieved higher F-measures on the same data set (91% and 88%). A possible reason for lower F-measure for the piezoelectric sensor is the mechanical stress caused by head movements and speech (as the sensor will touch the necklace). This limits the use of laryngeal movement-based swallowing detection methods in unrestricted free living.

7.4.4 Other Swallowing Sensors

Apart from monitoring the acoustic signals originating from swallowing sounds and laryngeal movements during swallowing, there are other physiological phenomena associated with swallowing that can be used for monitoring of swallowing events such as the movements of thyroid cartilage or by monitoring the apnea in breathing signals.

During normal swallowing the thyroid cartilage moves in a triangular path, that is, upward, forward, and then back to the original position where upward movement is related to the oral activity, whereas the forward movement is associated with the pharyngeal activity (Kandori et al. 2012). A swallowing detection system was proposed where two coils (an oscillation coil and detection coil) were placed on both sides of the thyroid cartilage (Kandori et al. 2012). The magnetic field produced by the oscillation coil causes inductive voltage in the detection coil. The passage of the bolus of food during swallowing causes movement of the thyroid cartilage, which results in the changes of the distance between the two coils and hence the variation in the induced voltage in the detection coil. The voltage changes that correspond to the changes in the distance between the two coils can be used for monitoring of swallowing. This system was used in conjunction with a swallowing sound monitoring system (microphone) and videoflourography. Peak times for points of videoflourographic waveform, distance between the coils of the magnetic system, and the swallowing sounds had an intraclass correlation coefficient of 0.9, which shows a strong agreement between all the three methods. This showed that the magnetic sensor-based approach could be used for accurate detection of swallowing events. Compared to acoustic-based approach, magnetic sensors are not influenced by the presence of environmental noises (acoustic); however, it is not clear that how the performance of such system will be in the presence of body movements.

Another approach for monitoring of swallowing relies on detection of swallowing apnea. Apnea or temporary cessation of breathing can be observed in normal breathing pattern when swallowing occurs. This observation is used in a system for monitoring the breathing patterns using a respiratory inductance plethysmography (RIP) belt (Dong and Biswas 2014). Characteristic apnea patterns were extracted from the sensor signal by using matched filters and decision tree classifier with time and frequency domain features. Data was collected from seven subjects while consuming liquids and performing other activities such as talking and upper body movements. Researchers collected data on both normal breathing cycles and breathing cycles with swallows. Their results suggest that with the proposed sensor system, the decision tree classifier obtained a true positive rate of up to 98% and false positive rate of as low as 1% for 10-fold cross validation. Although promising, the data was collected in a controlled environment where only liquid intake was considered with restricted body movements. A common limitation of this approach (like other swallowing-based approaches) is that the performance of such systems degrades with excessive body movements, for example, physical activity will increase the breathing rate of the individuals that will impact the performance of the classifier.

7.5 CHARACTERIZATION OF FOOD INTAKE FROM CHEWING AND SWALLOWING

Automatic detection of food intake is the first step toward the understanding of dietary intake and ingestive behavior. The chews and swallows are not only reliable indicators of food intake but also carry information about the physical properties of the foods being consumed, the number of distinct foods in an eating episode, the rate of ingestion, and mass and volume. This section provides a review of the methods and sensor systems used for characterization of food intake using chewing and swallowing.

7.5.1 Estimation of the Number of Foods and Food Type

Food science differentiates several classes of physical properties of foods, such as rheological, mechanical, thermal, dielectric, and others (Sahin and Sumnu 2006). The rheological properties play the significant role in how the food *feels* during the consumption. The rheological properties also define the chewing sounds and the pattern of chews and swallows during food consumptions. Intuitively this effect can be illustrated by the example of representative foods such as a peanut butter sandwich (a viscous, soft food that results in slow, prolonged chewing, and low-intensity chewing sound) and chips (a crisp, hard food that results in frequent, quick chewing, and a high-intensity chewing sound).

The impact of the rheological properties of the food on the chewing and swallowing patterns was explored in Lopez-Meyer et al. (2012) to determine the number of unique food items in a meal without identifying the specific foods consumed. A sample of 17 participants was recruited and asked to consume five different foods (cheese pizza, yogurt, apple, peanut butter sandwich, and water). Chewing and swallowing data was collected from each participant and three different features were computed for segmentation of the meal into different food types. The features were the time of each swallowing event, time to preceding swallow, that is, the time difference between the current and previous swallow, and the number of chews taken before a swallow, which shows the number of chews between two swallowing events. To account for the person-specific individuality of the chewing and swallowing patterns, the authors proposed using unsupervised clustering methods to group food types in a meal. The number of foods was estimated as the number of clusters formed by the agglomerative hierarchical clustering and affinity propagation (Frey and Dueck 2007). The agglomerative hierarchical clustering was able to estimate the number of food items with an accuracy of 95%, whereas the affinity propagation resulted in an accuracy of 90%. A limitation of the study was that the participants were required to consume the food items in a specific sequence and were not allowed to mix different foods together. Future research is required to test the feasibility of using the clustering approach for identifying foods (including mixed foods) in a more realistic, unrestricted meal.

Chewing sounds have been extensively applied in an attempt to recognize individual food types, typically for a limited set of several food items with varying properties. The chewing sounds collected by an earpad sensor were used to differentiate 19 food types based on the spectral content of the sound in Amft (2010). Chewing sound (captured through a miniature microphone in outer ear canal) recognition with Hidden Markov models was able to estimate the number of food items from chewing sounds with an accuracy of 79% in a study with 51 participants consuming seven different solid foods and a drink (Päßler et al. 2012). Hidden Markov models are probabilistic sequence models, which are used to compute the probability distribution of possible output sequences, given a sequence of observations. A similar approach is to collect acoustic data on chewing and swallowing events using a high fidelity microphone worn by individuals around their neck (Bi et al. 2015). Authors proposed a multistep process to identify chewing and swallowing events and to differentiate between seven different food types, using HMM and decision trees along with a sound library obtained from thousands of prerecorded chewing and swallowing events for different known food types. These and other similar studies often rely on a limited set of food items with distinctly different rheological (textural) properties. In general, most of the chewing sounds recognition may be utilized for the assessment of the food properties (e.g., Masaki Shuzo 2010) or the number of distinct foods in a meal. Most likely, the accuracy of food type classification will degrade severely with the increase in the number of food types and will not extend to the wide variety of foods encountered in daily life. Further studies are needed to verify the extensibility and potential limitations of such approach.

7.5.2 Estimation of Mass and Energy Intake from Chewing and Swallowing

Several studies explored the relationship between the number of chews and swallows and the ingested mass and energy intake. The study of Sazonov et al. (2009) used a limited data set collected from 20 participants consuming five fixed-size food items. As the first step, each swallow was attributed to the consumption of solids or liquids. Next, the ingested mass of solid food was estimated from the average of the product of the counts of chews and swallows and average mass per chew and swallow. The ingested mass of liquids was estimated as the product of the number of liquid swallows and the average mass per liquid swallow. The reported accuracy was 91% for estimation ingested solid mass and 83% for estimation of ingested liquid mass. Estimation of the bite mass from the chewing sounds was proposed in Amft et al. (2009). The study involved eight participants that performed a total of 504 bites of three food types. The bite mass estimated as a weighted average of *microstructure variables* such as the number of chews, chewing duration, and others. Using subject-dependent models, the estimation error varied from 19% to 31% for the three different food items (potato chips, lettuce, and apple).

A limitation of the studies (Amft et al. 2009; Sazonov et al. 2009) is a very restricted set of food items and the fixed serving size. The study of Fontana et al. (2015) extended the approach to estimation of the energy intake on a wide variety of cafeteria food items consumed with no service-size restrictions. Subject-dependent models were used to estimate the total mass ingested in a meal and the energy intake was computed by multiplying the total ingested mass by the average caloric density of the food. The energy estimation from the models was compared to the self-reported diet diaries and photographic food records. Weighed food records were used as gold standard. For the training meals, the estimated energy intake had a lower reporting error (16%) compared to 28% for diet diaries and 20% for photographic food records. For the validation meal, the energy intake estimated from chews and swallows was not significantly different from other methods. The models proposed in the studies of Sazonov et al. (2009); Fontana et al. (2015) are of a pilot nature and rely on the assumption of the stable average caloric density of the foods consumed by the individual (repeatable food choices). Further development of the method is possible to incorporate the imagery-based estimates of the caloric density, which may potentially improve the accuracy of the model. Existing research also suggests that the accuracy of self-reporting decreases (Goris et al. 2001) with the increase in the duration. Although the existing studies estimated the accuracy of the energy intake estimation from chews and swallows cross-sectionally, a longitudinal study may produce improved results. The counts of chews and swallows needed for such studies may be acquired from the wearable sensors (Sazonov et al. 2010; Kandori et al. 2012; Schultheiss et al. 2014; Farooq and Sazonov 2015).

7.6 FUTURE STUDIES

This chapter reviewed a variety of the sensors and methods reported in the recent literature. A number of common limitations of the existing studies and suggestion for the future work can be made based on this review.

Most of the reported work relies on the experiments conducted in the laboratory conditions. This is a necessary first step especially that the ground truth for ingestion is incredibly difficult to establish in the free-living conditions. However, the results obtained in the laboratory cannot be readily generalized to the community dwelling. For example, many sound-based studies limited the noise interference or did not allow the participants to talk or move because the artifacts from the interfering sources may negatively impact the performance of sound recognition. In general, the sensors used to detect chewing and swallowing also respond to other sources of excitation and need to be tested in the realistic conditions of daily life. In addition, all of the studies to date have been cross-sectional, rather than longitudinal. The assessment of the performance of the proposed methods over a reasonable observation period remains an open issue.

The capabilities and the limitations of the proposed methodologies also need to be clearly evaluated and understood. As an example, the recognition of food type based on the chewing sounds has been a popular topic, with many studies attempting to perform such differentiation. The chewing sound is a one-dimensional signal that carries significantly less information than, for example, an image of the food. The extensibility of chewing sound recognition to the full variety of foods encountered in the daily life is yet to be proven and need to be carefully assessed. Similarly, many of the studies report subject-dependent models that need to be calibrated to an individual. This fact is a good indicator of the tremendous intersubject variability rooted in different lifestyles, cultural backgrounds, and personal traits. The focus of future work should be the development of the methods that work well with or without minimal individual calibration.

In addition, with a few exception reviewed above, the major effort to date has been focused on the development of the method for detection of food intake. While this work is ongoing, the field of automatic characterization of the food intake remains largely unexplored. Chewing and swallowing present an appealing target for extraction of the metainformation about the ingestion process, such as timing, duration, rate of ingestion, and mass and energy intake. Fusion with other information sources, such as food imagery, may be needed to improve the accuracy. The real time about ingestion may also be used to inform just-in-time adaptive interventions that rely on sensor data to modify the eating behavior.

Finally, social acceptance of the proposed sensor modalities is a key issue for future adoption of the technology. An ideal sensor is imperceivable by the user and, when used for observation, minimally impacts the way people eat. The chewing and swallowing sensors naturally tend to locate in the neck and head regions, making them highly visible to the public. Some of the sensor solution, such as collar-like devices, are not well tolerated and may lead to noncompliance. Although a few of studies report on the user acceptance of the proposed sensing modalities, the future studies are needed regarding which form-factors and locations are tolerated and accepted by subjects.

7.7 CONCLUSIONS

The reliable, objective, and minimally invasive assessment of ingestion and diet is very promising, but remains an open issue. The emergence of wearable sensor technology stimulated tremendous growth in the development of sensors and devices for monitoring of food intake. A large portion, if not majority, of these devices rely on the behavioral and physiological phenomena, such as chewing and swallowing to detect and characterize food intake. Sensors such as microphones, accelerometers, piezoelectric, and magnetic sensors capture the information about events occurring during the ingestion to detect the eating episodes and to estimate the number of foods consumed, consumed mass, and energy intake. The development of the technology is driven toward miniaturization and advancements in reliability and functionality.

The chewing and swallowing are being actively explored as potential foundations for monitoring of the diet and ingestive behavior. A significant amount of open questions needs to be resolved before these sensors relying on chewing and swallowing may receive widespread acceptance in research and clinical practice.

REFERENCES

Amft, O. 2010. A wearable earpad sensor for chewing monitoring. In *Sensors, 2010 IEEE*, pp. 222–227. doi:10.1109/ICSENS.2010.5690449.

Amft, O., M. Kusserow, and G. Troster. 2009. Bite weight prediction from acoustic recognition of chewing. *IEEE Transactions on Biomedical Engineering* 56 (6): 1663–1672. doi:10.1109/TBME.2009.2015873.

Amft, O., M. Stäger, P. Lukowicz, and G. Tröster. 2005. Analysis of chewing sounds for dietary monitoring. In M. Beigl, S. Intille, J. Rekimoto, and H. Tokuda, eds., *UbiComp 2005: Ubiquitous Computing*. Lecture Notes in Computer Science, Vol. 3660. Berlin, Germany: Springer.

Amft, O., and G. Troster. 2009. On-body sensing solutions for automatic dietary monitoring. *IEEE Pervasive Computing* 8 (2): 62–70. doi:10.1109/MPRV.2009.32.

Barlow, S. M. 2009. Central pattern generation involved in oral and respiratory control for feeding in the term infant. *Current Opinion in Otolaryngology & Head and Neck Surgery* 17 (3): 187–193. doi:10.1097/MOO.0b013e32832b312a.

Bedri, A., A. Verlekar, E. Thomaz, V. Avva, and T. Starner. 2015a. A wearable system for detecting eating activities with proximity sensors in the outer ear. In *Proceedings of the 2015 ACM International Symposium on Wearable Computers*, pp. 91–92. ISWC'15. New York, NY: ACM. doi:10.1145/2802083.2808411.

Bedri, A., A. Verlekar, E. Thomaz, V. Avva, and T. Starner. 2015b. Detecting mastication: A wearable approach. In *Proceedings of the 2015 ACM on International Conference on Multimodal Interaction*, pp. 247–250. ICMI'15. New York, NY: ACM. doi:10.1145/2818346.2820767.

Bi, Y., M. Lv, C. Song, W. Xu, N. Guan, and W. Yi. 2015. AutoDietary: A wearable acoustic sensor system for food intake recognition in daily life. *IEEE Sensors Journal* 16 (3): 806–816. doi:10.1109/JSEN.2015.2469095.

Cadavid, S., M. Abdel-Mottaleb, and A. Helal. 2011. Exploiting visual quasi-periodicity for real-time chewing event detection using active appearance models and support vector machines. *Personal and Ubiquitous Computing* 16 (6): 729–739. doi:10.1007/s00779-011-0425-x.

Carruth, B. R., and J. D. Skinner. 2002. Feeding behaviors and other motor development in healthy children (2–24 months). *Journal of the American College of Nutrition* 21 (2): 88–96.

Chen, C. T., Y. L. Wang, C. A. Wang, M. J. Ko, W. C. Fang, and B. S. Lin. 2015. Wireless monitoring system for oral-feeding evaluation of preterm infants. *IEEE Transactions on Biomedical Circuits and Systems* 9 (5): 678–685. doi:10.1109/TBCAS.2015.2438031.

da Costa, S. P., L. van den Engel–hoek, and A. F. Bos. 2008. Sucking and swallowing in infants and diagnostic tools. *Journal of Perinatology* 28 (4): 247–257. doi:10.1038/sj.jp.7211924.

Damouras, S., E. Sejdic, C. M. Steele, and T. Chau. 2010. An online swallow detection algorithm based on the quadratic variation of dual-axis accelerometry. *IEEE Transactions on Signal Processing* 58 (6): 3352–3359. doi:10.1109/TSP.2010.2043972.

Dodds, W. J. 1989. The physiology of swallowing. *Dysphagia* 3 (4): 171–178. doi:10.1007/BF02407219.

Dong, B., and S. Biswas. 2014. Wearable sensing for liquid intake monitoring via apnea detection in breathing signals. *Biomedical Engineering Letters* 4 (4): 378–387. doi:10.1007/s13534-014-0149-8.

Drake, B. 1965. On the biorheology of human mastication: An amplitude-frequency-time analysis of food crushing sounds. *Biorheology* 3 (1): 21–31.

Effect of food texture on EMG profiles detected from masticatory muscle. 2016. Accessed June 27. https://www.jstage.jst.go.jp/article/jjps1957/37/6/37_6_1329/_article.

Engelen, L., A. Fontijn-Tekamp, and A. van der Bilt. 2005. The influence of product and oral characteristics on swallowing. *Archives of Oral Biology* 50 (8): 739–746. doi:10.1016/j.archoralbio.2005.01.004.

Ertekin, C., I. Aydoğdu, and N. Yüceyar. 1996. Piecemeal deglutition and dysphagia limit in normal subjects and in patients with swallowing disorders. *Journal of Neurology, Neurosurgery, and Psychiatry* 61 (5): 491–496.

Ertekin, C., M. Celik, Y. Seçil, S. Tarlaci, N. Kiyloğlu, and I. Aydoğdu. 2000. The electromyographic behavior of the thyroarytenoid muscle during swallowing. *Journal of Clinical Gastroenterology* 30 (3): 274–280.

Farooq, M., P. Chandler-Laney, M. Hernandez-Reif, and E. Sazonov. 2015a. Monitoring of infant feeding behavior using a jaw motion sensor. *Journal of Healthcare Engineering* 6 (1): 23–40. doi:10.1260/2040-2295.6.1.23.

Farooq, M., P. Chandler-Laney, M. Hernandez-Reif, and E. Sazonov. 2015b. A wireless sensor system for quantification of infant feeding behavior. In *Proceedings of the Conference on Wireless Health*, 16:1–16:5. WH'15. New York, NY: ACM. doi:10.1145/2811780.2811934.

Farooq, M., J. M. Fontana, A. Boateng, M. A. McCrory, and E. Sazonov. 2013. A comparative study of food intake detection using artifcial neural network and support vector machine. In *Proceedings of the 12th IEEE International Conference on Machine Learning and Applications (ICMLA'13)*, pp. 153–157. Miami, FL.

Farooq, M., J. M. Fontana, and E. Sazonov. 2014. A novel approach for food intake detection using electroglottography. *Physiological Measurement* 35 (5): 739. doi:10.1088/0967-3334/35/5/739.

Farooq, M., and E. Sazonov. 2015. Comparative testing of piezoelectric and printed strain sensors in characterization of chewing. In *2015 37th Annual International Conference of the IEEE Engineering in Medicine and Biology Society (EMBC)*, pp. 7538–7541. doi:10.1109/EMBC.2015.7320136.

Fontana, J. M., M. Farooq, and E. Sazonov. 2013. Estimation of feature importance for food intake detection based on random forests classification. In *2013 35th Annual International Conference of the IEEE Engineering in Medicine and Biology Society (EMBC)*, pp. 6756–6759. doi:10.1109/EMBC.2013.6611107.

Fontana, J. M., M. Farooq, and E. Sazonov. 2014. Automatic ingestion monitor: A novel wearable device for monitoring of ingestive behavior. *IEEE Transactions on Biomedical Engineering* 61 (6): 1772–1779. doi:10.1109/TBME.2014.2306773.

Fontana, J. M., J. A. Higgins, S. C. Schuckers et al. 2015. Energy intake estimation from counts of chews and swallows. *Appetite* 85 (February): 14–21. doi:10.1016/j.appet.2014.11.003.

Fontana, J. M., P. Lopez-Meyer, and E. S. Sazonov. 2011. Design of a instrumentation module for monitoring ingestive behavior in laboratory studies. In *Engineering in Medicine and Biology Society, EMBC, 2011 Annual International Conference of the IEEE*, pp. 1884–1887. doi:10.1109/IEMBS.2011.6090534.

Fontana, J. M., and E. S. Sazonov. 2013. Evaluation of chewing and swallowing sensors for monitoring ingestive behavior. *Sensor Letters* 11 (3): 560–565. doi:10.1166/sl.2013.2925.

Foster, K. D., A. Woda, and M. A. Peyron. 2006. Effect of texture of plastic and elastic model foods on the parameters of mastication. *Journal of Neurophysiology* 95 (6): 3469–3479. doi:10.1152/jn.01003.2005.

Frey, B. J., and D. Dueck. 2007. Clustering by passing messages between data points. *Science* 315 (5814): 972–976. doi:10.1126/science.1136800.

Goris, A. H. C., E. P. Meijer, and K. R. Westerterp. 2001. Repeated measurement of habitual food intake increases under-reporting and induces selective under-reporting. *British Journal of Nutrition* 85 (5): 629–634. doi:10.1079/BJN2001322.

Grassi, R., R. Farina, I. Floriani, F. Amodio, and S. Romano. 2005. Assessment of fetal swallowing with gray-scale and color doppler sonography. *American Journal of Roentgenology* 185 (5): 1322–1327. doi:10.2214/AJR.04.1114.

Guandalini, S., A. Dhawan, and D. Branski. 2015. *Textbook of Pediatric Gastroenterology, Hepatology and Nutrition: A Comprehensive Guide to Practice*. Cham, Switzerland: Springer.

Hiiemae, K., M. R. Heath, G. Heath, E. Kazazoglu, J. Murray, D. Sapper, and K. Hamblett. 1996. Natural bites, food consistency and feeding behaviour in man. *Archives of Oral Biology* 41 (2): 175–189. doi:10.1016/0003-9969(95)00112-3.

Imtiaz, U., K. Yamamura, W. Kong et al. Application of wireless inertial measurement units and EMG sensors for studying deglutition #x2014; preliminary results. In *2014 36th Annual International Conference of the IEEE Engineering in Medicine and Biology Society*, pp. 5381–5384. doi:10.1109/EMBC.2014.6944842.

Kalantarian, H., N. Alshurafa, T. Le, and M. Sarrafzadeh. 2015. Monitoring eating habits using a piezoelectric sensor-based necklace. *Computers in Biology and Medicine* 58 (March): 46–55. doi:10.1016/j.compbiomed.2015.01.005.

Kandori, A., T. Yamamoto, Y. Sano et al. 2012. Simple magnetic swallowing detection system. *IEEE Sensors Journal* 12 (4): 805–811. doi:10.1109/JSEN.2011.2166954.

Kinnunen, T., E. Karpov, and P. Franti. 2006. Real-time speaker identification and verification. *IEEE Transactions on Audio, Speech, and Language Processing* 14 (1): 277–288. doi:10.1109/TSA.2005.853206.

Lau, C., and I. Kusnierczyk. 2001. Quantitative evaluation of infant's nonnutritive and nutritive sucking. *Dysphagia* 16 (1): 58–67.

Lear, C. S. C., J. B. Flanagan, and C. F. A. Moorrees. 1965. The frequency of deglutition in man. *Archives of Oral Biology* 10 (1): 83–99. doi:10.1016/0003-9969(65)90060-9.

Lee Hi, W. E., A. E. Deibel, C. T. Glembin, and E. G. Munday. 1988. Analysis of food crushing sounds during mastication: Frequency-time studies. *Journal of Texture Studies* 19 (1): 27–38. doi:10.1111/j.1745-4603.1988.tb00922.x.

Li, C.-Y., Y.-C. Chen, W.-J. Chen, P. Huang, and H.-H. Chu. 2013. Sensor-embedded teeth for oral activity recognition. In *Proceedings of the 2013 International Symposium on Wearable Computers*, pp. 41–44. ISWC'13. New York, NY: ACM. doi:10.1145/2493988.2494352.

Lopez-Meyer, P., O. Makeyev, S. Schuckers, E. Melanson, M. Neuman, and E. Sazonov. 2010. Detection of food intake from swallowing sequences by supervised and unsupervised methods. *Annals of Biomedical Engineering* 38 (8): 2766–2774. doi:10.1007/s10439-010-0019-1.

Lopez-Meyer, P., S. Schuckers, O. Makeyev, J. M. Fontana, and E. Sazonov. 2012. Automatic identification of the number of food items in a meal using clustering techniques based on the monitoring of swallowing and chewing. *Biomedical Signal Processing and Control* 7 (5): 474–480. doi:10.1016/j.bspc.2011.11.004.

Makeyev, O., P. Lopez-Meyer, S. Schuckers W. Besio, and E. Sazonov. 2012. Automatic food intake detection based on swallowing sounds. *Biomedical Signal Processing and Control* 7 (6): 649–656. doi:10.1016/j.bspc.2012.03.005.

Makeyev, O., E. Sazonov, S. Schuckers et al. 2008. Recognition of swallowing sounds using time-frequency decomposition and limited receptive area neural classifier. In *Proceedings of AI-2008, The Twenty-Eighth SGAI International Conference on Innovative Techniques and Applications of Artificial Intelligence*, pp. 33–46. London, UK: Cambridge.

Makeyev, O., E. Sazonov, S. Schuckers, E. Melanson, and M. Neuman. 2007. Limited receptive area neural classifier for recognition of swallowing sounds using short-time fourier transform. In *Neural Networks, 2007. IJCNN 2007. International Joint Conference on*, 1601–1606. doi:10.1109/IJCNN.2007.4371197.

Masaki Shuzo, S. K. 2010. Wearable eating habit sensing system using internal body sound. *Journal of Advanced Mechanical Design Systems and Manufacturing* 4 (1): 158–166. doi:10.1299/jamdsm.4.158.

Matsuo, K., and J. B. Palmer. 2008. Anatomy and physiology of feeding and swallowing—Normal and abnormal. *Physical Medicine and Rehabilitation Clinics of North America* 19 (4): 691–707. doi:10.1016/j.pmr.2008.06.001.

Matsuo, K., and J. B. Palmer. 2009. Coordination of mastication, swallowing and breathing. *Japanese Dental Science Review* 45 (1): 31–40. doi:10.1016/j.jdsr.2009.03.004.

McFarland, D. H., and J. P. Lund. 1995. Modification of mastication and respiration during swallowing in the adult human. *Journal of Neurophysiology* 74 (4): 1509–1517.

Mioche, L., P. Bourdiol, J. F. Martin, and Y. Noël. 1999. Variations in human masseter and temporalis muscle activity related to food texture during free and side-imposed mastication. *Archives of Oral Biology* 44 (12): 1005–1012.

Miyaoka, Y., I. Ashida, S. Kawakami, Y. Tamaki, and S. Miyaoka. 2011. Generalization of the bolus volume effect on piezoelectric sensor signals during pharyngeal swallowing in normal subjects. *Journal of Oral Biosciences* 53 (1): 65–71. doi:10.1016/S1349-0079(11)80037-X.

Nishimura, J., and T. Kuroda. 2008. Eating habits monitoring using wireless wearable in-ear microphone. In *3rd International Symposium on Wireless Pervasive Computing, 2008. ISWPC 2008*, pp. 130–132. doi:10.1109/ISWPC.2008.4556181.

Ottenhoff, F. A. M., A. Van Der Bilt, H. W. Van Der Glas, F. Bosman, and J. H. Abbink. 1996. The relationship between jaw elevator muscle surface electromyogram and simulated food resistance during dynamic condition in humans. *Journal of Oral Rehabilitation* 23 (4): 270–279. doi:10.1111/j.1365-2842.1996.tb00852.x.

Passler, S., W. Fischer, and I. Kraljevski. 2012. Adaptation of models for food intake sound recognition using maximum a posteriori estimation algorithm. In *2012 Ninth International Conference on Wearable and Implantable Body Sensor Networks (BSN)*, pp. 148–153. doi:10.1109/BSN.2012.2.

Päßler, S., and W.-J. Fischer. 2014. Food intake monitoring: Automated chew event detection in chewing sounds. *IEEE Journal of Biomedical and Health Informatics* 18 (1): 278–289. doi:10.1109/JBHI.2013.2268663.

Päßler, S., M. Wolff, and W.-J. Fischer. 2012. Food intake monitoring: An acoustical approach to automated food intake activity detection and classification of consumed food. *Physiological Measurement* 33 (6): 1073–1093. doi:10.1088/0967-3334/33/6/1073.

Rahman, S. A., C. Merck, Y. Huang, and S. Kleinberg. 2015. Unintrusive eating recognition using google glass. In *Proceedings of the 9th International Conference on Pervasive Computing Technologies for Healthcare*, pp. 108–111. PervasiveHealth'15. ICST, Brussels, Belgium: ICST (Institute for Computer Sciences, Social-Informatics and Telecommunications Engineering). http://dl.acm.org/citation.cfm?id=2826165.2826181.

Sahin, S., and S. G. Sumnu. 2006. *Physical Properties of Foods*. Food Science Text Series. New York, NY: Springer. http://link.springer.com/10.1007/0-387-30808-3.

Sazonov, E., and J. M. Fontana. 2012. A sensor system for automatic detection of food intake through non-invasive monitoring of chewing. *IEEE Sensors Journal* 12 (5): 1340–1348. doi:10.1109/JSEN.2011.2172411.

Sazonov, E., O. Makeyev, S. Schuckers, P. Lopez-Meyer, E. L. Melanson, and M. R. Neuman. 2010. Automatic detection of swallowing events by acoustical means for applications of monitoring of ingestive behavior. *IEEE Transactions on Bio-Medical Engineering* 57 (3): 626–633. doi:10.1109/TBME.2009.2033037.

Sazonov, E., S. Schuckers, P. Lopez-Meyer et al. 2008. Non-invasive monitoring of chewing and swallowing for objective quantification of ingestive behavior. *Physiological Measurement* 29 (5): 525–541.

Sazonov, E., S. A. C. Schuckers, P. Lopez-Meyer et al. 2009. Toward objective monitoring of ingestive behavior in free-living population. *Obesity* 17 (10): 1971–1975.

Schmalz, M. S., A. Helal, and A. Mendez-Vasquez. 2009. Algorithms for the detection of chewing behavior in dietary monitoring applications. In 7444:74440E–74440E–11. doi:10.1117/12.829205.

Schultheiss, C., T. Schauer, H. Nahrstaedt, and R. O. Seidl. 2014. Automated detection and evaluation of swallowing using a combined EMG/bioimpedance measurement system. *The Scientific World Journal* 2014 (July): e405471. doi:10.1155/2014/405471.

Shaker, R., P. C. Belafsky, G. N. Postma, and C. Easterling, eds., 2013. *Principles of Deglutition*. New York, NY: Springer New York. http://link.springer.com/10.1007/978-1-4614-3794-9.

Stellar, E., and E. E. Shrager. 1985. Chews and swallows and the microstructure of eating. *The American Journal of Clinical Nutrition* 42 (5 Suppl): 973–982.

Taffoni, F., E. Tamilia, M. R. Palminteri et al. 2013. Ecological sucking monitoring of newborns. *IEEE Sensors Journal* 13 (11): 4561–4568. doi:10.1109/JSEN.2013.2271585.

Taki, M., K. Mizuno, M. Murase, Y. Nishida, K. Itabashi, and Y. Mukai. 2010. Maturational changes in the feeding behaviour of infants—A comparison between breast-feeding and bottle-feeding. *Acta Pædiatrica* 99 (1): 61–67. doi:10.1111/j.1651-2227.2009.01498.x.

Tate, G. S., G. S. Throckmorton, E. Ellis, and D. P. Sinn. 1994. Masticatory performance, muscle activity, and occlusal force in preorthognathic surgery patients. *Journal of Oral and Maxillofacial Surgery: Official Journal of the American Association of Oral and Maxillofacial Surgeons* 52 (5): 476–481; discussion 482.

Toyosato, A., S. Nomura, A. Igarashi, N. Ii, and A. Nomura. 2007. A relation between the piezoelectric pulse transducer waveforms and food bolus passage during pharyngeal phase of swallow. *Prosthodontic Research & Practice* 6 (4): 272–275. doi:10.2186/prp.6.272.

van den Braber, W., H. van der Glas, A. van der Bilt, and F. Bosman. 2004. Masticatory function in retrognathic patients, before and after mandibular advancement surgery. *Journal of Oral and Maxillofacial Surgery: Official Journal of the American Association of Oral and Maxillofacial Surgeons* 62 (5): 549–554.

Wang, S., G. Zhou, L. Hu, Z. Chen, and Y. Chen. 2015. CARE: Chewing activity recognition using noninvasive single axis accelerometer. In *Adjunct Proceedings of the 2015 ACM International Joint Conference on Pervasive and Ubiquitous Computing and Proceedings of the 2015 ACM International Symposium on Wearable Computers*, pp. 109–112. UbiComp/ISWC'15 Adjunct. New York, NY: ACM. doi:10.1145/2800835.2800884.

Wilding, R. J., and M. Shaikh. 1997. Muscle activity and jaw movements as predictors of chewing performance. *Journal of Orofacial Pain* 11 (1): 24–36.

Ye, X., G. Chen, and Y. Cao. 2015. Automatic eating detection using head-mount and wrist-worn accelerometers. In *2015 17th International Conference on E-Health Networking, Application Services (HealthCom)*, pp. 578–581. doi:10.1109/HealthCom.2015.7454568.

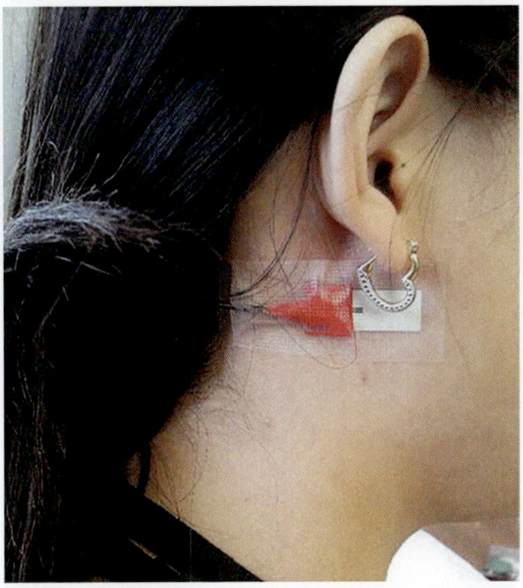

FIGURE 7.1 A piezoelectric strain sensor attached to the skin below the ear to capture jaw motion associated with chewing.

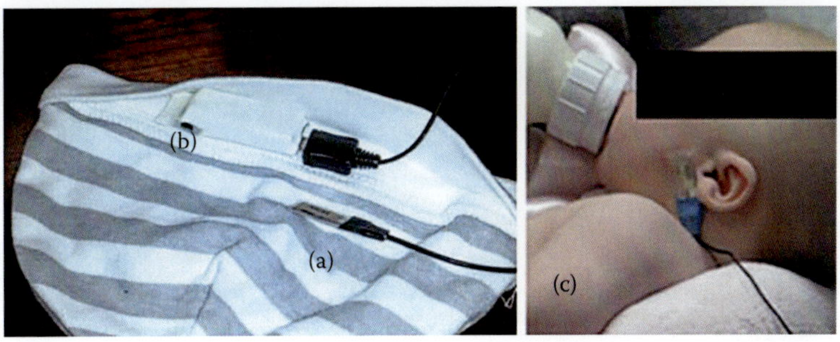

FIGURE 7.2 Jaw motion sensor in the infant's hat: (a) jaw motion sensor, (b) wireless data acquisition module, and (c) a bottle-fed infant wearing the sensor.

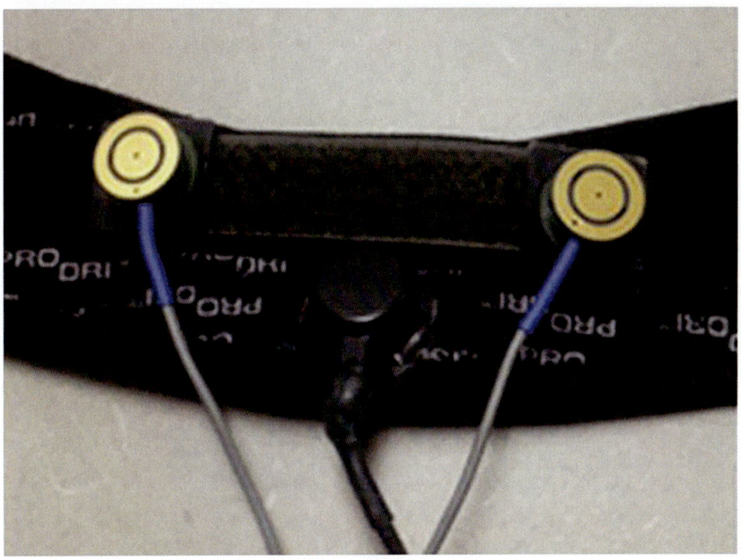

FIGURE 7.4 A collar with a microphone and electrodes for EGG device (Farooq et al. 2014).

FIGURE 7.6 Necklace with piezoelectric strain sensor for detection of swallowing. (From Kalantarian, H. et al., *Comput. Biol. Med.*, 58 [March], 46–55, 2015. doi:10.1016/j.compbiomed.2015.01.005.)

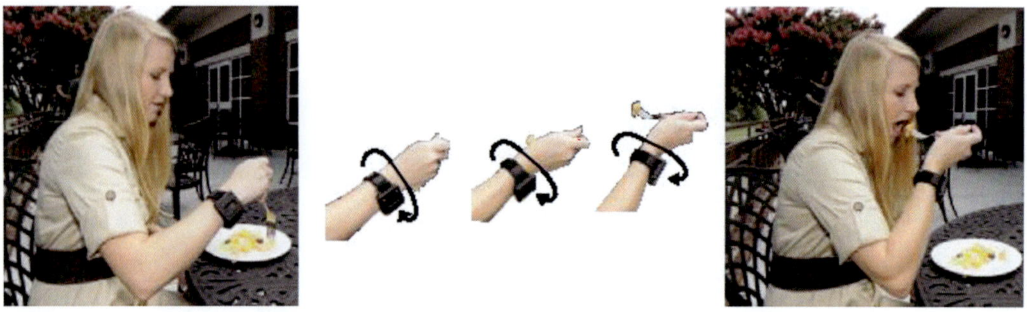

FIGURE 8.1 Depiction of wrist-roll motion that occurs during a bite.

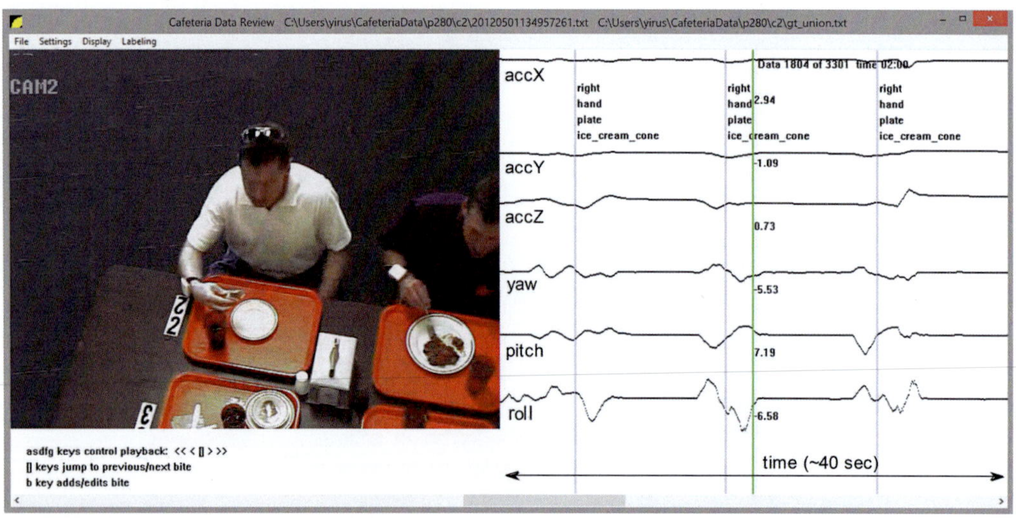

FIGURE 8.4 A custom program created for manual labeling of ground truth bites. The left panel shows the video and the right panel shows the wrist-motion tracking. Vertical purple lines indicate the times marked as bites, the vertical green line indicates the time currently displayed in the video. Variables (hand, utensil, container, and food) are identified for each bite.

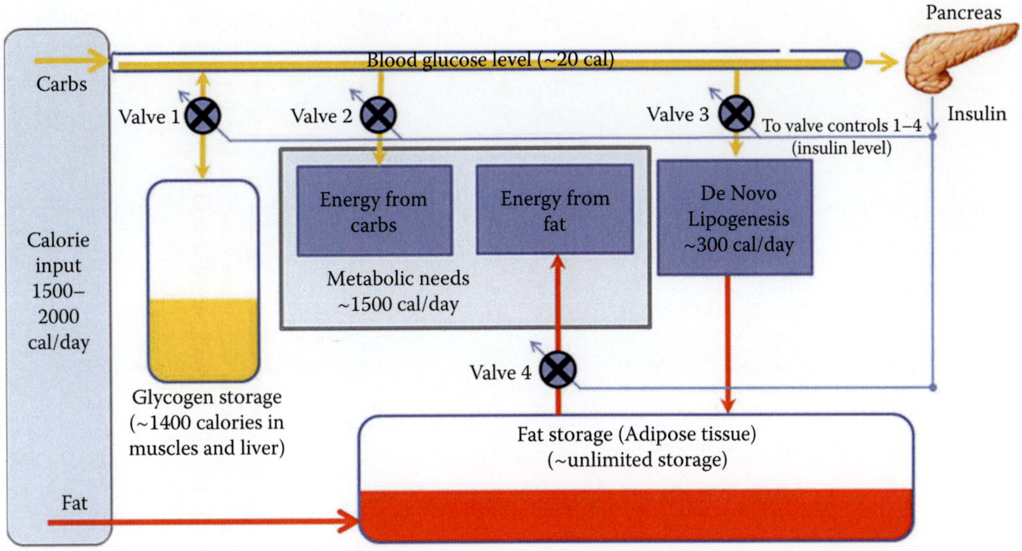

FIGURE 9.4 Closed-loop metabolic state prediction model emphasizing glucose control quantitatively as diet and exercise change. Four valves determine disposition of dietary intake and use of stored energy. Real-time measurement of O_2 consumed and CO_2 produced in exhaled breath will reveal the *settings* of valves 2–4. (Adapted from Candell, L. M. and G. A. Shaw, RER for the masses, *Presented at the annual conference of Recent Advances and Controversies in Metabolic Energy Measurement*, Tokyo, Japan, October 10–12, 2014.)

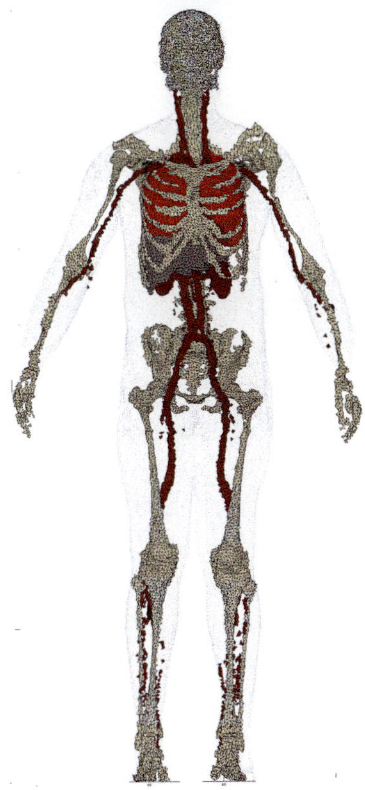

FIGURE 9.5 Individualized male avatar based on a body surface scan from the ANSUR II database (Gordon et al. 2014) using the method of Zientara and Hoyt (2016). Avatar surface (fine blue dots), selected internal anatomy (lungs, kidneys, heart, and liver; light brown), and cardiovascular system (dark and bright red) shown as 3D tetrahedral finite elements.

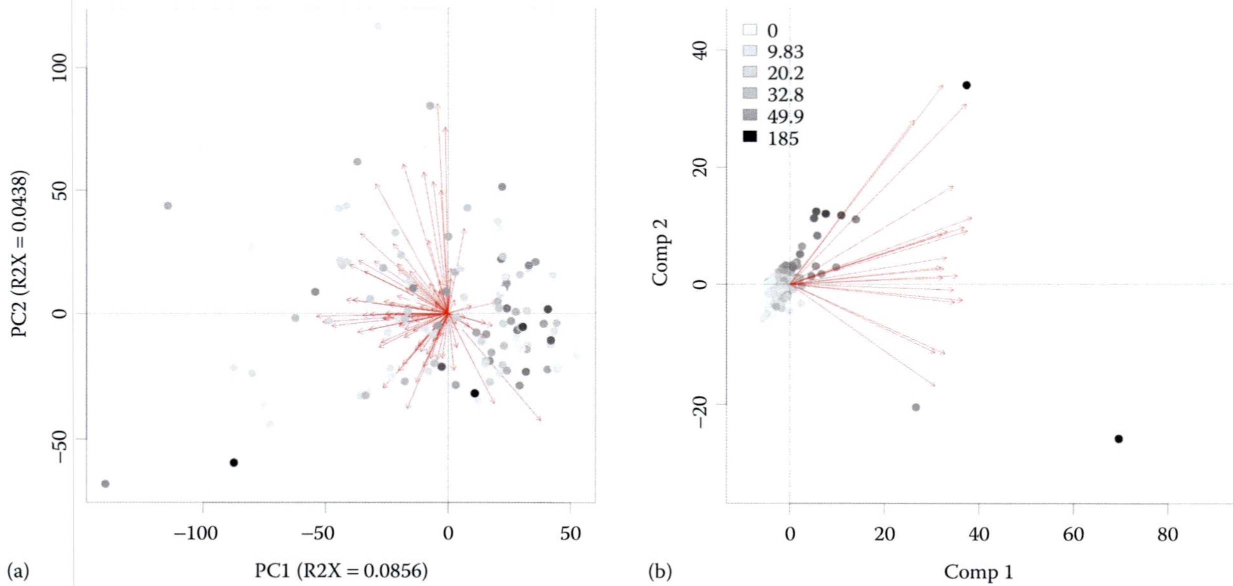

FIGURE 18.1 Biplots from (a) PCA and (b) PLS analyses. Observations are grey-scaled according to whole grain rye exposure (see legend) and variable loadings are shown as red arrows. The PCA loadings (100 randomly selected variables visualized in the figure) correspond to the main variability in the metabolomics data, which however is seemingly unrelated to the response variable as seen from the observation scores. The 20 top ranked variables in the PLS analysis, shortlisted in the original paper (Hanhineva et al. 2015), show a pattern in observation scores clearly associated with the response variable.

8 Bites as a Unit of Measurement

Adam Hoover, Ryan Mattfeld, and Eric Muth

CONTENTS

8.1 INTRODUCTION

This chapter considers the contribution of the measurement of bites to the quantification of dietary intake. A bite refers to the process of placing food or beverage into the mouth for consumption. In order to motivate the use of bites as a measurement unit it is important to discuss its potential strengths and weaknesses. On the strength side, a bite is a physical phenomenon familiar to everyone. Also, the act of taking a bite requires physical actions that can be observed during eating and thus measured in real time. A third strength is that taking a bite requires a conscious effort, and thus presents opportunities for quantifying behaviors. For example, the interbite interval (in seconds) can be measured and used to guide a person toward a slower eating rate (Scisco et al. 2011). As another example, the running count of total bites consumed can be measured and used to guide a person toward a dietary limit goal (Jasper et al. 2016). These feedbacks are enabled by the ability to measure bites in real time during consumption. However, the use of bites as a unit of measure raises many potential questions. The energy content of a bite varies depending on the energy density of the food and the size of bite consumed. Bites occur using a variety of hand, utensil, and container combinations. Eating habits can vary between nationalities and other demographics. This chapter describes evidence that bites can be physically detected and measured, and that an average relationship between number of bites and energy intake can be established at the meal level.

A secondary purpose of this chapter is to describe the engineering paradigm called empirical evaluation in computer vision (Bowyer and Phillips 1998), and its relationship to the development of new technologies for measuring dietary intake. As new technologies and new devices are offered to researchers and consumers there is a need to assess their capabilities in terms of accuracy, strengths, and limitations. This chapter will explain the role served by empirical evaluation in computer vision, and specifically how it can help provide context for evaluating the maturity and potential use of new devices in clinical and behavioral research. Two technologies are summarized in this chapter and can be viewed as case studies within this paradigm. The first uses a wrist-worn device to track a characteristic pattern of wrist motion associated with bites. The second tracks wrist motion all day in order to detect periods of eating activity (i.e., meals or snacks). After describing these technologies, a study on the relationship between number of bites and energy intake is reported. This chapter concludes with the advantages and disadvantages of bite-based measures and an assessment of the maturity of engineering research in quantifying dietary intake.

8.2 BACKGROUND ON THE APPLICATION OF COMPUTER VISION TO QUANTIFICATION OF DIETARY INTAKE

Existing tools for quantifying dietary intake include food diaries and recalls of the foods consumed (Chapter 1). However, these tools rely on self-report and have limitations, as discussed in Chapters 2, 4, and 10. The need for technologies to be developed to improve measures of dietary intake has been widely advocated within the dietetics community (McCabe-Sellers 2010; Schoeller et al. 2013; Thompson et al. 2010).

Computer vision is a discipline concerned with interpreting sensor data to extract symbolic meaning. The main goal of using computer vision is automation. It is important to understand that if a human cannot complete a task then it is unlikely that computer vision methods will be able to do so. The automation provided by computer vision serves two purposes: (1) it relieves the burden of manually completing a repetitive task and (2) it eliminates human subjectivity. Both these factors motivate the use of applying computer vision to the problem of quantifying dietary intake.

Several consumer-formatted approaches related to using computer vision to quantify dietary intake are being developed. One approach is to instrument utensils and containers used for eating. The HAPIfork (http://www.hapi.com) recently rebranded as the 10S fork (http://www.slowcontrol.com/en/10s-fork/) is an example of a utensil. It uses sensors to detect a closed electrical circuit between the hand holding the fork and the lips touching the tines to detect when a bite has been taken. The Pryme Vessyl (http://prymevessyl.com/) is an example of a container. It uses sensors to measure the liquid volume of a thermos and communicates with a cell phone app to track the daily amount of water consumed. Both of these are elementary applications of computer vision. The strength of these tools is that they tend to operate reliably within their application domain. However, because they do not require extensive research to develop, there is little to no published literature describing or testing the engineering principles of the computer vision technique. In fact, the techniques are usually held proprietary to facilitate a commercial advantage. Note that this is different from usability studies such as in Hermsen et al. (2016). A weakness of these tools is that they are limited in the types of consumption that can be measured. For example, an instrumented fork is only useful for foods that are eaten with a fork. Another weakness is that instrumented utensils and containers must be carried everywhere by an individual using them to track daily consumption and thus can impose a burden.

A second approach to using computer vision to quantify dietary intake is to instrument a table used for eating. The idea of using a table scale to continuously measure consumption was pioneered by a technique called the universal eating monitor (Kissileff et al. 1980). It measures grams consumed per unit time. It has been used in clinical and laboratory settings to test the effects of anorectic drugs (Kissileff et al. 1981) and behavior modifications on eating rate and total consumption (Hubel et al. 2006; Laessle et al. 2007; Westerterp-Plantenga 2000; Westerterp-Plantenga et al. 1990). It has also been used to measure variations in eating rate between the beginning and end of a meal to categorize subjects as typical (slowing down as a meal progresses), linear, or binging (speeding up as a meal progresses) (Dovey et al. 2009). Current research seeks to extend the approach. Instead of using a single scale to weigh everything, a cloth can be equipped with a fine grained pressure textile matrix and a weight sensitive tablet to recognize where intake actions such as cutting, scooping, or stirring occur, and thus identify individual containers or plate locations (Zhou et al. 2015). A tabletop touch screen has been investigated for the same purpose (Manton et al. 2016). The computer vision used in these techniques ranges from elementary to more sophisticated depending on the implementation. Under controlled conditions, for example, when eating a restricted meal under clinical observation, the computer vision needed to analyze the data is more basic than what is needed to analyze unrestricted eating in a natural environment. An advantage of an instrumented table is that it does not require specific utensils or containers to be used. A disadvantage is that it only measures eating activities occurring at the table.

A third approach to applying computer vision is wearables. Several positions on the human body can be instrumented to detect activities associated with eating (Troster and Amft 2009). Chapter 7 describes the biology and experiments with head-mounted sensors. The arms can be instrumented with motion sensors to detect patterns of limb motion associated with activities during eating, such as the use of cutlery and hand-to-mouth gestures (Amft et al. 2005; Junker et al. 2008; Troster and Amft 2008). In a simpler configuration, the wrist alone can be instrumented with motion sensors to detect patterns of wrist motion associated with eating (Amft 2010; Dong et al. 2009, 2012, 2014; Ramos et al. 2015; Thomaz et al. 2015). The computer vision for wearables is more difficult than for utensils or tables because it is a person being instrumented rather than a tool or location specifically intended for eating. An advantage to wearables is that they are naturally carried everywhere with a person and so can be used wherever they eat. They are independent of tables, utensils, or containers used. A disadvantage is that a person must be willing to wear them. Wrist-mounted sensors can be embodied in a device that resembles a common watch, and ear-mounted sensors can be embodied in glasses or an earpiece. Sensors located at other positions such as on the neck are not as easy to disguise.

A fourth approach to using computer vision involves the use of cameras. Researchers have investigated the automatic recognition of foods in images (He et al. 2015; Pouladzadeh et al. 2014; Zhu et al. 2015). Experiments are typically performed on ten to multiple tens of different food types with moderate to high accuracy in recognition. The computer vision in this approach is the most complex. It requires solving several problems that are known to be difficult for any computer vision problem including segmentation (for separating image areas corresponding to different foods), shape reconstruction (for portion estimation), and classification (for determining food identities). Recall that computer vision is intended to automate problems that can be accomplished manually by a human. Visual inspection of foods to determine identity is difficult for people, especially for mixed dishes and multiingredient items. It is reasonable to consider the typical human ability to recognize foods from images as an upper bound on what can be expected from computer vision research in this area.

All computer vision methods can be assessed using empirical evaluation (Bowyer and Phillips 1998). The process requires defining the problem, collecting a representative data set, *ground truthing* the data, and then using the ground truth to compare the accuracy of different computer vision algorithms. The ground truth is a manual marking of the data by one or more trained experts who carefully review the data and label it to identify the image regions or pieces of signal that are of interest. This evaluation process bears some resemblance to validation as used in the clinical or experimental sense of the word, but it is not the same thing. Its primary purpose is to enable incremental advances in engineering design by allowing different algorithms to be tested to determine which provides the best performance when compared to the human experts that labeled the data. The goal of 100% accuracy cannot always be achieved, but the failures are important as they motivate continued computer vision research to improve the methods.

The most important item in this evaluation process is the ground truth data set. Generating ground truth data sets is difficult, time consuming, and expensive. It requires carefully recording large amounts of data, building software tools to facilitate manual reviewing and labeling of the data, and trained experts to perform the labeling. When computer vision is first applied to a new problem, the data sets tend to be small and independently generated by different research groups who use them only to test their own algorithms. As work in a problem domain matures, larger data sets tend to get generated and shared, so that algorithm performance between groups can be evaluated and compared. Eventually some of these data sets become standards. The U.S. National Institute of Standards and Technology (http://www.nist.gov/) maintains a large number of these data sets, including standards used for fingerprint recognition, face recognition, and optical character recognition. Wikipedia provides a long list of many important data sets (https://en.wikipedia.org/wiki/List_of_datasets_for_machine_learning_research). The progression from small data sets independently developed in separate labs to large data sets settled on as standards is analogous to the increase in the expected number of participants in the phases of clinical research.

For automated analysis of dietary intake, data set generation is in its infancy. All the algorithms being developed are being tested on independent data sets generated by each research group. Partly this is due to the difficulty of generating large data sets that are representative of all eating activities during unrestricted free living. The generation of such data sets will require standardizing the problem definition. It will also be costly due to the difficulty of physically recording computer vision signals and obtaining ground truth about individual eating activities (i.e., meals) of large numbers of people over large periods of time. There are also academic pressures to publish and mine information from data sets before publicly sharing them, and business pressures to produce products. It can be expected that as research continues in this area, shared data sets will eventually emerge. The two case studies described in the next sections involve larger data sets than are typically generated in early proof-of-concept computer vision research and can be considered examples of the current state of the art.

8.3 WRIST-MOTION TRACKING TO MEASURE BITES

During eating, our group discovered that the wrist of a person undergoes a characteristic rolling motion that is indicative of the person taking a bite of food (Hoover et al. 2009). The concept is demonstrated in Figure 8.1. As food is picked up and brought toward the mouth, the wrist rotates. The rolling part of the motion is independent of whatever else the arm does. Anatomically, it can be observed that the comfortable range of rotary motion of the wrist facilitates the process of picking up food and delivering it to the mouth. Animals lacking this rotary ability, such as dogs, cats, and horses, do not use their limbs during eating; instead they eat by moving their head into areas containing food. Our method takes advantage of the rotary motion of the wrist exhibited by humans during eating to detect when a bite occurs.

The pattern of roll motion depicted in Figure 8.1 can be tracked using a wrist-mounted gyroscope (Dong et al. 2009). The relevant pattern is shown in Figure 8.2 where the x-axis is time and the y-axis is degrees per second as measured by the gyroscope. When the wrist is at rest the gyroscope reads zero and as it rotates in either direction the signal exhibits peaks. The computer vision needed to analyze this signal is to look for the peaks. Thresholds are used on the minimum peak magnitude,

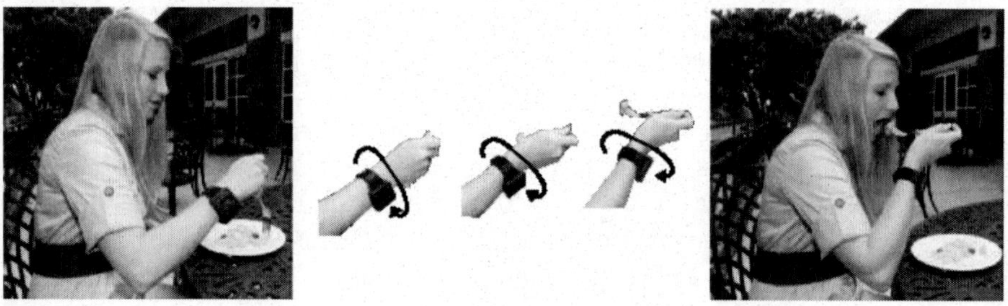

FIGURE 8.1 **(See color insert.)** Depiction of wrist-roll motion that occurs during a bite.

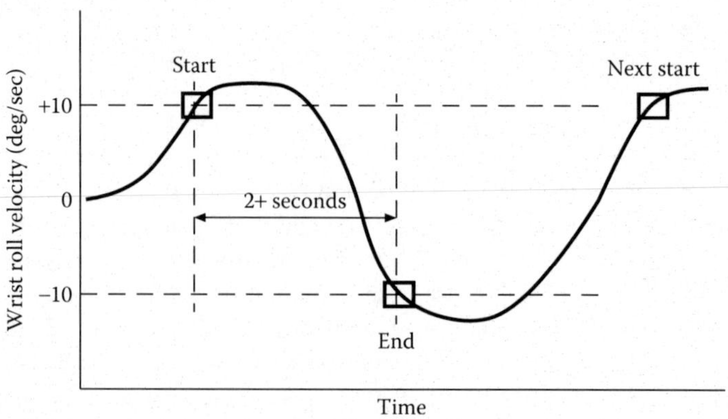

FIGURE 8.2 Tracking of wrist-roll motion using a gyroscope.

time between positive and negative peaks, and minimum time between successive bites. This algorithm can be executed on basic microcontroller hardware due to its low computational complexity. Our group is also exploring more sophisticated hidden Markov modeling of the signal that can recognize a wider variety of appearance of the signal but requires more sophisticated computing hardware due to its higher computational complexity (Ramos et al. 2015).

The following summarizes a set of proof-of-concept experiments from Dong et al. (2009) to test the accuracy of the method. The experiments were conducted in a laboratory setting. See http://www.youtube.com/watch?v=mLOBgBIXPUU for a video clip from the second experiment. Ground truth was established by video recording participants while they simultaneously wore wrist-motion tracking devices and then manually reviewing the video to mark times when bites were taken. In the first experiment, 139 meals from 51 subjects (scheduled for 3 sessions with some drop out) were recorded. Each meal consisted of waffles and was eaten with a fork. The sensitivity of the bite-counting method for counting bites taken during a meal was 94% (6% of actual bites were undetected) and the positive predictive value was 80% (about one false positive per five actual bites). Although the food and utensil were restricted to one type, this experiment provided evidence that the method worked across a large number of people. In a second experiment, 49 meals of 47 subjects (two subjects participated twice) were recorded. Food, beverage, and utensil choices were not restricted; each participant brought their own meal. Example meals included pizza, noodles and vegetables, chicken fingers, sandwiches, and a variety of drinks, eaten with hands and a variety of utensils. A total of 1,675 bites were consumed. The sensitivity of the bite-counting method for counting bites during a meal was 86% and the positive predictive value was 81%. This experiment provided evidence that the bite-counting method worked across a variety of foods and beverages. During these meals, participants spent on average 45% of their time doing things other than taking a bite, such as engaging in conversation, showing that the method is robust in the presence of confounding activities. In addition, all subjects wore two sensors: (1) an inexpensive micro-electro-mechanical systems (MEMS) gyroscope and (2) a more sophisticated magnetic, angular rate, and gravity (MARG) sensor. The experiment revealed that the MEMS gyroscope had the same performance as the more sophisticated MARG sensor.

8.3.1 Cafeteria Experiment

Once proof-of-concept had been established, the next step was to create a much larger ground truth data set. This is necessary to help determine what variables associated with bites cause variations in accuracy, which in turn drives new algorithm development. Our next data collection took place in the Harcombe Dining Hall at Clemson University, South Carolina. This location is important because it is a place where eating occurs naturally, as opposed to the initial proof-of-concept experiments that took place in a laboratory. The cafeteria seats up to 800 people and serves a large variety of foods and beverages from 10 to 15 different serving lines. Figure 8.3 shows an illustration and picture of our instrumented table (Huang 2013). It is capable of recording data from up to four participants simultaneously. Four digital video cameras in the ceiling (approximately five meters height) were used to record each participant's mouth, torso, and tray during meal consumption. A custom wrist-worn device containing MEMS accelerometers (STMicroelectronics LIS344ALH) and gyroscopes (STMicroelectronics LPR410AL) was used to record the wrist motion of each participant at 15 Hz. The three-axis accelerometer and three-axis gyroscope data were all recorded to help with future algorithm development in case additional axes were found to be useful for improving the algorithm. An Ohaus SP401 scale was embedded in the table provided below each tray to continuously measure tray weight. The scale is of course not related to tracking wrist motion. It was included because it allowed for bite-to-grams relationships to be explored in case they proved useful for determining why different types of bites might provide different sensitivities in detection. As will be shown in Section 8.5, the scale data also allowed us to begin development of an automatic bite-detection algorithm based on scale data. All the hardware were wired to the same computers

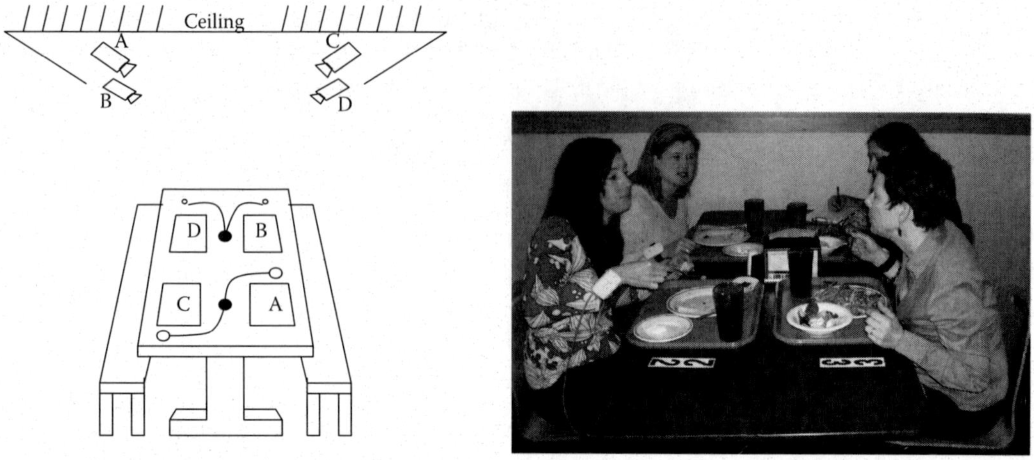

FIGURE 8.3 Cafeteria table instrumented for data collection. Each participant wore a custom-tethered device to track wrist motion. Below each tray is a scale to continuously measure tray weight.

and used timestamps for synchronization. All the data were smoothed using a Gaussian-weighted window of width 1 s and standard deviation of ⅔ s.

A total of 276 participants were recruited and each consumed a single meal (Salley et al. 2016). Note that this is a much larger number of participants compared to the earlier proof-of-concept experiments. In addition, because the setting changed to a more natural eating environment, the recording of raw signals is more complicated. Participants were free to choose any available foods and beverages. On sitting at the table to eat, an experimental assistant placed the wrist-motion tracking device on the dominant hand of the participant and interviewed them to record the identities of foods selected. The participant was then free to eat naturally. If additional servings were desired, the participant was instructed to notify the experimental assistant to assist with removing the wrist-motion tracker before moving through the cafeteria to obtain more food or beverage, returning to the table to begin a new segment of recording. Each such segment is referred to as a course, and meals had to subdivided into multiple courses (when applicable) to facilitate recording. Total usable data includes 271 participants, 518 courses with a range of 1–4, and an average of 1.8 courses per participant.

The creation of ground truth for this data set was difficult because it was so large and the variety of behaviors during bites so diverse. A total of 22 raters contributed more than 1,000 hours to the creation of ground truth. The video was manually reviewed to identify the time, food, hand, utensil, and container for each bite. These variables were considered in case they revealed differences in sensitivity for detecting bites, in which case they could drive further algorithm development. Figure 8.4 shows a custom program we built to facilitate the process. The left panel displays the video, whereas the right panel shows the synchronized wrist-motion tracking data. Keyboard controls allow for play, pause, rewind, and fast forward. The horizontal scroll bar allows for jumping throughout the recording and additional keyboard controls allow for jumping to previously labeled bites. A human rater annotates a course by watching the video and pausing it at times when a bite is seen to be taken, using frame-by-frame rewinding and forwarding to identify the time when food or beverage is placed into the mouth. Once the bite time is identified, the rater presses a key to spawn a pop-up window that allows the user to select from a list of foods recorded as having been eaten by the participant during the course, and a list of hand, utensil, and container options. The agreement of two raters was checked and in the case of disagreement was arbitrated by a third rater. A total of 24,088 bites were annotated in the ground truth. A total of 374 different food and beverage types were chosen by participants. Food and beverage names were taken from the menus of the cafeteria. Some foods are given the generic name of the food line from which they are served due to the heterogeneous mixture of ingredients that could be custom selected by the participant,

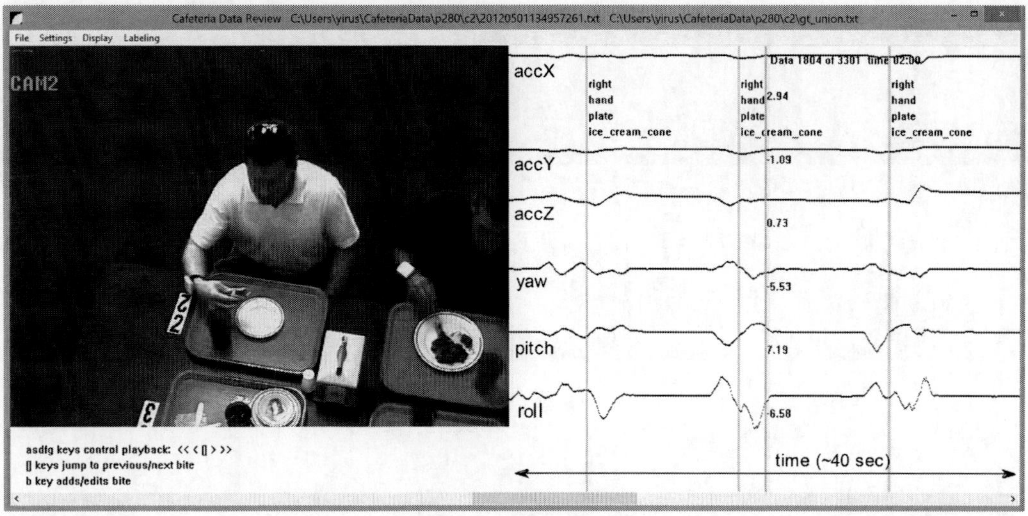

FIGURE 8.4 **(See color insert.)** A custom program created for manual labeling of ground truth bites. The left panel shows the video and the right panel shows the wrist-motion tracking. Vertical purple lines indicate the times marked as bites, the vertical green line indicates the time currently displayed in the video. Variables (hand, utensil, container, and food) are identified for each bite.

for example, from a salad bar. Example food identities include salad bar, shoestring French fries, Asian vegetables, pasta tour of Italy, cheese pizza, homestyle chicken sandwich, hamburger, custom sandwich, garlic bread sticks, fried shrimp, and grapefruit. Example beverage identities include whole milk, coca cola, water, sweet tea, coffee, and apple juice. Foods and beverages were served in four types of containers: plate, bowl, glass, and mug. Four different utensils were used: fork, spoon, chopsticks, and hand. Hand could be identified as left, right, or both.

Across all bites the sensitivity for counting bites taken during a meal was 75% with a positive predictive value of 89%. The algorithm parameters were originally trained using data recorded in a laboratory setting (Dong et al. 2012) in which the average eating rate was slower (n = 49, seconds per bite = 19.1 ± 6.4) compared to what was observed in the cafeteria setting (n = 271, seconds per bite = 14.7 ± 5.6). We therefore experimented with shortening the algorithm parameter controlling the minimum time between detections of bites to six seconds. With this value the algorithm produced 81% sensitivity with a positive predictive value of 83%, which is within a few percentage of the accuracy observed in the laboratory experiment. This observed difference in speed is currently driving new algorithm development to try to automatically detect time between bites and use that information to adjust algorithm parameters.

The next stage in development of data sets would be to repeat this type of experiment in different locations, for example, multiple homes or restaurants. It should also be repeated in multiple cultures, for example, Europe or Asia, to determine if cultural differences affect accuracy. This level of experimentation will require the involvement of multiple research groups. The methods outlined above provide the procedures by which the data can be collected, but it will require a tremendous amount of work to repeat them in a large variety of locations.

8.3.2 Detecting Eating Activities

A limitation of the bite-counting method using wrist-motion tracking is that it requires a person to turn the device on or off at the beginning and end of an eating activity. To overcome this we have been researching methods to automatically detect eating activities (entire periods of time in which a person is eating) in order to enable automatically turning our bite-counting method on or off (Dong et al. 2014). Figure 8.5 shows a plot of wrist-motion activity calculated as the

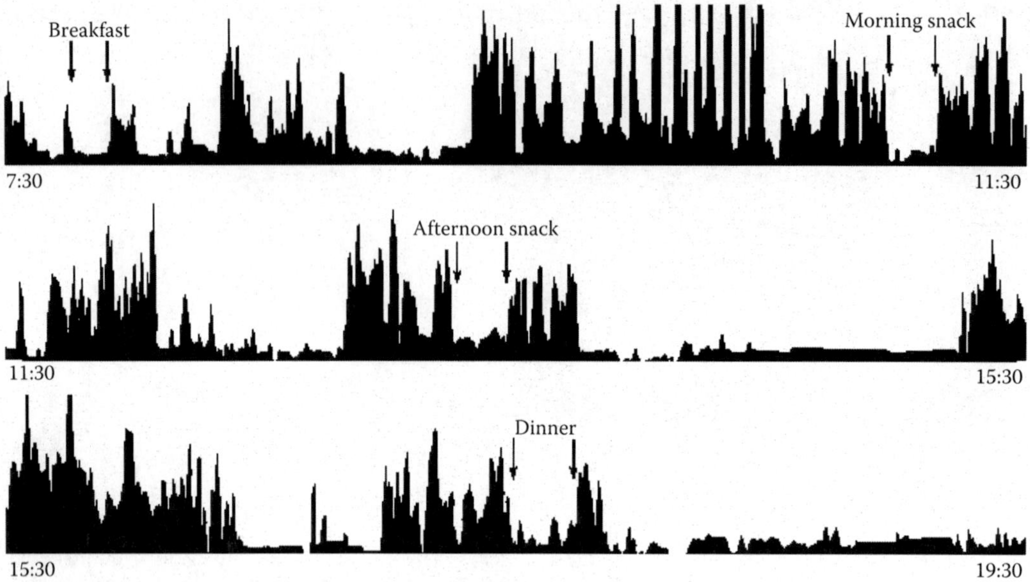

FIGURE 8.5 Plot of wrist-motion activity for an entire day. Manually logged eating activities are delineated. It can be seen that wrist motion peaks prior and subsequent to eating activities.

sum of acceleration across the three axes. The plot shows the measure for a person for a whole day. The meal and snack times manually logged by the person are labeled. It can be seen that the wrist-motion activity spikes both before and after eating for a few minutes each. This typically happens as a person prepares to eat, doing things like unwrapping foods, washing hands, setting dishes, and cleaning up. In between these periods, during the actual eating, wrist-motion activity tends to be moderate but much lower than the two spikes. We use this feature to segment the data and detect potential eating activities. Of course, other activities throughout the day can cause similar spikes, as shown in Figure 8.5. We therefore calculate additional features during each segment for final classification. For example, another feature we developed focuses on rotational motions that tend to be very high during eating activities. We tested our classifier on 43 participants over 449 total hours of data containing 116 periods of eating. Our results showed an accuracy of 81% for detecting eating at one second resolution in comparison to manually marked event logs of eating periods (Dong et al. 2014). This represents a proof-of-concept experiment. Additional algorithm development is ongoing that involves the collection of a much larger data set from 500 people.

8.4 BITES TO ENERGY INTAKE

This section describes two experiments to test the relationship between bites and ingested energy. The relationship was tested at the meal level, that is, the number of bites taken in a meal was compared against the energy (kcal) consumed.

The first experiment collected data in natural free living of 77 people over a two-week period (Scisco et al. 2014). To accomplish this, it was necessary to build custom hardware to facilitate data collection. The methods described in Section 8.3.1 for cafeteria data collection could not be applied as participants would be expected to eat at a variety of locations during natural free living. Therefore we manufactured the device shown in Figure 8.6. The device has a limited memory and does not store raw motion data. Instead, it executes the bite-counting algorithm and stores a time-stamped log of eating activities and associated bite counts. The left button turns the bite-counting algorithm on or off and is pressed at the beginning and end of an eating activity (e.g., meal or snack).

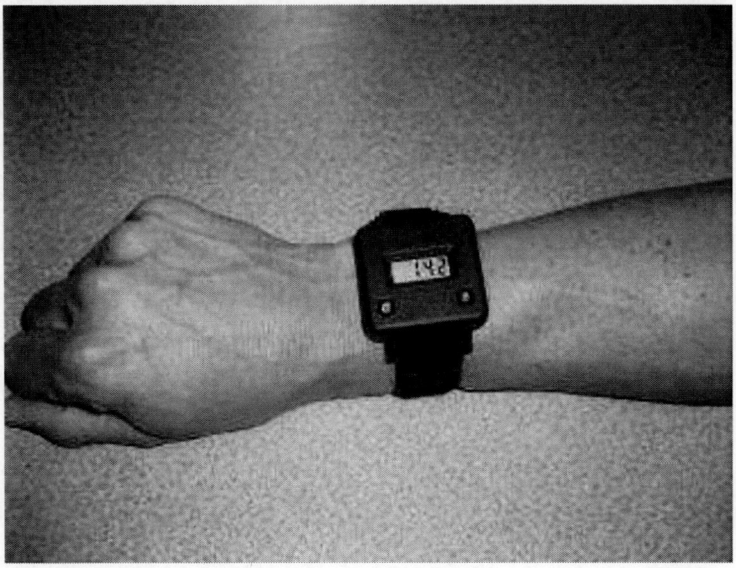

FIGURE 8.6 Bite-counter device used to collect free-living data.

When not in bite-counting mode the device functions as a standard watch. The right button allows reviewing the last recorded activity, total daily activity, and battery charge. The device was programmed to display "ON" during bite counting, instead of the actual count, to reduce potential influence on total intake.

Participants were asked to record every eating activity. To estimate energy intake, participants completed the National Cancer Institute's Internet-based automated self-administered 24-hour (ASA24) dietary recall each day. This is a self-reported measure that can suffer from all the problems associated with self-report including underestimation bias and noncompliance over long periods of time. However, participants were paid to motivate compliance and they were not dieting, they were simply instructed to record as many meals as possible. The use of the ASA24 enabled the collection of a large amount of data at little cost. A total of 2,975 eating activities were analyzed, an average of 39 per person. Figure 8.7 shows plots of data for two participants. The x-axis shows energy (kcal) and the y-axis shows bites; each data point is a single eating activity. The scatter of the points from

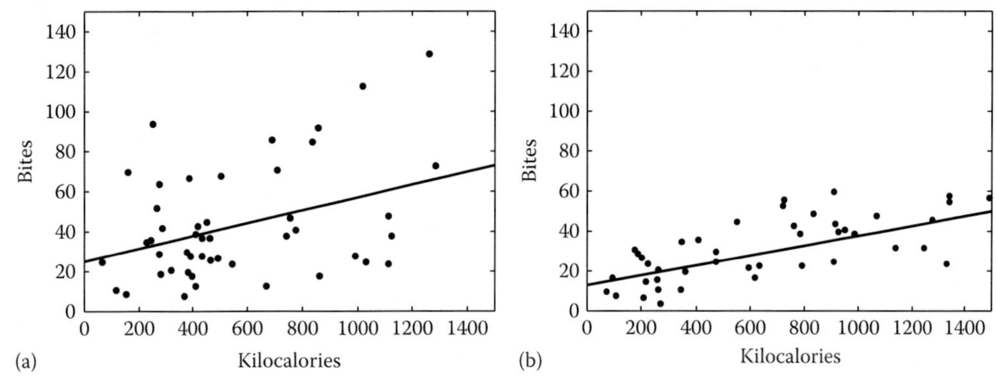

FIGURE 8.7 (a, b) Data for two participants wearing bite counters for two weeks. Each data point is a single eating activity (meal or snack).

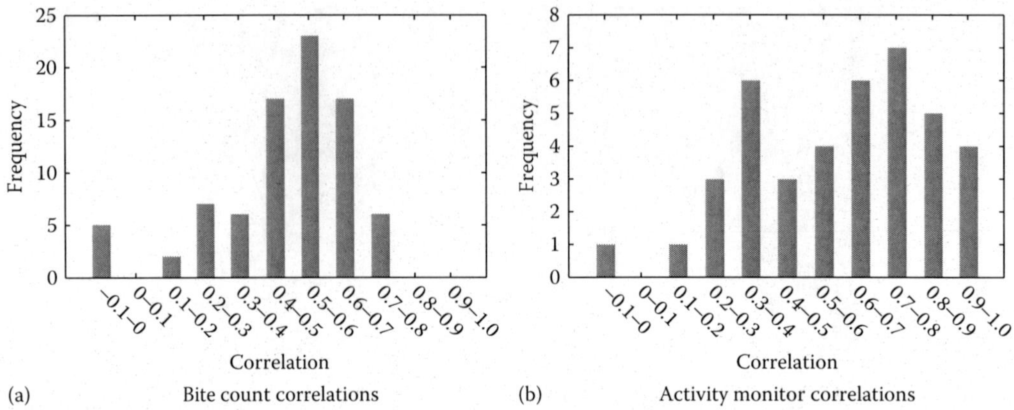

FIGURE 8.8 (a, b) A comparison of the correlations of bite count with energy intake versus physical activity monitors with energy expenditure.

the fitted line shows the effect of dietary variety on the energy per bite relationship. A data point high in bites and low in energy represents a low-energy dense meal such as a salad. A data point low in bites and high in energy represents a high-energy dense meal such as a cheeseburger.

The data in Figure 8.7a correlate at approximately 0.4 and the data in Figure 8.7b correlate at approximately 0.7. Figure 8.8a shows a histogram of the correlations of all 77 participants. The majority (86%) showed correlations above 0.4. For context, Figure 8.8b shows a histogram of correlations found in a metastudy evaluation of 41 studies that compared energy expenditure as measured by physical activity monitors to doubly-labeled water (Plasqui and Westerterp 2007). The comparison must be viewed with some caveats: each data point in Figure 8.8a represents a person, whereas each data point in Figure 8.8b represents a study; the correlations in Figure 8.8a are at the meal level, whereas the correlations in Figure 8.8b are at the daily level; the correlations in Figure 8.8a used doubly-labeled water, whereas the correlations in Figure 8.8b used a self-reported estimate. Nonetheless, the range of correlations in both histograms is similar. Just as counting steps can provide a moderate correlation to energy expenditure, counting bites may provide a moderate correlation to energy intake. The strength of the correlation depends on many variables including foods consumed and individual demographic variables. Section 8.6 discusses future work to continue to explore this relationship.

In a second experiment, we developed a formula to convert bites to energy intake as energy intake (kcal) = bites × kpb where kpb = kilocalories per bite (Salley et al. 2016). The value for kpb is determined individually according to gender, age, height, and weight, similar to the Mifflin St. Jeor formula (Mifflin et al. 1990). The formula for kpb was calculated using the 2,975 eating activities of the 77 participants described above in a multiple linear regression. These formulas were then tested on the cafeteria data set (271 meals) described in Section 8.3.1. Figure 8.9 shows the result. Each bar plots the distribution of estimates of energy intake. The actual energy intake was determined using direct observation by independent observers reviewing the videos and cafeteria database for serving sizes and energies. Each participant was asked to estimate their energy consumed; half were aided by a daily menu that listed serving sizes, and energies for foods served that day and half were unaided. Finally, the rightmost column plots the bite-derived estimate. It can be seen that participants underestimated energy intake in both conditions, but that the use of a menu reduced the underestimation. It can also be seen that the bite-derived estimates were more accurate on average than the participant estimates.

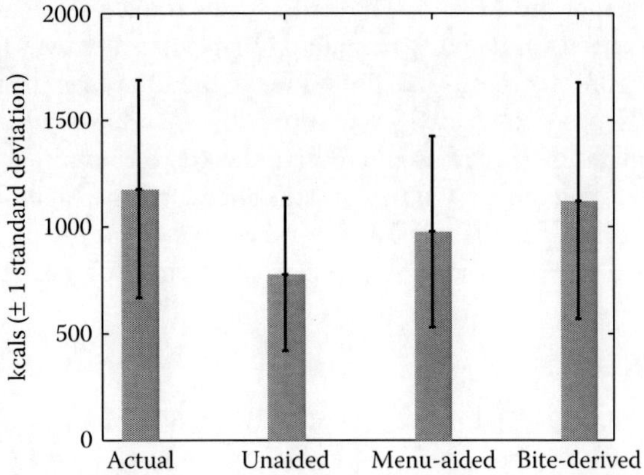

FIGURE 8.9 Comparison of distributions of estimates of kilocalories consumed for 271 meals in a cafeteria. Actual was calculated via direct observation, unaided was participant guesses, menu-aided was participant estimates using serving size and kilocalorie measures on a menu, and bite-derived was calculated using bites × kpb.

8.5 DISCUSSION

Numerous research groups are researching and developing tools to advance the measurement of energy intake. This chapter described a subset of this work related to the automatic measurement of bites. It also tried to provide context for how new engineering developments in this area can be evaluated in terms of their maturity by clinical and behavioral scientists. The field of computer vision plays a major role in the development of these technologies and follows a paradigm of developing ground truth data sets for evaluating algorithm performance. It is very time consuming and difficult to generate ground truth data sets. Currently, the field has not yet established large, publicly shared ground truth data sets of recordings of sensor signals during eating activities. These data sets will likely begin to emerge in the coming decade as more engineering groups get involved in this area, causing the need for standards to evaluate algorithm performance between groups.

Compared to other methods, the biggest advantage of sensors is their potential for real-time feedback. For example, sensors can be used to detect the pace of consumption while a person is eating and provide a cue to slow down (Scisco et al. 2014). Sensors can also be used to monitor portion consumption. For example, a cue can be initiated after a target number of bites has been reached, assisting a person with limiting total consumption during a meal (Jasper et al. 2016).

Sensors also have disadvantages. First, a person must be willing to wear one or more sensors, either during all waking hours or at least during eating activities, or eat at an instrumented table. Second, there is limited potential for determining the type of food consumed using sensors. Third, sensors have limited applicability for monitoring grazing. They are best suited for measuring meals, snacks, and other similar periods of contiguous eating, in which the sensor patterns associated with eating stand out most clearly. Fourth, sensors do not directly measure energy intake.

8.6 FUTURE WORK

Much more work remains to reach the maximum potential utility of bites as a dietary assessment measure. Currently, our group is repeating the first experiment described in Section 8.4 but this time using the remote food photography method (RFPM) (Chapter 6) to measure energy consumed during free living in place of the ASA24. The RFPM uses trained raters to analyze images of the foods consumed and has been validated against doubly-labeled water (Martin et al. 2012). It is interesting to note that in Figure 8.9 there is a slight underestimation bias in the bite-derived measure compared to the directly observed (actual) measure. This may have been due to underestimation bias in the ASA24 estimates of energy used to derive the kpb formula. The ongoing RFPM experiment will enable a three-way comparison that should offer further insight. Future experiments also need to be undertaken to analyze the relationship between bites at the daily and weekly level and compare them against measures of energy intake derived from doubly-labeled water.

8.7 CONCLUSION

Wearable sensors have the potential to offer new methods of dietary assessment that help extend research from the laboratory to the field. In order to maximize the utility of new sensors and the speed at which they can be developed and validated, large, standardized, *ground truth* data sets of eating-related activities need to be established. This chapter attempted to illustrate how this can be realized using the development of a bite-based sensor by a collaborative team of engineers and behavioral scientists applying computer vision best practices.

REFERENCES

Amft, O. 2010. A wearable earpad sensor for chewing monitoring. *IEEE Sens* 1:222–227.

Amft, O., H. Junker, G. Troster. (October 18–21, 2005). Detection of eating and drinking arm gestures using inertial body-worn sensors. Osaka, Japan: *IEEE International Symposium on Wearable Computers.*

Bowyer, K. and P. Phillips. 1998. *Empirical Evaluation Techniques in Computer Vision.* Los Alamitos, CA: IEEE Computer Society Press.

Dong, Y., A. Hoover, E. Muth. (November 1–4, 2009). A device for detecting and counting bites of food taken by a person during eating. Washington D.C.: *IEEE International Conference on Bioinformatics and Biomedicine.*

Dong, Y., A. Hoover, J. Scisco, E. Muth. 2012. A new method for measuring meal intake in humans via automated wrist motion tracking. *Appl Psychophysiol Biofeedback* 37:205–215.

Dong, Y., J. Scisco, M. Wilson, E. Muth, A. Hoover. 2014. Detecting periods of eating during free-living by tracking wrist motion. *IEEE J Biomed Health Inform* 18:1253–1260.

Dovey, T., D. Clark-Carter, E. Boyland, J. Halford. 2009. A guide to analysing universal eating monitor data: Assessing the impact of different analysis techniques. *Physiol Behav* 96:78–84.

He, H., F. Kong, J. Tan. 2015. DietCam: Multi-view food recognition using a multi-kernel SVM. *IEEE J Biomed Health Inform* 20:848–855.

Hermsen, S., J. Frost, E. Robinson, S. Higgs, M. Mars, R. Hermans. 2016. Evaluation of a smart fork to decelerate eating rate. *J Acad Nutr Diet* 116:1066–1068.

Hoover, A., E. Muth, Y. Dong. 2009. Weight control device using bites detection. United States Patent 8310348.

Huang, Z. 2013. An assessment of the accuracy of an automated bite counting method in a cafeteria setting, Thesis, Clemson University.

Hubel, R., R. Laessle, S. Lehrke, J. Jass. 2006. Laboratory measurement of cumulative food intake in humans: Results on reliability. *Appetite* 46:57–62.

Jasper, P., M. James, A. Hoover, E. Muth. 2016. Effects of bite count feedback from a wearable device and goal setting on consumption in young adults. *J Acad Nutr Diet* 116:1785–1793.

Junker, H., O. Amft, P. Lukowicz, G. Troster. 2008. Gesture spotting with body-worn inertial sensors to detect user activities. *Pattern Recognit* 41:2010–2024.

Kissileff, H., G. Klingsberg, T. V. Itallie. 1980. Universal eating monitor for continuous recording of solid or liquid consumption in man. *Am J Physiol* 238:R14–R22.

Kissileff, H., F. Pisunyer, J. Thornton, G. Smith. 1981. C-terminal octapeptide of cholecystokinin decreases food-intake in man. *Am J Clin Nutr* 34:154–160.

Laessle, R., S. Lehrke, S. Duckers. 2007. Laboratory eating behavior in obesity. *Appetite* 49:399–404.

Manton, S., G. Magerowski, L. Patriarca, M. Alonso-Alonso. 2016. The smart dining table: Automatic behavioral tracking of a meal with a multi-touch-computer. *Front Psychol* 11(7):142.

Martin, C. J., J. B. Correa, H. Han, H. Allen, J. Rood, C. Champagne, B. Gunturk, G. Bray. 2012. Validity of the remote food photography method (RFPM) for estimating energy and nutrient intake in near real-time. *Obesity* 20(4):891–899.

McCabe-Sellers, B. 2010. Advancing the art and science of dietary assessment through technology. *J Am Diet Assoc* 110:52–54.

Mifflin, M., S. St. Jeor, L. Hill, J. Scott, S. Daugherty, Y. Koh. 1990. A new predictive equation for resting energy expenditure in healthy individuals. *Am J Clin Nutr* 51:241–247.

Plasqui, G. and K. Westerterp. 2007. Physical activity assessment with accelerometers: An evaluation against doubly labeled water. *Obesity* 15:2371–2379.

Pouladzadeh, P., S. Shirmohammadi, R. Al-Maghrabi. 2014. Measuring calorie and nutrition from food image. *IEEE Trans Instrum Meas* 63:1947–1956.

Ramos, R., E. Muth, J. Gowdy, A. Hoover. 2015. Improving the recognition of eating gestures using intergesture sequential dependencies. *IEEE J Biomed Health Inform* 19:825–831.

Salley, J., A. Hoover, M. Wilson, E. Muth. 2016. A comparison between human and bite-based methods of estimating caloric intake. *J Acad Nutr Diet* 116:1568–1577.

Schoeller, D., D. Thomas, E. Archer et al. 2013. Self-report-based estimates of energy intake offer inadequate basis for scientific conclusions. *Am J Clin Nutr* 97:413–415.

Scisco, J., E. Muth, Y. Dong, A. Hoover. 2011. Slowing bite-rate reduces caloric consumption; an application of the bite counter device. *J Am Diet Assoc* 111:1231–1235.

Scisco, J., E. Muth, A. Hoover. 2014. Examining the utility of a bite-count based measure of eating activity in free-living human beings. *J Acad Nutr Diet* 114:464–469.

Thomaz, E., I. Essa, G. Abowd. (September 7–11, 2015). A practical approach for recognizing eating moments with wrist-mounted inertial sensing. Osaka, Japan: *International Conference on Ubiquitous Computing.*

Thompson, F., A. Subar, C. Loria, J. Reedy, T. Baranowski. 2010. Need for technological innovation in dietary assessment. *J Am Diet Assoc* 110:48–51.

Troster, G. and O. Amft. 2008. Recognition of dietary activity events using on-body sensors. *Artif Intel Med* 42:121–136.

Troster, G. and O. Amft. 2009. On-body sensing solutions for automatic dietary monitoring. *IEEE Pervasive Comput* 8:62–70.

Westerterp-Plantenga, M. 2000. Eating behavior in humans, characterized by cumulative food intake curves—A review. *Neurosci Biobehav Rev* 24:239–248.

Westerterp-Plantenga, M., K. Westerterp, N. Nicolson, A. Mordant, P. Schoffelen, F. ten Hoor. 1990. The shape of the cumulative food intake curve in humans, during basic and manipulated meals. *Physiol Behav* 47:569–576.

Zhou, B., J. Cheng, M. Sundholm et al. (March 23–27, 2015). Smart table surface: A novel approach to pervasive dining monitoring. St. Louis, Missouri: *IEEE International Conference on Pervasive Computing and Communications.*

Zhu, F., M. Bosch, N. Khanna, C. Boushey, E. Delp. 2015. Multiple hypotheses image segmentation and classification with application to dietary assessment. *IEEE J Biomed Health Inform* 19:377–388.

9 Direct and Indirect Measures of Dietary Intake

Use of Sensors and Modern Technologies

Holly L. McClung, Joseph J. Kehayias, Gary P. Zientara, and Reed W. Hoyt

CONTENTS

9.1 INTRODUCTION

A sensor in the broadest sense is an object or device that detects events or changes in its environment and provides a corresponding output. When used in biological applications, they are classified as biosensors. Included within this category are biosensors that can be used for various aspects of dietary assessment.

The United States currently leads the biosensor market, largely driven by interest in medical sensors due to an aging population with chronic diseases such as diabetes and heart related issues. This upsurge in biosensor interest reflects a recent shift in health care focus from chronic disease and end-of-life medical care to managing and maintaining a longer and healthful life. Future wearable biosensors help make this a reality by enabling the ability to detect, analyze, store, and transmit information for prevention, early detection and management of health, fitness, and disease states. More refined and integrated than current wearables, biosensors may be contact-based, implanted, ingested, or injected, and tend to mitigate the inconvenience, discomfort, and expense of regular medical monitoring (W. P. Carey Research 2015). Future medical biosensor development efforts are likely to focus on personalized assessment and feedback in free-living environments (Richinick 2015). In 2020 the global market for biosensors is projected to reach U.S.$ 21 billion with a 10% rate of growth primarily due to emerging technologies and a growing need and demand for new applications (Global Industry Analysts 2014).

9.2 BACKGROUND

9.2.1 What Is a Sensor?

A sensor in the broadest sense is an object or device that detects events or changes in its environment and provides a corresponding output. The term sensor is often used interchangeably with the term transducer. A transducer is a device that converts one form of energy to another, for example, sound to movement; temperature to electricity. For our purposes, we will consider a transducer to be the component of a more complex sensor or sensing system that allows for an observable response. Sensors may be divided into three types: *physical* sensors measure distance, mass, temperature, and so on (e.g., Avatar), *chemical* sensors (*chemosensors and chemoreceptors*) respond to chemical changes in their environment and *biosensors* in which a biochemical process is the source of the analytical signal (Hulanicki et al. 1991; Eggin 2004).

9.2.2 Sensor Function, Differentiation, and Classification

Sensors are composed of three main elements: the *recognition* component, *transducer* component, and *electronic* system. Figure 9.1 shows schematic diagram of a sensor. The recognition component, or receptor, is the part of the sensor where the chemical/biochemical information is transformed into a form of suitable energy for measurement. A key determinant of sensor sensitivity is the ability of the receptor to recognize a specific analyte of interest among the milieu of chemical and/or biological components it may be exposed to. Once recognized, analyte measurement is then performed by the transducer component, effectively quantifying the analyte–receptor interaction via a signal proportional to the presence of the target analyte. The transducer in effect makes use of a physical change that accompanies the reaction of interest, often relative to a reference or baseline signal to provide a usable signal. The electronic system will take the signal produced by the transducer and condition it via amplification, filtering, or other means, and provide a final output suitable for analysis, display, transmission, or storage.

A *chemosensor* is designed to detect a given analyte under well-defined conditions through a chemical reaction, and can be used for qualitative or quantitative determination of the analyte (Cattrall 1997). An analyte is simply the substance whose properties are of interest and are being

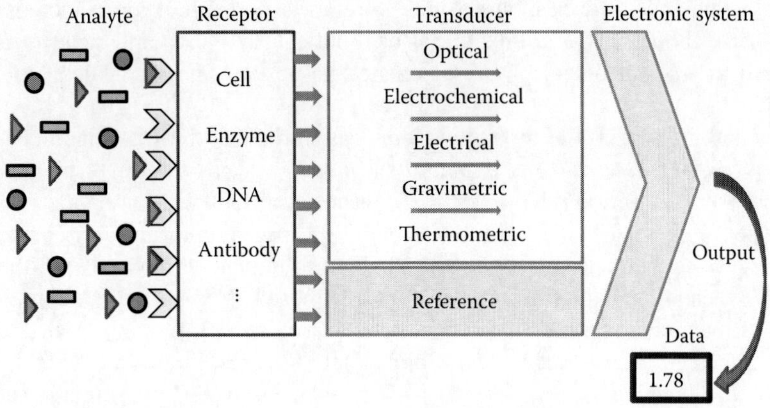

FIGURE 9.1 Schematic diagram of components of a sensor. Main components include the following: *receptor* specific to the *anlayte* of interest; *transducer* to determine the analyte presence (concentration); and *electronics* including a *reference* to condition the signal that produces the *output*, *data* display, or transmission.

quantified. Chemosensors may be classified by the operating principle of the transducer, such as: (1) *optical:* (transduction) where the receptor–analyte interaction is quantified through an optical phenomena (subdivided into absorbance, reflectance, luminescence, fluorescence, refractive index, or light scattering); (2) *electrochemical:* which transform the effect of the analyte–electrode interaction into an electrical phenomenon (further classified as voltammetric or potentiometric); (3) *electrical:* based on measurements where the signal originates from the change in electrical properties caused by the interaction of the analyte and receptor without electrochemical processes taking place; (4) *piezoelectric* or *piezoresistive:* transforming a change in mass or physical deformation into a measurement; or (5) *thermometric:* also referred to as *calorimetric* measurement of the heat effects (exothermic or endothermic) of a specific chemical reaction or absorption (or resorption) involving the analyte (Hulanicki et al. 1991; Eggin 2004; Banica 2012).

Biosensors are considered a subgroup of chemosensors, with the key difference being that the recognition of the target analyte is biological in nature. For example, enzymes or antibodies may be used as the bioreceptor element used to provide a measurable signal. Bioreceptors are often modeled after a biological system that interacts with a specific analyte or biomolecule of interest. A biosensor signal maybe amplified by coupling several enzymatic systems into a multisensor. For example, an auxiliary reaction (or primary reaction) actually detects the system of interest and produces an initial product. This initial product is then converted by a second indicator reaction that better communicates with the transducer and enables a more producible signal (Banica 2012). Biological analytes may range from tissues or cells down to more specific enzymes or antibodies. Biosensors are commonly classified by the particular bioreceptor used, such as antibody or antigen, enzymatic, nucleic acids, epigenetic modifications (e.g., changes in DNA structure), or cellular responses (e.g., monitoring membrane receptor response) (Vo-Dinh and Cullum 2000). Unique and fundamental to the function of a biosensor is the ability to attach the analyte (usually small molecules) to the sensor surface for measurement. Often this is accomplished through surface engineering, or functionalizing, to increase the affinity of the biological molecule to the biosensor surface usually through the addition of polymer coatings or multidimensional layers to chemically or physically entrap the biomolecule of interest (Banica 2012).

9.2.3 The History of Biosensors

Interest in chemosensors began in the 1970s when these new technologies were considered low-cost alternatives to laboratory chemical methods. Basic conductivity-sensing technologies evolved into more selective sensing alternatives such as thermal and piezoelectric sensors. Despite increased selectivity, slow chemosensor response times were an issue leading to an increased interest in optical sensor capabilities. The ability to perform optical, low cost, and rapid measurements of common chemical and biological processes created an upsurge in sensor technology exploration (Eggin 2004).

It was not until 1977 that the term *biosensor* was introduced by Cammann (1977), replacing terms such as *enzyme electrodes* or *biocatalytic membrane electrodes.* A more formal definition of a biosensor was introduced in 2001 (Thevenot et al. 2001). Momentum picked up in the late 1980s when a pen-sized meter for home blood glucose monitoring was commercialized by MediSense, Inc (Newman and Turner 2005). This led to implanted devices with the first success noted in 2007 when an implanted glucose biosensor (Freestyle Navigator™, Abbott Diabetes Care, Alameda, CA) remained operational for five consecutive days (Borgmann et al. 2011). Although implantables appeared ideal for accuracy of biological measures, they came with biocompatibility concerns. Specifically, biologic responses to sensor implantation (e.g., inflammation and encapsulation response) create an abnormal local environment and even device failure, precipitating removal, and replacement with additional risks and procedures (Kotanen et al. 2012). In contrast, wearable biosensors, including minimally-invasive blood glucose monitors, are a common use today and have less risk and a broader range of useful applications.

9.3 DIETARY INTAKE ASSESSMENT IN UNIQUE POPULATIONS

A better understanding and the utility of dietary intake in populations exposed extreme conditions, or subject to unusual dietary restrictions, may in turn lead to new techniques and technologies that prove practical for use by the general nutrition community. Both the military and National Aeronautics and Space Administration (NASA) nutrition programs have invested in new dietary intake assessment capabilities due to population-specific needs and constraints. Military and astronaut populations are unique in that their dietary intake and food systems are highly-regulated, consumption monitoring is feasible, and participant motivation is robust. In addition, strong leadership support exists for accurate dietary intake assessment before, during, and recovery from missions due in part to the documented benefits to both physical and psychological performance (Smith and Lane 1999b; Lieberman et al. 2002; McClung et al. 2009b). The developments of dietary intake sensors for these special populations can, however, create opportunities for broader scientific applications.

Nutritional issues affect the military community from both ends of the dietary intake spectrum, overconsumption and weight gain is prevalent in garrison (i.e., military base setting), and severe underconsumption leading to undesirable weight loss (in terms of degree and composition) in field or deployed scenarios (Baker-Fulco 1995). Total body weight loss can be significant during sustained field training exercises with reported fat-free mass (FFM) losses accounting for ~25% of weight loss (Hoyt et al. 2006). However, service members are increasingly aware that good nutrition is tightly linked to not only personal performance, for example, proper fueling is needed to improve and sustain strength with aerobic and cognitive performance (Lieberman et al. 2002; McClung et al. 2009b), but also to career goals, for example, physiological performance standards and body age- and sex-specific body fat standards must be met to be eligible for promotions or awards (U.S. Department of the Army 2006). Likewise, for the space program the performance and health of astronauts has been linked to energy imbalance and inadequate food intake (30%–40% below recommendations) during both short- and long-duration missions (Smith and Lane 1999a, 1999b; Smith et al. 2001). These aims are not dissimilar to those for athletic training and performance, and thus direct translation to studies in this population is anticipated.

To mitigate undesirable changes in muscle, fat, and bone mass, and the associated impact on health and mission performance the military and NASA have developed unique feeding to ensure nutritional well being of the individual, and ultimately improve the likelihood of successful missions. Military and space-based food systems consist primarily of preplanned, extended shelf-life, and packaged foods that follow population-specific dietary intake recommendations (Lane and

Feeback 2002; U.S. Department of the Army 2016). Space dietary intake is unique in that food is not only prepackaged, but often dietary intake patterns or meal plans are individualized and preselected, six or more months in advance for those missions where outside resupply is not available (Lane and Feeback 2002).

Programmed food systems may seem like a dream for a researcher dealing with the unstable and ever growing food environment of free-living test participants. However, working with a *captured audience* such as the military or space community comes with a host of challenges. For example, researcher access to test volunteers is often strictly controlled by the instructors in order to maintain training schedules; the time soldiers are given to consume their food is typically far too short; unprogrammed ration stripping, that is, removal of unwanted items, prior to missions to lighten the load carried; and research is a secondary concern of military leadership. During most data collection periods, time is a limiting factor and creative time management skills are often needed to incorporate appropriate dietary intake assessments with a minimal footprint. It is not unusual for daily interactions to be limited to <10 minutes per participant for the successful completion of dietary intake, anthropometric, and biochemical measures. In space flight, data collection may be even more complicated with biochemistry idiosyncrasies, data transfer, and communication delays making data intake and outcomes often a challenge, for example, water recycling complicates the use of doubly-labeled water (DLW) (Stein et al. 1999; Smith and Lane 1999b).

Nutrition researchers have developed expedient methods to accomplish dietary intake data collection under a variety of limitations, often utilizing a range of familiar tools and techniques. These include the use of multiple data collection stations, for example, rotation among surveys, anthropometric measurements, phlebotomy, food recall, and isotope dosing or collection, to manage work flow, and the use of practical methods such as food wrapper collection and the use of bar code scanning for dietary intake collection.

9.3.1 Capture of Individual Dietary Intake

Not unlike any controlled feeding trial, the military and NASA nutrition programs assess dietary intake through an array of traditional techniques and data collection methods. Both groups often utilize smaller, representative groups in laboratory or field settings to capture detailed individual dietary intake data that can be extrapolated to larger populations, for example, ground-based research extrapolated to extended-duration space flights or platoon-based research extended to company interventions. Historically, dietary intake is obtained using self-reported methods such as diet histories, food frequency questionnaires (FFQs) (Smith and Lane 1999b, 2001; Lutz et al. 2012), 24-hour recalls, food records or diaries ranging from three to seven days (Pikosky et al. 2008; McClung et al. 2009a), complemented by biochemical measures (Smith and Lane 1999b; McClung et al. 2009b) or as tools used in series (e.g., FFQ and biochemical data) (Smith et al. 2001; Pasiakos et al. 2011, 2015) to document caloric intake and macronutrient and/or specific micronutrient distributions and consumptions across available food items.

In the late 1990s, the Nutritional Biochemistry Laboratory at the Johnson Space Flight Center (Houston, TX) developed a population-specific computerized FFQ to aid in dietary intake assessment during space flight (Smith et al. 1998). Currently, International Space Station (ISS) crew members complete the FFQ weekly. Data files are telemetered to earth for analysis and reports are generated for the flight surgeon who focuses on key areas of concern, for example, water, calories, protein, iron, calcium, and sodium, and provides near real-time nutritional recommendations (Smith and Lane 1999b). An integrated *smart medical system* is being developed for use on the ISS or a future trip to Mars, is a nutritional status assessment system that integrates daily dietary intake, anthropometric measures, and biochemical measures (e.g., hydration status and micronutrient status) (Soller et al. 2002).

When strict dietary control is required, food records are typically analyzed using dietary software linked to standardized and custom-built databases. Examples include intervention studies

(Pikosky et al. 2008; Pasiakos et al. 2011, 2015), and individual nutritional status monitoring during a space flight (Smith et al. 2001). The value of accurate dietary intake assessment through the use of food records generally outweighs the high participant and researcher burden that accompanies their use; however, new technologies offer ways to streamline dietary intake data collection and analysis, particularly within the uniquely structured environments of the military and space programs. In 2009 the military compared self-reported energy intake data collection using a personal digital assistant (PDA)-based dietary intake program with traditional paper-and-pencil-based food records over a seven-day study were DLW served as the gold standard measure (McClung et al. 2009a). The PDA method was found to be as accurate as written records for estimating dietary intake in weight-stable individuals and was perceived to be less cumbersome and time consuming. Currently NASA is fielding a new software application for touch-screen tablets that monitors the dietary intake of individuals in real time. The dietary intake information input is linked to a population-specific nutrient database for analysis, with the output used by medical monitors to make point-in-time dietary corrections during flight (Smith and Zwart 2016).

9.3.2 Capture of Group Dietary Intake

Larger scale military nutrition studies are routinely conducted to collect more representative dietary intake samples or to capture detailed field data during specific trainings (e.g., basic training). The need to study test populations of ≥100 during short data collection periods have led nutrition researchers to rely heavily on dietary intake questionnaires such as the validated Block FFQ (Block et al. 1990). Questionnaires are tailored to capture specific intakes during selected time periods, for example, intermittently during 3-to-10 week training courses, often complemented by blood measures of nutritional status (e.g., iron and vitamin D) (Lutz et al. 2012).

Although accurate, use of direct observation and visual estimation (portions served and plate waste) for dietary assessment are even less well-suited for garrison-based research studies of dietary intake in dining facilities. Typically, service members dine simultaneously in large groups, are allowed little time to eat (<20 minutes), and researchers are required to minimize interference with test volunteers. To overcome these challenges, researchers studying military populations are increasingly using digital food photography (DFP) methods to augment traditional dietary intake assessment methods. The DFP method is rapid, reliable, accurate, and reduces the various demands on test participants and avoids other problems, for example, inaccuracies among data collectors, requirement for large number of food standards, and so on. (Williamson et al. 2003, 2004; Crombie et al. 2013). A relatively recent advance in DFP, the remote food photography method (Martin et al. 2009a, 2009b), may make quantification of dietary intake more accurate in a free-living environment, such as the military garrison setting.

Other techniques utilized extensively in the military, and to a limited degree by NASA include the use of the DLW method to monitor energy expenditure during periods of intense training, semi-starvation in remote environments where participants are inaccessible (Hoyt et al. 1991; Stein et al. 1999; Tharion et al. 2005). The DLW method has been used to estimate dietary intake during extreme trainings (Hoyt et al. 2006; Margolis et al. 2014a, 2014b), space flight (Stein et al. 1999), and intervention studies (Margolis et al. 2016). Researchers have also utilized ^{13}C and ^{15}N stable isotope methodology in the lab (Pasiakos et al. 2011, 2015) and field environment (Margolis et al. 2014b, 2016) to better understand protein metabolism.

9.3.3 Future Dietary Assessment in Special Populations

Military and space nutrition programs have unique requirements and constraints, but also offer opportunities to develop new dietary intake methods. As in the civilian sector, the need is to understand individual's dietary intake patterns, nutrient intakes, and deficiencies to support valid nutrition recommendations. Toward this end, new dietary intake assessment methods are being developed and validated that facilitates the timely collection of accurate and real-time consumption data with a minimal burden on test participants.

9.4 BIOSENSORS FOR DIETARY INTAKE ASSESSMENT

Biosensors detect biological components (analytes) of interest, commonly through optical or electrochemical detectors (Banica 2012). Biosensors, which are often capable of near real-time measurements of analytes (e.g., continuous glucose monitors), offer new ways to understand nutrient absorption, metabolism, and excretion, and individual requirements for energy and macronutrients and micronutrients.

9.4.1 Blood Biosensors

For decades blood analysis has been the gateway to a better understanding of nutritional biochemistry and status markers. Venous blood sampling typically requires trained venipuncturists, a sterile environment, and the funds and specialized equipment for analysis. The diabetic community is particularly burdened by the need for repeated blood sampling and analysis. To reduce this burden, bioengineers developed blood glucose biosensor systems, which are among the most common and most profitable biosensors in use today.

Typically, an amperometric sensor measures glucose concentration as a function of electrons released during glucose oxidase-mediated glucose oxidation (Newman and Turner 2005). A major early contribution to in vivo glucose sensing was the development of a needle-type enzyme electrode suitable for subcutaneous implantation (Shichiri et al. 1982). This work led to commercial electrochemical transcutaneous continuous glucose monitoring (CGM) systems (Newman and Turner 2005). The CGM systems have enabled diabetes management to evolve beyond traditional dietary intake assessment and periodic blood glucose measurements. For example, CGM has been used concurrently with written food records to map carbohydrate-specific dietary intake contributions and timing (Fabricatore et al. 2011), and to assess diet quality (Nansel et al. 2016). Traditional dietary intake measures used in concert with CGM has shown positive effects on long-term biomarkers, such as HbA1c, over the course of an 18-month diet quality study (Nansel et al. 2016). CGM has also been used in studies of cognitive performance under caloric deprivation (Lieberman et al. 2008), and to investigate carbohydrate metabolism during strenuous exercise (Sengoku et al. 2015).

Optical biosensor technologies provide various noninvasive spectrophotometric means of assessing physiological and nutritional status. For example, carotenoid (vitamin A) status, which is normally assessed in serum and tissue samples using high-performance liquid chromatography (HPLC) methods, can also be assessed by transcutaneous Raman spectroscopy (RS)-based photonic methods (Zidichouski et al. 2009). The RS method has been used concurrently with self-reported dietary intake to validate carotenoid consumption (Smidt and Burke 2004). The RS method of evaluating carotenoid status is attractive given it is fast, painless, cost effective, and ideal for spot measurements or periodic monitoring of large groups of test volunteers (Zidichouski et al. 2009).

Near infrared continuous wave spectroscopy (NIR) is commonly used to measure venous hemoglobin oxygen saturation (Mancini et al. 1994). NIR has also been validated for the continuous measurement of muscle oxygen saturation and pH, a measure of anaerobic metabolism (Ellerby et al. 2013). NASA has also explored the feasibility of measuring blood electrolyte chemistry (sodium, potassium, and calcium) with NIR and chemometric sensors (Soller et al. 2003).

9.4.2 Breath Sensors

Analysis of exhaled breath can offer rapid and noninvasive insights into nutritional status. Readily repeatable, breath analyses can also complement other diagnostics, for example, blood, urine, and tissue. Analytical techniques for breath include mass spectrometry, gas chromatography, photoacoustic spectroscopy, chemiluminescence, and nanotechnology-oriented techniques (Mathew et al. 2015). Introduction of nanotechnology into the field of breath analysis promises to increase sensor sensitivity and reduce response time, thanks in part to the increased surface area and active interfaces of the nanomaterials (Xu et al. 2014).

Elevated ketone concentrations in breath can reflect fasting, low carbohydrate diets, glycogen depletion with prolonged exercise, or untreated type 1 or (to a lesser degree) type 2 diabetes mellitus. Detection of breath acetone could be used as a screening tool for the early onset of type 1 diabetes or to help guide nutritional counseling (Toyooka et al. 2013). Recent crowd-funded product development efforts (Indiegogo 2016) include the *Breathometer*™ (2013), a sensor developed to detect blood alcohol concentration in exhaled breath through a smartphone plug-in breathalyzer device, and *Mint*™ (2014) a plug-in sensor from the same company (Breathometer, Burlingame, CA), that detects breath odor through measurement of volatile sulfur compounds (hydrogen sulfide, methyl mercaptan, and hydrogen disulfide). Another emerging technology is a low cost, hand held, and personalized metabolic sensor that measures oxygen consumption and carbon dioxide production from the volume and composition of exhaled breath (Candell et al. 2016). This personal metabolic sensor will enable periodic determinations of energy expenditure and metabolic fuel use (respiratory exchange ratio, RER) in free-living individuals.

Exhaled breath condensate, captured using a cold trap, can be analyzed for H_2O_2, a clinically-relevant marker of oxidative strain (Marek et al. 2009), and lactate concentration, an indicator of exercise intensity (%VO_{2max}) (Marek et al. 2010). This noninvasive method is promising, but needs further characterization and validation.

9.4.3 Smart Sensors

Normally, dietary intake assessment is cumbersome for both user and researcher alike. Methods that automate dietary intake data collection, and reduce or remove the need for manual data input, will help test volunteers stay engaged during protracted periods of dietary assessment. For example, body-worn sensors for energy expenditure have been shown to provide objective measurements with less user burden and cost than traditional energy balance methods (Westerterp 2009). Currently, most consumer-grade wearable devices track body motion and heart rate, and derive activity and energy expenditure estimates via wrist-worn accelerometry and photoplethysmography sensors (Tamura et al. 2014). Wearable technologies specifically relevant to dietary intake assessment include a device introduced to the dietetics community that detects eating periods through multisensory lanyards worn around the neck (Sun et al. 2010). Watch-like sensors for the wrist (Hoover, Chapter 8 this book) and smart eyeglasses, that is, Google Glass, recent preliminary studies have shown promise for both activity (Hernandez et al. 2014) and dietary intake monitoring (Zhang et al. 2016). The modification of existing commercially-available wearable sensor systems to enable dietary intake data collection, for example, through the addition of new algorithms, is likely to continue.

The extent to which automated capture of dietary intake assessment data are reliable and accurate depends on the determination of the food portion size and precise identification of the food type. Typically, quantifying food portions eaten have been challenging, with significant effort spent developing methods and technologies to avoid or limit user bias with respect to food volume and weight measures. Use of standardized tools, cards, and even anatomical measures (e.g., user's thumb), have been utilized as points of reference to increase accuracy of photographic estimations by trained individuals (Williamson et al. 2004) or by automated image processing (see Chapter 5). Preliminary smart kitchen studies have incorporated equipment and tables that are capable of recording the weight of food, with or without plates, and before and after meal consumption (Chang et al. 2006).

Determination of food type appears simple given that many packaged foods come with food identifiers that can be recorded and linked to nutrient component databases or food labels. For example, Universal Product Codes (UPC) (Stevens et al. 2010; Slining 2012) and Global Trade

Item Number (GTIN) can be scanned with hand-held scanners or smartphones to capture nutrition information. In addition, grocery store purchases can be automatically fed into a food record to quickly create records of individual dietary intake and limiting the time spent laboriously matching food consumed with database item (Brinkerhoff et al. 2011). However, not all food comes packaged with bar codes or food labels with nutrition facts. Other sensing and informatics-based research tools are needed to determine food type and nutritional composition, for example, by food classification-based acoustical sensing (Passler et al. 2012), or through the use of miniaturized NIR spectrometers that can scan food items and determine characteristic food matrix properties (Tellspec 2016).

Many new sensor technologies are stand-alone, wearable devices that serve a specific purpose, and utilize proprietary firmware, data transfer, and storage methods. Other approaches require the acquisition of new (kitchen) equipment or alterations of the individual's physical environment and require users to learn to operate and even troubleshoot such systems. Clearly, reliable, open-architected easy-to-use systems are needed, for example, sensors or multisensor systems linked to smartphones enabled with various software applications (Swan 2012).

9.5 INDIRECT ASSESSMENT OF DIETARY INTAKE

Changes in body composition are related to energy balance and food composition. This approach of evaluating dietary intake goes beyond body weight as an outcome; instead it uses changes in body composition. We study body composition by dividing body mass into compartments, either by function (e.g., bone, skeletal muscle, adipose tissue, and extracellular fluid) or by chemical composition (e.g., water, protein, triglycerides, bone mineral, and glycogen). We then study the balance among these compartments to monitor nutritional status, function, and physiological parameters, or even to measure the efficacy of nutritional and pharmacological interventions. Body composition also provides information on gender differences regarding fuel metabolism.

9.5.1 The Body Composition Approach

Even without an intervention or disease, the composition of the body changes naturally with age or under the effect of environmental and lifestyle factors such as climate, occupation, intensity, and type of exercise. Understanding the mechanisms that govern these changes will suggest ways of preserving or enhancing lean tissue and functional capacity. Most methods of measuring body composition are based on certain assumptions that may not always hold for all subjects especially when applied to diseased populations or under conditions of extreme environmental conditions or physical activity. For example, the derivation of FFM from the measurement of total body water (TBW) relies on the assumption that the hydration of FFM is constant (0.83). Similarly, measurement of total body fat by densitometry assumes constant physical densities for FFM and fat mass (FM). Slight deviations from the assumed constant values result in large errors (Kehayias 1993). As we evaluate energy intake, balance, and the efficacy of nutritional interventions, the techniques employed should have the ability to detect (rather than assume as constant) numerically small, but physiologically significant, changes in tissue composition.

The two most commonly used body composition models are: (1) the two-compartment model, where body mass is simply divided into fat and FFM; and (2) the four-compartment model, which explicitly includes the major components of FFM: protein, bone mineral, and water. We use the two-compartment model by estimating FFM with a physical measurement (such as body density by underwater weighing) and then deriving fat mass by subtraction from body weight. The four-compartment model and its variations provide us with closer monitoring of body composition. As recent measuring techniques improved, more detailed body composition models have been developed (Wang et al. 1992). We will introduce later the *quality of lean* principle and will discuss how this parameter provides the most sensitive tool for evaluating the long-term effect of dietary intake.

9.5.2 Total Body Potassium: Connecting Body Composition to Function

Total body potassium (TBK) is a measure of body cell mass, the metabolizing, and oxygen-consuming portion of FFM. TBK is closely related to skeletal muscle mass and its function (Fiatarone et al. 1994). The findings support the hypothesis that there is a systematic and continuous decline of cell mass and muscle throughout adult life. In most cases, this loss of muscle mass is accompanied by a decline of other components of lean tissue, such as protein and bone mineral. Both men and women lose on average 5% of their original TBK per decade. The systematic difference between men and women disappears when potassium content (TBK/weight) is used instead of absolute values. Figure 9.2 shows the results of total body potassium (TBK) measurements. However, when it comes to percentage body fat, there is a systematic gender difference as shown in Figure 9.3 that shows percentage body fat measured by neutron inelastic scattering. Hoyt et al. reported that female cadets under sustained exercise and at negative energy balance maintained significantly more fat-predominant fuel metabolism than male cadets (Hoyt et al. 2006).

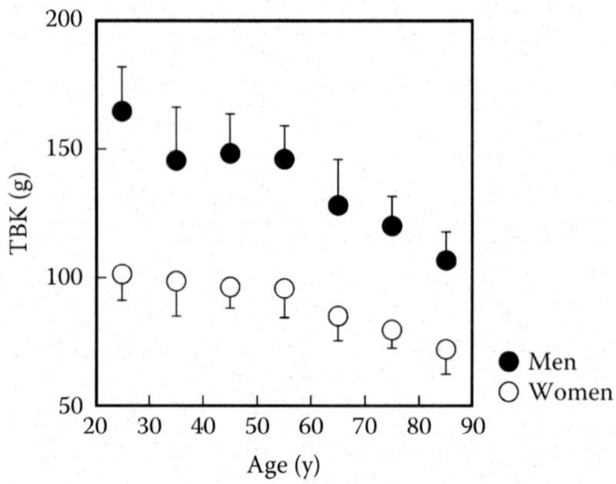

FIGURE 9.2 Total body potassium (TBK) measurements for men and women plotted as a function of age. The points represent the average TBK for each age group and the error bars one standard deviation. (Adapted from Kehayias, J. J. et al., *Am. J. Clin. Nutr.*, 66, 904–910, 1997.)

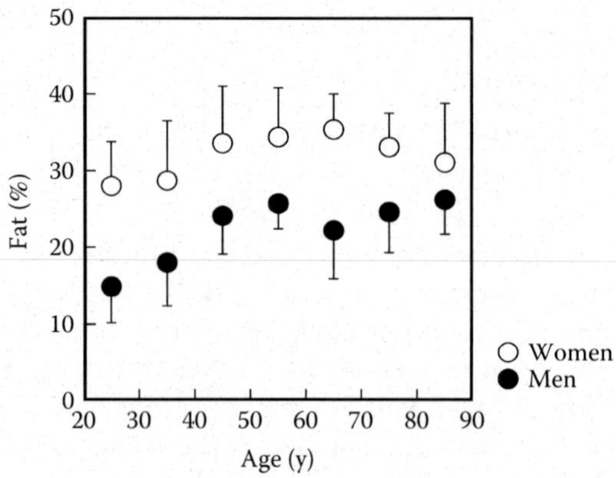

FIGURE 9.3 Percentage body fat measured by neutron inelastic scattering by decade of age in a healthy population. (Adapted from Kehayias, J. J. et al., *Am. J. Clin. Nutr.*, 66, 904–910, 1997.)

Historically, weight-for-height has been used as the macroscopic criterion of nutritional status. The availability of body composition devices, such as DEXA scanners, made the use of FFM as a measurable outcome of dietary intake and exercise possible. This approach protects against unwanted, even dangerous, and disproportional increases in body fat or significant losses of lean tissue. It is possible, however, to increase FFM without an increase in muscle, cell mass, or metabolism, by increasing extracellular fluid. The *quality of lean*, defined as the cell-mass content of FFM, represents the oxygen-consuming and metabolizing fraction of FFM. It is an indirect measure of skeletal muscle and the most sensitive body composition index of nutritional status. In practice, the *quality of lean* factor can be measured by TBK, assessed by whole body gamma counting of ^{40}K, and divided by FFM (Kehayias et al. 1997). A true measure of this quality factor was difficult until the introduction of direct techniques for measuring lean mass. This is because earlier models had to assume a constant composition of FFM or even derive lean mass directly from TBK. Acute changes in hydration status do not allow derivation of FFM from TBK but instead they require an independent FFM measurement, for example, from a DEXA scan.

TBK models have taught us that the important outcome of any dietary regimen is the maintenance or increase of the quality of lean mass. Goals for *optimum* quality of lean (or muscle mass) should be defined separately for men and women.

9.5.3 Field-Expedient Employment of Stable Isotopes

Direct assessment of intracellular mass is achieved by TBK measurements. This, however, is not practical in the field because of the type of equipment required for the measurement. A surrogate approach is the simultaneous use of two dilution methods: deuterated water (deuterated water) for TBW, and NaBr for extracellular water (ECW) space. Both methods have been used extensively for field measurements of TBW, hydration, and by subtraction of intracellular space (Schoeller et al. 1980; Miller and Cappon 1984; Prelack et al. 2005). Some of the very first applications of this approach were in the assessment of dietary intake under specific medical or physical conditions such as cystic fibrosis, severe burns, high intensity physical activity, and so on.

Measurement is simple: known amounts of D_2O and NaBr are administered together in solution by mouth or intravenously. After three hours, isotopic equilibrium is achieved in the water compartments. A biological sample is collected as plasma, serum, urine, sweat, or saliva and analyzed for D and Br. Specifically, deuterium-to-hydrogen (D/H) ratio can be analyzed by table-top cavity-enhanced laser resonance techniques, replacing isotope ratio mass spectrometry. Similarly, Br in plasma or urine can be analyzed by nondestructive energy dispersive X-ray fluorescence (XRF) (Zaichick 1998; Kehayias et al. 2012). New analytical techniques have improved the portability, cost, and precision of this approach.

9.5.4 Use of ^{13}C for Macronutrient Intake Modeling

The stable isotope ^{13}C may be useful in developing alternative ways for the evaluation of dietary intake, absorbed food composition, and can provide a valuable alternative to traditional methods (e.g., FFQ and food records). In 1988, Elia et al. proposed a model for short-term energy expenditure measurements based on breath CO_2 measurements following a 36-hour infusion of ^{13}C-labeled bicarbonate (Elia et al. 1988). Isotopic enrichment in breath correlated with activity levels (higher during sleep and lower during exercise). Although infusing methods are not practical and at the time some of the experiments were still performed with radioactive labels, this model and supporting experiments increased knowledge in modeling the active pools participating in the bicarbonate kinetics. Elia's model also documented the timing and magnitude of response of CO_2 production, the isotopic enrichments in relationship to energy expenditure, and established correlations between breath and urine CO_2. More recently, with the advancement of measuring $^{13}CO_2$ by mass spectrometry, bicarbonate studies have been expanded to include physiological responses to specific diets following oral administration of prelabeled nutrients. Larsson et al. have used this modern version of the bicarbonate method with dogs to study short-term metabolizable energy (ME) by measuring isotopic enrichment in breath CO_2. A mathematical model was used to calculate ME based on the composition of food intake (Larsson et al. 2015).

Protein intake and digestion was studied by Geboes and colleagues who used ^{13}C labeled amino acids that were incorporated into protein (Geboes et al. 2004). To simulate *natural* intake conditions the investigators prepared food using eggs from hens that were fed ^{13}C-leucine that resulted in the labeling of the egg white and yolk. Stable isotopes have been in the military environment to assess protein turnover and amino acid supplementation, typically by means of ^{13}C-leucine infusion methods (Pasiakos et al. 2011, 2015), and more recently using ^{15}N-glycine oral administration in the field (Margolis et al. 2016).

Another approach is to take into account the fact that some foods (such as corn products) are naturally enriched in ^{13}C isotope. Investigators took advantage of the observation that sweeteners based on corn syrup or sugar cane are naturally enriched in ^{13}C with respect to the natural abundance of this isotope (see Chapter 14). This is because they are using a unique photosynthetic pathway to incorporate CO_2 from the atmosphere to the plant molecules, which favors the heavier carbon isotope. The investigators proposed a biomarker for blood glucose, and sugar intake, based on breath ^{13}C enrichment.

Early metabolic models used body weight as the outcome of their predictions. Body composition provides a more useful description of the effect of dietary intake. We expect that metabolic models based on carbon kinetics are not restricted to breathe CO_2 and can be developed to include most of the significant processes of metabolism when combined with large molecular analysis in blood and urine. One of the goals of modeling, however, is to identify easy ways to measure biomarkers to best describe the intake needs of the individual. Another goal is to develop and validate personal metabolic sensors that can monitor metabolic processes and serve as behavior modifiers.

9.6 FUTURE DIRECTIONS AND CHALLENGES FOR WEARABLE SENSOR SYSTEMS IN DIETARY INTAKE ASSESSMENT

9.6.1 Predictive Modeling and Sensor Integration

Integrated health care is a forward-looking approach to managing various health and disease states that can include, for example, sociomedical aspects (e.g., conversational medical advice), wearable sensors, and Cloud services for content delivery and data security (Chouvarda et al. 2014). An integrated, common-platform approach enables interventions by the health care team, to include patient-related (e.g., self-management), professional-directed (e.g., provider education), and organizational (e.g., continuum of care and follow-up) care.

Integration of personal sensors, personalized predictive models, and individualized avatars, will provide new opportunities for nutritional assessment and biofeedback. Imagine dietary intake assessment free of the typical burdens associated with self-recorded food records, weekly meal interviews to guide meal patterns, or monitor personal fitness goals, and monthly blood work. Instead, personal sensors linked to individualized applications and models hosted on a smartphone will track your fuel metabolism by bites, exhalations, and steps, and teach you fundamental principles of nutrition and physiology. This idea is not far from fruition, Massachusetts Institute of Technology's Lincoln Laboratory (MIT LL) engineers in partnership with U.S. Army nutritionists, physiologists, and modelers are developing tools to make this part of the everyday norm. An austere model has been developed that provides a simplified but rigorous view of the dynamics of fuel metabolism. Figure 9.4 illustrates the metabolic model; a key goal is to streamline the model to the point it can be understood and used by the public (Candell and Shaw 2014). The Candell and Shaw model of fuel metabolism focuses on carbohydrate management, rather than body weight

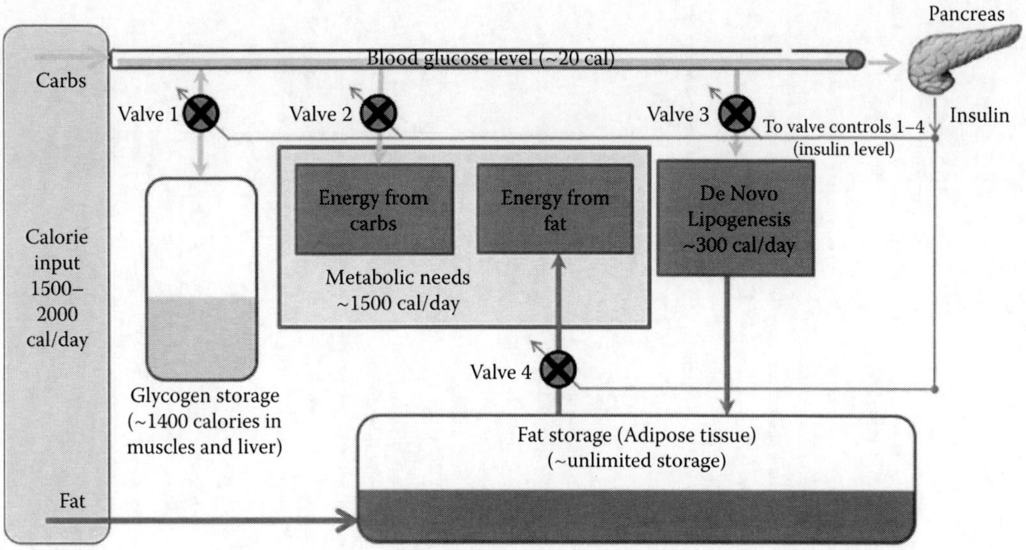

FIGURE 9.4 **(See color insert.)** Closed-loop metabolic state prediction model emphasizing glucose control quantitatively as diet and exercise change. Four valves determine disposition of dietary intake and use of stored energy. Real-time measurement of O_2 consumed and CO_2 produced in exhaled breath will reveal the *settings* of valves 2–4. (Adapted from Candell, L. M. and G. A. Shaw, RER for the masses, *Presented at the annual conference of Recent Advances and Controversies in Metabolic Energy Measurement*, Tokyo, Japan, October 10–12, 2014.)

management (e.g., Hall 2009). Clearly, modeling energy and macronutrient management relative to the timing, type, and portion of food consumed, will provide individualized near real-time insights into dietary and body fuel stores utilization.

To complement the metabolic fuel model (Candell and Shaw 2014), MIT LL in partnership with the U.S. Army, is developing a low cost, wearable, and indirect calorimeter that accurately measures gas exchange and energetics (Candell et al. 2016). This device will measure oxygen consumption (VO_2) and carbon dioxide (CO_2) production, enabling accurate determinations and modeling of metabolic fuel requirements, heat production, heat storage or thermal strain, and water requirements. This hand-held metabolic monitoring system will provide means of collecting data needed to inform the Candell and Shaw metabolic fuel model and better understand the energy costs and fuel requirements associated with sedentary and physically-demanding life styles.

9.6.2 Sensors' Key Role in Future-Enhanced Virtual Experiences

Dietary intake management is multifaceted (Banica 2012), and, as such, is a field that can take advantage of numerous sensor technologies and models to reveal personal health status, and advise users regarding behavioral modification. Adding sensing within synthetic (virtual) environments, those constructed by computer, creates enhanced participant experiences. Such environments can be exploited in dietary intake management particularly for individualized self-awareness. Table 9.1 introduces the use, strengths, and limitations of advanced technologies. Through demonstration and programmed communication, participants are involved in a more holistic, deeper fashion because synthetic environments convey experiences and information using all the senses. Experiences can then be designed to influence decision making, emotional balance, and psychological state—a powerful future tool.

Emerging technologies offer a wide range of possibilities for integrating sensors into complex ensembles of devices, computers, clothing, and environments for immersive simulations.

TABLE 9.1
Advanced Technologies: Usage, Strengths, and Limitations

Advanced Technology	Real-Time Measurements	Applications	Metrics for Success	Strengths and Benefits	Limitations
Synthetic environments and experiences (including virtual reality and virtual worlds); participant (e.g., trainee, test subject, medical patient)	Visually-recorded responses (if supervised); oral and physical participation	Behavioral modification; psychological and behavioral research experimentation; psychological counseling	DI management assessed versus individual goals promoted within the synthetic environment (e.g., to maintain or improve weight reduction); reduction in negative emotions; psychological status	Overall flexibility in environment and experience simulation design; safety; repeatability; cost savings versus tangible environments; highly controlled environment and experience; significant participant emotional involvement	Effectiveness dependent on believability of experience and environment; modest or no real-time changes to experience as compared to opportunities when sensors are used; personal/family issues not always addressed; software and technical costs
Synthetic environment; participant with wearable sensor	Physiological, chemical, and spatial sensing; visually-recorded responses of oral and physical participation	Behavioral modification; psychological and behavioral research experimentation; psychological counseling—with sensor-based participant involvement	DI modification as response to real-time sensor information and synthetic environment stimuli within the virtual experience; degree of self-control expressed; reduction in negative emotions; psychological status	Many opportunities for sensor integration; sensor feedback to participant and control mechanism; flexibility in environment and site simulation; safety; cost savings versus tangible environments; highly controlled environment and experience; significant participant emotional involvement	High technology delivery system costs; wearable sensor technology availability; complex software; environment simulation apparatus may be required; personal/family issues not always addressed
Synthetic environment, participant(s) with wearable sensor(s), animated avatar(s) of participant and possibly others	Physiological, chemical, and spatial sensing; visually-recorded responses of oral and physical participation; (possible) virtual sensing of avatar(s) through physiological simulations of avatar(s)	Behavior and reactions in group settings and complex environments; DI management training scenario; research experimentation; DI management through behavioral modification as caused by sensor feedback; improved self-image	DI management assessed versus goals; assessment of behavior in groups; degree of self-control expressed; reduction in negative emotions; psychological status	Virtual sensing can provide feedback to participant regarding training and environment; avatar's health status can affect participant reactions, emotions and behavior; highly believable experience when near real-time control of experience is possible; Avatars eliminate costs and logistics of multiple-person experiences; participant safety	Highly compute intensive; software and technical costs and support; sensing of health and spatial relationships must be tightly coupled with environment control mechanisms; Effectiveness dependent on believability of experience and environment; personal/family issues not always addressed

Note: DI, dietary intake.

Virtual-reality apparatus, synthetic training environments, and avatars, for example, can be joined with biosensors, chemosensors, and physiological sensors monitoring participants during a virtual experience. Together, these elements create an integrated immersive environment for an augmented virtual experience beyond the standard experiences of a *synthetic environment*, *virtual-reality* scenario, or a *virtual world* by providing data for information feedback into the scenario control system, as well as to the participant(s). Human participants can be, for example, test subjects, trainees, or patients undergoing therapy. The development of unobtrusive sensors for health monitoring in active real-life scenarios clearly benefits synthetic environment experiences.

The drive to develop the capability for synthetic environments is based on many practical reasons. First, the creation of a technology to create synthetic environments or virtual worlds provides extreme flexibility to simulate almost any environment that one can imagine. In addition, the experiences of participants in these synthetic environments enjoy the safety of their local surroundings. This well-controlled environment eliminates real-world hazards, offering a method for medically ethical studies. The experience (i.e., test, training, and therapy) in the synthetic environment has a higher likelihood to be effective because the environment and experience are so highly controlled and are not subject to random occurrences, for example, weather. Using synthetic environments simplifies testing and training site selection, personnel management, and logistics as participants would be able to experience a synthetic environment or virtual world in any number of locations globally. This provides a clear cost benefit by reducing expenditures for personnel management and care. Also, experiencing synthetic environments allows for repetition of an experience without risk of endangering participants.

Experiences in synthetic environments exploiting sensors are constructed using *content*, *delivery systems/mechanisms*, *monitoring*, and a *control system/mechanism* embodying an analytical engine. Surveillance may occur by an external viewer/operator or oracle/authority, who may also control aspects of the experience. The participant(s) are immersed within this technology.

The *content* of an augmented synthetic environment—either pregenerated/stored, created in real time or a mixture—refers to the input to the senses of sight, hearing, smell, taste, or touch. This may include media, climatic conditions such as temperature and humidity; smells; mechanical and electronic devices—including sensors; and the haptic feedback from devices. For example, the content of an augmented synthetic environment may consist of computer animation generated in real time from a hot climate synthetic environment for a specific training scenario with one human trainee, which can include an individualized avatar of the trainee participant, the main actor of the enhanced synthetic environment along with other animated avatars. The physiological state of the human trainee can be monitored by attached sensors, and the monitoring information may be used by the trainee, by the simulation control system, and/or by the external operator to alter the contents in the future.

Avatars can be a key content component of a virtual world or an experience in a synthetic environment. Finite element surface mesh-based avatars are used in movies and virtual world website animations. Emerging from research are individualized avatars as finite element tetrahedral (volume) mesh figures with completely labeled anatomy (Field 2016). Figure 9.5 shows current avatar technology. When avatars become available at high resolution and with more accurate anatomical representations, more realistic synthetic environment experiences can be created. *Virtual sensing*—simulating sensor output from the avatar—can be performed via rapid computer physiological simulation computations (Zientara and Hoyt 2016). For example, virtual sensing can enable monitoring of the physiological state of an avatar performing work or exercise. These simulations can reveal the health of avatars engaged in a synthetic experience that can be analyzed alongside the medical monitoring data from the real-world human participant.

Delivery systems/mechanisms for enhanced synthetic environment to the participants include: head mounted virtual reality goggles, computer screens, climatic chambers, and a variety of specialty simulators (e.g., airplane cockpit simulators), clothing, visualization systems, devices, computers, and electronics among others.

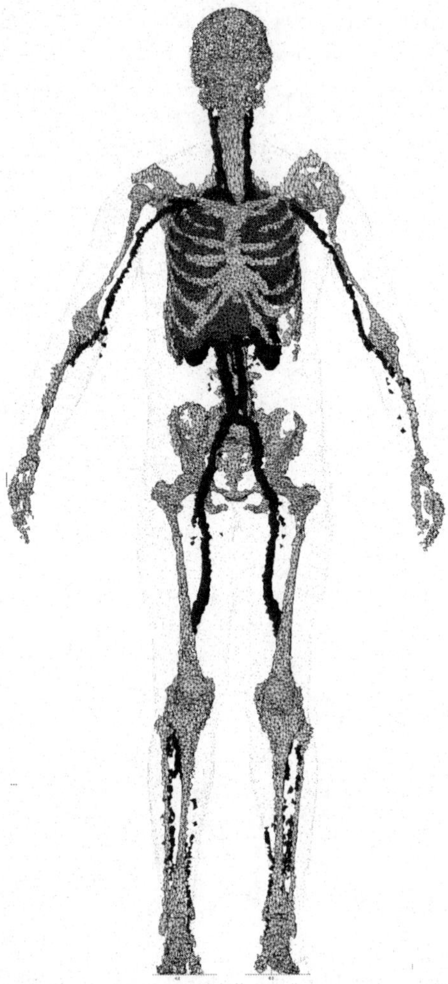

FIGURE 9.5 (See color insert.) Individualized male avatar based on a body surface scan from the ANSUR II database (Gordon et al. 2014) using the method of Zientara and Hoyt (2016). Avatar surface (fine blue dots), selected internal anatomy (lungs, kidneys, heart, and liver; light brown), and cardiovascular system (dark and bright red) shown as 3D tetrahedral finite elements.

Monitoring of a participant experiencing a synthetic environment may employ many types of sensing and tracking devices to acquire data regarding the medical status of the participant, as well as the conditions and contents of the environment. The value of these data is that it may be used for physical, medical, and environmental characterization, and aid in producing a continuous believable experience. Sensing or tracking data may include locations, orientations of participants and objects in the virtual world, and interactions among them or experienced by them, such as forces applied by the participants or experienced by the participants. Biosensors, chemosensors, and physiological sensors can monitor the physical or medical state of participants as an enhanced synthetic environment unfolds, as well as a characterization of his or her near environment.

The *control system* or *mechanism* in an augmented synthetic environment will include an analytical engine that will process the *fire hose* of streaming real-time data coming from participant-generated actions and commands, and from sensors and trackers to yield meaningful information critical to control the synthetic environment. Sensors can play a significant active role in the new paradigm of enhanced synthetic environments if they are tightly coupled with the control mechanism, rather than only working passively. Rapid transmission of sensor information from the participant and synthetic environment to the control mechanism is an exploitable technical advantage as it enables real-time or near real-time modifications to the content and the delivery systems of the synthetic environment.

Monitoring information can be used by the control mechanism to modify the virtual experience in real time, for example, to evoke a reaction from the participant. Feedback from the experience can be shown to a participant in order to attempt to alter his or her behavior. This has been proposed as a new form of psychological therapy (Riva et al. 2010, 2011). Participant's reactions can be elicited by overt stimulus, or through introspection, and with the help of an advisor, or not.

As the future unfurls, simulations will likely be designed to have human participants interact with other human participants, avatars, and robotics within a graphically rich high-technology environment. This undoubtedly will enable widespread applications of synthetic environments for personalized communications, testing, training, and therapies, and applications yet to be imagined.

9.6.3 Future Challenges

New networked sensor ecosystems are emerging, driven by the development of wireless, low-power, Internet of Things (IoT) technologies that allow sensors on everyday objects, for example, appliances or humans to be recognizable, readable, and controllable through the Internet (Swan 2012). Particularly in the area of health care, IoT technologies offer opportunities for innovation that go beyond the familiar Internet-connected computer, smartphone, and tablet devices. IoT-enabled sensors and applications will help individuals be responsible for managing their own health and health care. Examples include IoT-enabled tools to manage dietary intake, body composition, physical activity, physical fitness, and metabolic fuel use.

However, the development and use of IoT-enabled technologies for personal health care and other applications is impeded by the proprietary systems that companies commonly produce. Open architecture standards, in contrast, would allow IoT-enabled systems to share wireless protocols and data standards, and more readily accommodate new algorithms. Additional concerns include cyber-security of wireless IoT-enabled systems and devices, and the question of who owns and controls individual user data. In addition, information overload, poor data quality from consumer health and performance monitoring devices, and a blurring of the distinction between consumer devices and FDA-certified medical devices may also present challenges.

9.7 CONCLUSIONS

Current trends in biosensor development and use of hand-held devices (e.g., smartphones) will foster continued modernization and advancements in the field of dietary intake assessment. Existing intake methods have begun the transformation into automated, technologically-advanced data collection tools with the potential for open platform communication. Computerized questionnaires, automated food recognition, and computing systems (e.g., digital food photography) are becoming standard, soon the integration of *smart medical systems* (e.g., implanted, transcutaneous scan, or by breath) and body-worn sensors will further tailor-diet intake data to the individual. The addition of stable isotope methodology will allow for accurate and dependable body composition monitoring for a more profound understanding of personal nutrient utilization, and point-in-time recommendations.

As knowledge expands, so too will the challenges of education and information security. Industry awareness and new technology may solve the latter. Although virtual experiences may compliment or even replace traditional nutrition education tools and techniques of the professional. Avatars could provide powerful feedback through expert-led or self-taught training and therapies. Synthetic environments function as *safe* settings for active, real-life scenarios that serve to foster self-awareness and attempt to modify behaviors. The future is an individualized, holistic approach to the assessment of nutrient status.

REFERENCES

Baker-Fulco, C. J. 1995. Overview of dietary intakes during military exercises. In *Not Eating Enough*, ed. B. M. Marriott, pp. 121–149. Washington, DC: National Academy Press.

Banica, F. G. 2012. *Chemical Sensors and Biosensors: Fundamentals and Applications*. Chichester: John Wiley & Sons, LTD.

Block, G., M. Woods, A. Potosky, and D. Clifford. 1990. Validation of a self-administered diet history questionnaire using multiple diet records. *J Clin Epidemiol* 43(12):1327–1335.

Borgmann, S., A. Achulte, S. Neuebauer et al. 2011. Amperometric biosensors. In *Bioelectrochemistry: Fundamentals, Applications and Recent Development*, ed. R. Alkire, D. M. Kolb, and J. Lipkowski, pp. 1–56. Weinheim: Wiley-VCH GmbH & Co.

Brinkerhoff, K., P. Brewster, E. Clark et al. 2011. Linking supermarket sales data to nutritional information: An informatics feasiblity study. *AMIA Annu Symp Proc* 2011:598–606.

Cammann, K. 1977. Biosensors based on ion-selective electrodes. *Fresenuius J Anal Chem* 287(1):1–9.

Candell, L. M. and G. A. Shaw. 2014. RER for the masses. *Presented at the annual conference of Recent Advances and Controversies in Metabolic Energy Measurement*, Tokyo, Japan, October 10–12.

Candell, L. M., G. A. Shaw, and R. W. Hoyt. 2016. A nonlinear control model that challenges conventional thinking about obesity & type 2 diabetes. *Poster presented at the Metabolic Therapeutics Conference*, Tampa, FL, January 28–30.

Cattrall, R. W. 1997. *Chemical Sensors*. New York: Oxford University Press.

Chang, K., S. Lui, H. Chu et al. 2006. The diet-aware dining table: Observing dietary behaviors over a tabletop surface. In *Proceedings of the International Conference on Pervasive Computing*, pp. 366–382.

Chouvarda, I., N. Y. Philip, P. Natsiavas et al. 2014. WELCOME—Innovative integrated care platform using wearable sensing and smart cloud computing for COPD patients with comorbidities. *Conf Proc IEEE Eng Biol Soc* 2014:3180–3183.

Crombie, A. P., L. K. Funderburk, T. J. Smith et al. 2013. Effects of modified foodservice practices in military dining facilities on ad libitum nutritional intake of US army soldiers. *J Acad Nutr Diet* 113:920–927.

Eggin, B. R. 2004. *Chemical Sensors and Biosensors*. Chichester: John Wiley & Sons, LTD.

Elia, M., N. Fuller, and P. Murgatroyd. 1988. The potential use of labelled bicarbonate method for estimating energy expenditure in man. *Proc Nutr Soc* 47:247–258.

Ellerby, G. E. C., C. P. Smith, F. Zou et al. 2013. Validation of spectroscopic sensor for the continous, noninvasive measurement of muscle oxygen saturation and pH. *Physiol Meas* 34:859–871.

Fabricatore, A. N., C. B. Ebbeling, T. A. Wadden et al. 2011. Continuous glucose monitoring to assess the ecologic validity of dietary glycemic index and glycemic load. *Am J Clin Nutr* 94:1519–1524.

Fiatarone, M. A., E. F. O'Neill, N. D. Ryan et al. 1994. Exercise training and nutritional supplementation for physical frailty in very elderly people. *N Engl J Med* 330:1769–1775.

Field, K. 2016. USARIEM making 3-D soldier models. Accessed June 9, 2016. http://www.army.mil/article/165340/.

Geboes, K., B. Bammens, A. Luypaerts et al. 2004. Validation of a new test meal for a protein digestion breath test in humans. *J Nutr* 134:806–810.

Global Industry Analysts, Inc. 2014. Expanding medical and non-medical applications to stimulate growth in the global biosensor market. Accessed November 12. http://www.strategyr.com/MarketResearch/Biosensors_Market_Trends.asp.

Gordon, C., C. Blackwell, B. Bradtmiller et al. 2014. 2012 Anthropometric survey of U.S. Army personnel: Methods and summary statistics. Natick, MA: U.S. Army Soldier Systems Command. Technical Report: TR-15/007.

Hall, K. D. 2009. Predicting metabolic adaptation, body weight change, and energy intake in humans. *Am J Physiol Endocrinol Metab* 298:E449–E466.

Hernandez, J., Y. Li, J. M. Rehg et al. 2014. Bioglass: Physiological parameter estimation using a head mounted wearable device. *EAI International Conference of Wireless Mobile Communication and Healthcare*, pp. 55–58.

Hoyt, R. W., T. E. Jones, R. Schwartz et al. 1991. Doubly labeled water measurement of human energy expenditure during strenuous exercise. *J Appl Physiol* 71:16–22.

Hoyt, R. W., P. K. Opstad, A. H. Haugen et al. 2006. Negative energy balance in male and female rangers: Effects of 7 d of sustained exercise and food deprivation. *Am J Clin Nutr* 83:1068–1075.

Hulanicki, A., S. Glab, and F. Ingman. 1991. Chemical sensors: Definitions and classifications. *Pure Appl Chem* 63(9):1247–1250.

Indiegogo. 2016. Breathometer mint for breath quality and hydration. Accessed June 15. https://www.indiegogo.com/projects/breathometer-mint-for-breath-quality-and-hydration#/.

Kehayias, J. J. 1993. Aging and body composition: Possibilities for future studies. *J Nutr* 123:454–458.

Kehayias, J. J., M. A. Fiatarone, H. Zhuang, and R. Roubenoff. 1997. Total body potassium and fat: Relevance to aging. *Am J Clin Nutr* 66:904–910.

Kehayias, J. J., S. M. L. Ribeiro, A. Shahan et al. 2012. Water homeostasis, frailty and cognitive function in the nursing home. *J Nutr Health Aging* 16(1):35–39.

Kotanen, C. N., G. M. Francis, S. Carrara, and A. Guiseppi-Elie. 2012. Implantable enzyme amperometric biosensors. *Biosens Bioelectron* 35(1):14–26.

Lane, H. W. and D. L. Feeback. 2002. History of nutrition in space flight: Overview. *Nutrition* 18:797–804.

Larsson, C., Ø. Ahlstrøm, R. B. Jensen et al. 2015. The oral ^{13}C bicarbonate technique for measurement of short-term energy expenditure of sled dogs and their physiological response to diets with different fat:carbohydrate ratios. *J Nutr Sci* 4:e32

Lieberman, H. R., C. M. Falco, and S. S. Slade. 2002. Carbohydrate administration during a day of sustained aerobic activity improves vigilance, as assessed by a novel ambulatory monitoring device, and mood. *Am J Clin Nutr* 76:120–127.

Lieberman, H. R., C. M. Caruso, P. J. Niro et al. 2008. A double-blind, placebo-controlled test of 2 d of calorie deprivation: Effects on cognition, activity, sleep, and interstitial glucose concentrations. *Am J Clin Nutr* 88:667–676.

Lutz, L. J., J. P. Karl, J. C. Rood et al. 2012. Vitamin D status, dietary intake, and bone turnover in female soldiers during military training: A longitudinal study. *J Inter Soc Sports Nutr* 9(1):38.

Mancini, D. M., L. Bolinger, H. Li et al. 1994. Validation of near-infrared spectroscopy in humans. *J Appl Physiol* 77:2740–2747.

Marek, E., J. Volke, P. Platen et al. 2009. H_2O_2 release and acid-base status in exhaled breath condensate at rest and after maximal exercise in young and healthy subjects. *Eur J Med Res* 14:134–139.

Marek, E., J. Volke, I. Hawener et al. 2010. Measurements of lactate in exhaled breath condensate at rest and after maximal exercise in young and healthy subjects. *J Breath Res* 4:1–8.

Margolis, L. M., A. P. Crombie, H. L. McClung et al. 2014a. Energy requirements of US Army Special Operation Forces during military training. *Nutrients* 6(5):1945–1955.

Margolis, L. M., N. E. Murphy, S. Martini et al. 2014b. Effects of winter military training on energy balance, whole-body protein balance, muscle damage, soreness, and physical performance. *App Physiol Nutr Metab* 39(12):1395–1401.

Margolis, L. M., N. E. Murphy, S. Martini et al. 2016. Effects of supplemental energy on protein balance during 4-d Arctic military training. *Med Sci Sports Exerc* 48:1604–1612. PMID:27054679.

Martin, C., S. Kaya, and B. Gunturk. 2009a. Quantification of food intake using food image analysis. *Conf Proc IEEE Eng Med Biol Soc* 1:6869–6872.

Martin, C., H. Han, S. Coulon et al. 2009b. A novel method to remotely measure food intake of free-living individuals in real-time: The remote food photography method. *Br J Nutr* 101(3):446–456.

Mathew, T. L., P. Pownrai, S. Abdulla et al. 2015. Technologies for clinical diagnosis using expired human breath analysis. *Diagnostics* 5(1):27–60.

McClung, H. L., L. D. Sigrist, T. J. Smith et al. 2009a. Monitoring energy intake: A hand-held personal digital assistant provides accuracy comparable to written records. *J Acad Nutr Diet* 109:1241–1245.

McClung, J. P., J. P. Karl, S. J. Cable et al. 2009b. Randomized, double-blind, placebo-controlled trial of iron supplementation in female soldiers during military training: Effects on iron status, physcial performance, and mood. *Am J Clin Nutr* 90:124–131.

Miller, M. E. and C. J. Cappon. 1984. Anion-exchange chromatographic determination of bromide in serum. *Clin Chem* 30(5):781–783.

Nansel, T. R., L. M. Lipsky, and A. Liu. 2016. Greater diet quality is associated with more optimal glycemic control in a longitudinal study of youth with type 1 diabetes. *Am J Clin Nutr* 104(1):81–7.

Newman, J. D. and A. P. F. Turner. 2005. Home blood glucose biosensors: A commercial perspective. *Biosens Bioelectron* 20(12):2435–2453.

Pasiakos, S. P., H. L. McClung, J. P. McClung et al. 2011. Leucine-enriched essential amino acid supplementation during moderate steady state exercise enhances postexercise muscle protein synthesis. *Am J Clin Nutr* 94:809–818.

Pasiakos, S. P., H. L. McClung, L. M. Margolis et al. 2015. Human muscle protein synthetic responses during weight-bearing and non-weight-bearing exercise: A comparative study of exercise modes and recovery nutrition. *PLoS One* 10(10):e0140863.

Passler, S., M. Wolff, and W. J. Fischer. 2012. Food intake monitoring: An acoustical approach to automated food intake activity detection and classification of consumed food. *Physiol Meas* 33(6):1073–1093.

Pikosky, M. A., T. J. Smith, A. Grediagin et al. 2008. Increased protein maintains nitrogen balance during exercise-induced energy deficit. *Med Sci Sports Exerc* 40(3):505–512.

Prelack, K., J. Dwyer, R. Sheridan et al. 2005. Body water in children during recovery from severe burn injury using a combined tracer dilution method. *J Burn Care Rehabil* 26(1):67–74.

Richinick, M. 2015. Obama seeks $215 million for personalized medicine. *MSNBC*, January 30. Accessed June 23, 2016. http://www.msnbc.com/msnbc/obama-seeks-215-million-personalized-medicine.

Riva, G., R. Raspelli, D. Algeri et al. 2010. Interreality in practice: Bridging virtual and real worlds in the treatment of posttraumatic stress disorders. *Cyberpsychol Behav Soc Netw* 13(1):55–65.

Riva, G., B. K. Wiederhold, F. Mantovani, and A. Gaggioli. 2011. Interreality: The experiential use of technology in the treatment of obesity. *Clin Prac Epidemiol Ment Health* 7:51–61.

Schoeller, D. A., E. van Santen, D. W. Peterson et al. 1980. Total body water measurement in humans with O-18 and H-2 labeled water. *Am J Clin Nutr* 33:2686–2693.

Sengoku, Y., K. Nakamura, H. Ogata et al. 2015. Continuous glucose monitoring during a 100-km race: A case study in an elite ultramarathon runner. *Int J Sports Physiol Perform* 10(1):124–127.

Shichiri, M., R. Kawamori, R. Yamaski et al. 1982. Wearable artifical endrocrine pancreas with needle-type glucose biosensor. *Lancet* ii:1129–1131.

Slining, M. 2012. Linking data sources. In *Proceedings of the National Nutrient Bank Conference*, Houston, TX, March 25–28.

Smidt, C. R. and D. S. Burke. 2004. Nutritional significance and measurement of carotenoids. *Curr Topics Nutraceut Res* 2:79–91.

Smith, S. M., G. Block, B. L. Rice et al. 1998. A food frequency questionnaire for use during space flight: A ground-based evaluation. *FASEB J* 12:A526 [abstract no. 3057].

Smith, S. M. and H. W. Lane. 1999a. Gravity and space flight: Effects on nutritional status. *Curr Opin Clin Nutr Metab Care* 2:335.

Smith, S. M. and H. W. Lane. 1999b. Nutritional biochemistry of space flight. *Life Support Biosph Sci* 6(1):5–8.

Smith, S. M., J. E. Davis-Street, B. L. Rice et al. 2001. Nutritional status assessement in semiclosed environments: Ground-based and space flight studies in humans. *J Nutr* 131(7):2053–2061.

Smith, S. M. and S. Zwart. 2016. NASA's International Space Station (ISS) Fitness Tracker (FIT) App to monitor daily dietary intake. *Personal communication*, May 19.

Soller, B. R., M. Cabrera, S. Smith et al. 2002. Smart medical systems with application to nutrition and fitness in space. *Nutrition* 18:930–936.

Soller, B. R., J. Favreau, and P. O. Idwasi. 2003. Investigation of electrolyte measurement in diluted whole blood using spectroscopic and chemometric methods. *Appl Spectro* 57(2):146–151.

Stein, T. P., M. J. Leskiw, M. D. Schluter et al. 1999. Energy expenditure and balance during spaceflight on the space shuttle. *Am J Physiol* 276(6):R1739–R1748.

Stevens, J., M. Bryant, L. Wang et al. 2010. Exhaustive measurement of food items in the home using a universal product code scanner. *Public Health Nutr* 14(2):314–318.

Sun, M., J. D. Fernstrom, W. Jia et al. 2010. A wearable electronic system for objective dietary assessment. *J Am Diet Assoc* 110(1):45–47.

Swan, M. 2012. Sensor Mania! The Internet of Things, wearable computing, objective metrics and the quantified self 2.0. *J Sens Actuator Netw* 1:217–253.

Tamura, T., Y. Maeda, M. Sekine, and M. Yoshida. 2014. Wearable photoplethysmographic sensors—Past and present. *Electronics* 3:282–302.

Tellspec. 2016. Tellspec Inc. is a data company that provides predictive intelligence about food. Accessed June 15. http://tellspec.com/howitworks/.

Tharion, W. J., H. R. Lieberman, S. J. Montain et al. 2005. Energy requirements of military personnel. *Appetite* 44:47–65.

Thevenot, D. R., K. Toth, R. A. Durst et al. 2001. Electrochemical biosensors: Recommended definitions and classifications. *Anal Lett* 34(5):635–659.

Toyooka, T., S. Hiyama, and Y. Yamada. 2013. A prototype portable breath acetone analyzer for monitoring fat loss. *J Breath Res* 7:1–8.

U.S. Department of the Army. 2006. *AR 600-9: The Army Weight Control Program*. Washington, DC: Department of the Army.

U.S. Department of the Army. 2016. *AR 40-25: Nutrition Menu Standards for Human Performance Optimization*. Washington, DC: Department of the Army.

Vo-Dinh, T. and B. Cullum. 2000. Biosensors and biochips: Advances in biological and medical diagnostics. *Fresenius J Anal Chem* 366(6–7):540–551.

Wang, Z. M., N. R. Pierson, and S. B. Heymsfield. 1992. The five-level model: A new approach to organizing body-composition research. *Am J Clin Nutr* 56:19–28.

Westerterp, K. 2009. Assessment of physical activity: A critical appraisal. *Eur J Appl Physiol* 105(6):823–828.

Williamson, D. A., H. R. Allen, P. D. Martin et al. 2003. Comparison of digital photography to weighed and visual estimation of portion sizes. *J Acad Nutr Diet* 103:1139–1145.

Williamson, D. A., H. R. Allen, P. D. Martin et al. 2004. Digital photography: A new method for estimating food intake in cafeteria settings. *Eat Weight Disord* 9:24–28.

W. P. Carey Research. 2015. Consumer wearables: Biosensors and health care. April 29. Accessed June 2, 2016. http://research.wpcarey.asu.edu/health/consumer-wearables-biosensors-and-health-care/.

Xu, H., Y. Wei, L. Zhu et al. 2014. Bifunctional magnetic nanoparticles for analysis of aldehyde metabolites in exhaled breath of lung cancer patients. *J Chrmotogr* 1324:29–35.

Zaichick, V. 1998. X-ray fluorescence analysis of bromide for the estimation of extracellular water. *Appl Raiat Isot* 49(12):1665–1669.

Zhang, R., S. Berhart, and O. Amft. 2016. Diet eyeglasses: Recognizing food chewing using EMG and smart eyeglasses. *Presented at the Annual Conference of Body Sensor Networks*, San Francisco, CA, June 13–16.

Zidichouski, J. A., A. Mastaloudis, S. J. Poole et al. 2009. Clinical validation of a noninvasive, Ramen spectroscopic method to assess carotenoid nutritional status in humans. *J Am Coll Nutr* 28(6):687–693.

Zientara, G. and R. W. Hoyt. 2016. Individualized avatars with complete anatomy constructed from the ANSUR II 3D antrhopometric database. *Int J Digtal Human* (in review).

10 Use of Doubly-Labeled Water Measured Energy Expenditure as a Biomarker of Self-Reported Energy Intake

Dale A. Schoeller and David B. Allison

CONTENTS

10.1 INTRODUCTION

Energy, or more correctly potential chemical energy, is provided by diet in the form of three macronutrients, carbohydrate, fat, and protein, and are the fuels used by the body to provide energy for life. These macronutrients comprise the vast majority of the dry mass of the diet, and thus, dietary energy is the key property of the diet. It follows from this that the assessment of dietary energy intake is essential to both clinical and human nutrition, and many survey tools have been developed, including weighed food records, 24-hour dietary recalls, diet histories, and semiquantitative food frequency questionnaires for this task.

The survey tools used for assessing dietary energy intake, assess metabolizable energy intake, which is the fraction of gross chemical energy intake that is available to the body to fuel metabolism (Merrill and Watt 1973). The gross energy is the energy released as heat when the food is combusted to carbon dioxide, water, and nitrogen gas. The gross energy of foods, however, is not all that is available to the body as metabolic fuel. Approximately 5%–10% of energy substrates we consume are not absorbed by the body and thus lost in feces. The fraction that is absorbed is termed the digestible energy. In addition, an even smaller portion of absorbed energy cannot be metabolized by the body and is instead excreted in urine as compounds such as urea, ammonia, and short chain fatty acids. The difference between gross energy consumed and the energy lost in feces and urine is termed metabolizable energy, and it is the metabolizable energy that is tabulated in the food composition tables and used to calculate the energy content of the diet as part. It is notable that for any one food, the difference between metabolizable energy and gross energy will vary slightly between persons and within persons across time and conditions (Heymsfield and Pietrobelli 2011).

Hence, metabolizable energy contents of foods are not fixed values, though they are typically treated as such for most practical purposes. Commonly, however, metabolizable energy is simply referred to as dietary energy.

The energy consumed by an individual each day is quite variable and the coefficient of variation is estimated to be 26% (Beaton et al. 1983). As such a single day's estimate of dietary energy is an imprecise estimate of the average or usual energy intake of an individual that must be averaged over five to six days with care to sample weekday and weekends to obtain a precision of less than 10%. Among individuals who are weight stable and not increasing or decreasing body energy stores as evidenced by changes in weight and body composition, the usual energy intake is equal to energy expenditure. Children, pregnant women, and lactating women are exceptions to the equality of habitual energy intake and energy expenditure, because of the surfeit of energy intake to support childhood growth, pregnancy, or lactation. These surfeits, averaged across periods of high and low daily energy demand, are estimated to be 25, 190, and 450 kcal/d, respectively (Food and Nutrition Board 2005).

With these three major exceptions, however, habitual or usual energy intake is the average dietary energy intake to maintain body weight with no change in body composition and is equal to energy expenditure in nonpregnant or nonlactating adults. Energy expenditure is therefore a biomarker or alternate means of assessing dietary energy intake. Energy expenditure, however, has not been easily measured with accuracy and precision until the doubly-labeled water method became available to measure human energy expenditure. Using doubly-labeled water, energy expenditure could be measured and averaged over 7–14 days with a precision of about 5% and thus provide a biological marker of usual energy intake for weight maintenance (Schoeller 2002).

10.2 BACKGROUND

The doubly-labeled water method was pioneered and validated in small animals by Nathan Lifson and coworkers over 60 years ago (Lifson et al. 1955). The method was based on the observation that the oxygen atoms in carbon dioxide were in equilibrium with the oxygen in water (Lifson et al. 1949). Thus it was reasoned that if a loading dose of water labeled with ^{2}H and ^{18}O, the ^{2}H and ^{18}O elimination kinetics could be used to calculate an animal's carbon dioxide production. The elimination of the ^{2}H in body water results from the daily ingestion of beverages containing water, food moisture, and metabolic water that dilute the excess tracer from the loading dose followed by excretion of water to maintain a steady state. The elimination of ^{18}O from body water is not only by the dilution and excretion of water, but also by dilution from the oxygen atoms in carbon dioxide produced during respiration (Figure 10.1). This can be stated mathematically using a single pool mathematical model. The elimination rate of ^{2}H from the total body water pool times the pool size is directly proportional to water flux, and the elimination rate of ^{18}O from the total body water times the pool size is directly proportional to the sum of water flux and twice the carbon dioxide flux (Figure 10.1). The carbon dioxide flux is doubled because there are two atoms of oxygen in each carbon dioxide molecule for each atom of oxygen in water.

One of the primary advantages of the doubly-labeled water method is that it essentially converts body water into a metabolic recorder. The practical aspect of this is that collection of the labeled carbon dioxide as it exits the body is not required, but rather one can measure the amount of label in body water at the start and end of the metabolic period and calculate how much has been lost. Thus, the subject is able to perform their normal daily functions without significant restrictions for specimen collection. These tracers are stable isotopes and thus not radioactive that would introduce safety concerns. Although a stable isotope, ^{2}H in body water at high concentrations does result in altered enzyme kinetics, the concentration of the ^{2}H used for the doubly-labeled water method, is orders of magnitude below the lowest concentrations at which the health effects have been identified for $^{2}H_2O$ (Jones and Leatherdale 1991). No adverse health effects of $H_2{}^{18}O$ have been reported and the doubly-labeled water method is considered safe for use in humans of all ages (Jones and Leatherdale 1991).

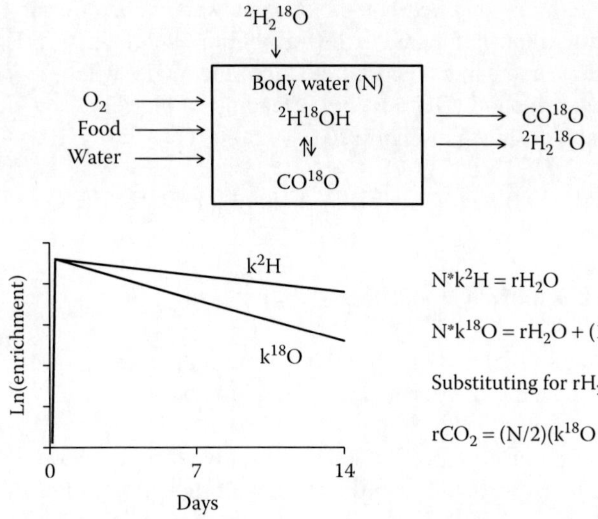

$N^*k^2H = rH_2O$

$N^*k^{18}O = rH_2O + (1/2)rCO_2$

Substituting for rH_2O:

$rCO_2 = (N/2)(k^{18}O - k^2H)$

FIGURE 10.1 The single pool model of Lifson (1966). Following a dose of doubly-labeled water, the tracers rapidly distribute across body water. The deuterium tracer is diluted by water influx and eliminated as water. The oxygen tracer is not only diluted by water influx, but also by equilibration with carbon dioxide and is eliminated as both. The area between the two elimination curves is proportional to carbon dioxide production.

When first validated by Lifson and coworkers (1955), the doubly-labeled water method was impractical for use in humans. The measurements of the abundances of ^2H and ^{18}O by mass spectrometry were not very precise necessitating the use of still safe, but higher tracer abundances. In particular, it required doses of ^{18}O that were cost prohibitive for human use. In addition, the isotope ratio analyses available in the 1950s and 1960s were time consuming and thus laboratory throughput was very limited. Lifson and coworkers (1975) estimated that the cost of the first validation of the doubly-labeled water method was about of \$10,000 (2016 dollars) for each rat. The accelerated development of isotope ratio mass spectrometers included increased precision and automated operation that began in the 1970s and continues till today and has reduced the analysis to a few hours per subject and the cost of the dose of doubly-labeled water to an adult to about \$300 (2016 dollars). The cost of the dose, however, will vary with supply and demand of $H_2{}^{18}O$ and as increased 200%–300% during two periods in the past 30 years when ^{18}O water was in short supply (Schoeller 1999).

Recently, cavity ring-down spectroscopy, which has the same precision as isotope ratio mass spectrometry and thus is an alternative to isotope ratio mass spectrometer, has been validated. Cavity ring-down spectroscopy, however, is limited by design problems for the inlet that leads to long analysis times for any specimen other than water (Thorsen et al. 2011). Should these design problems be solved, cavity ring-down spectroscopy will become a viable alternative that will further reduce analysis time and the need for extensive operator training.

Since its invention, the doubly-labeled water method had been extensively shown to be accurate in small animals, although a slightly different calculation has been used because of differences in the influence of isotope fractionation and isotope incorporation into urea and molecules other than water (Schoeller 1984; Speakman 1997).

The validity of the doubly-labeled water method is based on four assumptions, all of which have been found to be true or at least sufficiently close to true so as not to worsen the accuracy and precision of the doubly-labeled water method (Lifson 1966; Speakman 1997). These assumptions arise from the well-understood assumptions of steady-state kinetics. Two of these assumptions, distribution of the tracer only in the total body water pool and elimination of products having the same stable isotope tracer abundances as body water, however, are not perfect. The tracers undergo very rapid exchange with labile hydrogens in protein and labile oxygens in phosphate, carboxyl, or similar moieties, and the elimination of carbon dioxide and water vapor are subject to isotope fraction, but the exchange is very rapid compared to the tracer elimination kinetics. These deviations are consistent across individuals and can be incorporated into the model without violating the validity of the doubly-labeled water method (DeLany 1997) leading to the following similar, but slightly more complex model than that shown in Figure 10.1.

$$rCO_2 = (TBW/2.078)(1.01\ k_O - 1.04\ k_H) - 0.0246rH_2O_f \qquad (10.1)$$

where:

rCO_2 is the rate of carbon dioxide production
TBW is total body water
k are the elimination rates of ^2H and ^{18}O from body water in pools/d
rH_2O_f is the rate of isotopically fractionated water loss in moles/d

This can be expressed in far more simple, but less informative form as shown below, because loss of water vapor in breath is proportional to carbon dioxide production in most, but not all human studies (Schoeller et al. 1986a).

$$rCO_2 = 0.455(TBW)(1.007\ k_O - 1.045\ k_H) \qquad (10.2)$$

where:

TBW is total body water in moles
k_O and k_H are the oxygen and hydrogen elimination rates in pools/d
rH_2O_f is the rate of water lost through fractionated routes in moles/d

A recent paper proposes a slight modification due a refinement in the ratio of the ^2H and ^{18}O dilution spaces (Sagayama et al. 2016). The other assumptions have been shown to be valid except under unusual circumstances that result in a few restrictions on the use of the doubly-labeled water method. The assumption that the abundances of these stable isotopes in food and water entering the body are constant, restricts use of the method to individuals who do not travel between areas of distinctly different hydrological conditions because the abundances of ^2H and ^{18}O in water is subject to natural variations as a function of latitude and elevation (Taylor 1974; Horvitz and Schoeller 2001). As such, subjects are usually asked not to travel more than 300 km during the metabolic period and subjects are excluded if they have recently or expect to receive a half liter or more of intravenous fluids. The assumption of no tracer reentry of the organism during the metabolic period limits the use among astronauts and other individuals who consume recycled water unless special precautions are taken (Gretebeck et al. 1997).

As indicated above the doubly-labeled water method measures the subject's carbon dioxide production. To convert carbon dioxide production to energy expenditure, the Weir equation (1949) is used, but in a form below that has been rearranged to express energy expenditure as a function of carbon dioxide production.

$$TEE(kcal/d) = (22.4 * 4.18/1000)(1.1 + 3.9/RER) \qquad (10.3)$$

where:

22.4 is the liters in a mole of gas at standard temperature and pressure
4.18/1000 is the conversion of kcal to MJ.

This equation requires knowledge of the RER (respiratory exchange ratio = V_{CO2}/V_{O2}) for the metabolic period. While on a minute-to-minute as well as an hour-to-hour basis, respiratory exchange ratio is highly variable, variability is reduced when averaged over a period of one to two weeks

(Smith et al. 2000), and is not limiting for the method. Indeed, the RER can be estimated from one to two week diet because the body generally uses close to 100% of the macronutrients consumed and thus use of a food quotient to estimate a respiratory exchange ratio should not introduce an error in total energy expenditure (TEE) of greater than 1%–2% (Black 1986). Only in cases when energy intake and respiration are significantly out of balance are adjustments for changes in body composition required for the accuracy of this calculation (Black et al. 1986). Under some unusual conditions, this conversion can be the largest source of systematic error for the measurement of TEE by doubly-labeled water. The ratio has a range of 0.70–1.0 and each error of 0.01 units in estimating the ratio relative to the true value introduces a nearly 1% error in TEE. Thus, for most applications, a population mean food quotient can be used because the between-person standard deviation is less than 0.02 units (Black 1996). In practice, however, there will be a few individuals at the extremes of the distribution. A very low fat diet of 10% of energy as fat, 15% protein, and 75% carbohydrate would have a predicted ratio of 0.95 and a very high fat diet of 60% fat, 25% protein, and 15% carbohydrate would have an ratio of 0.74; an 8% and 13% difference, respectively, in the calculated TEE compared to the estimate using a ratio of 0.86. The respiratory ratio chosen for calculation of energy expenditure for a doubly-labeled water study may be based on the diet prescribed for the participant or from the estimate of the actual diet consumed be it a special diet or a general Western diet that gives a value of 0.86 (Neuhouser et al. 2008). For infants and individuals on unusual diets, clinical studies using diets of different macronutrient composition, or populations not consuming a Western diet, adjustments to the respiratory exchange ratio used to calculate TEE are recommended (Black et al. 1986).

10.3 DOUBLY-LABELED WATER METHODOLOGY

The doubly-labeled water method is considered the most objective and accurate measure of total energy expenditure for humans outside of a laboratory setting (Jéquier and Schutz 1988; Speakman 1997). On account of this, there exist several detailed resources on how to use the doubly-labeled water method. Despite this, however, there is no single protocol because the method is flexible with regard to timing of specimen collection and exact dose. Once the doubly-labeled water dose has been administered and the specimens collected, however, the method is highly dependent on the precision of the measurements of the abundances of ^{2}H and ^{18}O in water. Although dedicated isotope ratio mass spectrometry labs can attain the required precision, it has been demonstrated that in laboratories when they first attempted analyses for doubly-labeled water often did not perform well and could not reach a 5% precision for TEE (Roberts et al. 1995). As indicated above, a potential alternative to the use of isotope ratio mass spectrometry is cavity ring-down spectroscopy (Thorsen et al. 2011). The use of cavity ring-down spectroscopy, however, was limited because exchange of hydrogen between sequential specimens reduces accuracy and increases analysis time.

The need for highly precise measures of isotopic abundance is due to the need to minimize the dose of ^{18}O water, the cost of which during the past 30 years has varied between U.S. $2 and $10 per gram of 10 atom percent ^{18}O water. At these costs, the price of a typical adult dose of 75 g has been between U.S. $150 and $750. Current costs for this 75 g dose in 2016 are about U.S. $300. The relationship between dose of labeled water, length of metabolic period, and theoretical precision of the doubly-labeled water method has been described elsewhere (Schoeller 1983). In an experienced laboratory, the method, however, provides nearly optimal precision at doses of around 1.8 g of 10 atom percent excess ^{18}O water and 0.12 g of 99 atom percent $^{2}H_2O$ and measurement period equal to 1.5 half lives of the labeled water that is between 4 and 21 days, with shorter periods being optimal when water turnover is higher and 14–21 days when water turnover is slow (Schoeller 1983; Speakman 1997).

Typical protocols have been extensively described along with their theoretical bases. Methodological details will not be repeated here because they are extensively and clearly detailed elsewhere. Two such manuals have been prepared by and published by the International Atomic Energy Agency (Prentice 1990; IAEA 2009). These manuals describe both the theory of the method as well as practical details about clinical and laboratory procedures. A second detailed source book has been published that describes both human and animal applications of the method (Speakman 1997). A recent paper has reviewed means of adapting the method and its calculations in the face of unusual circumstances such as changes in baseline isotopic abundances and interferences from residual isotopic enrichments during serial measurements of TEE (Bhutani et al. 2015).

10.4 VALIDITY OF DOUBLY-LABELED WATER

The DLW method has been validated against both respiratory gas analysis and measured dietary intake plus change in energy balance (Table 10.1). The validations against whole room respiratory gas analysis have been performed for periods of three to five days. The whole room respiratory gas analysis systems offer high precision, but require that the subjects remain in the room for nearly 24 hour each day and thus are generally limited to less than five days. Validations of shorter than three days, are limited by the increasing imprecision of the doubly-labeled water method that arises because the difference between the isotope abundances of the tracers in body water between the start and end of the metabolic period are small and thus closer to the precision of the instrumental measurement of the isotope ratio (Schoeller 1983). An exception to this occurred when energy expenditure was elevated due to extreme exercise (Stein et al. 1987). Validations have been performed using measured dietary energy intake plus change in body energy stores for periods of up

TABLE 10.1
Validations of the Doubly-Labeled Water Method for Measuring Total Daily Energy Expenditure (TDEE)

Subjects (n)	Criterion	Mean Difference	Precision (CV)	Reference
Using Single Pool Lifson Equation and Adjusted to Two Pool Model				Lifson (1966); Schoeller et al. (1986a)
Adults (4)	I/B	–0.90%	5.60%	Schoeller and van Santen (1982)
Adult male (2)	IC	–5.50%	3.50%	Westerterp et al. (1984)
Adult male (1)	IC	–7.60%	–	Klein et al. (1984)
Adults (5)	IC	2.90%	7.80%	Schoeller and Webb (1984)
Using 2-Pool Model Individual or Fixed Ratio 1.03 or 1.034				Schoeller (1986); Coward et al. (1985)
Adults (4)	IC	2.00%	–	Coward et al. (1985)
Adult (6)	IC	2.00%	7.60%	Schoeller et al. (1986a)
Adult IBD on TPN (5)	I/B	3.30%	5.90%	Schoeller et al. (1986b)
Premature infants (4)	IC	–1.40%	4.80%	Roberts et al. (1986)
Infants (9)	IC	–0.90%	6.20%	Jones et al. (1987)
Exercising adults (5)	IC	1.40%	3.90%	Westerterp et al. (1988)
Infants (8)	IC	–1.00%	7.00%	Jones et al. (1988)
Adults (4)	IC	0.60%	1.00%	Seale et al. (1990)
Lean and obese adults (8)	IC	2.50%	5.80%	Ravussin et al. (1991)
Cold-field exercise (23)	I/B	–1.70%	3.40%	Hoyt et al. (1991)
Neonates (8)	IC	–4.50%	6.00%	Westerterp et al. (1991)
Preterm infants (12)	IC	0.30%	19.10%	Jensen et al. (1992)
College swimmers (8)	I/B	0.70%	14.40%	Jones and Leitch (1993)
Adults (9)	IC	1.60%	4.50%	Seale et al. (1993)
Adults on TPN (6)	IC	2.40%	1.70%	Pullicino et al. (1993)
Weighted average (131)		0.0%[a]		
Geometric mean		–	8.2% (median 5.8%)	

[a] This value used the 1.034 dilution space ratio of Racette et al. (1994) in adults. Using the recently published value of Sagayama et al. (2016) for adults while maintaining to 1.02 value for neonates, reduces the accuracy to –1.2%.

TABLE 10.2
Reproducibility of the Doubly-Labeled Water Method for Measuring Total Daily Energy Expenditure (TDEE) in Adults

Sample Size	Interval	Within Subject CV[a]	ICC[b]	Reference
9	6 months	11.20%	0.69	Haggarty et al. (1994)
6	6 months	7.80%	–	Schoeller and Hnilicka (1996)
24	2 weeks	5.10%	0.95	Trabulsi et al. (2003)
111	6 months	8.40%	0.72[c]	Neuhouser et al. (2008)
42	15 months	13.6%[c]	0.71[d]	Moshfegh et al. (2008)
53	6 months	–	0.88	Arab et al. (2011)

[a] Within subject coefficient of variation.
[b] Intraclass correlation coefficient (*r*).
[c] Ln $\log_e$ transformed.
[d] TEE read from published graph.

to two weeks. Overall the doubly-labeled water method has been demonstrated to have a relative accuracy of 1%–2% and a median precision of 6%, although two validation studies displayed much worse precision (Table 10.1). These validations have included males and females, activity levels from sedentary to highly active, a wide range of BMIs, and a wide range of adult ages.

As indicated above, interlaboratory validation studies demonstrated that isotopic abundance analysis can be a critical source of imprecision of the doubly-labeled water method, and that only the experienced and well-calibrated laboratories were capable of performing isotope abundance measurements with a precision and accuracy necessary to measure energy expenditure with an accuracy of 1%–2% and precision of 5%–6% (Roberts et al. 1995). With the use of a carefully standardized laboratory assay, the precision of the doubly-labeled water analyses can be maintained at 5% over two to five years (Wong et al. 2014). In addition, as shown in Table 10.2, within individual reproducibility studies have demonstrated that duplicate measures of energy expenditure in an individual have a relative precision of 5% when the interval between measures was two weeks. For longer intervals of six months or more, precision varied between 8% and 19%. Part of this may be seasonal variation in TEE (Haggarty et al. 1994; Plasqui and Westerterp 2004), although based on the unusually frequent high and low TEE values reported for the two studies with limited precision (Haggarty et al. 1994; Moshfegh et al. 2008) and limited experience with doubly-labeled water laboratory it may be that the analytical precision may not have been optimal in those two studies (Roberts et al. 1995). This demonstrates the importance of a laboratory to demonstrate precision by analyzing multiple, blinded aliquots from the same set of doubly-labeled water urine specimens.

10.5 APPLICATION OF THE DOUBLY-LABELED WATER BIOMARKER TO ASSESS ACCURACY OF SELF-REPORTED ENERGY INTAKE

The first reported use of doubly-labeled water as a biomarker to test the accuracy of self-reported energy intake was by Prentice et al. (1986). Thirteen healthy weight (22 + 2 kg/m^2) and nine obese women (33 + 5 kg/m^2) were studied as outpatients. The dietary intake was measured for seven day in the healthy weight group and 14 days in the obese group by self-recorded diary with all foods and beverages being weighed. The reported energy intakes in obese group were not significantly different from that among the healthy weight group (6.7 and 7.9 MJ/d), a pattern the authors discussed as being expected in previous studies. The biomarker energy expenditures measured by doubly-labeled water, however, were found to be greater in the obese group than in the healthy weight group (10.2 vs. 8.0 MJ/d, $p < 0.001$), which at that time the TEE difference was discussed as a surprising finding.

Thus, self-recorded dietary intake was only 67% of habitual energy intake for weight maintenance in the obese, 98% in the healthy weight. On account of both body weight and total body water decreased over the study period, it was reported that about half of the 33% underestimate of habitual energy intake was due to undereating as evidenced by loss of body energy stores and half due to an underrecording of dietary energy intake. Thus, this first energy expenditure biomarker assessment of self-reported dietary intake came to three important conclusions, which as indicated below have been confirmed by additional reports. Self-reported dietary intake was prone to underestimating of energy intake, the underestimate was far larger in the obese group, and the underestimate could be due to underreporting of actual intake or undereating. It interesting to note that, although this study by Prentice et al. was the first to use doubly-labeled water as a biomarker it was not the first biomarker study to raise concern about underestimates of dietary intake using self-reporting instruments. Warnold et al. (1978) compared self-reported protein intake against a urinary nitrogen biomarker of protein intake, and found that obese adults systematically underreported protein intake by about 50%. Based on this, Warnold et al. questioned the accuracy of the food diaries, but there were few studies to follow-up on these findings and it was the paper by Prentice et al. that appears to have initiated the current efforts to investigate the accuracy of self-reported dietary and which led to many of the developments reported in the other chapters of this book.

A second highly informative study comparing self-reported, weighed dietary diaries against doubly-labeled water was performed by Livingstone et al. (1990). Thirty one adults were studied using a stratified design to include a wide range of energy intakes in both men and women. Selection was based on a prior weighed dietary diary measurement. The important observation in this study was that individual self-reported energy intakes between the first weighed record were highly correlated with that of the second ($r = 0.79$) and covered the same range (study 1: 4.3–11.7 MJ/d vs. study 2: 4.2 vs. 11.4 MJ/d) in the same individuals. The underestimates of habitual energy intake compared to energy expenditure were large among the lowest tertile of self-reported energy intake (–34%), moderate in between the lower and upper tertile (–18%), and negligible above the upper tertile (–2%). Thus, these results suggested that much of error in the self-reported energy intake was not random within each subject but systematic, and that reproducibility of self-reported energy intake should not be taken as an evidence of accuracy nor can it be assumed that the errors would average out across a large sample population (Black et al. 1991). Even when intakes were recorded in household units without weighing using diet diaries, underreporting was observed and was due to both undereating and underreporting, and undereating was the predominant source of systematic error (Goris et al. 2000).

The systematic nature of underreporting of dietary energy intake was confirmed in the OPEN study, which was the first large ($n = 482$) dietary instrument validation study (Schatzkin et al. 2003; Trabulsi et al. 2003). Duplicate self-reported dietary intakes were again found to provide highly reproducible group means. In the OPEN study, two 24-hour recalls and two food frequency assessments were performed in each of the participants. Among women, the geometric mean for first and second 24-hour recall were 8.0 and 7.6 MJ/d (–5% difference) and the first and second food frequency questionnaire were 6.3 and 5.9 MJ/d (–7% difference). The geometric mean for TEE, however, was 9.5 MJ/d. Among men, the geometric mean for first and second 24-hour recall were 10.5 and 10.2 MJ/d (–3% difference) and the first and second food frequency questionnaire (FFQ) were 7.8 and 7.4 MJ/d (–5% difference). The geometric mean for TEE, however, was 11.5 MJ/d. For the entire group, the average 24-hour recall derived energy intakes were 16% less than energy expenditure and hence usual energy intake, and the average FFQs derived energy intakes were 34% less than usual energy intake indicating a larger underestimate of habitual energy intake.

Including those early reports, 43 studies have been identified by literature searches using combined terms of doubly-labeled water and diet, most of which have been included in reviews (Hill and Davis 2001; Trabulsi and Schoeller 2001). Of these, 36 were conducted using nonrandom sample of 100 or fewer subjects and thus could be considered to reflect what has or probably would occur when diet information is collected as part of a clinical research study. Many of these 36 studies compared more than one dietary assessment instrument against doubly-labeled water for a total of 45 attempted validations of energy intake against habitual energy intake from doubly-labeled water. Of these 45 comparisons, seven differences were less than 7% and 38 differences were underestimates of between 11% and 59%. Thus, the overwhelming finding was that self-reported energy intakes represented underestimates of usual energy intake. Weighing each study equally, there was no difference between the underestimates of three major dietary assessment instruments (diary −21 ± 15%, recall −15 ± 9%, and food frequency questionnaire −16 ± 8%). The large differences between the amounts of underreporting were probably due to the BMI status of the participants. Specifically, many studies reported greater underestimates as a function of increasing BMI, and the distributions of BMI among the participants varied between studies thus resulting in different percentage underestimates (Westerterp and Goris 2002). Of note, the magnitude of underestimates of energy intake observed for diet recall instruments and diary instruments are not dissimilar despite the fact that one group of self-reporting instruments is retrospective, whereas the other is prospective. As such, reporting error cannot simply be explained by a failure to remember a consumed food item (Westerterp and Goris 2002). Moreover, the few weighed diary studies also display a not so dissimilar degree of underestimates of dietary energy indicating that reporting error cannot be simply explained by underestimates of serving size but must also involve some items not being reported or consumed. A possible explanation would be that subjects either consciously or unconsciously interchange these diverse mechanisms to arrive at an underreported energy intake, but further research is needed to test this hypothesis.

There have been five studies performed that included between 263 and 544 participants, which were conducted by investigators experienced in epidemiology, and which quasi-randomly recruited participants from general populations (Freedman et al. 2014). These studies therefore probably represent results that would be expected from typical large epidemiologic studies of diet and disease. Energy intakes were underestimated in all five studies, with the underestimates from the food frequency questionnaires (−28%) exceeding those from the 24-hour recalls (−16%). Moreover, the correlations between self-reported energy intake and energy expenditure were between $r = 0.12$ and 0.42 depending on the dietary instrument and number of replicate dietary assessments. These correlations indicate that the reporting errors were highly variable between individuals and thus not due to a consistent proportional error across individuals.

Most of the above studies were performed in the United States or Europe. Underestimating of usual energy intake has been reported in Japan (Okubo et al. 2008), Africa (Orcholski et al. 2015), and Brazil (Ferriolli et al. 2010) and thus it is likely to be a problem that has been encountered on a global basis.

Underestimating of dietary energy intake has also been observed in children, although, the results are highly influenced by age. The age effect was first reported by Livingstone et al. (1992) who performed an age-stratified comparison of reported energy intake and usual energy intake by doubly-labeled water in 78 children aged 5–18 years. Using weighed dietary records, reported energy intakes prepared by a parent were not different from usual energy intake measured by doubly-labeled water for children between seven and nine years of age. Underestimates of energy intake increased slightly at age 12. Beginning at age 15 years, underestimates became large and differences exceeded −30% at age 18 years. A very different pattern was observed using detailed diet histories, with diet histories tending to slightly over estimate (12%) expenditure through age seven years and equaling expenditure during adolescents. In a 2010 review, Burrows et al. (2010) reported that this age effect for weighed or household common unit, dietary diaries observed parental or child reports that were accurate up through age 11. Thereafter, underestimates of energy intake were common and large (−20% or worse). Even when group means for the younger children were accurate, regression of reported energy intake on energy expenditure indicated that individual errors were large and thus not individually precise (Livingstone et al. 1992).

10.6 CONCLUSION

The use of doubly-labeled water measured energy expenditure as a biomarker of dietary energy intake has been extensive and continues. It is recognized as an accurate and quantitative biomarker for usual energy intake, although cost limits it use in large studies. Recently it has been an important tool for assessing the accuracy of new technology for the dietary assessment in free-living individuals, including many of the techniques reported in this text such as computer assisted dietary recall (Chapter 3), photographic diet diaries (Chapter 4), and swallowing and chewing monitors (Chapter 5). The doubly-labeled water is also an important research biomarker for studies designed to understand errors in the dietary assessment instruments (Chapter 2).

The literature utilizing doubly-labeled water as a biomarker of usual energy intake demonstrates that self-report-based estimates of energy intake have generally been underestimates of energy intake and that the degree of underestimation can be large; particularly among individuals with BMI exceeding 30 kg/m^2. The number of variables related to the degree of underestimation is not limited to BMI, however, making simple error corrections difficult and the correlation between energy intakes by self-report and DLW are very low with an average sharing far less than 25% of their variance.

In conclusion, the use of the doubly-labeled water biomarker for usual energy intake has demonstrated that the traditional dietary instruments of diet diaries, diet recalls, and food frequency questionnaires are subject to systematic underreporting. As such, none of these instruments can be used to accurately measure energy intake. On account of the increased underreporting of energy intake with increasing BMI repeatedly documented for these traditional dietary instruments demonstrates that dietary instruments cannot be used to study energy balance among obese persons or, for that matter, to monitor dietary compliance to energy restriction (Lissner et al. 2009). Indeed, given this information, a large group of investigators have stated that self-report-based estimates of energy intake should no longer be used when seeking to test hypotheses of the association between health of other outcomes and actual intake (Dhurandhar 2015).

10.7 FUTURE RESEARCH AND CONCLUSIONS

Perhaps one the most important current and future applications of the energy biomarker will be in use with other biomarkers to understand and possibly model or calibrate dietary instruments as discussed in this book (Chapter 13). By combining urinary nitrogen as a biomarker for protein intake with the energy biomarker, it has been demonstrated that protein, and thus foods with low protein density, are underreported more than average or high-protein density foods (Freedman et al. 2014). It has also been demonstrated using urinary potassium as a biomarker of fresh food consumption, that low potassium foods are underreported more than high potassium foods (Freedman et al. 2015). Further efforts to better identify foods that are over and underestimated are required to fully enable investigators to accurately link dietary intakes to disease (Chapter 13). As described elsewhere in this book, new approaches are being developed for which preliminary studies indicate less systematic error and thus they hold promise for future diet studies.

ACKNOWLEDGMENT

Dr. Allison's effort was supported in part by NIH grant P30DK056336. Dr. Schoeller's effort in development of the doubly-labeled water method was previously supported by NIH grants RO1DK30031 and P30DK26678. The opinions expressed are those of the authors and do not necessarily represent those of the NIH or any other organization.

REFERENCES

Arab, L., C.-H. Tseng, A. Ang, P. Jardack. 2011. Validity of a multipass, web-based, 24-hour self-administered recall for assessment of total energy intake in Blacks and Whites. *Am J Epidemiol* 174:1256–1265.

Beaton, G. H., J. Milner, V. McGuire, T. F. Feather, J. A. Little. 1983. Source of variance in 24-hour dietary recall data: Implications for nutrition study design and interpretation. Carbohydrate sources, vitamins, and minerals. *Am J Clin Nutr* 37:986–995.

Bhutani, S., N. Racine, T. Shriver, D. A. Schoeller. 2015. Special considerations for measuring energy expenditure with doubly labeled water under atypical conditions. *J Obes Weight Loss Ther* 5(Suppl 5):002.

Black, A. E., A. M. Prentice, W. A. Coward. 1986. Use of food quotients to predict respiratory quotients for the doubly-labelled water method of measuring energy expenditure. *Hum Nutr Clin Nutr* 40:381–391.

Black, A. E., G. R. Goldberg, S. A. Jebb, M. B. Livingstone, T. J. Cole, A. M. Prentice. 1991. Critical evaluation of energy intake data using fundamental principles of energy physiology: 2. Evaluating the results of published surveys. *Eur J Clin Nutr* 45:583–599.

Burrows, T. L., R. J. Martin, C. E. Collins. 2010. A systematic review of the validity of dietary assessment methods in children when compared with the method of doubly labeled water. *J Am Diet Assoc* 110:1501–1510.

Coward, W. A., A. M. Prentice, P. R. Murgatroyd et al. 1985. Measurement of CO2 and water production rates in man using ^{2}H and ^{18}O labeled H_2O. In A. J. H. van Es, ed. *European Nutrition Report 5; Human Energy Metabolism*. The Haugue, the Netherlands: CIP-gegevens Koninklijke Bibliotheek.

DeLany, J. P. 1997. Doubly labeled water for energy expenditure. In S. J. Carlson-Newberry and R. B. Costello, eds. *Emerging Technologies for Nutrition Research: Potential for Assessing Military Performance Capability*. Institute of Medicine (US) Committee on Military Nutrition Research; Washington, DC: National Academies Press (US).

Dhurandhar, N. V., D. A. Schoeller, A. W. Brown et al. 2015. Energy balance measurement: When something is not better than nothing. *Int J Obesity* (*Lond*) 39:1109–1113.

Ferriolli, E., K. Pfrimer, J. V. Moriguti et al. 2010. Under-reporting of food intake is frequent among Brazilian free-living older persons: A doubly labelled water study. *Rapid Commun Mass Spectrom* 24:506–510.

Food and Nutrition Board. 2005. *Dietary Reference Intakes for Energy, Carbohydrate, Fiber, Fat, Fatty Acids, Cholesterol, Protein, and Amino Acids* (*Macronutrients*). Washington, DC: National Academy of Science, Institute of Medicine.

Freedman, L. S., J. M. Commins, J. E. Moler et al. 2014. Pooled results from 5 validation studies of dietary self-report instruments using recovery biomarkers for energy and protein intake. *Am J Epidemiol* 180:172–88.

Freedman, L. S., J. M. Commins, J. E. Moler et al. 2015. Pooled results from 5 validation studies of dietary self-report instruments using recovery biomarkers for potassium and sodium intake. *Am J Epidemiol* 181:473–487.

Goris, A. H., M. S. Westerterp-Plantenga, K. R. Westerterp. 2000. Undereating and underrecording of habitual food intake in obese men: selective underreporting of fat intake. *Am J Clin Nutr* 71:130–134.

Gretebeck, R. J., D. A. Schoeller, R. A. Socki et al. 1997. Adaptation of the doubly labeled water method for subjects consuming isotopically enriched water. *J Appl Physiol* 82:563–570.

Haggarty, P., G. McNeill, M. K. Abu Manneh, L. Davidson. 1994. The influence of exercise on the energy requirements of adult males in the UK. *Br J Nutr* 72:799–813.

Heymsfield, S. B. and A. Pietrobelli. 2011. Individual differences in apparent energy digestibility are larger than generally recognized. *Am J Clin Nutr* 94:1650–1651

Hill, R. J. and P. S. Davies. 2001. The validity of self-reported energy intake as determined using the doubly labelled water technique. *Br J Nutr* 85:415–430.

Horvitz, M. A. and D. A. Schoeller. 2001. Natural abundance deuterium and 18-oxygen effects on the precision of the doubly labeled water method. *Am J Physiol Endocrinol Metab* 280:E965–E972.

Hoyt, R. W., T. E. Jones, T. P. Stein et al. 1991. Doubly labeled water measurement of human energy expenditure during strenuous exercise. *J Appl Physiol* 71:16–22.

IAEA. 2009. Assessment of body composition and total energy expenditure in humans using stable isotope techniques, IAEA Human Health Series No 3., IAEA, Vienna. http://www-pub.iaea.org/MTCD/publications/PDF/Pub1370_web.pdf (accessed August, 2016).

Jensen, C. L., N. F. Butte, W. W. Wong, J. K. Moon. 1992. Determining energy expenditure in preterm infants: Comparison of 2H(2)18O method and indirect calorimetry. *Am J Physiol* 263:R685–R692.

Jéquier, E. and E. Schutz. 1988. Stable isotopic methods for measuring energy expenditure. Classical respirometry and the doubly-labelled-water ($^{2}H_2{}^{18}O$) method: Appropriate applications of the individual or combined techniques. *Proc Nutr Soc* 47:219–225.

Jones, P. J., A. L. Winthrop, D. A Schoeller et al. 1988. Evaluation of doubly labeled water for measuring energy expenditure during changing nutrition. *Am J Clin Nutr* 47:799–804.

Jones, P. J. H. and C. A. Leitch. 1993. Validation of doubly labeled water for measurement of caloric expenditure in collegiate swimmers. *J Appl Physiol* 74:2909–2914.

Jones, P. J. H., A. L. Winthrop, D. A. Schoeller et al. 1987. Validation of doubly labelled water for assessing energy expenditure in infants. *Pediatr Res* 21:242–246.

Jones, P. J. and S. T. Leatherdale. 1991. Stable isotopes in clinical research: Safety reaffirmed. *Clin Sci (Lond)* 80:277–280.

Klein, P. D., W. P. T. James, W. W. Wong et al. 1984. Calorimetric validation of the doubly labelled water method for determination of energy expenditure in man. *Hum Nutr Clin Nutr* 38C:95–106.

Lifson, N., G. B. Gordon, R. McClintock. 1955. Measurement of total carbon dioxide production by means of D2O18. *J Appl Physiol* 7:704–710.

Lifson, N., W. S. Little, D. G. Levitt, R. M. Henderson. 1975. D2 18O (deuterium oxide) method for CO_2 output in small mammals and economic feasibility in man. *J Appl Physiol* 39:657–664.

Lifson, N. 1966. Theory of use of the turnover rates of body water for measuring energy and material balance. *J Theor Biol* 12:46–74.

Lifson, N., C. B. Gordon, M. B. Visscher, A. O. Nier. 1949. The fate of utilized molecular oxygen and the source of the oxygen of respiratory carbon dioxide, studied with the aid of heavy oxygen. *J Biol Chem* 180:803–811.

Lissner, L., R. P. Troiano, D. Midthune et al. 2007. OPEN about obesity: Recovery biomarkers, dietary reporting errors and BMI. *Int J Obes (Lond)* 31:956–961.

Livingstone, M. B., A. M. Prentice, J. J. Strain et al. 1990. Accuracy of weighed dietary records in studies of diet and health. *BMJ* 300:708–712.

Livingstone, M. B., A. M. Prentice, W. A. Coward et al. 1992. Validation of estimates of energy intake by weighed dietary record and diet history in children and adolescents. *Am J Clin Nutr* 56:29–35.

Merrill, A. L. and B. K. Watt. 1973. Energy value of foods, basis and derivation. Human Nutrition Research Branch, Agricultural Research Service, USDA. (Agriculture Handbook no. 74.) U.S. Government Printing Office Washington, D.C.

Moshfegh, A. J., D. G. Rhodes, D. J. Baer et al. 2008. The US Department of Agriculture Automated Multiple-Pass Method reduces bias in the collection of energy intakes. *Am J Clin Nutr* 88:324–332.

Neuhouser, M. L., L. Tinker, P. A. Shaw et al. 2008. Use of recovery biomarkers to calibrate nutrient consumption self-reports in the Women's Health Initiative. *Am J Epidemiol* 167:1247–1259.

Okubo, H., S. Sasaki, H. H. Rafamantanantsoa, K. Ishikawa-Takata, H. Okazaki, I. Tabata. 2008. Validation of self-reported energy intake by a self-administered diet history questionnaire using the doubly labeled water method in 140 Japanese adults. *Eur J Clin Nutr* 62:1343–1350.

Orcholski, L., A. Luke, J. Plange-Rhule et al. 2015. Under-reporting of dietary energy intake in five populations of the African diaspora. *Br J Nutr* 113:464–472.

Plasqui, G. and K. R. Westerterp. 2004. Seasonal variation in total energy expenditure and physical activity in Dutch young adults. *Obes Res* 12:688–694.

Prentice, A. M. 1990. The doubly labelled water method for measuring energy expenditure. Technical recommendations for its use in humans. IAEA, Vienna. https://humanhealth.iaea.org/HHW/Nutrition/TotalEnergyExpenditure/DLW_nahres4.pdf (accessed August, 2016).

Prentice, A. M., A. E. Black, W. A. Coward et al. 1986. High levels of energy expenditure in obese women. *Br Med J (Clin Res Ed)* 292:983–987.

Pullicino, E., W. A. Coward, M. Elia. 1993. Total energy expenditure in intravenously fed patients measured by the doubly labelled water technique. *Metabolism* 42:58–64.

Racette, S. B., D. A. Schoeller, A. H. Luke, K. Shay, J. Hnilicka, R. F. Kushner. 1994. Relative dilution spaces of 2H- and 18O-labeled water in humans. *Am J Physiol* 267:E585–E590.

Ravussin, E., I. Harper, R. Rising, C. Bogardus. 1991. Energy expenditure by doubly labeled water: Validation in lean and obese subjects. *Am J Physiol* 261:E402–E409.

Roberts, S. B., W. Dietz, T. Sharp. G. E. Dallal, J. O. Hill. 1995. Multiple laboratory comparison of the doubly labeled water technique. *Obes Res* 3 Suppl 1:3–13.

Roberts, S. B., W. A. Coward, K. H. Schlingenseipen, V. Nohra, A. Lucas. 1986. Comparison of the doubly labeled water ($^2H_2{}^{18}O$) method with indirect calorimetry and a nutrient balance study for simultaneous determination of energy expenditure, water intake, and metabolizable energy intake in preterm infants. *Am J Clin Nutr* 44:315–322.

Sagayama, H., Y. Yamada, N. M. Racine, T. C. Shriver, D. A. Schoeller, DLW Study Group. 2016. Dilution space ratio of 2H and 18O of doubly labeled water method in humans. *J Appl Physiol* 120:1349–1354.

Schatzkin, A., V. Kipnis, R. J. Carroll et al. 2003. A comparison of a food frequency questionnaire with a 24-hour recall for use in an epidemiological cohort study: Results from the biomarker-based Observing Protein and Energy Nutrition (OPEN) study. *Int J Epidemiol* 32:1054–1062.

Schoeller, D. A. and J. M. Hnilicka. 1996. Reliability of the doubly labeled water method for the measurement of total daily energy expenditure in free-living subjects. *J Nutr* 126:348S–354S.

Schoeller, D. A., E. Ravussin, Y. Schutz, K. J. Acheson, P. Baertschi, E. Jéquier. 1986a. Energy expenditure by doubly labeled water: Validation in humans and proposed calculation. *Am J Physiol* 250:R823–R830.

Schoeller, D. A. and E. van Santen. 1982. Measurement of energy expenditure in humans by doubly labeled water method. *J Appl Physiol* 53:955–957.

Schoeller, D. A. and P. Webb. 1984. Five-day comparison of the doubly labeled water method with respiratory gas exchange. *Am J Clin Nutr* 40:153–158.

Schoeller, D. A., R. F. Kushner, P. J. Jones. 1986b. Validation of doubly labeled water for measuring energy expenditure during parenteral nutrition. *Am J Clin Nutr* 44:291–298.

Schoeller, D. A. 1984. Use of two point sampling for the doubly labeled water method. *Hum Nutr Clin Nutr* 38C:477–480.

Schoeller, D. A. 1983. Energy expenditure from doubly labeled water: Some fundamental considerations in humans. *Am J Clin Nutr* 38:999–1005.

Schoeller, D. A. 1999. The shortage of O-18 water. *Obes Res* 7:519.

Schoeller, D. A. 2002. Validation of habitual energy intake. *Public Health Nutr* 5:883–888.

Seale, J. L., J. M. Conway, J. J. Canary. 1993. Seven-day validation of doubly labeled water method using indirect room calorimetry. *J Appl Physiol* 74:402–409.

Seale, J. L., W. V. Rumpler, J. M. Conway, C. W. Miles. 1990. Comparison of doubly labeled water, intake balance, and direct- and indirect-calorimetery methods for measuring energy expenditure in adult men. *Am J Clin Nutr* 52:66–71.

Smith, S. R., L. de Jonge, J. J. Zachwieja et al. 2000. Fat and carbohydrate balances during adaptation to a high-fat. *Am J Clin Nutr* 71:450–457.

Speakman, J. R. 1997. *Doubly Labelled Water: Theory and Practice.* London: Chapman & Hall.

Stein, T. P., R. W. Hoyt, R. G. Settle, M. O'Toole, W. D. Hiller. 1987. Determination of energy expenditure during heavy exercise, normal daily activity, and sleep using the doubly-labelled-water ($^2H_2$ ^{18}O) method. *Am J Clin Nutr* 45:534–539.

Taylor, H. P. 1974. The application of oxygen and hydrogen isotope studies to problems of hydrothermal alteration and ore deposition. *Econ Geol* 69:843–883.

Thomas, D. M., C. Bouchard, T. Church et al. 2012. Why do individuals not lose more weight from an exercise intervention at a defined dose? An energy balance analysis. *Obes Rev* 13:835–847.

Thorsen, T., T. Shriver, N. Racine, B. A. Richman, D. A. Schoeller. 2011. Doubly labeled water analysis using cavity ring-down spectroscopy. *Rapid Commun Mass Spectrom* 25:3–8.

Trabulsi, J. and D. A. Schoeller. 2001. Evaluation of dietary assessment instruments against doubly labeled water, a biomarker of habitual energy intake. *Am J Physiol Endocrinol Metab* 281:E891–E899.

Trabulsi, J., R. P. Troiano, A. F. Subar et al. 2003. Precision of the doubly labeled water method in a large-scale application: Evaluation of a streamlined-dosing protocol in the Observing Protein and Energy Nutrition (OPEN) study. *Eur J Clin Nutr* 57:1370–1377.

Warnold, I., G. Carlgren, M. Krotkiewski. 1978. Energy expenditure and body composition during weight reduction in hyperplastic obese women. *Am J Clin Nutr* 31:750–763.

Weir, J. B. 1949. New methods for calculating metabolic rate with special reference to protein metabolism. *J Physiol* 109:1–9.

Westerterp, K. R. and A. H. Goris. 2002. Validity of the assessment of dietary intake: Problems of misreporting. *Curr Opin Clin Nutr Metab Care* 5:489–493.

Westerterp, K. R., J. O. de Boer, W. H. M. Saris, P. F. M. Schoffelen, F. Ten Hoor. 1984. Measurement of energy expenditure using doubly labeled water. *Int J Sports Med* 5:S74–S75.

Westerterp, K. R., F. Brouns, W. H. M. Saris, F. Ten Hoor. 1988. Comparison of doubly-labelled water with respirometry at low activity and high activity levels. *J Appl Physiol* 65:53–56.

Westerterp, K. R., H. N. Lafeber, E. J. Sulkers, P. J. Sauer. 1991. Comparison of short term indirect calorimetry and doubly labeled water method for the assessment of energy expenditure in preterm infants. *Biol Neonate* 60:75–82.

Wong, W. W., S. B. Roberts, S. B. Racette et al. 2014. The doubly labeled water method produces highly reproducible longitudinal results in nutrition studies. *J Nutr* 144:777–783.

11 Biomarker for Energy Intake
Resting Energy Expenditure and Physical Activity

Klaas R. Westerterp

CONTENTS

11.1 INTRODUCTION

Energy intake is a function of energy requirement as determined by resting energy expenditure and activity-induced energy expenditure. Resting energy expenditure can be measured or predicted with an equation based on weight, height, age, and gender of the subject. Activity-induced energy expenditure (AEE) can be estimated with an accelerometer for movement registration. The validity of these measures for estimating energy intake is presented with respirometry assessed resting energy expenditure and doubly-labeled water assessed total energy expenditure as a reference. Prediction of resting energy expenditure based on subject sex, weight, height, and age with equations is applicable for most subjects. For some ethnicities or for subjects with an exceptional physical activity level like athletes, measurement of body composition is indicated; to use a fat-free mass-based equation. For activity-induced energy expenditure, a valid sensor for body movement should allow explaining most of the variation as observed with the reference; doubly-labeled water assessed total energy expenditure.

11.2 BACKGROUND

Adult subjects maintain a balance between their energy intake and energy expenditure. The energy store of the body does not fluctuate much, as shown by the constancy of body weight and body composition. Thus, energy intake can be estimated from energy requirement to maintain energy balance, that is, to maintain body weight and body composition. Energy requirement is a function of energy expenditure for body maintenance or resting energy expenditure (REE) and of physical activity or AEE. Total energy expenditure (TEE) consists of three components: REE, AEE, and the thermic effect of food (TEF) also known as diet-induced energy expenditure, where the latter is the smallest component (Figure 11.1). The thermic effect of food is determined by food intake

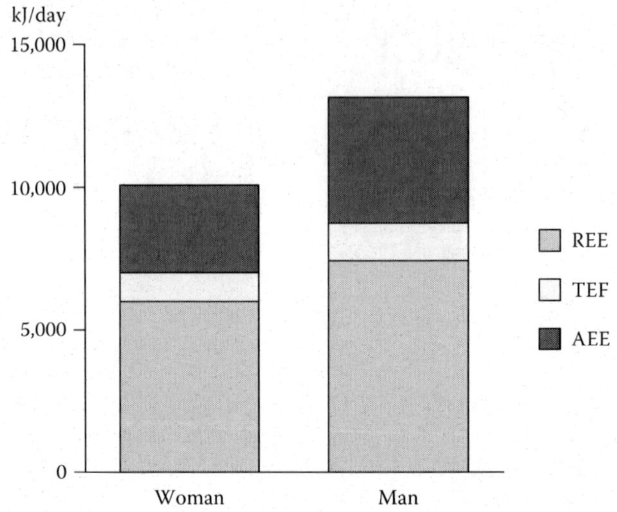

FIGURE 11.1 Total energy expenditure and the components resting energy expenditure (REE), thermic effect of food (TEF), and activity-induced energy expenditure (AEE), for a typical adult woman and man.

and food composition. Reported TEF values for separate nutrients are 0%–3% for fat, 5%–10% for carbohydrate, 20%–30% for protein (Tappy 1996), and 10%–30% for alcohol (Acheson 1993). In healthy subjects with a mixed diet, TEF represents about 10% of the total amount of energy ingested over 24 hours. When a subject is in energy balance, where intake equals expenditure, TEF can be estimated as a fixed proportion of 10% of total energy expenditure (Westerterp 2004).

The reference technique for the assessment of REE and AEE is indirect calorimetry. In indirect calorimetry the energy production is calculated from oxygen consumption, carbon dioxide production, and urine-nitrogen loss. The basis of the calculation is the gaseous exchange and energy release from the metabolized carbohydrate, fat, and protein. For the measurement of REE, a subject is observed under standard conditions excluding TEF and AEE. For the measurement of AEE, the subject is observed under free-living conditions. Then, total energy expenditure is measured with the doubly-labeled water technique, based on the measurement of carbon dioxide production as described in the foregoing chapter, and AEE is derived from total energy expenditure adjusted for REE. Measurements of REE with indirect calorimetry allowed the development of prediction equations for REE based on subject characteristics including height, age, weight, and gender. The alternative for doubly-labeled water assessed AEE is an accelerometer for body movement registration as validated with doubly-labeled water assessed AEE as a reference.

The current chapter describes the assessment of energy intake based on the assessment of REE and AEE, where REE is estimated with a prediction equation and AEE is derived with a doubly-labeled water validated accelerometer for movement registration. Theory of measure, validation, protocol, and application are presented for the two components REE and AEE separately, followed by the combination of the two, as biomarker for energy intake.

11.3 RESTING ENERGY EXPENDITURE

11.3.1 Measured Resting Energy Expenditure

The energy expenditure for body maintenance or REE is, generally, the largest component of TEE (Black et al. 1996). It is defined as the energy expenditure to maintain and preserve the integrity of vital functions. The measurement of REE under the basal conditions must meet four conditions: (1) the subject is awake, (2) is measured in a thermoneutral environment to avoid energy expenditure for the maintenance of body temperature, (3) is fasted long enough to eliminate TEF, and (4) is in rest to eliminate AEE. To perform accurate measurement of REE, one usually adopts an in-patient protocol. A subject stays overnight in the research facility where food intake and physical activity are strictly controlled, and REE is measured directly after waking up in the morning. A 10–12 hour fast before REE measurement is the accepted procedure to eliminate TEF. Thus, when REE is measured at 7.00 a.m., subjects should be fasted from about 8.00 p.m. the day before. High intensity exercise should be prevented on the day before REE measurement. An outpatient protocol, where subjects are transported by car or public transport to the laboratory after spending the night at home produces sufficiently reproducible results when subjects are carefully instructed and behave accordingly (Adriaens et al. 2003).

A typical protocol for a REE measurement with a ventilated-hood system takes 30 minutes. To eliminate effects of subject habituation to the testing procedure, the respiratory measurements over the first 10 minutes are discarded and the following 20 minutes are used to calculate REE. The criterion for the chosen time interval is the reproducibility of the calculated REE value. Longer measurements tend to result in higher values because subjects become restless. Reproducibility of REE measurement is influenced by the within-machine variability of the ventilated-hood system (Adriaens et al. 2003). Calibration procedures include standard gases covering the span of the oxygen and carbon dioxide analyzers, and a standard volume for calibration of the hood ventilation. Overall performance of a ventilated-hood system can be checked with methanol burning (Schoffelen et al. 1997). Methanol 99.8% is combusted by using a gas burner, placed on a calibrated balance under the hood. The methanol burner is set at a burning rate equivalent to the oxygen consumption and carbon dioxide production of the average subject and the burning time is comparable to a typical hood measurement.

Resting energy expenditure is usually compared between subjects by standardizing to an estimate of metabolic body size, where fat-free mass is the main predictor (Cunningham 1991). The reliable way of comparing REE data is by regression analysis. REE should never be divided by the absolute fat-free mass value, because the relationship between energy expenditure and fat-free mass has an axes intercept that is significantly different from zero (Figure 11.2). The smaller the fat-free mass the higher the REE/kg, and thus the REE per kg fat-free mass is on average higher in women with a lower fat-free mass compared to men. When fat-free mass, fat mass and gender are included as covariates in a regression analysis; gender does not come out as a significant contributor to the explained variation (Westerterp 1999).

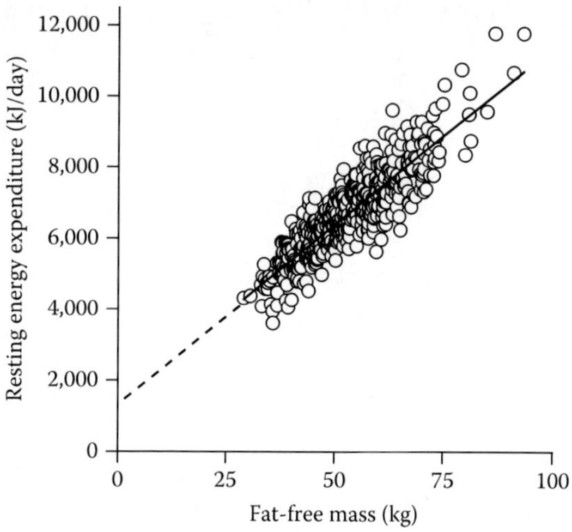

FIGURE 11.2 Resting energy expenditure plotted as a function of fat-free mass with the linear regression showing a significant positive intercept. (From Speakman, J. R. and K. R. Westerterp. *Am. J. Clin. Nutr*, 92, 826–834, 2010.)

11.3.2 Prediction of Resting Energy Expenditure

Although fat-free mass is the main predictor of REE, fat-free mass is not routinely used for the prediction of REE as measurement of fat-free mass is complicated. Thus prediction equations for REE were developed, based on subject sex, weight, height, and age. The most commonly used prediction equations are Harris Benedict (Harris and Benedict 1919), Food and Agricultural Organization (FAO/WHO/UNU 1985), Owen (Owen et al. 1986 and 1987), and Mifflin (Mifflin et al. 1990). An equation to predict REE from subject sex, weight, height, and age is based on an estimation of fat-free mass from subject sex, weight, height, and age, as shown below.

To simulate changes in body weight in response to a change in energy intake or physical activity, we developed a prediction equation for REE from subject sex, weight, height, and age with data from 190 subjects, of which 105 women, body mass 38 to 215 kg, height 1.30 to 2.05 m, and age 19 to 95 year (Westerterp et al. 1995). First fat-free mass (FFM) was predicted from subject sex, weight, height, and age, and then the FFM was used in a prediction equation for REE:

$$\text{For women: FFM (kg)} = 0.218 \text{ BM (kg)} + 27.392 \text{ height (m)} - 0.074 \text{ age (y)} - 12.5;\ r^2 = 0.81;$$

$$\text{For men: FFM (kg)} = 0.292 \text{ BM (kg)} + 34.009 \text{ height (m)} - 0.105 \text{ age (y)} - 18.4;\ r^2 = 0.79.$$

The equations show that women have relative less FFM per kg body mass than men, taller subjects have more FFM, and FFM decreases with increasing age.

As shown below, prediction equations for REE from subject sex, weight, height, and age are similar to prediction equations for FFM. Here, the Harris Benedict equation and the Owen equation were derived by combining the original data for women and men. The WHO-prediction uses separate equations for women and men and for subjects aged 18–30, 30–60, and over 60 year.

Harris Benedict: 231 subjects, 99 women, 6 with BMI > 30 kg/m^2, age 18–74 year;

$$\text{REE (kJ/d)} = 49.0 \text{ BM (kg)} + 23.5 \text{ height (m)} - 23.4 \text{ age (y)} + 448 \text{ Sex} - 218;\ r^2 = 0.80;$$

Owen: 104 subjects, 44 women, 32 with BMI > 30 kg/m^2, age 18–82 year;

$$\text{REE (kJ/d)} = 33.7 \text{ BM (kg)} + 18.5 \text{ height (m)} - 14.9 \text{ age (y)} + 1200 \text{ Sex} + 590;\ r^2 = 0.71;$$

Mifflin: 498 subjects, 247 women, 234 with BMI > 30 kg/m^2, age 19–78 year;

$$\text{REE (kJ/d)} = 41.8 \text{ BM (kg)} + 26.2 \text{ height (m)} - 20.6 \text{ age (y)} + 695 \text{ Sex} - 674;\ r^2 = 0.71;$$

Where for women Sex = 0 and for men Sex = 1.

WHO: 4814 subjects, 1239 women, including adults of different weight for height, age > 18 year.

Women	18–30 year	REE (kJ/d) = 55.6 BM (kg) + 1397.4 height (m) + 146
	30–60 year	REE (kJ/d) = 36.4 BM (kg) – 104.6 height (m) + 3619
	>60 year	REE (kJ/d) = 38.5 BM (kg) + 2665.2 height (m) – 1264
Men	18–30 year	REE (kJ/d) = 64.4 BM (kg) – 113.0 height (m) + 3000
	30–60 year	REE (kJ/d) = 47.2 BM (kg) + 66.9 height (m) + 3769
	>60 year	REE (kJ/d) = 36.8 BM (kg) + 4719.5 height (m) – 4481

The four mostly used REE prediction equations as presented earlier have been evaluated extensively. Prediction accuracy is usually defined as the percentage of individuals in the study group whose REE is predicted to within plus or minus 10% of the measured REE (Frankenfield et al. 2005; Neelemaat et al. 2012; Rao et al. 2012; Siervo et al. 2014; Ten Haaf and Weijs 2014; Weijs 2008; Weijs and Vansant 2010). The more an individual shares characteristics with the group of people from whom the equation was developed the better the estimate. The Harris Benedict and Mifflin equation provided accurate estimates over a wide range of body mass index. Above body mass index 45 kg/m^2, the WHO equation should not be used in this extremely obese group (Weijs and Vansant 2010). The Mifflin equation was associated with the largest error in subjects older than 60 year (Siervo et al. 2014). In older and malnourished patients, the WHO equation came out best (Neelemaat et al. 2012).

As an example, Figure 11.3 presents the frequency distribution for percentage differences between predicted REE with the Harris Benedict, WHO, Owen, and Mifflin equation and

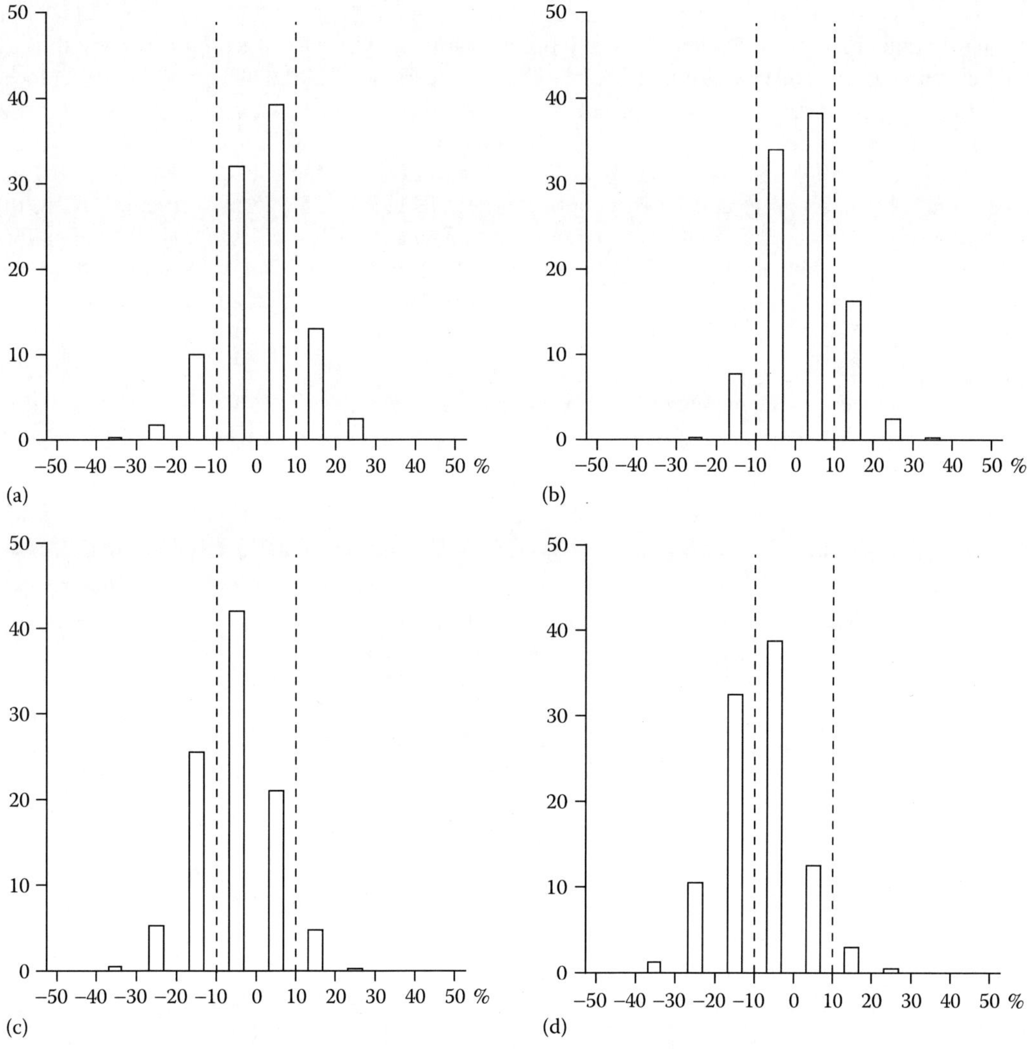

FIGURE 11.3 Frequency distribution for percentage differences between predicted and measured resting energy expenditure for 529 subjects (From Speakman, J. R. and K. R. Westerterp. *Am. J. Clin. Nutr*, 92, 826–834, 2010.): (a) prediction equation Harris et al. (1919); (b) prediction equation WHO/FAO/UNU (1985); (c) prediction equation Owen et al. (1986, 1987); and (d) prediction equation Mifflin et al. (1990).

measured REE for 556 subjects, 234 women, 95 with BMI > 30 kg/m^2, age 18 to 96 year (data from Speakman and Westerterp 2010). In this reference group, covering a large age span, the Harris Benedict and WHO equation did best with more than 70% of the subjects within the defined range of plus or minus 10% of the measured REE, whereas with the Mifflin equation this accuracy level was reached for only 50% of the subjects. The Mifflin equation was derived from a population with nearly 50% obese subjects and apparently does not apply for this reference population with less than 20% obese subjects.

Additional subject characteristics to take into account for the prediction of REE are ethnicity affecting body build and special behavior affecting body composition, like athletic performance. Clear examples of differences in body build affecting the accuracy of the generally Caucasian-based prediction equations for REE are Asian Indians. Where the Owen and Mifflin equation were appropriate for the prediction of REE in Chinese (Rao et al. 2012), Caucasian-based prediction equations systematically overestimate REE in South Asian Indians (Wouters-Adriaens et al. 2008; Wulan et al. 2010). South Asian Indians have a higher percentage of body fat at a lower body mass index compared to Caucasians. The high body fat percentage at low body mass index can be partly explained by differences in body build, that is, differences in trunk-to-leg-length ratio and differences in slenderness and muscularity (Deurenberg et al. 2002). There are no differences in REE between South Asian Indians and Caucasians after adjustment for differences in fat-free mass (Wouters-Adriaens et al. 2008). Thus, South Asian Indians are an exceptional population requiring a population-specific equation for predicting REE from subject sex, weight, height, and age. Similarly, muscularity and fat-free mass can be affected by athletic performance. Then, the Cunningham equation (Cunningham 1991) based on fat-free mass is the first choice for the prediction of REE. Alternatively, there are equations based on subject sex, weight, height, and age for athletes like an equation derived from REE measurement in a group of young recreational athletes (Ten Haaf and Weijs 2014):

90 subjects, 37 women, age 18–35 year;

$$\text{REE (kJ/d)} = 49.940\ \text{BM (kg)} + 2459.1\ \text{height (m)} - 34.0\ \text{age (y)} + 799\ \text{Sex} + 122;$$

Where for women Sex = 0 and for men Sex = 1.

A special case, where even the Cunningham equation based on fat-free mass does not produce accurate estimates for REE are elite athletes. Sjödin et al. (1996) measured in top international cross-country skiers that REE values were 16% higher than that predicted from the weight-based WHO equation and 12% higher than in sedentary fat-free mass matched control subjects.

Prediction equations for REE based on subject sex, weight, height, and age are applicable for most subjects, where for some situations measurement of body composition is indicated as described. An example where indirect calorimetry measurement of REE shows a systematic difference with predicted values is in subjects after weight loss. Underfeeding induces a reduction in REE below-predicted values, as based on the new body composition reached after underfeeding-induced weight loss (Camps et al. 2013; Major et al. 2007; Rosenbaum et al. 2008; Schwartz et al. 2012). The REE reduction, adjusted for changes in body composition, ranges between 5% and slightly more than 10% of the initial value, depending on time interval after the intervention. Van Gemert et al. observed an average reduction of 12% at three months after the start of weight loss and of 6% when weight loss was maintained for more than three years (Van Gemert et al. 1998).

In summary, prediction of REE based on subject sex, weight, height, and age with equations as presented from Harris and Benedict and from WHO/FAO/UNU give estimates within 10% for most subjects. For some ethnicities like South Asian Indians or subjects with an exceptional physical activity level like athletes, measurement of body composition is indicated; to use a fat-free mass-based equation. In some situations, only an indirect calorimetry measurement guarantees an estimate within 10%.

11.4 ACTIVITY ENERGY EXPENDITURE

11.4.1 Measured Activity Energy Expenditure

Activity energy expenditure is the most variable component of TEE. Activity energy expenditure is determined by body movement and by the mass moved, that is, body mass, where body movement or physical activity level varies between individuals. The indicated method for the assessment of body movement in daily life is a motion sensor. Motion sensors for the assessment of body movement recently evolved from mechanical devices like pedometers to miniature electronic triaxial accelerometers for movement registration (Plasqui et al. 2013; Plasqui and Westerterp 2007).

Accelerometer-based activity monitors quantify physical activity by measuring the acceleration of the human body during movement. Accurate and unobtrusive measurements of physical activity are achieved only when the accelerometer has certain physical and technical characteristics in terms of dimensions, weight, and amount of information processed and recorded. Considering that there is a trade-off between the energy consumption, portability, and performance (quality or quantity of information collected) of an activity monitor, the design largely determines the degree of accuracy and unobtrusiveness. Accurate measurements of physical activity can be achieved when the acceleration signal from the body is collected at a frequency sufficient to ensure that the full range of human motions is captured. The frequency content of the acceleration of the body during physical activity varies according to the measurement location. At the waist level, 95% of the variability of the acceleration signal can be determined by harmonics with 10 Hz. The amplitude does not exceed 6 g in magnitude (1 g = 9.8 m/s^2) (Antonsson and Mann 1985; Bouten et al. 1997). Thus, according to the Nyquist theorem, an accurate accelerometer should collect acceleration data at a sampling frequency of 20 Hz. Furthermore, it should be able to process the acceleration signal to filter out noise and extract relevant characteristics from the acceleration pattern so as to describe physical activity.

Nowadays, there are many accelerometer-based sensors for monitoring body movement available, including sensors incorporated in body worn devices like wristwatches and smartphones. The capability of motion sensors to predict AEE is usually tested during standardized activities like walking and running on a treadmill under laboratory conditions, generally showing good performance (Plasqui and Westerterp 2007). The ultimate test is a validation over one or more weeks under free-living conditions, where AEE is derived from simultaneous measurement of TEE with doubly-labeled water in combination with a measurement of REE. Then, only a few usable sensors remain so far (Westerterp 2014). Many devices only produce values for total energy expenditure based on subject sex, weight, height, age, and measured body acceleration. Unfortunately, the validity of this method is questionable. The outcome often is a function of proprietary equations. Thus, the user has no information on the equation for REE adopted, based on subject sex, weight, height, and age. In addition, only for a few devices, the correlation of the outcome for total energy expenditure with the measured doubly-labeled water assessed total energy expenditure as reference is driven by measured body acceleration (Plasqui and Westerterp 2007).

11.4.2 Prediction of Activity Energy Expenditure

Approaches to prediction of AEE can be broken into two categories: those that estimate a multiple of REE and those that estimate AEE directly. The former way of expressing the physical activity level of a subject is by expressing TEE as a multiple of REE, adjusting TEE for differences in maintenance metabolism (PAL = TEE/REE; WHO/FAO/UNU 1985). The limits of TEE are established at around 1.2 REE for nonambulatory subjects to 4.5 REE for elite endurance athletes (Black et al. 1996). FAO/WHO/UNU classified the physical activity level of a subject in three categories (FAO/WHO/UNU 2004). The physical activity for sedentary and light activity lifestyles ranges between 1.40 and 1.69, for moderately active or active lifestyles between 1.70 and 1.99, and for vigorously active lifestyles between 2.00 and 2.40. Thus, for most subjects REE is the largest component of

TEE as stated already in the section on REE. Activity energy expenditure is only one third of TEE at a PAL of 1.70, the value for a moderately active subject, as shown below by examples of a typical woman and man.

Woman: BM 65 kg; H 1.70 m; age 40 y; TEE 9.9 MJ/d; REE 5.8 MJ/d; TEF 1.0 MJ/d, and thus AEE 3.1 MJ/d;

Man: BM 75 kg; H 1.80 m; age 40 y; TEE 12.6 MJ/d; REE 7.4 MJ/d; TEF 1.3 MJ/d, and thus AEE 3.9 MJ/d.

Variation in AEE between subjects can be partly explained by body mass, age, and sex. As mentioned above, AEE is higher for the same body movement in a subject with a higher body mass. Thus, AEE is generally higher in heavier subjects. Activity energy expenditure peaks when adult weight is reached and then gradually declines with increasing age. Thus, it seems that AEE is highest during the reproductive years (Heitmann et al. 2012). In addition, there is a systematic sex difference. Mainly due to the fact that men are generally larger than women, AEE tends to be higher in men than in women while differences in PAL, adjusting TEE of differences in body size, are negligible (Black et al. 1996). Body mass, age, and sex are included as significant parameters in a prediction equation for AEE as derived from data from 556 subjects, 234 women, 95 with BMI > 30 kg/m^2, age 18–96 year (data from Speakman and Westerterp 2010).

$$\text{AEE (kJ/d)} = 24.3\ \text{BM (kg)} - 39.4\ \text{age (y)} + 1172\ \text{Sex} + 3064;\ r^2 = 0.38;$$
$$\text{Where for women Sex} = 0 \text{ and for men Sex} = 1.$$

Comparing the prediction of REE and AEE from subject characteristics, the explained variation for AEE is only half the explained variation for REE. Body movement, as the main determinant of AEE, has to be included to get to more accurate predictions of AEE.

Prediction equations for AEE from motion sensor-assessed body movement should have an explained variation HIGHER than 40%, what can be reached already by including subject body mass, age, and sex in the equation as shown above. Thus, out of 11 sensors validated under free-living conditions, most showed a poor performance and only three showed sufficient validity (Westerterp 2014). They explained 50%–70% of the variation in AEE, still leaving at least 30% of the variation unexplained. Further improvement will be reached with monitors allowing the assessment of activity type. The relation between body movement and AEE is different for activities like walking and cycling. Identification of activity types improved the assessment of AEE compared with overall unidentified body movement only (Bonomi and Westerterp 2012; Van Hees et al. 2009).

Physical activity and thus AEE shows short-term variation, from day-to-day, and long-term variation in relation to season. In a climate with adverse weather conditions in winter and more friendly weather in summer, subjects are more homebound, and thus less physically active in winter, and more outgoing and thus more physically active in summer (Figure 11.4). In the Netherlands, physical activity level was lower in winter with ambient temperatures around 5°C than in summer with ambient temperatures around 20°C (Plasqui and Westerterp 2003). However, seasonal differences were small and the higher AEE in summer was partly compensated by a lower REE and thus total energy expenditure was not significantly different between seasons, as observed in a study on the effects of seasonal changes in physical activity and energy requirements in the United Kingdom (Haggarty et al. 1994).

In summary, prediction equations based on subject weight and body movement can explain 50%–70% of the variation in AEE, when choosing a valid sensor for body movement.

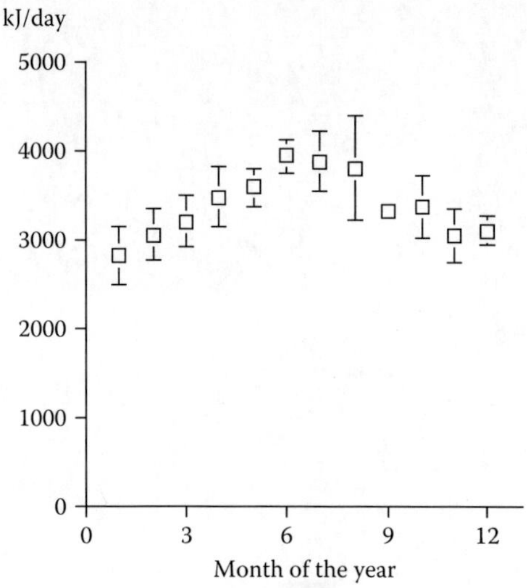

FIGURE 11.4 Annual pattern of activity-induced energy expenditure of a typical individual living in the Netherlands as derived from an accelerometer for movement registration. Monthly values are averages with SD over a seven-year interval.

11.5 APPLICATIONS OF ENERGY EXPENDITURE BIOMARKERS TO DIET ASSESSMENT

One of the applications of resting energy expenditure and activity energy expenditure as biomarker of intake is the identification of underreporters of energy intake (EI) with the Goldberg cut off (Goldberg et al. 1991). Reported intake is expressed as a multiple of REE and the ratio should be higher than the PAL value for a sedentary lifestyle. However, excluding underreporters by excluding subjects reporting unlikely low intake does not imply that reported intake figures of the remaining subjects are valid. When a subject with an REE of 6 MJ/d reports an intake of 10 MJ/day, the EI/REE figure is with a value 1.7 well above the value of 1.4 to 1.6 for a sedentary lifestyle. When the subject has an active lifestyle with a PAL of 2.0, energy intake should be approximately two times REE and equals 12 MJ/day. Thus, a reported intake of 10 MJ/day for this subject is 20% lower than their habitual intake. The indicated method to identify not only underreporters but also misreporters, in general, is by comparing EI/REE with a PAL figure derived from REE and AEE.

A first evaluation of reported energy intake by using calculated expenditure from REE and AEE was reported by Goris et al. (2001). The study was performed in 24 subjects, 12 women, age 55–65 year, and body mass index 19 to 33 kg/m^2. Energy intake was reported with a seven-day dietary record and energy expenditure was calculated from REE estimated with the WHO equation (FAO/WHO/UNU 1985) and AEE estimated with a doubly-labeled water validated accelerometer. Simultaneously, energy expenditure was measured with the doubly-labeled water method, as described in the foregoing chapter, as a reference. Reported intake of 9.0 $\pm$ 2.1 MJ/day was significantly lower than calculated expenditure of 10.8 $\pm$ 1.7 MJ/d ($p < 0.001$) and measured energy expenditure of 11.3 $\pm$ 2.3 MJ/day ($p < 0.001$). The percentage of underreporting derived from calculated and measured energy expenditure was similar, showing the validity of the combination of estimated REE and AEE as a biomarker of energy intake.

An example of the application of calculated energy expenditure to evaluate estimated energy intake is a study in Danish children (Biloft-Jensen et al. 2013). Energy intake, recorded with web-based dietary assessment software, was not different from calculated expenditure at group level. At the individual level, 20% of the children were classified as overreporters and 20% were classified as underreporters. Interestingly, underreporters were more likely than both acceptable

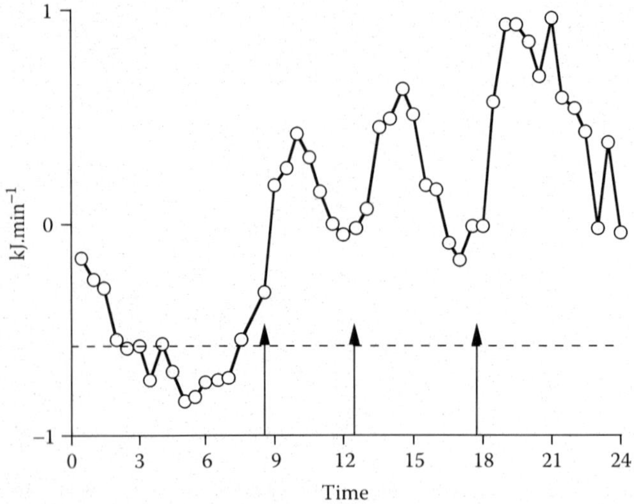

FIGURE 11.5 The diurnal pattern of resting energy expenditure including the thermic effect of food minus the level of basal metabolic rate (dotted line); subjects followed a protocol with fixed meal times (arrows) and were allowed to sleep from 23.00 to 07.00. (From Westerterp, K. R. *Nutr Metab*, 1, 5, 2004.)

and overreporters to report that illness, due to a flu epidemic, affected eating during the recording period. Similarly, overreporting might reflect real overeating to compensate for a flu-induced reduction of intake just before the reporting interval. Thus, reported intake might be different from calculated expenditure for other reasons than misreporting as well. However, in healthy subjects, intake can be assumed to match expenditure over observation intervals of minimally one week (Edholm et al. 1955).

A point of discussion is whether energy intake estimated from REE and AEE as described includes TEF, as third component of total energy expenditure. The thermic effect of food peaks after meals during the active part of the day and is excluded from REE when REE is measured or predicted as described above. Resting energy expenditure including TEF shows a diurnal pattern, increasing above basal metabolic rate at breakfast and decreasing again below basal metabolic rate after midnight when asleep (Figure 11.5). In adults, overnight metabolic rate is equivalent to basal metabolic rate or REE as measured or estimated with the prediction equations described. Overnight metabolic rate is lower than basal metabolic rate in children and higher in older adults (Wouters-Adriaens et al. 2006). Valenti at al. (2016) showed an inverse association of overnight metabolic rate with quality sleep, explaining the age-related increase in overnight metabolic rate. Physical activity as measured with a doubly-labeled water validated accelerometer for movement registration includes TEF when activity is expressed as PAL. Thus all components of energy expenditure are included in the estimate of energy intake when measured or estimated REE is multiplied with an accelerometer derived PAL value.

11.6 CONCLUSIONS

The assessment of energy intake with REE and AEE as biomarkers of dietary energy intake requires minimal information. REE is measured or estimated from an equation based on subject sex, weight, height, and age; AEE is measured with a doubly-labeled water validated accelerometer for movement registration. The easiest way to estimate habitual energy intake for weight maintenance from REE and AEE is the application of a device producing values for total energy expenditure based on subject sex, weight, height, age, and measured body acceleration. Unfortunately, the validity of most devices is questionable. The outcome often is a function of proprietary equations. Thus, the user has no information on the equation for REE adopted, based on subject sex, weight, height, and age. When estimation energy intake from subject sex, weight, height, age, and body acceleration, it is suggested to estimated REE and AEE separately by choosing the REE prediction equation, optimal for the population of study, in combination with measured AEE with a valid sensor for body movement.

REFERENCES

Acheson, K. J. 1993. Influence of autonomic nervous system on nutrient-induced thermogenesis in humans. *Nutrition* 9: 373–380.

Adriaens, M. P., P. F. Schoffelen, and K. R. Westerterp. 2003. Intra-individual variation of basal metabolic rate and the influence of daily habitual physical activity before testing. *Br. J. Nutr.* 90: 419–423.

Antonsson, E. K. and R. W. Mann. 1985. The frequency content of gait. *J. Biomech.* 18: 39–47.

Biloft-Jensen, A., M. F. Hjorth, E. Trolle et al. 2013. Comparison of estimated energy intake using web-based dietary assessment software with accelerometer-determined energy expenditure in children. *Food Nutr. Res.* December, 17: 57. doi:10.3402/fnr.v57i0.21434.

Black, A. E., W. A. Coward, T. J. Cole, and A. M. Prentice. 1996. Human energy expenditure in affluent societies: An analysis of 574 doubly-labelled water measurements. *Eur. J. Clin. Nutr.* 50: 72–92.

Bonomi, A. G. and K. R. Westerterp. 2012. Advances in physical activity monitoring and lifestyle interventions in obesity: A review. *Int. J. Obes.* 36: 167–177.

Bouten, C. V., K. T., Koekoek, M. Verduin, R. Kodde, and J. D. Janssen. 1997. A triaxial accelerometer and portable data processing unit for the assessment of daily physical activity. *IEEE T Biomed Eng* 44: 136–147.

Camps, S. G., S. P. Verhoef, and K. R. Westerterp. 2013. Weight loss, weight maintenance and adaptive thermogenesis. *Am. J. Clin. Nutr.* 97: 990–994.

Cunningham, J. J. 1991. Body composition as a determinant of energy expenditure: A synthetic review and a proposed general prediction equation. *Am. J. Clin. Nutr.* 54: 963–969.

Deurenberg, P., M. Deurenberg-Yap, and S. Guricci. 2002. Asians are different from Caucasians and from each other in their body mass index/body fat per cent relationship. *Obes. Rev.* 3: 141–146.

Edholm, O. G., J. G. Fletcher, E. M. Widdowson, and R. A. McCance. 1955. The energy expenditure and food intake of individual men. *Br. J. Nutr.* 9: 286–300.

Food and Agricultural Organization/World Health Organization/United Nations University. 1985. Energy and protein requirements. Report of a joint FAO/WHO/UNU expert consultation. World Health Organization technical report series 724. Geneva, Switzerland.

Food and Agricultural Organization/World Health Organization/United Nations University. 2004. Human energy requirements. Rome: FAO Food and nutrition report series 1.

Frankenfield, D., L. Roth-Yousey, and C. Compher, 2005. Comparison of prediction equations for resting metabolic rate in healthy non obese and obese adults: A systematic review. *J. Am. Diet. Assoc.* 105: 775–789.

Goldberg, G. R., A. E. Black, S. A. Jebb et al. 1991. Critical evaluation of energy intake data using fundamental principles of energy physiology: 1. Derivation of cut-off limits to identify under-recording. *Eur. J. Clin. Nutr.* 45: 569–581.

Goris, A. H., E. P. Meijer, A. Kester, and K. R. Westerterp. 2001. Use of a triaxial accelerometer to validate reported food intakes. *Am. J. Clin. Nutr.* 73: 549–553.

Haggarty, P., G. McNeill, M. K. Manneh, L. Davidson, E. Milne, G. Duncan, and J. Ashton. 1998. The influence of exercise on the energy requirements of adult males in the UK. *Br. J. Nutr.* 72: 799–813.

Harris, J. A. and F. G. Benedict. 1919. A biometric study of basal metabolism in man. Publication no. 297. Washington, DC: Carnegie Institute of Washington.

Heitmann, B. L., K. R. Westerterp, R. J. Loos et al. 2012. Obesity: Lessons from evolution and the environment. *Obes. Rev.* 13: 910–922.

Major, G. C., E. Doucet, P. Trayhurn, A. Astrup, and A. Tremblay. 2007. Clinical significance of adaptive thermogenesis. *Int. J. Obes.* 31: 204–212.

Mifflin, M. D., S. T. St Jeor, L. A. Hill, B. J. Scott, S. A. Daugherty, and Y. O. Koh. 1990. A new predictive equation for resting energy expenditure in healthy individuals. *Am. J. Clin. Nutr.* 51: 241–247.

Neelemaat, F., M. A. Van Bokhorst-de van der Schueren, A. Thijs, J. C. Seidell, and P. J. Weijs. 2012. Resting energy expenditure in malnourished older patients at hospital admission and three months after discharge: Predictive equations versus measurements. *Clin. Nutr.* 31: 958–966.

Owen, O. E., E. Kavle, R. S. Owen et al. 1986. A reappraisal of the caloric requirements in healthy women. *Am. J. Clin. Nutr.* 44: 1–19.

Owen, O. E., J. L. Hollup, D. A. D'Alessio et al. 1987. A reappraisal of the caloric requirements of men. *Am. J. Clin. Nutr.* 46: 875–885.

Plasqui, G., A. G. Bonomi, and K. R. Westerterp. 2013. Daily physical activity assessment with accelerometers: New insights and validation studies. *Obes. Rev.* 14: 451–462.

Plasqui, G. and K. R. Westerterp. 2003. Seasonal variation in total energy expenditure and physical activity in Dutch young adults. *Obes. Res.* 12: 688–694.

Plasqui, G. and K. R. Westerterp. 2007. Physical activity assessment with accelerometers: An evaluation against doubly labeled water. *Obesity* 15: 2371–2379.

Rao, Z., X. Wu, B. Liang, M. Wang, and W. Hu. 2012. Comparison of five equations for estimating resting energy expenditure in Chinese young, normal weight healthy adults. *Eur. J. Med. Res.* 17: 26.

Rosenbaum, M., J. Hirsch, D. A. Gallagher, and R. L. Leibel. 2008. Long-term persistence of adaptive thermogenesis in subjects who have maintained a reduced body weight. *Am. J. Clin. Nutr.* 88: 906–912.

Schoffelen, P. F., K. R. Westerterp, W. H. Saris, and F. Ten Hoor. 1997. A dual-respiration chamber system with automated calibration. *J. Appl. Physiol.* 83: 2064–2072.

Schwartz, A., J. L. Kuk, G. Lamothe, and E. Doucet. 2012. Greater than predicted decrease in resting energy expenditure and weight loss: Results from a systematic review. *Obesity* 20: 2307–2310.

Siervo, M., S. Bertoli, A. Battezzati et al. 2014. Accuracy of predictive equations for the measurement of resting energy expenditure in older subjects. *Clin. Nutr.* 33: 613–619.

Sjödin, A. M., A. H. Forslund, K. R. Westerterp, A. B. Andersson, J. M. Forslund, and L. M. Hambraeus. 1996. The influence of physical activity on BMR. *Med. Sci. Sports Exerc.* 28: 85–91.

Speakman, J. R. and K. R. Westerterp. 2010. Associations between energy demands, physical activity and body composition in adult humans between 18 and 96 y of age. *Am. J. Clin. Nutr.* 92: 826–834.

Tappy, L. 1996. Thermic effect of food and sympathetic nervous system activity in humans. *Reprod Nutr Dev* 36: 391–397.

Ten Haaf, T. and P. J. Weijs. 2014. Resting energy-expenditure prediction in recreational athletes of 18–35 years: Confirmation of Cunningham equation and an improved weight-based alternative. *PLoS One* 9: e108460.

Valenti, G., A. G. Bonomi, K. R. Westerterp. 2016. Quality sleep is associated with overnight metabolic rate in healthy older adults. *J Gerontol A Biol Sci Med Sci.* 72 (4): 567–571.

Van Gemert, W. G., K. R. Westerterp, J. M. Greve, and P. B. Soeters. 1998. Reduction of sleeping metabolic rate after vertical banded gastroplasty. *Int. J. Obes.* 22: 343–348.

Van Hees, V. T., R. C. Van Lummel, and K. R. Westerterp. 2009. Estimating activity-related energy expenditure under sedentary conditions using a tri-axial seismic accelerometer. *Obesity* 17: 1287–1292.

Westerterp, K. R. 1999. Energy metabolism: Human studies. In: Tarnopolski. M. (ed) *Nutritional implications of gender differences in metabolism*. CRC Press, Boca Raton, FL: 249–264.

Westerterp, K. R. 2004. Diet induced thermogenesis. *Nutr Metab.* 1: 5.

Westerterp, K. R. 2014. Reliable assessment of physical activity in disease: An update on activity monitors. *Cur. Opin. Clin. Nutr. Metab. Care* 17: 401–406.

Westerterp, K. R., J. Donkers, E. W. Fredrix, and P. Boekhoudt. 1995. Energy intake, physical activity and body weight: A simulation model. *Br. J. Nutr.* 73: 337–347.

Weijs, P. J. 2008. Validity of predictive equations for resting energy expenditure in US and Dutch overweight and obese class I and II adults aged 18–65 years. *Am. J. Clin. Nutr.* 88: 959–970.

Weijs, P. J. and G. A. Vansant. 2010. Validity of predictive equations for resting energy expenditure in Belgian normal weight to morbid obese women. *Clin. Nutr.* 29: 347–351.

Wouters-Adriaens, M. P. and K. R. Westerterp. 2006. Basal metabolic rate as a proxy for overnight energy expenditure: The effect of age. *Br. J. Nutr.* 95: 1166–1170.

Wouters-Adriaens, M. P. and K. R. Westerterp. 2008. Low resting energy expenditure in Asians can be attributed to body composition *Obesity* 16: 2212–2216.

Wulan, S., K. R. Westerterp, and G. Plasqui. 2010. Ethnic differences in body composition and the associated metabolic profile: A comparative study in Asians and Caucasians. *Maturitas* 65: 315–319.

12 Dynamic Modeling of Energy Expenditure to Estimate Dietary Energy Intake

Diana Thomas and Vincent W. Antonetti

CONTENTS

12.1 INTRODUCTION

Researchers in human physiology and nutrition have accumulated a significant amount of quantitative knowledge regarding how changes in energy balance impact body weight. Theoretically, when more energy is consumed than burned through exercise and normal daily activities, excess calories are stored leading to increased body weight and when less energy is consumed than burned, weight loss should be observed. This is known as the conservation of energy principle, referred to as the first law of thermodynamics.

Eventually human physiology and nutrition data was integrated with the first law of thermodynamics through mathematical modeling (Antonetti 1973a, Hall 2006, Thomas et al. 2009). The integrated approach resulted in the quantitative prediction of weight change. Presently, several validated thermodynamic mathematical models of human weight change are available as web-based applications that have resulted in widespread use (Casazza et al. 2013, Hall et al. 2011). These models basically address the question: How long will it take an individual to lose a certain amount of weight while on a given restriction on energy intake (EI) or a change in physical activity expenditure?

The same models can be applied to the inverse question: What must the dietary intake be of an individual in order to lose a certain amount of weight in a given amount of time? The answer to this question provides an accurate and objective estimate of energy intake between body weights. An equally important use of the models that is central to this book is that the developed equations can also be used to calculate energy intake that can be used as a biomarker for dietary energy (Prentice 2017).

Efforts to predict weight change from a change in energy intake or energy expenditure date as far back as 1952 when Max Wishnofsky (1952) assumed a linear formulation for weight prediction. Specifically, Wishnofsky used the best available weight loss data available at the time and determined the energy content of weight change to be approximately 3500 kcal/lb (7700 kcal/kg) (Thomas et al. 2014). Wishnofsky used this result to predict weight loss. For example, if initial weight is 70 kg and an individual is decreasing energy intake by 500 kcal/d, then predicted weight loss after one day (represented by $W(1)$) is $W(1) = 70 - (500/7700) \cdot 1 = 69.9$ kg. Similarly at five days, we have $W(5) = 70 - (500/7700) \cdot 5 = 69.7$. At seven days, or one week, $W(7) = 70 - (500/7700) \cdot 7 = 69.5$, resulting in a predicted weight change of 0.45 kg or 1 lb. Written as a formula, Wisnofsky's model, commonly referred to as the 3500 kcal rule is expressed as

$$W(t) = W_0 - \frac{1}{7700} \Delta EBt \tag{12.1}$$

where:

$W(t)$ represents weight in kg on day t
W_0 is the baseline weight (kg)
ΔEB is the change in energy balance (typically thought of as a deficit) in kcal/day

Wishnofsky's model is based on data from short-term very low calorie experiments in study populations comprised largely of obese women. But even after Wishnofsky was able to estimate a value for the energy content of weight change (namely 3500 kcal/lb), there remained a substantial problem. A plot of weight loss versus time using Wishnofsky's 3500 kcal rule demonstrates a constant never-ending weight loss, and in the extreme leads to nonsensical predicted body weights of zero and even negative values (Allison et al. 2014, Antonetti 1973a, Hall et al. 2011, Thomas et al. 2014). Obviously, Wishnofsky's equation is invalid for long-term predictions. The often used rule of a pound lost for each 3500 kcal reduction in cumulative energy intake is appropriate only for short-term projections of weight change. Thus to predict weight change over longer time durations another approach is required. The better approach is to use time-dependent thermodynamic models. A time-dependent thermodynamic model represents a differential equation derived using the first law of thermodynamics. The differential equation is a model that contains both $W(t)$ and the speed of weight change expressed as (dW/dt) in one formula. The *solution* to the differential equation is a graph or trajectory of $W(t)$. In most cases, the solution of a differential equation cannot be expressed as a formula, referred to as a *closed form*, like the 3500 kcal rule in Equation 12.1. In these cases, the solution is numerically approximated using algorithms programmed into special software (see http://pbrc.edu/research-and-faculty/calculators/weight-loss-predictor/ and https://www.niddk.nih.gov/health-information/health-topics/weight-control/body-weight-planner/Pages/bwp.aspx). Therefore, these more complex models were initially not as attractive as the 3500 kcal Rule that requires no special software or simulations. To overcome the challenge of delivering model results to nonmathematicians, researchers in the past have printed tables containing personalized predictions (Antonetti 1973b), or more recently supplied web-based applications (Hall et al. 2011, Thomas et al. 2013).

In the next section, we develop a new thermodynamic model that has and that can be expressed in closed form similar to the manner the 3500 kcal Rule is conveyed.

12.2 WEIGHT CHANGE MODELS BASED ON THE FIRST LAW OF THERMODYNAMICS

In 1973, Vincent Antonetti realized that as an individual loses weight, the energy needed to maintain the individual's lower weight also decreases (Antonetti 1973a). In addition, if the individual's restricted energy intake (EI) and physical activity does not change over the course of the intervention, the individual's calorie deficit (change in energy balance) must decrease in magnitude as weight is lost. The result is that the individual's weight loss trajectory slows and eventually plateaus at a new lower weight.

Applying the preceding thought experiment, the conclusions of Wishnofsky's work and the first law of thermodynamics, Antonetti constructed the first weight change predictive model. Development of the model began with an energy balance equation applied to humans:

$$\text{ES} = \text{EI} - \text{TEE} \tag{12.2}$$

where:

ES is the rate of change in energy stores
EI is the rate (or speed) of energy intake
TEE is the rate of total energy expenditures all measured in kcal/d

Antonetti applied Wishnofsky's estimate of the energy content of weight change, (3500 kcal/lb = 7700 kcal/kg), to model ES

$$\text{ES} = 7700\frac{dW}{dt} \tag{12.3}$$

Recall that (dW/dt) represents the rate of change or speed of weight loss. Then Antonetti model split total energy expenditure (TEE) into several key components:

$$\text{EE} = \text{TEM} + \text{EEPA} + \text{REE} \tag{12.4}$$

The thermic effect of meals (TEM) that represents the rate of heat energy generated by the body when an individual eats was modeled as TEM = 0.10 EI (Antonetti 1973a). Physical activity energy expenditures (EEPA) was represented as a direct proportion of body weight, EEPA = K_aW. The activity coefficient K_a is the sum of the calories expended per kg of body mass multiplied by the time spent at each activity. As such a detailed calculation for most individuals is impractical, Antonetti adapted Taylor and Pye's more realistic approach and determined the values for the activity coefficient, K_a, shown in Table 12.1 by assuming 16 active hours per day.

At the time the Antonetti model was being developed, resting energy expenditure (REE) or the energy required to maintain organ functions at rest was thought to be best represented by assuming that it was dependent on body surface area (Bruen 1930, Du Bois and Du Bois 1989) with body surface area expressed as $W^{0.425}$. However, this introduces weight not as a linear variable, but rather as power function. The power function causes the resulting differential equation to be nonlinear. As described earlier, the solution to the differential equation in this case cannot be expressed simply as a formula for $W(t)$ (like the 3500 kcal Rule in Equation 12.1). After input of an individual's height, baseline weight, age, and gender, solving this model requires special algorithms and software to generate a personalized graph of weight loss (Antonetti 1973b).

TABLE 12.1
List of Activity Coefficients (K_a) for Varying Levels of Activity

Activity Level	Activity Coefficient (K_a)
Sedentary	8.10
Light	9.50
Moderate	12.7
Vigorous	17.7
Severe	27.1

To eliminate this complexity, we replaced the REE formula with the validated and widely used Mifflin–St. Jeor regression equations for REE (Mifflin et al. 1990): $\text{REE} = 9.99W + B$, where $B = 6.25H - 4.92A - 161$ for females, and $B = 6.25H - 4.92A + 5$ for males.

Now we substitute the expressions for TEM, EEPA, and REE into Equation 12.2:

$$\underbrace{7700\frac{dW}{dt}}_{\text{ES}} = \text{EI} - \underbrace{\left(\overbrace{0.10\text{EI}}^{\text{TEM}} + \overbrace{K_aW}^{\text{EEPA}} + \overbrace{9.99W + B}^{\text{REE}}\right)}_{\text{EE}} \tag{12.5}$$

Setting $K_1 = (1-\alpha)\text{EI} - B$ and $K_2 = K_a + 9.99$, transforms Equation 12.5 into an easily solved first-order linear differential equation (Stewart 2011):

$$7700\frac{dW}{dt} = K_1 - K_2W \tag{12.6}$$

We can express the solution of modified Antonetti model, hereafter referred to as the Antonetti–Thomas model, as a formula:

$$W(t) = \frac{K_1}{K_2} + Ce^{-\frac{K_2}{7700}t} \tag{12.7}$$

where $C = W_0 - (K_1/K_2)$. Although Formula 12.7 is more complicated than the 3500 kcal Rule (Equation 12.1), it does not need special software or approximation algorithms to calculate weight on any given day. To calculate an individual's weight at any time during an intervention, simply enter the baseline weight W_0 values for K_1 and K_2 and the value of t (desired day of intervention you need to see a prediction) into Formula 12.7.

There are a fair number of other energy balance models that vary in the breakdown of the three terms in the energy balance equation (ES, EI, EE), or differ in their application such as modeling the dynamics of starvation (Alpert 1979, Kozusko 2001, Speakman and Westerterp 2013, Westerterp et al. 1995). For example, the Hall model (Hall 2010) refines ES and EE by the three macronutrient stores, carbohydrates, fats, and proteins and their oxidations while the Flatt model (Flatt 2004) combines ES as carbohydrates and fats. The Westerterp model (Westerterp et al. 1995) and the Thomas model (Thomas et al. 2011) separates ES into lean and fat compartments. The Westerterp model defines the lean–fat relationship using a piecewise defined function, whereas the Thomas model utilizes a nonlinear algebraic relationship. Although the models qualitatively describe different weight loss mechanisms, the models do not differ significantly in their ability to predict weight change over periods of time (Thomas et al. 2014). In fact, simulations indicate they have similar predictions after one month till time increases to weight plateau (~1.5 years). But in contrast to the newly proposed Antonetti–Thomas model, all existing thermodynamic models require approximating solutions using computer simulations.

As the solution to the Antonetti–Thomas model can be expressed as a formula, any variable in the formula can be solved using algebra, or *back calculated*. There are two main advantages of applying the Antonetti–Thomas model for the purpose of back-calculating energy intake. The first is that energy stores are formulated in terms of body weight that is easily measured and often the main outcome variable in weight change interventions. The second, and more important advantage, is that the solution can be expressed as a formula (Equation 12.7) without requiring repeated measures of body weight, special software, and computer simulation. We demonstrate in the next section using examples how to apply Equation 12.7 to back-calculate energy intake directly.

12.3 CALCULATION OF ENERGY INTAKE FROM ENERGY BALANCE MODELS

As energy balance models can predict weight change from a change in EI or TEE, we can theoretically solve for EI given information about weight change. This concept is analogous to expressing solutions in terms of different variables in algebraic equations. For example, $y = mx + b$ can be solved in terms of x:

$$y = mx + b$$

$$\text{subtract } b \text{ from both sides} \Rightarrow y - b = mx$$

$$\text{divide both sides by } m \Rightarrow \frac{(y-b)}{m} = x$$

In this manner, we have *back-calculated* x in terms of y. However, the fundamental difference in this discussion is that the *variable* we are interested in EI is housed within a differential equation that yields an entire trajectory of values instead of a single value.

There are several existing approaches for extracting a variable from a differential equation. The first was proposed by Thomas (Thomas et al. 2010) who applied a *shooting* method. The second approach by Hall (Hall and Chow 2011) applied a linearization of the equation. Both methods work in theory but are relatively complex to implement. The simpler Antonetti–Thomas model can be applied to directly solve for EI, which provides a closed formula for immediate direct calculation. Thus, after some algebra, Equation 12.7 can be rearranged to yield EI:

$$\text{EI} = \frac{(-K_2W_0 - B)e^{-\frac{tK_2}{\gamma}} + K_2W_f + B}{\left(e^{-\frac{tK_2}{\gamma}} - 1\right)(\alpha - 1)} \quad (12.8)$$

where:

$K_2 = K_a + 9.99$

K_a is the activity level (kcal/kg/d) from Table 12.1

$B = 6.25H - 4.92A - 161$ for females, and $B = 6.25H - 4.92A + 5$ for males

A is the age (years)

H is the height (cm)

t is the time on intervention (days)

α is the percentage of EI contributing to TEF, $\alpha = 0.10$

γ is the energy content of weight change, $\gamma = 7700$ kcal/kg

W_0 is the baseline weight (kg)

W_f is the final weight (kg)

12.4 APPLICATIONS OF THE ANTONETTI–THOMAS ENERGY INTAKE MODEL

We first compared our EI intake model to individual data from Racette et al. (1995) to EI measured as the sum of changed body energy stores and doubly-labeled water energy expenditures (de Jonge et al. 2007). After removal of a subject who substantially increased total daily energy expenditures during the intervention, the mean absolute error between actual and predicted was 237 ± 100 kcal/d. This agreement is consistent with results reported using existing models (Hall and Chow 2011, Sanghvi et al. 2015, Thomas et al. 2010).

We will now apply Equation 12.7 to determine how long will it take for an individual to lose a certain amount of weight when restricting EI. This is the type of problem that most available thermodynamic models have addressed to date.

Example 2.1

Determine how long it will take for a 30-year-old female weighing 85 kg to lose 15 kg on a target EI of 1200 kcal/d. She weighs 85 kg at baseline, is 165 cm tall, and her activity level is considered light.

In this case we use the solution of Equation 12.7 to back-calculate an expression t. We note that the steps to arrive at the below formula for t only require algebra but are involved and are omitted:

$$t = \left(-\frac{7700}{K_2}\right)\ln\left(\frac{K_1 - K_2 W_f}{K_1 - K_2 W_0}\right) \tag{12.9}$$

We first need to determine K_1 and K_2:

For females: $B = 6.25H - 4.92\text{A} - 161 = 6.25(165) - 4.92(30) - 161 = 722.65$

$$K_1 = (1-\alpha)\,\text{EI} - B = (1-0.10)1200 - 722.65 = 357.35$$

From Table 12.1, $K_a = 9.50$ kcal/kg/day for light activity

$$K_2 = K_a + 9.99 = 9.50 + 9.99 = 19.49$$

Substituting into Equation 12.9 yields $t = 100.7$ days.

Next we apply our new model to determine EI of an individual in order to lose a certain amount of weight in a given amount of time.

Example 2.2

Determine the target energy intake for a 30-year-old female who weighs 85 kg and desires to lose 15 kg in 100 days. She is 165 cm tall and her activity level is considered light.

To calculate EI, we need to substitute the parameter values into Equation 12.8:

$$\text{EI} = \frac{(-K_2W_0 - B)e^{-\frac{tK_2}{\gamma}} + K_2W_f + B}{\left(e^{-\frac{tK_2}{\gamma}} - 1\right)(\alpha - 1)}$$

where:

t is the time on intervention (days), $t = 100$
α is the percentage of EI contributing to TEM, $\alpha = 0.10$
γ is the energy content of weight change, $\gamma = 7700$ kcal/kg
W_0 is the baseline weight, $W_0 = 85$ kg
W_f is the final weight, $W_f = 70$ kg
A is the age (years), $A = 30$ years
K_a is the activity level, $K_a = 9.50$ kcal/kg/d (from Table 12.1 for light activity)
H is the height (cm), $H = 165$ cm

Substituting these values, we arrive at

$$K_2 = K_a + 9.99 \Rightarrow K_2 = 9.50 + 9.99 = 19.49$$

$B = 6.25H - 4.92A - 161$ for females. Then $B = 6.25(165) - 4.92(30) + 5 = 722.65$

TABLE 12.2
Energy Intake (kcal/d) for a 30-Year-Old Woman That Achieved 15 kg of Weight Loss for Various Values of Baseline Weight, Height, and Activity Levels

Baseline Weight (kg)	Height (cm)	Activity Level (K_a)	EI kcal/d
80 kg	160	Light	1050
	180	Light	1190
	160	Moderate	1302
	180	Moderate	1441
90 kg	160	Light	1264
	180	Light	1403
	160	Moderate	1554
	180	Moderate	1694
100 kg	160	Light	1483
	180	Light	1622
	160	Moderate	1803
	180	Moderate	1939

Finally, substituting into Equation 12.8 results in

$$\mathrm{EI} = \frac{(-19.49\cdot 85 - 722.65)\,e^{-\frac{100\times 19.49}{7700}} + 19.49\cdot 70 + 722.65}{\left(e^{-\frac{100\times 19.49}{7700}} - 1\right)(0.10 - 1)} = 1191\ \text{kcal/d}$$

In summary, the female in this example's target energy intake would have to be 1191 kcal/d to achieve a 15 kg weight loss by 100 days.

The advantage of directly computing the parameter values for a closed form solution as in Example 12.2 is that we can not only directly calculate EI objectively from measured weights between any two given points in time for an individual patient, but we can also examine the effect of changing gender, age, or height on the resulting EI. For example, one can easily compute EI for the same data in Example 12.2, switching only the gender to male, to arrive at a higher target EI (1339 kcal/d).

To illustrate how EI can vary as a function of baseline weight, height, and activity level for a 30-year-old woman who wants to lose 15 kg in 100 days, we provide back-calculated EI in Table 12.2 for a sample of different parameter values. As observed from Table 12.2, considering only this set of variables, baseline weight has the most significant impact on EI.

12.5 ADVANTAGES AND LIMITATIONS

In this chapter, we derived a new thermodynamic energy balance model and showed how the new model can be used to calculate EI. We used our new model rather than any of several existing more complex models because the existing models would have required a back-calculation from energy balance weight change prediction models using methods such as shooting or estimating the rate of change of body weight with respect to time through repeated weight measures. Instead, here we provide an uncomplicated alternative method that relies on our simple validated thermodynamic model that can be directly expressed in closed form. Finally, we provide several straightforward examples of how the new model can be used to compute EI between body weights.

The main limitation of using any model is that some of the model parameters are estimated using measured data (such as doubly-labeled water measured EE, dual energy X-ray absorptiometry-measured body composition). These measurements are not without error and the error is encompassed within model's prediction. Specifically, there exist two types of error. There is the measurement error in determining TEE and REE that are inherently encompassed within the data derived model terms. In addition, there exists physiologic error due to interindividual variation. Using a Bland Altman analysis, Thomas et al. (Thomas et al. 2010) estimated error between actual and predicted EI as -48 ± 226 kcal/d at six months. Similar error estimates are provided by Hall et al. (Hall and Chow 2011) who used confidence interval calculations during model development. Despite this limitation, the windows of error are known and can be taken into account when utilizing the models to estimate EI.

12.6 CONCLUSIONS AND FUTURE DIRECTIONS

The Antonetti–Thomas energy balance model provides a simple, flexible, and scalable method to apply at either the individual or population-wide level to calculate EI objectively.

REFERENCES

Allison, D. B., D. M. Thomas, S. B. Heymsfield. 2014. Energy intake and weight loss. *JAMA* 312 (24):2687–2688. doi:10.1001/jama.2014.15513.

Alpert, S. S. 1979. A two-reservoir energy model of the human body. *Am J Clin Nutr* 32 (8):1710–1718.

Antonetti, V. W. 1973a. The equations governing weight change in human beings. *Am J Clin Nutr* 26 (1):64–71.

Antonetti, V. W. 1973b. *The Computer Diet: A Weight Control Guide*. New York, M. Evans: distributed by Lippincott.

Bruen, C. 1930. Variation of basal metabolic rate per unit surface area with age. *J Gen Physiol* 13 (6):607–616.

Casazza, K., K. R. Fontaine, A. Astrup et al. 2013. Myths, presumptions, and facts about obesity. *N Engl J Med* 368 (5):446–454. doi:10.1056/NEJMsa1208051.

de Jonge, L., J. P. DeLany, T. Nguyen et al. 2007. Validation study of energy expenditure and intake during calorie restriction using doubly labeled water and changes in body composition. *Am J Clin Nutr* 85 (1):73–79. doi:85/1/73[pii].

Du Bois, D. and E. F. Du Bois. 1989. A formula to estimate the approximate surface area if height and weight be known. 1916. *Nutrition* 5 (5):303–311; discussion 312–313.

Flatt, J. P. 2004. Carbohydrate-fat interactions and obesity examined by a two-compartment computer model. *Obes Res* 12 (12):2013–2022. doi:10.1038/oby.2004.252.

Hall, K. D. 2006. Computational model of in vivo human energy metabolism during semistarvation and refeeding. *Am J Physiol Endocrinol Metab* 291 (1):E23–E37. doi:10.1152/ajpendo.00523.2005.

Hall, K. D. 2010. Predicting metabolic adaptation, body weight change, and energy intake in humans. *Am J Physiol Endocrinol Metab* 298 (3):E449–E466. doi:10.1152/ajpendo.00559.2009.

Hall, K. D. and C. C. Chow. 2011. Estimating changes in free-living energy intake and its confidence interval. *Am J Clin Nutr* 94 (1):66–74. doi:10.3945/ajcn.111.014399.

Hall, K. D., G. Sacks, D. Chandramohan et al. 2011. Quantification of the effect of energy imbalance on bodyweight. *Lancet* 378 (9793):826–837. doi:10.1016/S0140-6736(11)60812-X.

Kozusko, F. P. 2001. Body weight setpoint, metabolic adaption and human starvation. *Bull Math Biol* 63 (2):393–403. doi:10.1006/bulm.2001.0229.

Mifflin, M. D., S. T. St Jeor, L. A. Hill, B. J. Scott, S. A. Daugherty, Y. O. Koh. 1990. A new predictive equation for resting energy expenditure in healthy individuals. *Am J Clin Nutr* 51 (2):241–247.

Prentice, A. 2017. *Advances in the Assessment of Dietary Intake*. Boca Raton, FL: Taylor & Francis Group/CRC Press.

Racette, S. B., D. A. Schoeller, R. F. Kushner, K. M. Neil, K. Herling-Iaffaldano. 1995. Effects of aerobic exercise and dietary carbohydrate on energy expenditure and body composition during weight reduction in obese women. *Am J Clin Nutr* 61 (3):486–494.

Sanghvi, A., L. M. Redman, C. K. Martin, E. Ravussin, K. D. Hall. 2015. Validation of an inexpensive and accurate mathematical method to measure long-term changes in free-living energy intake. *Am J Clin Nutr* doi:10.3945/ajcn.115.111070.

Speakman, J. R. and K. R. Westerterp. 2013. A mathematical model of weight loss under total starvation: Evidence against the thrifty-gene hypothesis. *Dis Model Mech* 6 (1):236–251. doi:10.1242/dmm.010009.

Stewart, J. 2011. *Calculus: Early Transcendentals*. 7th ed. Belmont, CA: Brooks/Cole, Cengage Learning.

Thomas, D. M., A. Ciesla, J. A. Levine, J. G. Stevens, C. K. Martin. 2009. A mathematical model of weight change with adaptation. *Math Biosci Eng* 6 (4):873–887. doi:10.3934/mbe.2009.6.873.

Thomas, D. M., M. C. Gonzalez, A. Z. Pereira, L. M. Redman, and S. B. Heymsfield. 2014. Time to correctly predict the amount of weight loss with dieting. *J Acad Nutr Diet* 114 (6):857. doi:10.1016/j.jand.2014.02.003.

Thomas, D. M., C. K. Martin, S. Heymsfield, L. M. Redman, D. A. Schoeller, J. A. Levine. 2011. A simple model predicting individual weight change in humans. *J Biol Dyn* 5 (6):579–599. doi:10.1080/17513758.2010.508541.

Thomas, D. M., C. K. Martin, S. Lettieri et al. 2013. Can a weight loss of one pound a week be achieved with a 3,500 kcal deficit? commentary on a commonly accepted rule. *Int J Obes* doi:10.1038/ijo.2013.51.

Thomas, D. M., D. A. Schoeller, L. A. Redman, C. K. Martin, J. A. Levine, S. B. Heymsfield. 2010. A computational model to determine energy intake during weight loss. *Am J Clin Nutr* 92 (6):1326–1331. doi:10.3945/ajcn.2010.29687.

Westerterp, K. R., J. H. Donkers, E. W. Fredrix, P. Boekhoudt. 1995. Energy intake, physical activity and body weight: A simulation model. *Br J Nutr* 73 (3):337–347.

Wishnofsky, N. 1952. Caloric equivalents of gained or lost weight. *Metabolism* 1 (6):554–555.

13 Use of Intake Biomarkers in Nutritional Epidemiology

Ross L. Prentice

CONTENTS

13.1 INTRODUCTION

There have been more than 40 years of study by multiple research groups of associations between self-reported intake of foods and nutrients and subsequent chronic disease incidence. In spite of strong suggestions from animal feeding studies, international correlational analyses, and migrant studies, and in spite of consistent analytic epidemiologic findings of elevated risks of cardiovascular diseases, major cancers and diabetes among overweight and obese persons, rather few diet and disease associations have been identified by expert panels convened to review the worldwide evidence (World Cancer Research Fund/American Institute for Cancer Research 1997; 2007; World Health Organization 2003). An important limitation of the self-report approach, whether based on food frequency questionnaires (FFQs), food records or diaries (e.g., four-day food records), or dietary recalls (e.g., one or more 24-hour dietary recalls) is the inability to adequately measure total energy intake (Neuhouser et al. 2008; Prentice et al. 2011; Subar et al. 2003). In fact, self-reported energy not only is highly variable, but it also has systematic biases, with overweight and obese persons tending to substantially underestimate energy intake. Hence, there are major limitations to nutritional epidemiology as currently practiced. The incorporation of objective intake measures, usually referred to as intake biomarkers, into the nutritional epidemiology research agenda has potential for a fresh and penetrating study of a broad range of dietary intake associations with subsequent chronic disease risk.

13.2 BACKGROUND

A dietary biomarker can be defined as an intake measure that is equal to the nutritional variable under study, plus random *noise* that depends neither on the value of the targeted nutritional variable nor on other pertinent study subject characteristics. An intake measure satisfying this *classical* measurement error model can anchor the nutritional variable assessment. Analytic methods are available to properly estimate disease associations with the targeted dietary variable under this classical measurement model. Both the plausibility of the classical measurement error model and the efficiency of disease association estimates will be enhanced if the biomarker measurement error variance is small relative to that for the targeted dietary variable. Most biomarkers to date arise either from measures in urine of metabolites produced when nutrient or foods are expended, referred to as recovery biomarkers, or from blood concentrations of nutrients, referred to as concentration biomarkers (Kaaks 1997). One of the most important of the former type is a doubly-labeled water (DLW) recovery biomarker (Schoeller 1999) of short-term energy expenditure, which for persons in energy balance provides an objective measure of short-term energy consumption.

Energy overconsumption over the lifespan is likely to be a major driver of obesity and chronic disease risk, and any comprehensive approach to nutritional epidemiologic research needs to accommodate this crucial aspect of diet. The DLW biomarker shows considerably higher energy expenditure for overweight and obese persons, compared to normal weight persons, in various U.S. populations (Arab et al. 2010; Moshfegh et al. 2008; Mossavar-Rahmani et al. 2015; Neuhouser et al. 2008; Prentice et al. 2011; Subar et al. 2003). However, large epidemiologic studies using self-reported dietary assessments have typically not been able to detect a higher energy intake by overweight and obese persons. The properties and application of the DLW biomarker, which assesses the energy expenditure needed to maintain body weight, will be described in detail elsewhere in this volume. Its logistics and cost are such that it is not practical for application to the tens of thousands of study subjects in a typical epidemiologic cohort study, nor can it be assessed using stored biospecimens. Hence, applications to date have involved the use of DLW, often in conjunction with other nutritional biomarkers, in specifically designed biomarker studies of a few hundred persons (Arab et al. 2010; Moshfegh et al. 2008; Mossavar-Rahmani et al. 2015; Neuhouser et al. 2008; Prentice et al. 2011; Subar et al. 2003). If such biomarker studies are embedded in larger epidemiologic cohorts with recorded clinical outcomes (Mossavar-Rahmani et al. 2015; Neuhouser et al. 2008; Prentice et al. 2011), one has the potential to use the biomarker data to correct self-report assessments for measurement error throughout study cohorts, as will be elaborated below, leading to association studies of enhanced reliability.

Perhaps because of the inability to adequately measure energy consumption by self-report, most cohort study reporting has focused on ratio of nutritional measures. However, nutrient density, or closely-related nutrient residual measures, may also be subject to systematic and random reporting biases when based on self-report. For example, 24-hour urinary nitrogen (UN) provides an established biomarker of total protein intake (Bingham 2003). By comparison to the ratio of the UN to the DLW biomarker, one sees that self-report assessments tend to overestimate the fraction of energy from protein (Freedman et al. 2014; Neuhouser et al. 2008; Prentice et al. 2011), indicating that protein intake is underestimated to a lesser extent than is energy as a whole. Also, 24-hour urinary sodium is regarded as a useful intake biomarker of sodium consumption. The ratio of self-reported sodium to self-reported total energy has systematic bias with magnitude dependent on the respondent's body mass index (BMI, defined as weight in kilograms divided by the square of height in meters) (Huang et al. 2013). These examples are among the few nutritional density variables with established biomarkers. They indicate that it would not be prudent to assume that self-reported estimates of dietary ratio measures are free of systematic bias.

The use of biomarkers in nutritional epidemiology, either directly through application to stored biospecimens from cases and controls nested within cohort studies, or indirectly through the calibration of self-report assessments using biomarker substudies, provides a major future research pathway for strengthening nutritional epidemiology research. However, as just noted, there are few nutrients or foods having an established intake biomarker. The development of biomarkers for additional nutritional variables, typically requiring well-controlled human feeding studies, poses an urgent challenge to the population science research community. These opportunities and needs will be elaborated in the remainder of this chapter, while often using the DLW biomarker as a specific example.

13.3 THEORY OF THE BIOMARKER APPROACH TO NUTRITIONAL EPIDEMIOLOGY

13.3.1 Classical Biomarker Measurement Model

The terminology *biomarker* is used ubiquitously in biomedical research, often simply referring to a correlate of some other variable of interest, such as a disease occurrence or disease recurrence. However, as mentioned above, a biomarker for the intake of a nutrient or food will typically need to adhere to a classical measurement model to be useful. For example, consider log-transformed average daily energy intake over, say, the preceding decade, denoted by z, as a nutritional variable of interest. Log transformation may simplify measurement error modeling, and often yields variables that are approximately normally distributed. A biomarker, denoted by w, for this nutritional variable will be defined as one satisfying the classical measurement model

$$w = z + e \tag{13.1}$$

where the error term e is random noise; that is, e has mean zero and depends neither on the targeted quantity (z) or on other study subject characteristics $v = (v_1, v_2, \ldots)'$, such as age, gender, race/ethnicity, socioeconomic status, or BMI. The log-transformed DLW biomarker of energy expenditure mentioned above meets these criteria with only a small error component over the short term (e.g., during the two-week application period for the DLW method), and possibly during the longer term as well, in populations free of major secular trends in dietary habits. Note, however that the magnitude of error component in Equation 13.1 will become larger if the targeted dietary variable is defined to average over a longer time period. A potential biomarker will typically be stronger, and the plausibility of Equation 13.1 will be greater, if the variance of the error component is small relative to that for the targeted nutritional variable. Although there is no firm criterion for the magnitude of this variance ratio, values of 50% or greater are consistent with an efficient biomarker. Specimen collections at multiple time points during cohort follow-up can yield potential biomarkers having relatively smaller error components. Repeat biomarker measurement over time for individual study subjects is also valuable for studying biomarker properties.

To be specific, continue with log-transformed daily average energy intake over the preceding decade, hereafter referred to simply as energy intake, as an example of a targeted variable; and with corresponding log-transformed DLW over a 14-day biomarker application period randomly selected in the 10-year period, hereafter referred to as DLW energy, as potential biomarker. Model (Equation 13.1) implies that the DLW energy can *anchor* the energy assessment in the sense that the average DLW energy values for an independent sample of subjects in a study population will approach that for the targeted energy intake as the sample size becomes large. The requirement that the error in Equation 13.1 is independent of study subject characteristics is also crucial. For example, self-report dietary assessments (e.g., 24-hour dietary recalls) of energy consumption tend to involve large negative measurement errors among overweight and obese persons (Prentice et al. 2011), with much smaller errors among normal weight persons, precluding (log transformed) self-reported energy intake from satisfying Equation 13.1.

A biomarker adhering to Equation 13.1 will be particularly useful for disease association analyses and its measurement error variance is small relative to that for the targeted dietary variable. However, the efficiency of the association analysis will be lower when this variance ratio is larger, and the modeling assumptions related to Equation 13.1 may then be more difficult to justify. These topics require careful consideration in application, particularly if the targeted dietary variable is defined as a function of intake over a time period of years or decades, while the biomarker derives from short-term intake over a few days, weeks, or months.

13.3.2 Direct Use of Biomarkers in Disease Association Analyses

Epidemiologic cohort studies typically require tens of thousands of study subjects to generate sufficient disease incidence, and sufficient precision in association analyses, with a few years of cohort follow-up. When exposure variables, or potential confounding variables, involve expensive analysis of stored biospecimens, the association analysis is typically conducted in a nested case-control (Prentice and Breslow 1978; Thomas 1977) or case-cohort (Prentice 1986; Self and Prentice 1988) mode. Here, exposure variable refers to a dietary variable of interest. A corresponding confounding variable is one that is associated with both this targeted dietary variable and the risk of the chronic disease under study and that, therefore, may need to be included in the disease risk model to avoid bias in estimating the dietary variable's influence on disease risk. Depending on the dietary variable definition, it may not be possible to classify some characteristics, such as study subject BMI, as exposure variables or as potential confounding variables, and association analyses may need to be presented and interpreted under each classification.

For disease-association analyses, a logistic-regression (Prentice and Pyke 1979) or Cox-regression (Cox 1972) analysis may be carried out to contrast biomarker values between cases and noncases (Prentice 1986; Prentice and Breslow 1978; Self and Prentice 1988; Thomas 1977) in the study sample. Standard tests of the null hypothesis of no association between incidence of the study disease and the targeted dietary exposure, based on hazard ratio (or odds ratio) modeling as a function of the individual's preceding biomarker history will be valid under the classical measurement model (Equation 13.1). However, estimated hazard ratios as a function of the biomarker assessment will typically be attenuated or otherwise distorted, compared to estimators that would arise if the targeted dietary variable could be directly measured and modeled. To adjust these estimates, so that they are appropriate to the targeted dietary variable, one requires an estimate of the ratio of the measurement error variance to that for the targeted variable in Equation 13.1.

Estimates of this variance ratio may be derived using a reliability subsample in which repeat biomarker values are obtained separated in time from the original biomarker measurements. For example, for a targeted energy intake one could obtain DLW measurements in a biomarker subset of a study cohort at cohort enrollment, with repeat DLW measurements six months later for some fraction of the subsample. Denoted by $w_1 = z + e_1$, the original biomarker measurement, and by $w_2 = z + e_2$, the replicate biomarker measurement for members of the reliability. If e_1 and e_2 are statistically independent, then

$$2\hat{\text{v}}\text{ar}(w_1 - w_2)/\{\hat{\text{v}}\text{ar}(w_1 + w_2) - \hat{\text{v}}\text{ar}(w_1 - w_2)\} \tag{13.2}$$

provides a simple estimator of the variance ratio in question. This quantity can be used to obtain a suitably adjusted estimator of disease risk association parameters as a function of z. However, for practical reasons, the time interval between replicate biomarker measurements is typically short, whereas the error term in Equation 13.1 may reflect the difference between a short-term biomarker and an actual intake over a time period of several years. Under these circumstances, errors for the replicate measurements for reliability subsample subjects can be expected to be positively correlated, perhaps strongly so.

One response to this dilemma is to alter the time period in targeted dietary variable definition to correspond more closely to the available biomarker data. For example, measurement errors based on biomarker assessments separated by about six months may be approximately independent if the targeted dietary variable is defined in terms of dietary intake over, say, the preceding year. The association of disease risk with a dietary variable defined in this manner may also reflect some aspect of earlier dietary intakes, if these tend to track over time for individuals. Alternatively, and better, repeat biomarker measurements over the more lengthy time that may be relevant to disease risk could be obtained, including a longitudinal reliability subsample, supporting the modeling of measurement error correlations, and subsequent adjustment of disease association parameter estimates, for a dietary variable defined over a longer time period. This type of longitudinal biomarker subsample has evidently not yet figured into nutritional epidemiology conduct and reporting.

As noted above, a challenging topic in nutritional epidemiology concerns the role of BMI, and other anthropometric or body fat distribution measures, in disease association analyses. These measures may substantially reflect energy imbalance over the lifespan, and are quite valuable in suggesting important influences of diet and/or physical activity on chronic disease risk. However, including BMI as a potential confounding variable in diet and disease association analyses, as is frequently done, may result in overcontrol, and to failure to identify important diet and disease associations (Hu 2008). For example, suppose that a wealthy donor made it possible to apply the DLW assessment to every person in a large epidemiologic cohort of weight-stable persons for the purpose of studying energy consumption in relation to coronary heart disease (CHD) incidence over a single year of follow-up. The investigators carry out a standard analysis, carefully controlling for multiple CHD risk factors, and find a strong positive association. A skeptic then comes along and says, *But you did not control for baseline BMI.* Upon including BMI in the disease risk regression model, the DLW association with CHD is much reduced and no longer significant. It may be tempting to then conclude that energy intake is not a CHD risk factor, only appearing to be in the first analysis because of confounding by BMI. However, even though only short-term energy consumption was measured in this study, the strong positive association in the first analysis may be due to tracking of individual study subject actual energy consumption over the years or decades prior to cohort enrollment. BMI may provide a strong correlate of the relevant long-term energy consumption for study subjects, so that the second analysis is essentially examining the CHD association with very recent energy consumption after controlling for energy consumption over a much longer time period that may be relevant to disease risk. It is difficult to know what to expect from the later analysis that includes BMI in the disease risk model. Certainly the CHD association with the individual's preceding energy consumption history as a whole would often be a highly relevant analytic target, much more so than very recent energy consumption in isolation.

Of course BMI at study enrollment is not a perfect correlate of energy consumption over the preceding decades that may be relevant to disease risk. For example, there is a genetic component to BMI, and BMI may be substantially influenced by lifetime activity patterns. An analytic way forward is to exclude BMI from the disease risk model, while including other relevant variables, such as genetic and activity factors, for confounding control.

Including BMI, or other measures of body fat deposition, in the disease risk model can also be expected to lead to overcontrol for other nutritional variables, such as fat intake, that are strongly correlated with total energy intake.

13.3.3 Biomarker Calibrated Intake in Disease Association Analyses

In contrast to the above hypothetical illustration, some established nutritional biomarkers, including the DLW energy expenditure biomarker, are impractical for application to an entire epidemiologic cohort, and also cannot be obtained from stored biospecimens for application in a case-control or case-cohort mode. However, some of the advantages of objective intake measures adhering to Equation 13.1 may be able to be achieved through the conduct of a biomarker study of moderate size in which both the biomarkers of interest and corresponding self-reported dietary intake are assessed. Preferably, the biomarker study would consist of a representative sample of the much larger epidemiologic cohort, but biomarker studies that are more specialized or that are external to the larger cohort may also have value. Consider a biomarker adhering to Equation 13.1 and let q denote a self-report assessment of targeted dietary variable on the same set of study subjects. Typically, the self-report would be based on a machine-readable FFQ, though food record/food diary or 24-hour dietary recall methods are sometimes used. Similar to the previously mentioned biomarkers, the self-report tools target intake over a few days or months, a time period that is short relative to the time period of potential importance to chronic disease risk. A statistical model for the self-report assessment needs to allow it to be relocated and rescaled relative to the targeted intake, and to additionally incorporate the possibility of systematic bias relative to one or more study subject characteristics coded as $v = (v_1, v_2,...)'$. A model (Prentice et al. 2002) for the observed can then be written as

$$w = z + e \tag{13.3}$$

$$q = \beta_0 + \beta_1 z + \beta_2 v + \varepsilon$$

with error term for the self-report that is independent of the targeted dietary variable given the modeled study subject characteristics. A joint normality specification for this error variable and the targeted dietary variable, given the modeled characteristics then leads to a linear model

$$E(z;q,v) = \alpha_0 + \alpha_1 q + \alpha_2 v \tag{13.4}$$

for the conditional expectation of the targeted dietary variable, with coefficients (α_0, α_1, α_2) that are functions of the regression coefficients and of the conditional variance for the self-report measurement error in Equation 13.3.

Our biomarker requirements imply that the measurement error in Equation 13.1 is independent of the targeted dietary variable and of other pertinent study subject characteristics, so that this measurement error will also be independent of the self-report assessment, provided the two error variables in Equation 13.3 are statistically independent. This latter independence may seem quite natural given the very different sources of error that attend a biomarker versus corresponding self-report assessment. However, correlation between these error variables can be expected if the targeted dietary variable is defined in terms of intake over a lengthy period of time, whereas the two error variables derive from assessments over similar short-time periods, because of temporal variations in actual intakes. Hence, to proceed with association analysis, it may be necessary to again redefine the targeted dietary variable in terms of short-term recent intake (e.g., over a one-year period), so that independence of the error terms is plausible. Under this independence assumption, one has

$$E(w;\, q,\, v) = E(z;\, q,\, v) \tag{13.5}$$

and coefficients in Equation 13.4 can be estimated by linear regression of biomarker values on self-report assessments and other modeled study subject characteristics, from which *calibrated* intake estimates

$$\hat{z} = \hat{\alpha}_0 + \hat{\alpha}_1 q + \hat{\alpha}_2 v \tag{13.6}$$

can be calculated for individuals in the epidemiologic cohort, from each person's self-reported dietary intake and modeled characteristics. These calibrated intakes can be thought of as estimates of a study subject's short-term dietary intake, that derive from corresponding self-reported intake after correcting the self-report assessment for random noise as well as for systematic bias related to the modeled study subject characteristics. Their utility will depend on the validity of the model (Equation 13.3) that needs to include all pertinent variables related to systematic bias in the self-report assessment, and will be reduced if the variances for the error terms in Equation 13.3 are not small relative to the variance of targeted dietary variable. As an example of calibration equation development, one can consider average energy intake over the preceding one-year period as targeted dietary variable, with DLW assessments at random two-week periods during the year as biomarker, and FFQ average daily energy intake over the same one-year period as the self-report measure. Calibration equations for the targeted energy variable would arise by linear regression of log-transformed DLW energy on log-transformed FFQ energy and such study subject characteristics as age, ethnicity, and BMI. Calibrated energy estimates for cohort members then derive from each individual's self-reported energy and other modeled characteristics using the fitted linear model (Equation 13.6).

Application of the calibrated estimates involves substituting these for the corresponding targeted dietary variables in standard hazard ratio (or odds ratio) regression models, while using a resampling procedure (Shaw and Prentice 2012; Sugar et al. 2007) to estimate the variance of the estimated regression coefficient to acknowledge random variation in the calibration equation coefficients in Equation 13.6. The regression coefficient for the calibrated intake estimate essentially estimates the disease association parameter linking the targeted dietary variable to disease risk, under the above measurement error modeling assumptions. This simple regression-calibration approach to disease association analysis typically involves little bias and has high statistical efficiency for relatively rare outcomes (e.g., <10% during cohort follow-up) (Shaw and Prentice 2012; Sugar et al. 2007). More complex analytic methods may be needed for common outcomes (Carroll et al. 2006). A biomarker substudy that includes longitudinal biomarker and self-report data over several years has potential for reliable association analyses for a dietary target involving intakes over a longer time period.

This issue of correlation between the biomarker measurement error and the random error component in the self-report model in Equation 13.3 may seem esoteric, but it is central to adequate measurement error correction in association analyses. Some authors use a second self-report assessment in place of the biomarker in Equation 13.3, typically based on one or more dietary recalls. However, even if the dietary target is defined in terms of recent short-term intake, the measurement properties of the second self-report assessment will often depend on study subject characteristics such as BMI, in contrast to biomarker requirements under Equation 13.1. Specifically, measurement errors for differing self-report assessments are likely to be positively correlated, strongly so for a dietary variable such as recent energy consumption, where overweight and obese persons tend to substantially underestimate short-term energy intake whether self-reporting uses food frequencies, food records, or dietary recalls (Prentice et al. 2011).

13.4 BIOMARKER EVALUATION AND DEVELOPMENT

13.4.1 Biomarker Evaluation

Established nutritional biomarkers mainly derive from the recovery in urine of metabolites resulting from nutrient utilization. For example, about 81% of protein expenditure appears in the urine in the form of UN giving 24-hour UN/0.81 as an objective biomarker of daily total protein intake (Bingham 2003). This and other recovery biomarkers, including 24-hour urinary sodium and potassium for sodium and potassium intake, have been developed in human feeding studies, often carried out in metabolic wards with carefully controlled dietary intake over a period of days or weeks. For example, for 24-hour urinary sodium excretion as a biomarker for corresponding sodium intake was studied in a human feeding study that included randomly fluctuating intake. A correlation of 0.55 was found between actual intake and this potential biomarker (Luft et al. 1982). A more recent study, involving constant sodium intake over some months among astronaut trainees, revealed important rhythmicities of 24-hour sodium excretion having durations longer than 24 hours, perhaps reflecting temporal variations in sodium tissue retention (Rakova et al. 2013), helping to explain the higher correlations in the earlier study when multiple days of 24-hour sodium were used as a sodium intake biomarker (Luft et al. 1982). Taken together, these studies provide moderate support for 24-hour sodium excretion as a biomarker of corresponding short-term sodium intake, with multiple days of excretion able to reduce the variance of the error term in Equation 13.1.

The DLW biomarker of total energy intake has been validated in carefully controlled feeding studies, and provides an excellent biomarker of short-term energy consumption for weight-stable study subjects (Schoeller 1999). As detailed elsewhere in this volume (Chapter 10), this biomarker requires the administration of a drink containing ^{2}H and ^{18}O water, with related metabolites recovered in the urine over a subsequent two-week period. None of the dietary self-report approaches have been shown to yield a usable estimate of total energy consumption, so the availability of these energy consumption biomarkers is crucial to nutritional epidemiology. For example, the amount of energy-containing foods consumed may be among the most important drivers of chronic disease risk (Zheng et al. 2014). The DLW biomarker differs from the others mentioned above in that it requires an *intervention* and, hence, cannot be derived from the analysis of stored biospecimens. The development of additional valid and precise objective measures of energy consumption, perhaps using device-related measures of activity, could provide novel and useful research tools.

13.4.2 Biomarker Development

The list of established nutritional biomarkers is short. Objective measures for many additional dietary variables are needed for the biomarker approach to offer a fresh and penetrating look at lifetime dietary patterns in relation to chronic disease risk. Human feeding studies provide the natural context for the development of novel biomarkers for nutrients or foods. This is a high priority research area and the impressive set of potential novel biomarkers, beyond those mentioned above, described in this volume, attests to the recognition of this need by the nutrition, biochemistry, and biomedical device research communities.

Our Women's Health Initiative (WHI) (Anderson et al. 2003; WHI Study Group 1998) research group has recently conducted a 153-women feeding study among WHI women living in its Seattle area for the express purpose of novel nutritional biomarker development. Although most feeding studies administer a small number of specific diets that differ in the intake levels for a specific dietary variable under study, the design of the WHI study attempted to approximate each woman's usual diet, to potentially support biomarker identification for many nutrients or foods. Also, choosing a diet that approximates each study subject's usual diet facilitates stability in urine and blood analyte concentrations by the end of the two-week feeding period, and retains the prestudy intake variation in the study population in provided diets. Usual intake was assessed using four-day food records, with adaptation based on known biases and on in-depth personal interview to ascertain additional detail concerning food choices and dietary patterns. All food and drink was provided by the Human Nutrition Lab at the Fred Hutchinson Cancer Research Center in Seattle, Washington. Women came to the lab two to three times per week during the feeding period to eat one meal on site and to take home food for other meals. Nutrients and foods consumption was assessed by entering all provided food and drink, adjusted for food not consumed or unplanned consumption and for alcohol intake (not provided), into the Nutrient Data System for Research (Nutrition Coordinating Center, version 2010, University of Minnesota, Minneapolis). Potential biomarkers for nutrients or foods are those that are able to explain a substantial fraction of the variation in consumption over the two-week feeding period based on blood and 24-hour urine specimens collected at the end of this period. The DLW and UN measures of energy and protein consumption provide benchmarks for the evaluation of potential novel biomarkers. A paper describing details of the study design and implementation of this feeding study, along with the evaluation of blood concentrations as biomarkers for various micronutrients and fatty acids, has recently appeared (Lampe et al. 2016). This work somewhat extends the set of known nutritional biomarkers by including blood concentration biomarkers for several micronutrients, but the need for biomarkers for many additional dietary variables remains. In the near future, the novel biomarkers identified in this study will be used to evaluate the relationship between the associated nutritional variable and chronic disease incidence in WHI cohorts, using FFQ nutrient intake estimates in a regression calibration mode.

For the development of further nutritional biomarkers our research group is currently examining the potential of urine and blood metabolomic profiles for additional novel biomarker development. This work involves both targeted and global platforms (LC–MS, NMR, and GC–MS) capable of estimating concentrations of about 1000 metabolites. Analyses, using modern high-dimensional data analytic methods for model development and validation are currently underway. Several investigators outside of WHI are also using this feeding study resource for nutritional biomarker identification and evaluation.

The feeding study design seems essential for biomarker development: Even though intake of foods or nutrients will be ascertained with some error in feeding study contexts (e.g., due to limitations of nutrient databases), estimated intakes will plausibly adhere to a classical measurement model (Equation 13.1) in a well-conducted feeding study, without systematic biases.

Potential biomarkers that correlate with their corresponding estimated intake about as strongly as do established biomarkers, such as DLW and UN, then have justification for use. Of course, it would not at all be advisable for biomarker development exercises to use self-reported intakes, with their associated systematic biases, in place of feeding study intake estimates for biomarker development.

Additional human feeding studies are needed to contribute to nutritional biomarker development, and for the study of such issues as biomarker transportability between study populations.

13.5 APPLICATION OF CALIBRATED INTAKE ESTIMATES IN THE WOMEN'S HEALTH INITIATIVE

Applications to date of the use of nutritional biomarkers in Women's Health Initiative cohorts have used the biomarker-calibration approach described in Section 13.3.3 though further applications are planned that will use biomarkers based on blood and urine metabolites in the case-control mode described in Section 13.3.2 without any reliance on dietary self-report data.

Table 13.1 shows estimated calibration equation coefficients (Equation 13.6) for log-transformed total energy, based on data from a Nutrition and Physical Activity Assessment Study (NPAAS)

TABLE 13.1
Calibration Equations from Regression of Log-Transformed Doubly-Labeled Water Energy Consumption on Log(self-report)

	Food Frequency				4DFR[a]				24HR[a]			
Variable	β	SE	R^2	Adj R^{2b}	β	SE	R^2	Adj R^2	β	SE	R^2	Adj R^2
Intercept	7.614	0.009			7.597[c]	0.009			7.607[c]	0.009		
Log(self-report Energy Consumption)	0.054[c]	0.017	3.8	6.5	0.161[c]	0.028	7.8	13.3	0.101[c]	0.026	2.8	4.8
Body Mass Index	0.013[c]	0.001	26.9	45.9	0.013[c]	0.001	27.0	46.0	0.013[c]	0.001	28.7	48.9
Age	−0.010[c]	0.001	9.7	16.5	−0.009[c]	0.001	8.4	14.3	−0.009[c]	0.001	9.1	15.5
Black	−0.023	0.019			−0.024	0.018			−0.024	0.018		
Hispanic	−0.062[c]	0.021	1.3	2.2	−0.065[c]	0.020	1.5	2.6	−0.063[c]	0.020	1.5	2.6
Other minority	−0.041	0.040			−0.039	0.038			−0.038	0.039		
(Total)[d]			41.7	71.1			44.7	76.2			42.1	71.8

Source: Prentice, R. L. et al. *Am J Epidemiol* 174, 591–603, 2011.

[a] 4DFR, four-day Food Record; 24HR, 24-Hour Recall.

[b] Adjusted R^2 is corrected for biomarker temporal variability, that is, R^2 divided by log (biomarker) correlation from reliability substudy.

[c] Coefficient different from zero at the $p = 0.05$ level of significance.

[d] Total percentage of variation explained by all variables. R^2 values for specific variables derive from regression of each variable individually, with subsequent rescaling to add to the total regression R^2. R^2 values for race/ethnicity arise from white, black, Hispanic, and other minority classification.

(Prentice et al. 2011) conducted from 2007–2010 among 450 women in the WHI Observational Study, a prospective cohort study that includes 93,676 postmenopausal women enrolled at 40 U.S. Clinical Centers during 1994–1998. As the biomarker study was conducted some years after enrollment, one needs to make the additional assumption that calibration equations formed by linear regression of biomarker values on concurrent dietary self-report data and other study subject characteristics are applicable to dietary self-report data and other measurements taken at the time of cohort enrollment. Separate calibration equations are shown for each of a FFQ, a four-day food record (4DFR), and 24-hour dietary recalls (24HRs) as the dietary self-report assessment. The FFQ assessment targeted the three-month period prior to the biomarker study, the 4DFR assessment took place within the two-week biomarker protocol, whereas the 24HR assessment was based on three 24-hour dietary recalls conducted during or within a few weeks following the two-week protocol. The targeted daily total energy intake, in Equation 13.3, for both the DLW and self-report assessments can be taken to be the log-transformed short-term average daily energy from, say, three months prior to until nine months following the biomarker study protocol period. Note that only a tiny fraction of the log-DLW assessment variation is *explained* by the FFQ (3.8%), or the 24HR (2.8%) assessments, while it is also small (7.8%) for the 4DFR, even though this latter assessment took place within the DLW protocol period. However, a much larger fraction (about 27%) of the log-DLW variation is explained by BMI. The positive association between BMI and DLW makes no use of clinical outcome data and primarily reflects greater energy consumption by overweight and obese women compared to normal weight women (Neuhouser et al. 2008; Prentice et al. 2011; Subar et al. 2003) and also reflects the very substantial underreporting of energy, using any of these self-report approaches, among overweight and obese women. With this calibration contribution from BMI, along with smaller contributions from age and ethnicity, more than 40% of the variation in log-DLW energy among women can be explained by these calibration equations. Moreover, actual energy intake varies somewhat across two-week time periods (within the targeted one-year time window) that could be used for DLW assessments. The entire NPAAS protocol was repeated about six months later for a 20% subsample of the 450 women. Relative to the one-year target for log-transformed energy intake, the measurement errors in the initial and replicate DLW assessments are plausibly independent, in which case the fraction of variation in the targeted log-energy intake, rather than in the log energy for just the initial two-week protocol period, can be obtained by dividing the R^2 values by the biomarker correlation between the paired assessments in the reliability subsample (Prentice et al. 2011). These adjusted R^2 values suggest that the calibrated estimate in Equation 13.6 can explain more than 70% of the variations among women for the targeted log-energy variable.

Corresponding calibration equations were also developed (Neuhouser et al. 2008; Prentice et al. 2011) for protein and for the ratio of protein to energy where, once again, the targeted dietary variable can be thought of as log-transformed average daily intake over a recent time period of approximately one year of duration.

The NPAAS study also included the development of calibration equations for total activity-induced energy expenditure (AEE) (Neuhouser et al. 2013). The biomarker used for this work was the logarithm of the difference between the same DLW energy expenditure biomarker minus resting energy expenditure as assessed using indirect calorimetry. Calibration equations were developed for each of an Arizona Activity Frequency Questionnaire (AAFQ), a seven-day Physical Activity Recall (PAR), and a physical activity construct based on WHI questionnaire data that included leisure activity assessment within a Personal Habits Questionnaire (PHQ). As shown in Table 13.2, the self-report assessments were only able to explain a modest fraction (3%–8%) of the biomarker AEE variation, increasing to 10%–24% after allowing for variation in AEE within the targeted one-year period. However, these R^2 values increase to 20%–25% on including BMI, age, and race in the calibration equations, with adjusted R^2 values in the 68%–80% range, suggesting a good ability of calibrated AEE values, using any of the self-report tools, to estimate average daily AEE over a targeted one-year period (Neuhouser et al. 2013).

TABLE 13.2
Calibration Equations from Regression of Log-Transformed Biomarker-Calibrated Activity-Related Energy Expenditure (AREE) on Log(self-reported AREE)

	AAFQ[a]			PAR[a]			PHQ[a]		
Variable	**β**	**R^2**	**Adj R^{2b}**	**β**	**R^2**	**Adj R^2**	**β**	**R^2**	**Adj R^2**
Intercept	4.789			5.353			5.632		
Log(self-report AREE)	0.273	7.6	24.0	0.184	4.8	15.1	0.153	3.4	10.7
Age	–0.018	9.1	28.7	–0.024	8.6	27.1	–0.022	8.5	26.8
Race		2.4	7.6		2.3	7.2		2.9	9.1
Black	0.086			0.075			0.095		
Hispanic	0.108			0.141			0.055		
Body Mass Index	0.015	6.1	19.2	0.012	5.7	18.0	0.015	7.0	22.1
TOTAL		25.2	79.4		21.5	67.8		21.8	68.7

[a] AAFQ, Arizona Activity Frequency Questionnaire; PAR, seven-day Physical Activity Recall; PHQ, Personal Habits Questionnaire.

[b] Adjusted R^2 is corrected for biomarker temporal variability, that is, R^2 divided by log(biomarker) correlation from reliability substudy.

These calibrated estimates of energy consumption and AEE have recently been associated with the risks of cardiovascular disease, cancer, and diabetes in the WHI Observational Study cohort, using the Cox model regression-calibration approach outlined above with log-hazard ratio function that depended linearly and simultaneously on log-transformed energy intake and log-transformed AEE. Table 13.3, adapted from Zheng et al. (2014), shows some of the key findings from these analyses. Hazard ratios and 95% confidence intervals are shown for 20% increments in energy consumption and in AEE based on FFQ and PHQ assessments (uncalibrated), and based on calibrated energy and AEE assessments. The former analyses suggest weak or nonexistent associations for either energy or AEE. In contrast, analyses based on calibrated energy and calibrated AEE, with their biomarker-based measurement error corrections to the self-report assessments, indicate that energy consumption and activity may be major drivers of the risk of cardiovascular diseases, major cancers, and diabetes mellitus. For example, from the diabetes analyses, one could project that a 20% reduction in energy consumption in conjunction with a 20% increase in AEE could reduce risk by about seven fold!

Table 13.3 analyses (Zheng et al. 2014) include disease-specific confounding factors in the disease risk model, but do not include BMI. The HR estimates based on calibrated exposures become unstable if BMI is added to the disease risk model, reflecting the limited exposure signal from the self-reported energy and AEE assessments over the one-year period in question, conditional on BMI at the beginning of the one-year period. Without including BMI in the disease risk model one has the benefit of BMI, beyond self-reported energy intake, in estimating short-term energy intake, whereas opening up the possibility of residual confounding if the additional modeled variables by themselves do not provide adequate confounding control. As noted above, this is an important interpretational topic that is largely independent of dietary self-report measurement error, and one that will require longitudinal biomarker, diet, and confounding variable data collection for fuller resolution. Aside from the BMI issue, it should be remembered that all the analyses discussed rely on observational data, so that careful attention to confounding control is necessary for useful interpretation even aside from measurement error in dietary exposure variables.

TABLE 13.3
Hazard Ratio Estimates for 20% Increments in Energy Consumption and in Activity-Related Energy Expenditure, without and with Calibration to Correct for Self-Report Measurement Error Based on Chronic Disease Occurrence from Baseline (1994–1998) through September 30, 2010, in the WHI[a] Observational Study

		Uncalibrated				Calibrated			
		Energy		AREE[a]		Energy		AREE	
Disease Category	Cases	HR[a]	95% CI[a]	HR	95% CI	HR	95% CI	HR	95% CI
CHD[b]	1660	1.00	0.98, 1.02	0.99	0.97, 1.01	1.57	1.19, 2.06	0.78	0.65, 0.95
Ischemic Stroke	1136	0.98	0.96, 1.01	0.99	0.97, 1.01	1.55	1.14, 2.10	0.78	0.64, 0.94
Heart Failure	780	1.04	1.01, 1.08	0.97	0.95, 1.00	3.51	2.12, 5.82	0.57	0.41, 0.79
Total CVD[c]	4212	1.00	0.99, 1.01	1.00	0.99, 1.01	1.49	1.23, 1.81	0.83	0.73, 0.93
Breast Cancer[d]	3798	1.01	0.99, 1.02	1.00	0.99, 1.01	1.47	1.18, 1.84	0.82	0.71, 0.96
Colon Cancer	677	1.00	0.96, 1.03	1.00	0.97, 1.03	1.86	1.18, 2.93	0.83	0.66, 1.04
Uterine Cancer	584	1.08	1.04, 1.12	1.01	0.98, 1.05	2.72	1.44, 5.13	0.77	0.49, 1.21
Total Cancer[e]	9227	1.01	1.00, 1.02	0.99	0.99, 1.00	1.43	1.17, 1.73	0.84	0.73, 0.96
Diabetes	6494	1.06	1.04, 1.07	1.01	1.00, 1.02	4.17	2.68, 6.49	0.60	0.44, 0.83

[a] WHI, Women's Health Initiative; AREE, Activity-related Energy Expenditure; HR, hazard ratio; CI, confidence interval
[b] CHD, coronary heart disease, composed of myocardial infarction (MI) and CHD death
[c] CVD, total cardiovascular disease, composed of MI, CHD death, coronary artery bypass graft, and percutaneous coronary intervention, heart failure, ischemic stroke, and hemorrhagic stroke
[d] Invasive breast cancer
[e] Total invasive cancer (excluding nonmelanoma skin cancer)

13.6 STRENGTHS AND WEAKNESSES OF THE BIOMARKER APPROACH TO NUTRITIONAL EPIDEMIOLOGY

The biomarker approach to dietary assessment opens the possibility of a fresh and reliable look at a broad range of nutritional epidemiology associations. Limitations to exercising this approach in the many cohorts used to develop nutritional epidemiology reports derived from the rather few cohorts having a nutritional biomarker substudy, the few cohorts that have stored urine in addition to blood specimens, and the very few cohorts that have collected and stored biospecimens and other necessary data over a sufficient longitudinal period to reflect the time period that may be relevant to chronic disease risk. Another limitation pertains to the rather few foods or nutrients for which there is an established biomarker that plausibly adheres to the classical measurement model (Equation 13.1). Also, some available biomarkers (e.g., 24-hour sodium excretion) may incorporate considerable noise as measures of corresponding intake, reducing their value for efficient data analysis. Biomarkers that derive from stored biospecimens have particular value for efficient cohort study analyses. An accurate and precise biomarker of total energy consumption from stored specimens would be of great value in nutritional epidemiology.

In the absence of ability to use stored biospecimens only, the biomarker approach can rely on a biomarker substudy that includes both biomarker and self-report assessments. This approach requires a usable *signal* for the targeted nutritional variate from both the biomarker and the self-report. Studies to date suggest that calibration coefficients may be quite similar (Freedman et al. 2014 and 2015) among diverse biomarker study populations, perhaps bypassing the need for each cohort to have its own biomarker substudy.

In spite of the limitations and challenges listed above, there is great potential for the development of reliable information on diet and chronic diseases, and on diet and chronic conditions, through the incorporation of objective dietary exposure assessments. Such developments will be most useful if carried out in conjunction with corresponding objective assessments of physical activity and its components (e.g., duration, intensity, and lifetime history), especially because self-report data for physical activity also incorporate important measurement error. The appropriate use of objective measures of diet and physical activity deserves a very high priority in the biomedical research and public health research communities, especially as the obesity epidemic continues unabated in an ever-widening segment of the world population.

ACKNOWLEDGMENTS

This chapter was partially supported by NIH grants CA53996 and CA119171. The author would thank Drs. Marian Neuhouser, Lesley Tinker, Johanna Lampe, Ying Huang, Cheng Zheng, and Dan Raftery for their valuable roles in the WHI research summarized above, and to thank Sheri Greaves for assistance in manuscript preparation.

REFERENCES

Anderson, G., Manson, J., Wallace, R. et al. 2003. Implementation of the women's health initiative study design. *Ann Epidemiol* 13:5–17.

Arab, L., Wesseling-Perry, K., Jardack, P. et al. 2010. Eight self-administered 24-hour dietary recalls using the internet are feasible in African Americans and Whites: The energetics study. *J Am Diet Assoc* 110:857–864.

Bingham, S. A. 2003. Urine nitrogen as a biomarker for the validation of dietary protein intake. *J Nutr* 133:921S–924S.

Carroll, R. J., Ruppert, D., Stefanski, L. A., Crainiceanu, C. P. 2006. *Measurement Error in Nonlinear Models: A Modern Perspective*, Second edition. Boca Raton, FL: Chapman and Hall.

Cox, D. R. 1972. Regression analysis and life tables (with discussion). *J R Statist Soc B* 34:187–220.

Freedman, L. S., Commins, J. M., Moler, J. E. et al. 2014. Pooled results from five validation studies of dietary self-report instruments using recovery biomarkers for energy and protein intake. *Am J Epidemiol* 180:172–188.

Freedman, L. S., Commins, J. M., Moler, J. E. et al. 2015. Pooled results from five validation studies of dietary self-report instruments using recovery biomarkers for potassium and sodium intake. *Am J Epidemiol* 181:473–487.

Hu, F. B. 2008. *Obesity Epidemiology*. New York: Oxford University Press.

Huang, Y., Van Horn, L., Tinker, L. F. et al. 2013. Measurement error corrected sodium and potassium intake estimation using 24-hour urinary excretion. *Hypertension* 63:238–244.

Kaaks, R. J. 1997. Biochemical markers and additional measurements in studies of the accuracy of dietary questionnaire measurements: Conceptual issues. *Am J Clin Nutr* 65 (suppl 4):S1232–S1239.

Lampe, J. W., Huang, Y., Neuhouser, M. L. et al, 2016. Dietary biomarker evaluation in a controlled feeding study in women from the women's health initiative cohort. *Am J Clin Nutr* doi: 10.3945/ajcn.116.144840.

Luft, F. C., Fineberg, N. S., Sloan, R. S. 1982. Estimating dietary sodium intake in individuals receiving a randomly fluctuating intake. *Hypertension* 4:805–808.

Moshfegh, A. J., Rhodes, D. G., Baer, D. J. et al. 2008. The U.S. department of agriculture automated multiple-pass method reduces bias in the collection of energy intakes. *Am J Clin Nutr* 88:324–332.

Mossavar-Rahmani, Y., Shaw, P. A., Wong, W. W. et al. 2015. Applying recovery biomarkers to calibrate self-report measures of energy and protein in the Hispanic Community Health Study/Study of Latinos. *Am J Epidemiol* 181:996–1007.

Neuhouser, M. L., Tinker, L., Shaw, P. A. et al. 2008. Use of recovery biomarkers to calibrate nutrient consumption self-reports in the women's health initiative. *Am J Epidemiol* 167:1247–1259.

Neuhouser, M. L., Di, C., Tinker, L. F. et al. 2013. Physical activity assessment: biomarkers and self-report of activity-related energy expenditure in the WHI. *Am J Epidemiol* 177:576–585.

Prentice, R. L. and Breslow, N. E. 1978. Retrospective studies and failure time models. *Biometrika* 65:153–158.

Prentice, R. L. and Pyke, R. 1979. Logistic disease incidence models and case-control studies. *Biometrika* 66:403–412.

Prentice, R. L. 1986. A case-cohort design for epidemiologic cohort studies and disease prevention trials. *Biometrika* 73:1–11.

Prentice, R. L., Sugar, E., Wang, C. Y., Neuhouser, M., Patterson, R. E. 2002. Research strategies and the use of biomarkers in studies of diet and chronic disease. *Public Health Nutr* 5:977–984.

Prentice, R. L., Mossavar-Rahmani, Y., Huang, Y. et al. 2011. Evaluation and comparison of food records, recalls and frequencies for energy and protein assessment using recovery biomarkers. *Am J Epidemiol* 174:591–603.

Rakova, N., Jüttner, K., Dahlmann, A. et al. 2013. Long-term space flight simulation reveals infradian rhythmicity in human Na(+) balance. *Cell Metab* 17:125–131.

Schoeller, D. A. 1999. Recent advances from application of doubly-labeled water to measurement of human energy expenditure. *J. Nutr* 129:1765–1768.

Self, S. G. and Prentice, R. L. 1988. Asymptotic distribution theory and efficiency results for case-cohort studies. *Ann Statist* 16:64–81.

Shaw, P. A., and Prentice, R. L. 2012. Hazard ratio estimation for biomarker-calibrated dietary exposures. *Biometrics* 68:397–407.

Subar, A. F., Kipnis, V., Troiano, R. P. et al. 2003. Using intake biomarkers to evaluate the extent of dietary misreporting in a large sample of adults: the OPEN study. *Am J Epidemiol* 158:1–13.

Sugar, E. A., Wang, C. Y., Prentice, R. L. 2007. Logistic regression with exposure biomarkers and flexible measurement error. *Biometrics* 63:143–151.

Thomas, D. C. 1977. Addendum to 'Methods for Cohort Analysis: Appraisal by Application to Asbestos Mining' by F. D. K. Liddell, J. D. McDonald, and D. C. Thomas. *J Roy Statist Soc A* 140:469–491.

Women's Health Initiative Study Group. 1998. Design of the Women's Health Initiative Clinical Trial and Observational Study (R. Prentice, Writing Group Chair). *Control Clin Trials* 19:61–109.

World Cancer Research Fund/American Institute for Cancer Research (WCRF/AICR). 1997. Food, Nutrition and the Prevention of Cancer: A Global Perspective. Washington, DC: American Institute for Cancer Research.

World Cancer Research Fund/American Institute for Cancer Research (WCRF/AICR). 2007. Food, Nutrition, Physical Activity, and the Prevention of Cancer: A Global Perspective. Washington, DC: American Institute for Cancer Research.

World Health Organization. Diet, nutrition and the prevention of chronic diseases: Report of a joint WHO/FAO expert consultation. 2003. *World Health Organ Tech Rep Ser* 916:i–viii, 1–149.

Zheng, C., Beresford, S. A. A., Van Horn, L. et al. 2014. Simultaneous association of total energy consumption and activity-related energy expenditure with cardiovascular disease, cancer, and diabetes risk among postmenopausal women. *Am J Epidemiol* 180:526–523.

14 Stable Isotopic Biomarkers of Diet

Sarah H. Nash and Diane M. O'Brien

CONTENTS

We are stardust, billion year old carbon

Joni Mitchell

14.1 INTRODUCTION

Natural abundance stable isotope ratios have only recently attracted attention from nutritional epidemiologists for their potential as biomarkers of dietary intake. These markers, initially developed by geologists, have been used extensively by animal ecologists (del Rio et al. 2009; Gannes et al. 1998) and archaeologists (Macko et al. 1999; Richards et al. 2003; Richards et al. 2005; Schoeninger and Moore 1992) to understand diet in animal and prehistoric populations, respectively, during the past several decades. During the past several years, however, these measures have begun to attract interest as candidate biomarkers of dietary intake for use in nutritional epidemiology studies.

The utility of stable isotope dietary biomarkers exists because the stable isotope ratios of carbon, nitrogen, oxygen, hydrogen, and sulfur vary in the foods we eat and the water we drink, and these variations are incorporated into body proteins and other molecules. These ratios can then be measured in a variety of biological sample types, including whole blood, serum, hair, and nails, to indicate dietary intake.

In this chapter, we will introduce the theoretical concepts underlying the use of stable isotope ratios as dietary biomarkers. As most epidemiologists are likely to be unfamiliar with natural abundance stable isotope ratios, we first provide a description of nomenclature and analytical methodologies, including considerations for sample handling and preparation. We then discuss how stable isotope ratios have been validated for use as dietary biomarkers, including free-living populations and controlled settings. Finally, we describe potential applications of these biomarkers in an epidemiologic setting and provide examples of some early work demonstrating the utility of stable isotope dietary biomarkers to population-based studies.

14.2 BACKGROUND

14.2.1 What Are Stable Isotopes?

Biological molecules are built from atoms of several elements: principally, carbon, nitrogen, oxygen, and hydrogen. Each of these elements exists in nature as a mixture of different isotopes, or atoms with the same number of protons but different numbers of neutrons. Although isotopes can be radioactive, many are stable. Carbon, for example, has two stable isotopes, ^{13}C and ^{12}C, and one radioactive isotope, ^{14}C. In nature, the ratios of heavy to light isotope vary slightly but consistently among types of plants and animals, according to their biology and their sources of each element. These measureable natural differences in stable isotope ratios have the potential to provide powerful biomarkers for the dietary sources of an element.

The stable isotopes of the common elements in biological molecules are given in Figure 14.1. By convention, isotopes with lower atomic mass are termed *light*, whereas those with higher atomic mass are termed *heavy*. As the table in Figure 14.1 illustrates, it is common in nature for one isotope to be more abundant than the others; for example, just over 1% of Earth's carbon exists as ^{13}C, whereas almost 99% exists as ^{12}C (only a very small fraction is the radioisotope ^{14}C). For all of the elements given in this table, the light isotope is by far the most abundant. It is important to note that this table refers to the naturally occurring abundance of each isotope. Molecular tracers that have been artificially enriched in heavy isotopes have a long history of use in nutritional biochemistry and epidemiology; for example, to measure protein turnover, metabolic substrate utilization, and the rate of energy expenditure by the doubly labeled water technique. However, this chapter focuses on natural abundance stable isotope ratios, and will not discuss further the use of stable isotope tracers.

14.2.2 Expressing Stable Isotope Ratios: The Delta Value

Natural variations in stable isotope ratios, although very consistent, are also very small: typically occurring at the fourth, fifth, or even sixth decimal place (Schoeller 1999). As relative abundance can be measured more accurately than absolute abundance, stable isotope ratios in nature are always

Element	Stable isotopes	Abundance (%)[a]
Hydrogen	^{1}H (H)	99.985
	^{2}H (D)[b]	0.015
Carbon	^{12}C	98.892
	^{13}C	1.108
Nitrogen	^{14}N	99.635
	^{15}N	0.365
Oxygen	^{16}O	99.759
	^{17}O	0.037
	^{18}O	0.204
Sulfur	^{32}S	95.0
	^{33}S	0.75
	^{34}S	4.21
	^{36}S	0.014

[a]Abundance does not sum to 100% because radioisotopes are omitted from this table.
[b]D = deuterium, ^{2}H:^{1}H is often given as D:H.

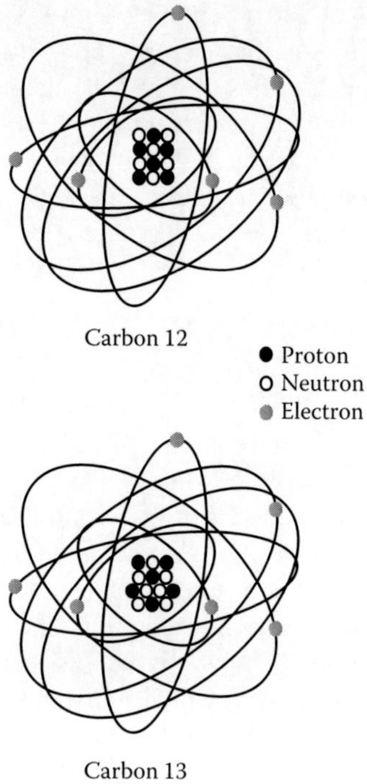

FIGURE 14.1 Naturally occurring stable isotopes of the light elements common in biological molecules, and their global abundances.

measured relative to standards and are expressed in units of relative abundance. These *delta values* (δ) are expressed in units of permil (‰) relative abundance (Macko et al. 1999; Richards et al. 2003; Richards et al. 2005; Schoeninger and Moore 1992). A value of zero permil indicates that the sample has the same ^{13}C:^{12}C as the reference material; thus, zero is a reference point not unlike temperature on the Fahrenheit or Celsius scales, rather than a null value. As an example of stable isotope notation, the carbon isotope ratio ($^{13}C/^{12}C$) is expressed as a $\delta^{13}C$ value, calculated as follows:

$$\delta^{13}C(‰) = ([{}^{13}C/{}^{12}C_{sample} - {}^{13}C/{}^{12}C_{standard}]/{}^{13}C/{}^{12}C_{standard}) \times 1000 \qquad (14.1)$$

The standard for carbon stable isotope analysis is a limestone (Vienna Pee Dee Belemnite, V-PDB), which rather unfortunately contains more ^{13}C than most living organisms, causing the $\delta^{13}C$ values of humans and their foods to be negative.

14.2.3 Isotopic Fractionation

Molecules bearing different isotopes have slightly different masses, which can cause them to have slightly different kinetic properties. For example, the lighter molecules may diffuse more rapidly, or may have slightly faster rates of chemical reactions. These effects can cause the isotope ratio of a product to differ from that of its source, depending on whether the product is derived from the faster moving molecules that react first or from the slower moving molecules that are left behind, and other features of the pathway such as where branch points and rate-limiting steps occur (Hayes 2001; Schoeller 1999). The summed effect of these rate differences, termed *isotope fractionation*, is that isotope ratios differ predictably throughout the environment as elements move through cycles of physical, chemical, and biological reactions. These characteristic differences can provide powerful isotopic *signatures* for specific environmental sources of an element. Isotope fractionation in the environment is the process that generates variability in stable isotope ratios among foods, and it is this variability that forms the basis for that the biomarkers described in this chapter.

14.3 MEASUREMENT OF STABLE ISOTOPE RATIOS

14.3.1 Analysis

Stable isotope ratios are measured using isotope ratio mass spectrometry (IRMS). Note that this is a distinct analysis from molecular scanning or *organic* mass spectrometry, with which many nutrition scientists are already familiar. A full description of IRMS is beyond the scope of this chapter; instead, we refer interested readers to a recent review on this topic by O'Brien (2015). However, there are several key points worth mentioning. First, IRMS requires that biological samples be converted to gas; thus, this is a destructive analysis. However, the amount of sample required is small: typically analyses require only 0.2–0.5 mg of dry sample, depending on the isotope(s) measured, which translates to 2–5 μL of plasma, serum, or red blood cells (RBC). In addition, most IRMS systems measure both the carbon and nitrogen isotope ratios in a single analysis.

14.3.2 Practical Considerations: Choice, Handling, and Preparation of Samples for Analysis

Many dietary biomarkers are molecules that originate from foods of interest, and are subsequently incorporated into, and measured in, specific tissues. Natural abundance stable isotope ratios differ from such measures in that it is the atoms of a given element that are measured, not the molecules in which those atoms occur. This has several important implications for the choice of the biological sample for analysis, and the sample handling.

First, because the light elements (carbon, nitrogen, oxygen, hydrogen, and sulfur) are present in all human tissues, stable isotope ratios can be measured in most sample types available in epidemiologic studies for dietary biomarker analysis. This includes whole blood, red blood cells, serum, plasma, and urine, as well as perhaps less common sample types, such as fingerstick blood, hair, and fingernails. The choice of sample for analysis will depend on availability of samples, as well as the time period one wishes the biomarker to reflect. Tissues differ in their rates of molecular turnover, or how rapidly molecules are catabolized and replaced from dietary sources. Tissues with more rapid turnover will reflect the isotope ratio of diet consumed more recently, whereas tissues with slower turnover will integrate the isotope ratio of the diet over a longer period of time. For example, red blood cells turn over more slowly than plasma, and thus are expected to reflect diet integrated over a longer period. A full discussion of elemental turnover and its implications for stable isotope analysis and interpretation is given later in this chapter.

Second, because stable isotope ratio measurements require the elements in samples be combusted to gases (CO_2 for carbon, N_2 for nitrogen), they are not affected by alterations to molecular structure that may occur before analysis. The only exception to this is where one is interested in the stable isotope ratio of specific molecules, such as amino acids or fatty acids, as described later. This means that the requirements for sample storage are less stringent than for other dietary biomarker analyses, such as vitamin C (Jenab et al. 2009), and enables the use of stored specimens that have undergone freeze–thaw cycles. Preservation additives do have the potential to alter stable isotope ratios, as the atoms contributed by the additive may have different isotope ratios than the sample. Although this effect is likely to be small due to the small number of additive atoms relative to those in the sample, it should be formally tested. Studies have found no effect of ethylenediaminetetraacetic acid (EDTA), polymerized acrylamide resin, or sodium fluoride additives on the $\delta^{13}C$ and $\delta^{15}N$ of blood samples (Kraft et al. 2008; Wilkinson et al. 2007), and there was a marginally nonsignificant effect of sodium fluoride on plasma $\delta^{13}C$ (Kraft et al. 2008).

It has also been demonstrated that autoclaving has no effect on blood $\delta^{13}C$ or $\delta^{15}N$ values (Wilkinson et al. 2007). This is important, as autoclaving can reduce the risk of blood borne pathogen contamination associated with human biospecimens. Samples are typically dried before stable isotope analysis and enclosed in a crushed, but not sealed, tin capsule. To remove the chance of pathogen exposure, we have implemented a process by which liquid samples are aliquoted into tin capsules and autoclaved prior to being dried, crushed, and analyzed (Nash et al. 2013).

14.3.3 Choice of Analyzing Complete Tissue Samples, or Isolating Specific Molecules

Most nutrition epidemiology studies to date that have utilized stable isotope dietary biomarkers have measured the stable isotope ratio of the total sample carbon or nitrogen (Davy et al. 2011; Nash et al. 2013; Yeung et al. 2010). This analysis is informally referred to as *bulk* analysis. However, several more recent studies have begun to explore associations of diet with specific molecules, such as amino acids (Choy et al. 2013), and plasma glucose (Cook et al. 2010; Nash et al. 2014b).

Each type of analysis has strengths and limitations that must be considered. Bulk analyses are relatively straightforward, automated and inexpensive, and have shown efficacy in reflecting intake of dietary patterns (Nash et al. 2012), or broad food groups such as sugar (Davy et al. 2011; Nash et al. 2013; Yeung et al. 2010) and animal protein (Petzke et al. 2005a; Petzke et al. 2005b). However, associations of bulk isotope ratios with foods of interest have the potential to be confounded by intake of other foods (Nash et al. 2013; Yeung et al. 2010), a challenge that we discuss later in this chapter (see Section 14.6.3). The stable isotope ratios of isolated molecules may provide more specific indicators of dietary intake, but analyses are usually more expensive and involved. This is because they require isolating the specific molecules using gas chromatography or other techniques before measurement of the stable isotope ratio using IRMS (Popp et al. 2007; Silfer et al. 1991; Takano et al. 2010). The derivatization of molecules for gas chromatography also adds atoms of differing isotope ratios, which, when converted to CO_2 along with the analyte, must be accounted for in analyses (Corr et al. 2007; O'Brien et al. 2002; Silfer et al. 1991).

14.4 HOW DO STABLE ISOTOPE RATIOS VARY AMONG FOODS?

Knowledge of how and why isotopes vary in our foods is important to understanding the strengths and limitations of stable isotope dietary biomarkers. In this section, we discuss the primary reasons that stable isotope ratios of the light isotopes vary in the environment and our foods.

14.4.1 Carbon

Ultimately, the carbon in our food is derived from atmospheric CO_2, which is captured (fixed) in organic molecules by plants through the process of photosynthesis. Differences in plant photosynthetic physiology result in differences in the proportion of the heavy isotope of carbon that plants fix. A full discussion of these processes is given in the recent review by O'Brien (2015). Importantly, most commonly consumed plants fall into one of two categories: C3 plants, which fix less of the heavy isotope and have $\delta^{13}C$ values ranging from –22‰ to –31‰, and C4 plants, which fix more of the heavy isotope and are relatively isotopically enriched (–17‰ to –10‰; Figure 14.2) (Farquhar et al. 1989; O'Leary 1988). A third group of plants utilize the crassulacean acid metabolism (CAM) photosynthetic pathway; these plants, including pineapple, agave, and prickly pear, have similar $\delta^{13}C$ values to C4 plants (Farquhar et al. 1989) but not heavily consumed in the U.S. diet.

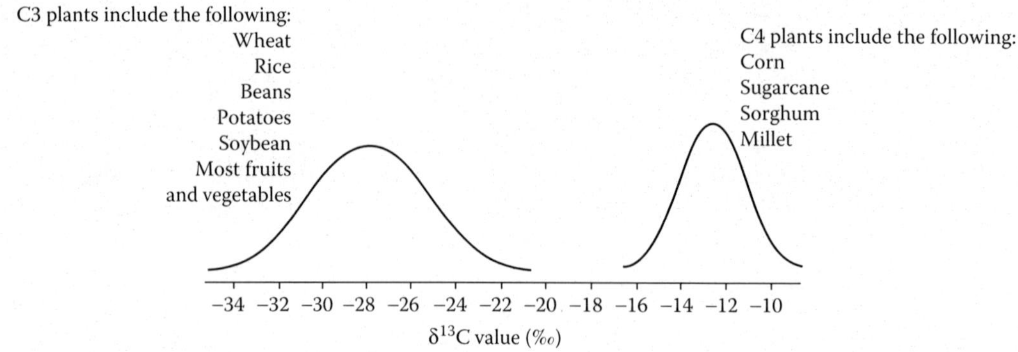

FIGURE 14.2 The distribution of $\delta^{13}C$ values in C3 and C4 plants.

The majority of plants in the U.S. food supply, including wheat, rice, beans, and most fruits and vegetables, utilizes the C3 photosynthetic pathway (Minagawa 1992; Nakamura et al. 1982; Schoeller et al. 1980; Schoeller et al. 1986). There are two notable exceptions: corn and sugarcane, both of which are C4 plants (Jahren et al. 2006). In the United States, much of the consumption of these two plants is in the form of added sugars. It is estimated that added sugars represent approximately 16% of total energy consumed in the United States (Johnson and Yon 2010; Marriott et al. 2010); of these, 78% are derived from cane sugar or corn-based syrups (Haley 2013). For this reason, there has been substantial interest in developing a carbon isotope biomarker of sugar consumption; this will be discussed in detail later in this chapter. However, it should be noted that C4 plant consumption varies globally. For example, the European food system is far less reliant on corn and cane sugar-based sweeteners, as sugar beet, a C3 plant, is produced and consumed more than sugarcane (Polet 2015). Therefore, the carbon isotope ratio will not act as a biomarker of sugar consumption in Europe, or other food systems where consumption of C4 sweeteners is low. In addition to corn and sugarcane, other C4 plants include millet and sorghum.

These variations in plant stable isotope ratios are transferred to consumers, human and animal, which can result in elevated isotope ratios elsewhere in the food chain. Most importantly to the U.S. diet, commercially raised livestock are heavily fed on corn, which gives U.S. store-purchased meat and dairy products an elevated $\delta^{13}C$ value (~−15‰ to −20‰) (Jahren and Kraft 2008; Nardoto et al. 2006; Nash et al. 2012). This feature of the U.S. agricultural system is the basis for observed associations of tissue $\delta^{13}C$ and animal protein consumption among U.S. populations (Nash et al. 2013; Yeung et al. 2010). As both C4 sugars and animal proteins are important contributors to the U.S. diet, refining $\delta^{13}C$ biomarkers to differentiate carbon contributed by carbohydrates versus carbon contributed by proteins is a priority for researchers who are interested in using this biomarker in U.S. study populations (see Section 14.6.3). In contrast, European commercial livestock consume a more heavily C3-based diet, which is reflected in the stable isotope ratios of EU meat products (Camin et al. 2007; Schmidt et al. 2005). However, due to a small average increase of approximately 1‰ that occurs between diet and consumer (McCutchan et al. 2003), these food items still have higher $\delta^{13}C$ values than the plants they consume. For this reason, there has been some interest in carbon stable isotope biomarkers of animal protein intake in European study populations (Patel et al. 2014; Petzke et al. 2005a; Petzke et al. 2005b).

Finally, although the majority of variation in food carbon isotope ratios derives from differences in photosynthetic pathway, there are also some differences observed between plants in terrestrial versus marine environments (Mook et al. 1974). Marine $\delta^{13}C$ values typically fall in the range of ~−17‰ to −20‰, intermediate between C3 and C4 plants. These elevated carbon isotope ratios occur throughout the marine foodweb, including among commercially important fish species such as salmon, cod, tuna, and pollock (Nash et al. 2012; O'Brien 2015). On average, the U.S.

population consumes 1–2 servings of fish per week (U.S. Environmental Protection Agency 2014), which is likely a small source of carbon relative to other C3 and C4 sources, such as wheat and corn. However, where populations consume a large proportion of energy from fish and other marine foods, then their carbon stable isotopes will reflect this marine intake (Nash et al. 2012).

14.4.2 Nitrogen

Tissue nitrogen is almost entirely derived from dietary protein; thus, the nitrogen isotope ratio is a potential biomarker of protein intake. The nitrogen isotope ratio varies among dietary protein sources for two reasons: variation in the isotope ratio of soil nitrogen and trophic level enrichment. Again, a complete review of these processes can be found in O'Brien (2015); however, we will briefly discuss trophic level enrichment, as it is this process that primarily drives variation in food nitrogen isotope ratios, in turn, leading to the potential of the nitrogen isotope ratio to indicate animal protein and/or fish intake.

Nitrogen isotope ratios increase with each step (trophic level) in the food chain. When animals consume plants (or other animals), they incorporate this dietary nitrogen into their tissue amino acids. This process is constant and results in the generation and excretion of waste nitrogen. Nitrogen excretion heavily favors the lighter isotope, resulting in consumer $\delta^{15}N$ values that are on average 3‰–4‰ higher than their diet (Minagawa and Wada 1984; Steele and Daniel 1978). As this increase in $\delta^{15}N$ occurs at each step in a food chain, top predators can have very high $\delta^{15}N$ values. For example, polar bears have demonstrated $\delta^{15}N$ values as high as 20‰ (Bentzen et al. 2007).

In general, commercial livestock are herbivores (cows, chickens, turkey, pigs, etc.), with nitrogen isotope ratios only slightly higher than conventionally grown crops: 2‰–4‰ (Huelsemann et al. 2013; Nash et al. 2012; O'Brien *pers. comm.*). The major exception to this are commonly consumed fish species, such as tuna, salmon, cod, pollock, and halibut. Marine food chains tend to be longer than terrestrial food chains, and these fish species are highly predatory. Consequently, their $\delta^{15}N$ values are markedly elevated relative to other animal protein sources, at approximately ~10‰ to 20‰ (Huelsemann et al. 2013). Thus, nitrogen isotope ratios have the potential to indicate both meat and fish intake, with meat having a smaller effect on tissue $\delta^{15}N$ values but (generally) consumed more frequently, and fish having a smaller effect but consumed less frequently (O'Brien 2015). Disentangling the effect of meat and fish intake on tissue $\delta^{15}N$ will be of key in the continued development of the nitrogen isotope biomarker; a particularly promising approach may be to comeasure tissue sulfur isotope ratios, as described in Section 14.4.3.

14.4.3 Sulfur

The sulfur isotope ratio ($\delta^{34}S$) has the potential to provide a useful biomarker of marine food intake, as $\delta^{34}S$ values differ markedly between marine and terrestrial food sources. Importantly, there is a little isotopic fractionation as plants assimilate sulfur into biological compounds, and $\delta^{34}S$ values do not exhibit a systematic increase with trophic level (McCutchan et al. 2003); thus, $\delta^{34}S$ values have the potential to provide a biomarker of marine intake that is independent of trophic level. Although this potential has been relatively under explored by nutritional epidemiologists, if measured in combination with $\delta^{15}N$ values, $\delta^{34}S$ may help distinguish the individuals with high meat intake from those with high fish intake.

14.4.4 Oxygen and Hydrogen

Although the stable isotope ratios of oxygen ($\delta^{18}O$) and hydrogen (δD) may be familiar to nutritional epidemiologists because of their use as enriched tracers in the doubly labeled water technique, the potential of naturally occurring variations in these isotope ratios has been relatively underexplored.

Unlike the other isotopes discussed, oxygen and hydrogen stable isotope ratios vary on the continental scale. The H and O in both drinking water and foods ultimately derive from precipitation, and because heavier isotopes condense into precipitation more readily than do lighter isotopes, successive precipitation events exhibit progressively lighter isotope ratios. These processes generate distinct patterns of $\delta^{18}O$ and δD values across the United States in drinking water, plants, animals, and humans (Bowen et al. 2007; Ehleringer et al. 2008).

Few human studies have utilized $\delta^{18}O$ and δD ratios; those that have, have largely been forensic or anthropologic in focus. For example, hair $\delta^{18}O$ and δD ratios have been used to develop predictive models of models for residence based on geographic differences in drinking water and dietary stable isotope ratios (Ehleringer et al. 2008). They have also been used to differentiate local and nonlocal residents (Podlesak et al. 2012), as well as to illustrate the use of a global *supermarket diet* (Thompson et al. 2010). Another study showed that beard hair $\delta^{18}O$ and δD values were sensitive to global travel, with values clearly changing with migration between the United Kingdom and the United States (O'Brien and Wooller 2007). Finally, in combination with C, N, and/or S isotope ratios, oxygen and hydrogen isotope ratios have also been used in forensic applications to identify location of origin for unidentified individuals (Bartelink et al. 2014; Fraser et al. 2006; Katzenberg and Krouse 1989).

However, the utility of these markers is presently diminished, due to unresolved complications. First, tissue O and H can derive from both diet and drinking water, yet the precise contribution from each is unknown. Studies suggest that, although the contribution of diet to tissue O and H is high, up to 31% of hydrogen in human hair may be derived from drinking water (Ehleringer et al. 2008; Sharp et al. 2003). Furthermore, the U.S. population consumes a decidedly nonlocal, supermarket-based diet, with food items grown or raised globally. Thus, the $\delta^{18}O$ and δD values of one's diet may differ from that of local drinking water (Chesson et al. 2008). Finally, a fraction of sample H is known to interact with water vapor in the air, which may lead to challenges in interpretability for δD values (Chesson et al. 2009), though recently developed techniques may address this issue (Sauer et al. 2009). Thus, although there is a potential for oxygen and hydrogen-stable isotope ratios to indicate dietary intake, these biomarkers will remain undeveloped until additional research can address some of these concerns. For this reason, the rest of this chapter will focus only on stable isotope biomarkers of diet using C, N, and S.

14.5 HOW ARE STABLE ISOTOPE VARIATIONS TRANSFERRED TO CONSUMERS?

All carbon, nitrogen, and sulfur in our bodies come from our diet. Unlike other molecules that can serve as dietary biomarkers, there are no other sources of these elements that can be incorporated directly into human tissues, and we cannot create elements endogenously. However, there are several factors that may affect elements within specific molecules within tissues, resulting in tissue isotope ratios that may not perfectly correspond to those of the diet.

14.5.1 Elemental Turnover

Our bodies are in a constant state of elemental turnover. Carbon, nitrogen, and sulfur are lost from our tissues, through processes such as excretion (feces, urine), expiration (respired CO_2), or growth (hair, nails, sloughed cells) and replaced with these elements from our diet. In the protein-rich tissues commonly used for isotopic analysis, the rate of elemental turnover should be closely linked to the rate of protein turnover. In general, the rate of elemental turnover does not directly correlate with tissue or whole-body metabolic rate (Carleton and Martinez del Rio 2005; Kraeer et al. 2014). The exception to this is where protein turnover and metabolic rates are linked, for example, during periods of intense exercise training (Pikosky et al. 2006).

The rate of protein/elemental turnover differs between tissues; therefore, stable isotope biomarkers may reflect different periods of time, depending on the turnover of the tissue analyzed.

For example, red blood cells (RBC) undergo 85% replacement approximately every 96 days (Cohen et al. 2008). Thus, stable isotope ratios measured in RBC will reflect dietary intake over the preceding ~3 months. Conversely, stable isotope ratios measured in blood plasma or serum are likely to reflect dietary intake over a slightly shorter period of time; although this has not yet formally been tested. Other measurements may be even more short term, for example, the carbon isotope ratio of breath is known to reflect dietary intake over the preceding hours (Schoeller et al. 1980; Whigham et al. 2014), and $\delta^{13}C$ isolated from one specific molecule in plasma, plasma glucose, was shown to alter quickly in response to an isotopically labeled meal (Cook et al. 2010). Carbon and nitrogen isotope ratios measured in urine and feces were shown to reflect dietary intake after a controlled feeding period of eight days (Kuhnle et al. 2013), suggesting that these sample types may provide additional short-term biomarkers of intake. Other biological samples, such as bone (O'Connell et al. 2001) or tooth (Ambrose 1990), have particularly slow turnover rates and can be used to indicate dietary intake over the very long term; however, these samples cannot be easily collected for epidemiologic studies.

It is particularly important to be aware of how elemental turnover may affect tissue stable isotope ratios if an individual or population is undergoing dietary change, for example, in a dietary intervention. In this case, the tissue will need to have sufficient time for complete turnover (a process known as isotopic equilibration) before the isotope ratio of the new diet is fully reflected in that tissue.

Finally, it is important to note that some structures, including hair and nail, do not turn over once synthesized. Instead, their isotope ratios will reflect dietary intake at the time of synthesis, a feature that can be used to the investigator's advantage. For example, hair grows at a rate of approximately 1 cm/month; thus hair sampled close to the scalp will reflect more recent dietary intake, whereas sampling further down the hair may reflect diet months (or years!) in the past. Fine-grained, sequential sampling along the hair may be used to construct detailed dietary histories (Huelsemann et al. 2009; Santamaria-Fernandez et al. 2009), and homogenization and analysis of larger sections of hair will indicate diet over its respective period of growth. Nails grow at an approximate rate of 0.4 cm/month (Yaemsiri et al. 2010); thus, samples clipped from the end of the fingernail will reflect dietary intake several months prior.

14.5.2 Diet to Tissue Routing

Another consideration in the use of stable isotope ratios as dietary biomarkers is how elements from foods of interest are incorporated into the tissues measured. Although elements cannot be synthesized *de novo* by the body, molecules may be broken down by the body and their constituent C, N, and S may be incorporated into tissues in a different form than supplied by the diet. Nitrogen and sulfur are primarily supplied by dietary proteins, and incorporated into tissue proteins (Ambrose and Norr 1993; Schwarcz 1991). Thus, they demonstrate protein to protein routing, meaning that both $\delta^{15}N$ and $\delta^{34}S$ values of protein-based tissues reflect the isotope ratios of dietary proteins. In contrast, carbon can be supplied by dietary protein, carbohydrates, or lipids. Although it was once assumed that carbon in tissue proteins reflects only that from dietary proteins (Ambrose and Norr 1993), both animal and human studies have shown that nonprotein carbon can provide substantial contribution to carbon-rich tissues such as RBC (Choy et al. 2013; Nash et al. 2013; Podlesak and McWilliams 2006). This indicates that there may be substantial *de novo* synthesis of nonessential amino acids from nonprotein sources.

It is for this reason that the carbon isotope ratio measured in protein-rich tissues such as RBC (Nash et al. 2013), whole blood (Davy et al. 2011), and plasma/serum (Nash et al. 2014b; Yeung et al. 2010) has shown associations with dietary sugar intake. If the carbon in these tissues was routed exclusively from dietary protein, then one would expect no association with dietary sugars intake, except where confounded by the intake of ^{13}C-enriched proteins such as commercial meats. Yet, in multiple study populations, tissue $\delta^{13}C$ has demonstrated independent associations with both sugar and commercial meat intake (Nash et al. 2013; Yeung et al. 2010), suggesting that carbon from both dietary carbohydrates and proteins is routed to tissue proteins. Further supporting this hypothesis, one study found strong associations of dietary sugars intake with RBC alanine, which is a nonessential amino acid closely linked with glucose metabolism (Choy et al. 2013). Approaches such as this, which maximize the routing of elements from foods of interest to the tissue or molecules measured, will be particularly valuable in the further development of stable isotope dietary biomarkers.

14.5.3 Physiologic Effects

Although the primary determinant of tissue stable isotope ratios is diet, there are physiological effects on the distribution of isotopes throughout the body that are important to keep in mind when using stable isotope biomarkers. As has been found in a large number of animal studies (del Rio et al. 2009), tissues vary predictably in their stable isotope ratios; for example, hair $\delta^{13}C$ values are 2‰ higher than plasma and RBC $\delta^{13}C$ values, whereas plasma and hair $\delta^{15}N$ values are 1.5‰ higher than RBC $\delta^{15}N$ values (Kraft et al. 2008; Nash et al. 2009; Nash et al. 2014b). Tissue isotope ratios also differ consistently from the diet. For example, in a study of 11 participants (five males, six females) that consumed a controlled diet for 30 days, O'Connell and colleagues (2012) demonstrated that RBC $\delta^{15}N$ values were 3.5‰ higher than those of the diet. This difference is termed the diet to tissue fractionation, and is particularly pronounced for $\delta^{15}N$ values, due to preferential excretion of ^{14}N relative to ^{15}N (Minagawa and Wada 1984; Steele and Daniel 1978). Among excreted biospecimens, one controlled feeding study showed that urine $\delta^{13}C$ and $\delta^{15}N$ values were approximately 1.5‰ elevated and depleted relative to diet, respectively, whereas both fecal $\delta^{13}C$ and $\delta^{15}N$ values were lower than those of the diet (Kuhnle et al. 2013). Although diet-tissue fractionation and between-tissue variations do not affect the potential of stable isotope ratios as biomarkers of diet, it is important for researchers to account for these differences when comparing the results of studies conducted with different tissues.

Many animal studies have shown that the diet to tissue fractionation for nitrogen can be influenced by nitrogen balance (Lee et al. 2012), with the $\delta^{15}N$ values of tissues synthesized during periods of negative nitrogen balance tending to be elevated and the $\delta^{15}N$ values of tissues synthesized during positive nitrogen balance tending to be reduced. In human studies, this *weight loss* effect has been demonstrated in early stage pregnant women who lost significant weight from morning sickness (Fuller et al. 2004; Fuller et al. 2005) and anorexia nervosa patients (Hatch et al. 2006). Conversely, the $\delta^{15}N$ of tissues synthesized during positive nitrogen balance are reduced, as has been demonstrated in pregnant women who are gaining weight (Fuller et al. 2004) and recovering anorexia patients (Mekota et al. 2006). Hair $\delta^{15}N$ values were also shown to be markedly lower in patients with liver cirrhosis, relative to matched healthy controls (Petzke et al. 2006). Studies utilizing $\delta^{15}N$ as a dietary biomarker should exclude participants who are pregnant, undergoing rapid weight gain or loss, or have a diagnosis of chronic liver disease.

The primary physiological influence on tissue $\delta^{13}C$ values is the lipid content of the tissue. Lipids have $\delta^{13}C$ values that are typically 5‰–7‰ lower than tissue protein or carbohydrate (Sweeting et al. 2006), due to discrimination against ^{13}C by an enzyme involved in lipid biosynthesis (DeNiro and Epstein 1977). This effect can be avoided by lipid-extracting samples, which has the unfortunate side effect of altering $\delta^{15}N$ and $\delta^{34}S$ values also, or by mathematically correcting tissue $\delta^{13}C$ using the C:N ratio, an accurate proxy for tissue lipid content (Oppel et al. 2010; Post et al. 2007). The effect of lipid on tissue $\delta^{13}C$ is likely to be minimal for tissues with low lipid content, such as RBC, nails, and hair, but could potentially affect tissues such as serum or plasma, if their lipid content was high.

Finally, as sulfur isotope dietary biomarkers are less well developed than those of carbon or nitrogen, physiologic effects on $\delta^{34}S$ values have been relatively little explored. However, given that diet to tissue fractionation averages ~0‰ (McCutchan et al. 2003), any such effects are likely to be minimal. There is limited evidence to suggest that, like $\delta^{15}N$ values, diet to tissue fractionation may increase for $\delta^{34}S$ when dietary protein intake is inadequate (Richards et al. 2003); however this assertion is based on a very small number of observations. Further research is necessary to better understand potential physiologic effects on tissue $\delta^{34}S$ values.

14.6 VALIDATING STABLE ISOTOPE RATIOS AS BIOMARKERS OF DIETARY INTAKE

14.6.1 Isotopic Surveys of Food Items

Understanding the complete *isotopic landscape* of the foods that a population consumes is critical to understanding the efficacy and utility of stable isotope biomarkers within that population. In Section 14.4, we discussed the theoretical underpinnings of stable isotope variations among the foods that we eat. Here, we discuss the findings from several studies that surveyed carbon and nitrogen stable isotope ratios of a variety of commonly consumed food items.

The first isotopic surveys of contemporary human food items were conducted in the 1980s to provide context for early studies of dietary effects on breath CO_2 (Schoeller et al. 1980). This and subsequent studies in the late 1980s and early 1990s confirmed marked differences in $\delta^{13}C$ values of plant foods based on their photosynthetic pathway (Minagawa and Wada 1984; Nakamura et al. 1982; Schoeller et al. 1980; Schoeller et al. 1986). Foods such as rice, wheat, vegetables, and many fruits exhibited lower $\delta^{13}C$ values (~−30‰ to −20‰), whereas corn, millet, and cane sugar exhibited higher $\delta^{13}C$ values (~−15‰ to −10‰). These studies also confirmed that both animal products (meat, eggs, and dairy) and fish had intermediate $\delta^{13}C$ values (−13‰ to −15‰), whereas $\delta^{15}N$ values were intermediate among animal products (5‰ to 8‰) and high for fish (9‰ to 18‰). These patterns were relatively consistent across foods purchased in Germany, Japan, and the United States; however, animal products from Germany did exhibit lower $\delta^{13}C$ values than those from the United States, reflecting lower use of corn for animal feed in the European agricultural system.

For carbon isotope ratios, recent focus has shifted to providing additional measurements of corn-based foods, particularly sweeteners. As with the earliest food surveys, this reflects the interests of the field more broadly: for some time, nutrition epidemiologists have sought to improve measurement of dietary sugars intake (Jenab et al. 2009; Tasevska et al. 2005; Tasevska et al. 2014), and the past five years has seen increasing interest in the potential for tissue carbon isotope ratios to provide such as biomarker (Jahren et al. 2014; O'Brien 2015). Recently, Jahren and colleagues (2006) examined the $\delta^{13}C$ values of corn products, cane sugar, fruits, vegetables, and plant-based proteins. This study corroborated findings of earlier studies by showing large differences in $\delta^{13}C$ value between C3-based foods (~−25‰) and C4-based foods (~−10‰), the latter category including two sweeteners common in the U.S. food system: corn syrup and sugarcane.

Further studies have provided information on the carbon isotope ratios of other isotopically enriched foods, including fish and marine mammals (Nash et al. 2012; Wilkinson et al. 2007), and animal products (Nash et al. 2012). Examining food items collected from an Alaska Native population that still practices subsistence, Nash and colleagues (2012) showed that fish and marine mammals exhibited intermediate $\delta^{13}C$ values (ca. −21‰), whereas terrestrial subsistence foods (i.e., wild-harvested game meats) had lower values reflective of their C3-based diet (ca. −25‰). In contrast, commercially purchased animal products exhibited higher $\delta^{13}C$ values (ca. −17‰), indicative of the use of corn feed in the U.S. agricultural system (Nardoto et al. 2006; Nash et al. 2012). Similar findings have also been demonstrated for U.S.-purchased fast food items (Chesson et al. 2008; Jahren and Kraft 2008).

Toward the development of the nitrogen isotope biomarker of animal protein intake, a recent study summarized the literature to provide the most comprehensive database of historical and modern food $\delta^{15}N$ values (Huelsemann et al. 2013). This study documented consistent patterns of $\delta^{15}N$ values among types of food types, with $\delta^{15}N$ values increasing predictably between plants, animal products (beef, pork, chicken, dairy), and fish. However, a survey of food and fingernail stable isotopes in Brazil showed that $\delta^{15}N$ values were higher throughout the Brazilian foodweb, relative to the United States, likely due to differences in the agricultural fertilizers used (Nardoto et al. 2006).

Thus, studies of the carbon and nitrogen isotope ratios of food items support relatively consistent and predictable patterns among the foods that we eat. This is particularly true for populations that consume a *supermarket diet*, which appears relatively isotopically consistent across continents. However, differences observed globally highlight the importance of validating stable isotope ratios in the country, and ideally the population, in which they are to be used.

Finally, very little research has been conducted to examine whether cooking or food preparation substantially modifies stable isotope ratios. One study showed that cooking and food preparation showed very minor effects on the nitrogen stable isotope ratios of fish and marine mammals (T. O'Hara *pers. comm.*), and another indicated that neither cooking nor fermentation affected the carbon or nitrogen isotope ratios of yeasted buns and cookies (Bostic et al. 2015). As discussed for sample handling and storage, one would not expect food isotope ratios to change during food preparation, unless atoms are either added to or removed from the food. Most importantly, changes in the stable isotope ratios introduced by food preparation are likely to be very small, relative to the variations naturally observed due to photosynthesis, trophic level, or other environmental processes discussed in Section 14.4.

14.6.2 Controlled Feeding Studies

Controlled feeding studies provide a critical opportunity to test the validity of dietary biomarkers without reliance on error-prone dietary self-report. Unfortunately, only a few human studies have tested the performance of stable isotope biomarkers in a controlled setting, although there have been numerous controlled feeding studies in animals (Martínez del Rio et al. 2009).

The primary focus of early controlled (human) feeding studies was methodological, examining length of time for dietary changes to be incorporated in tissue stable isotope ratios (Huelsemann et al. 2009), as well as diet-tissue fractionation (O'Connell et al. 2012). In one such study, four participants (two male, two female) underwent an isocaloric change from their habitual diets (C3-based, terrestrial proteins) to a controlled diet based mainly on C4 plants and marine proteins (Huelsemann et al. 2009). All participants showed an increase in both hair $\delta^{13}C$ and $\delta^{15}N$ values during the duration of the study; however, none reached a new isotopic steady state (i.e., they did not achieve isotopic equilibrium with their diet). Furthermore, these increases were observed earlier for hair $\delta^{15}N$ than for $\delta^{13}C$, suggesting that dietary N is incorporated into hair proteins more rapidly than dietary C. This finding is in contrast to several animal studies that demonstrated either no difference, or more rapid incorporation for dietary C (Bahar et al. 2014; Braun et al. 2013; Carleton and Martinez del Rio, 2005). As previously mentioned in Section 14.5.3, O'Connell and colleagues showed that RBC $\delta^{15}N$ values were 3.5‰ higher than diet, a finding that is in good agreement with those from animal studies (McCutchan et al. 2003).

Since this time, several additional studies have examined associations of tissue $\delta^{13}C$ and $\delta^{15}N$ with dietary intake variables of interest. In Europe, the focus of these studies has been isotopic biomarkers of animal protein intake. In a study sample of 14 UK participants (six male, eight female), Kuhnle and colleagues (2013) showed that the $\delta^{13}C$ and $\delta^{15}N$ values of urine and feces provide a short-term biomarker of meat and fish intake. In this controlled crossover design, each dietary treatment lasted eight days, and four participants also completed a vegetarian control period. Relative to the vegetarian controls, urine and feces $\delta^{15}N$ values were elevated in both the meat and fish treatments; however, $\delta^{13}C$ values were only elevated in the fish treatment. Furthermore, there were no dietary effects on whole blood isotope ratios, likely because of the short-term nature of the dietary intervention. The second study was a semicontrolled crossover design, in which Petzke and colleagues randomized 14 female participants to either add a daily serving of pork to, or omit meat from, their habitual diet for a period of four weeks per treatment (Petzke and Lemke 2009). The authors found no effect of the dietary interventions on either hair or plasma $\delta^{13}C$ or $\delta^{15}N$ values, and only a minimal effect on urine carbon and nitrogen isotope ratios. This finding was unexpected; although four weeks should have been sufficient to observe dietary effects on plasma, hair, and urine isotope ratios, it is possible that the duration of dietary treatments were too short. Alternatively, as the design was only semicontrolled, participants may have compensated for the impact of treatment on their habitual diet in such a way that effects of the treatment on tissue isotope ratios were obscured.

Despite the interest in the carbon isotope biomarker of sugar intake in the United States, only one study to date has examined the potential of this marker in a controlled setting. Cook and colleagues (2010) conducted a seven-day crossover feeding study where five participants (three men, two women) were fed three weight maintaining diets, which differed in the proportion of energy derived from C4 sugars. The $\delta^{13}C$ of plasma glucose was then measured at two-hour intervals throughout the day. They found a strong positive relationship between proportion of energy from C4 sugars and postprandial plasma glucose $\delta^{13}C$ values; however, there was no association with fasting plasma glucose $\delta^{13}C$ values.

14.6.3 Stable Isotope Validation in Observational Studies

Although controlled feeding studies are the gold standard study design for assessing the validity of potential dietary biomarkers, validation studies in free-living populations are also important, as they provide key insight into the effectiveness of these markers in populations with varying diet composition. However, because such studies rely on self-reported measures of intake for comparison, it should be understood that associations presented will likely be attenuated from their *true* associations, due to the high error associated with dietary self-report (Bingham et al. 2003; Kipnis et al. 2003; Subar et al. 2003). Fortunately, these errors are uncorrelated with those inherent in stable isotope measurements—something that is not true when validating self-reported measures against each other (Ocke and Kaaks 1997).

In European observational studies, the carbon and nitrogen isotope ratios have both been explored as biomarkers of animal protein intake. The first such study was performed by Petzke and colleagues (Petzke et al. 2005a), who showed that *bulk* hair $\delta^{13}C$ and $\delta^{15}N$ easily differentiated omnivores, ovo-lacto vegetarians, and vegans in a subset of participants of the VERA (Verbundstudie Ernährungserhebung und Risikofaktoren-Analytic; Nutrition Survey and Risk Factor Analysis) Study (n = 126), which oversampled participants reporting vegetarian and vegan dietary patterns. Intake was assessed using a seven-day food diary, and reported animal protein intake demonstrated moderate associations with hair $\delta^{13}C$ and $\delta^{15}N$ in a dual-isotope linear regression model ($R^2 = 0.31$, $P < 0.01$). The authors also examined hair amino acid $\delta^{13}C$ and $\delta^{15}N$, but found no ability to discriminate between categories of animal protein intake across sex and isotope-specific categories. Unfortunately, this study did not examine associations separately with fish intake.

More recently, Patel and colleagues (2014) examined associations of serum $\delta^{13}C$ and $\delta^{15}N$ with dietary correlates (estimated using a food frequency questionnaire [FFQ]) in a case-cohort study (n = 718) nested within the European prospective investigation into cancer and nutrition (EPIC)-Norfolk cohort. Fish protein intake showed moderate positive associations with serum $\delta^{13}C$ ($r = 0.22$) and $\delta^{15}N$ ($r = 0.20$). Animal protein intake was not associated with serum $\delta^{13}C$, but showed weak significant correlations with serum $\delta^{15}N$ (dairy protein $r = 0.11$, meat protein $r = 0.09$). These associations are weaker than those observed in the VERA study, likely due to differences in error associated with the measures of dietary self-report.

There has been relatively little work to validate stable isotope biomarkers of animal protein intake in U.S. study populations. Stable isotope biomarkers of fish intake have been validated in a Yup'ik study population in Southwest Alaska. This Alaska Native population has a strong cultural tradition of subsistence harvesting, including fishing, hunting, and gathering wild plants. Marine intake is high, but variable, in this population, accounting for an average of 15% energy intake (Bersamin et al. 2006; Bersamin et al. 2007). In the Center for Alaska Native Health Research Study, Nash and colleagues (2012) demonstrated strong associations of RBC $\delta^{15}N$ with self-reported (combined four 24-hour recall and one three-day food diary) fish and marine mammal intake ($r = 0.52$, $n = 260$). Furthermore, even stronger observations were observed with two independent biomarkers of marine food intake, the RBC *n*-3 fatty acids eicosapentaenoic and docosahexaenoic acid for both hair and RBC $\delta^{15}N$ (all $r > 0.8$; O'Brien et al. 2009).

In contrast, there has been a considerable interest in validating the carbon isotope biomarker of sugars intake among U.S. populations. Self-reported sugar intake suffers high error (Tasevska et al. 2014), and an objective biomarker would improve our understanding of the role of sugar in obesity, diabetes, and other chronic diseases of interest. Yeung and colleagues (2010) examined associations of serum $\delta^{13}C$ values with self-reported intake in 186 participants of the Atherosclerosis Risk in Communities (ARIC) study. This study, which oversampled at the low and high ends of self-reported sugar-sweetened beverage (SSB) consumption, demonstrated moderate associations with SSB consumption ($r = 0.18$) and no association with intake of total sugars, fructose, or glucose. Indeed, the strongest associations were observed with animal protein ($r = 0.28$) and animal fat ($r = 0.37$) intake. Shortly thereafter, Davy and colleagues (2011) examined associations of whole blood $\delta^{13}C$ with self-reported added sugar (four-day food record) and SSB (beverage questionnaire) in a community-based sample of 60 adult Virginians. Whole blood $\delta^{13}C$ was positively associated with intake of added sugars ($r = 0.37$) and SSB ($r = 0.35$). Unfortunately, associations with animal protein intake were not assessed. A further study by the same authors assessed associations in a community-based sample of 224 participants of the Talking Health trial (Hedrick et al. 2016), which aims to reduce SSB consumption among medically underserved Virginians. This study found that the Healthy Eating Index, SSB, and added sugars intake (estimated using an FFQ) all predicted whole blood $\delta^{13}C$; however, associations with animal protein intake were not assessed.

Several studies have explored various ways of addressing the lack of specificity associated with the carbon isotope biomarker of sugars. The first method used $\delta^{15}N$ values in an attempt to control for the confounding effects of animal protein intake on $\delta^{13}C$ values. In a sample of 68 Yup'ik participants, Nash and colleagues (2013) showed independent associations of RBC $\delta^{13}C$ with self-reported (four weekly 24-hour recalls) total sugar, meat, and fish intake. To reduce the confounding effect of fish intake on RBC $\delta^{13}C$ values, they included $\delta^{15}N$ as a covariate in a two-isotope predictive model of sugar intake; this model strongly predicted intake of total sugars ($r^2 = 0.48$), as did similar models using hair and plasma (Nash et al. 2014b). Unfortunately, due to the high and varying intakes of fish among Yup'ik people, $\delta^{15}N$ values were related to fish intake only in this study population. Therefore, the inclusion of $\delta^{15}N$ did not control for the confounding effects of meat intake on $\delta^{13}C$ values. In a subset of participants in the PREMIER dietary intervention trial ($n = 144$), Fakhouri and colleagues (2014) demonstrated modest improvements in associations of serum $\delta^{13}C$ with SSB intake (two 24-hour recalls) after adjusting for serum $\delta^{15}N$. The association of serum $\delta^{15}N$ and animal protein intake in this study population is presumed, but was not reported. In contrast, a third study in a sample of Southwest Virginian adults showed no difference between a single- or dual-isotope model of self-reported (three 24-hour recalls or one four-day food diary) added sugar or sugar-sweetened beverage intake (Hedrick et al. 2015). In this study sample, $\delta^{13}C$ values showed independent associations with sugar and animal protein intake; however, $\delta^{15}N$ values were not associated with animal protein intake. Thus, although there is a potential for the $\delta^{15}N$ value to provide a convenient way of improving the carbon isotope biomarker of sugar intake, its effectiveness will vary based on the underlying association of $\delta^{15}N$ and animal protein intake in the study population of interest.

The second method to increase the specificity of the carbon isotope biomarker of sugars in U.S. populations is to measure $\delta^{13}C$ values of molecules that preferentially incorporate carbon from dietary sugars. Two recent studies have taken this approach. The first, by Cook and colleagues (2010), described in detail in Section 14.6.2, showed a strong association of postprandial plasma glucose $\delta^{13}C$ with sugar intake in a seven-day randomized crossover feeding study. However, there was no association of fasting plasma glucose $\delta^{13}C$ and dietary sugars intake, a finding that was confirmed in an observational study by Nash et al. (2014b). In this study of 68 Yup'ik adults, the $\delta^{13}C$ value of fasting plasma glucose was not associated with habitual total or added sugars intake. Thus, fasting plasma glucose may provide a more specific marker of sugars intake in the very short term but does not appear to integrate sugar carbon over a longer period. In the same sample of 68 Yup'ik adults mentioned earlier, Choy and colleagues (2013) proposed an alternative approach. Here, the authors measured the carbon isotope ratios of RBC amino acids and examined associations with self-reported fish, meat, and sugars intake. They found a strong association ($r = 0.7$) of SSB intake with the δ^{13}C value of RBC alanine but no association with fish or meat intake. Alanine is metabolically connected to glucose intake via the glucose–alanine cycle (Perriello et al. 1995; Waterhouse and Keilson 1978), and may reflect a long-term record of average blood glucose δ^{13}C values in both RBC and hair (Choy et al. 2013). Additional research is required to confirm this finding in other, nonnative, U.S. populations.

14.7 APPLICATIONS OF STABLE ISOTOPE DIETARY BIOMARKERS

Although large-scale and controlled-setting validation of stable isotope dietary biomarkers is ongoing, a small number of studies have used these biomarkers to explore questions of interest to nutrition epidemiologists. Currently, these studies fall into two categories: those that examine associations with disease outcomes or surrogate (intermediate) endpoints, and those that use the marker to identify dietary changes.

14.7.1 Diet-Disease/Risk Factor Associations

RBC δ^{13}C and δ^{15}N values have been used to examine associations of both sugar and marine food intake with obesity, and risk factors for diabetes and cardiovascular disease in a cross-sectional Yup'ik study population (Nash et al. 2014a; O'Brien et al. 2014). It is important to note upfront that the findings of these studies were almost perfectly opposite, because intake of these foods is strongly negatively associated among Yup'ik people (Nash et al. 2012). Longitudinal studies will be necessary to disentangle the potentially independent effects of sugar and marine food consumption on these risk factors; however, despite this complication, these studies remain a powerful demonstration of the utility of stable isotope biomarkers in epidemiologic studies. The first of these studies showed that, in a sample of 1076 Yup'ik participants, the dual-isotope marker of sugar intake was not associated with measures of obesity (BMI and waist circumference), but did exhibit positive associations with blood pressure, triglycerides, and leptin and inverse associations with cholesterol (total, LDL, and HDL), and leptin (Nash et al. 2014a). The second, in a very similar study sample ($n = 772$), showed that RBC δ^{15}N values were positively associated with cholesterol (total, LDL, HDL), apolipoprotein A-I, and insulin-like growth factor binding protein-3 (IGFBP-3), and inversely associated with triglycerides, adiponectin, and blood pressure (O'Brien et al. 2014). Importantly, this second study also demonstrated that RBC δ^{15}N values captured the same diet-risk factor associations that had been previously observed with established biomarkers of marine intake, RBC EPA and DHA (docosahexaenoic acid) (Makhoul et al. 2010). Furthermore, because nitrogen isotope ratios were available on a larger sample of participants, the authors were able to detect associations of marine intake with blood pressure and adiponectin that were previously unknown. In a follow-up study, the authors showed that the association of RBC δ^{15}N values with blood pressure was modified by sex, adiposity, and hypertension status (Beaulieu-Jones et al. 2015).

In the same Yup'ik study population, RBC δ^{15}N values have been extensively used to examine whether fish and marine mammal intake modifies associations of genetic markers with obesity and metabolic traits (Aslibekyan et al. 2013; Aslibekyan et al. 2014; Klimentidis et al. 2014; Lemas et al. 2012; Lemas et al. 2013; Vaughan et al. 2015). For example, Lemas and colleagues (2012) showed that marine intake modified associations of several *CPT1A* SNPs with HDL-cholesterol and Apo A1, but not body composition, and Klimentidis and colleagues (2014) found that marine intake did not modify associations of type 2 diabetes SNPs with glycemic traits such as fasting glucose and HbA1c. In a smaller subsample of this same study population,

Aslibekyan and colleagues (2014) demonstrated very different patterns of gene methylation between high and low consumers of marine food.

In addition to this diabetes research, investigators have also used RBC $\delta^{15}N$ values to investigate other chronic disease questions of interest among Alaska Native populations. For example, Fohner and colleagues (2016) showed that several SNPs, marine diet, and characteristics such as body mass index (BMI), gender, season, and location of residence were all associated with serum 25(OH)D_3 concentrations in this Yup'ik study population. Additionally, Chi and colleagues (2015) used the dual-isotope biomarker of sweeteners intake to demonstrate positive associations of sugar intake with tooth decay in a sample of Yup'ik children ($n = 51$). The same associations were not observed with self-reported intake, highlighting the potential limitations associated with error-prone self-reported measures, relative to objective biomarkers.

Finally, a recent case-cohort nested within the EPIC-Norfolk study examined associations of serum $\delta^{13}C$ and $\delta^{15}N$ values with incident type 2 diabetes (Patel et al. 2014). In this study, $\delta^{15}N$ values showed positive associations with diabetes incidence (hazard ratio [HR] per tertile of $\delta^{15}N$: 1.23 [95% CI 1.09, 1.38]). Conversely, $\delta^{13}C$ values showed associations with fish intake only and were inversely associated with diabetes incidence (HR 0.74 [95% CI 0.65, 0.83]). Because of these disparate patterns of association, Patel and colleagues concluded that animal protein intake may lead to increased risk of diabetes, whereas fish intake may reduce risk. If this is indeed the case, the association with $\delta^{15}N$ may actually underestimate the risk associated with animal protein intake, as this biomarker reflects both animal protein and fish intake.

14.7.2 Monitoring Dietary Changes

Stable isotope biomarkers have the potential to be useful tools with which to observe dietary changes, if the diets of interest are isotopically distinct. To date, there are two areas in which this potential has been explored: documenting nutritional transitions and monitoring adherence to dietary interventions.

Nash and colleagues (2012) used stable isotope biomarkers to examine the nutrition transition among Yup'ik people living in rural Southwest Alaska. As described in Section 14.6.3, RBC $\delta^{15}N$ provided a biomarker of traditional food intake, whereas after correction for marine mammal intake, RBC $\delta^{13}C$ values showed strong correlations with nontraditional food intake ($r = 0.46$). Importantly, these isotopic biomarkers varied with key demographic and cultural factors: traditional food intake was higher in females, older participants, participants who reported high adherence to a traditional Yup'ik way of life, and participants who lived in coastal communities located farther from the regional hub. Another, very recent study among Yup'ik people examined the nitrogen isotope ratios of archived serum samples dating back to the 1960s, and showed significant decreases in traditional marine food intake each decade between the 1960s and 1990s, at which point it stabilized (O'Brien 2016). This was accompanied by similar decreases in serum 25-hydroxy-vitamin D_3 concentrations, as fish and marine mammals are a key source of vitamin D in the Yup'ik diet.

Although the above studies used stable isotope biomarkers to explore the nutrition transition in circumpolar communities, Nardoto and colleagues (2011, 2016) have used these markers to investigate dietary change in a distinctly different setting: Amazonian populations in Brazil. The authors measured fingernail $\delta^{15}N$ and $\delta^{13}C$ values of people living along Brazil's Solimões River and showed that consumption of nontraditional, C4-based foods increased with community size and proximity to urban centers. The authors concluded that increasing urbanization in the Brazilian Amazon may lead to substantial changes in food intake, including increased intake of nontraditional, C4-based foods. Although this study, as well as most of those conducted in the circumpolar north were cross-sectional and cannot directly speak to dietary changes among these populations, they do illustrate the potential of stable isotope biomarkers to distinguish traditional and nontraditional food consumers in populations experiencing the nutrition transition.

Finally, two studies have demonstrated the potential of stable isotope biomarkers in monitoring adherence to dietary interventions. In a subset of 144 participants of the PREMIER trial, an 18 month behavioral intervention trial in which participants changed their SSB consumption, Fakhouri and colleagues (2014) found that even modest reductions (4–6 fl oz/d) in SSB consumption were associated with slightly reduced serum $\delta^{13}C$ values. After adjustment for potential confounding variables, including animal protein intake, a reduction of intake equivalent to one serving a day (12 fl oz) was associated with a small (0.17‰) decrease in serum $\delta^{13}C$ values. Furthermore, patterns of $\delta^{13}C$ values over the course of the trial were very similar to patterns of reported SSB intake. Similar decreases were observed in fingerstick blood samples of 155 adults randomized to an intervention to reduce SSB consumption among rural Southwest Virginian adults, relative to matched controls (Davy et al. 2016). Although observed changes in $\delta^{13}C$ values were small, these studies do provide some evidence to support the use of stable isotope biomarkers to provide an objective record of adherence to dietary interventions, specifically those to reduce sugar intake.

14.8 ADVANTAGES AND LIMITATIONS

As with any dietary assessment method, stable isotope biomarkers have several advantages, as well as limitations. First, there are several biological features of stable isotope ratios that make them attractive candidate dietary biomarkers. Unlike other concentration biomarkers, stable isotope ratios are little affected by endogenous metabolic processes. This is particularly true for the nitrogen isotope ratio, as tissue nitrogen tends to be directly routed from dietary protein (see Section 14.5.2). Although tissue carbon isotope ratios do have the potential to incorporate dietary carbon from protein, carbohydrates, and/or fats, studies measuring the carbon isotope ratios of specific amino acids have shown compound specific approaches have great potential to distinguish carbon from different macronutrient sources (Choy et al. 2013). Another advantage is the ability of stable isotope markers to indicate diet over different time periods, because of differences in turnover rates between tissues and other sample types (e.g., serum, RBC and hair, as discussed in Section 14.5.1). Thus, studies incorporating stable isotope dietary biomarkers should determine which time period would be most appropriate for the question of interest, and choose sample type for analysis accordingly.

There are also several practical advantages to stable isotope dietary biomarkers. Key to the utility of these markers is their ability to be measured in a wide variety of sample types. As mentioned earlier, this gives the ability to provide dietary insights using whatever samples may be available from epidemiologic cohorts. Because stable isotope ratios can be measured in hair and nails, it also means that samples can be collected noninvasively and without the need for specialized personnel. This may be of particular interest to researchers working with potentially vulnerable populations, such as children, or those who are interested in the potential for study subjects to collect their own samples for analysis: with adequate instruction and training, study participants might reasonably be able to collect hair samples and mail back to study coordinating centers. Furthermore, because stable isotope ratios are not affected by processes that may alter molecular structure, they do not require specialized sample handing or storage, except that which may be required for biohazard control (e.g., for blood, plasma, or serum samples). This includes the process of freeze–thaw, so that stable isotope ratios can be measured in archived biological samples, including those that have been in long-term storage. Sample preparation and analysis is comparatively easy and inexpensive, and is available at many academic and commercial laboratories.

One of the primary considerations in the use of stable isotope dietary biomarkers is the need to understand the underlying isotopic ecology of the diet. This is well illustrated by differences in dietary associations with tissue carbon isotope ratios between the United States and Europe. In this chapter, we have discussed the extensive use of corn and sugarcane-based sweeteners in the U.S. agricultural system, and how this differs from Europe, where the primary sweetener is sugar beet (Polet 2015). It is this difference that results in the carbon isotope ratio demonstrating associations with sugar intake in U.S. study populations, but not in Europe. An additional example comes from the lack of specificity associated with the carbon isotope biomarker of sugar intake. Again, we have discussed how tissue $\delta^{13}C$ demonstrates associations with both sugar, and animal protein intake (Nash et al. 2013; Yeung et al. 2010), due to the elevated $\delta^{13}C$ values associated with both commercial meat products and fish (Jahren and Kraft 2008; Nash et al. 2012). Furthermore, it is unknown how the carbon isotope biomarker of sugars will perform in populations that consume a substantial proportion of energy from corn, and foods produced from whole corn. Thus, stable isotope biomarkers require further testing and validation in populations with a diverse range of diets to fully understand how these biomarkers are affected by dietary composition.

14.9 FUTURE STUDIES

Although stable isotope ratios have demonstrated great potential as dietary biomarkers, additional validation studies are necessary for these biomarkers to reach their full potential. Particularly lacking in the current literature are studies that examine the validity of stable isotope biomarkers in a controlled setting, particularly because comparisons against self-reported dietary intake are subject to reporting error. Controlled feeding studies will be the most appropriate design to explore several outstanding questions, including how long it takes for dietary C and N to be incorporated into different biological sample types, the degree of diet-tissue fractionation for both C and N, and most importantly, whether we can quantify dose-response relationships between changes in dietary intake and tissue stable isotope ratios. This is particularly critical for the carbon stable isotope biomarker of corn and cane sugar-based sweetener intake, given the high level of recent interest in this candidate biomarker. Regardless of focus, all controlled feeding studies will need to be of adequate duration to ensure that isotopic equilibration has occurred. As turnover of many commonly available tissues/molecules occurs slowly, reflecting intake over a period of weeks or months, controlled diets that replicate and maintain participants' usual intake may provide the most realistic approach.

Population-based studies of diet-biomarker associations will provide complimentary information to that from controlled feeding studies; thus, a validation strategy that includes both study designs should be employed. Specifically, population-based research among diverse populations will be necessary to evaluate how stable isotope biomarkers are affected by differences in diet composition. For example, RBC $\delta^{15}N$ values were strongly associated with fish intake with a Yup'ik study population with high marine intake (Nash et al. 2009; Nash et al. 2012; O'Brien et al. 2009), but did not show associations with either meat or fish in a sample of Southwest Virginian adults consuming a U.S. supermarket diet (Davy et al. 2011). Furthermore, the validity of the carbon isotope biomarker of sugar intake has not been assessed in a population that has high consumption of products made from whole corn, despite serum $\delta^{13}C$ values showing weak associations with whole corn intake in a study population residing in Maryland (Yeung et al. 2010). Importantly, these population-based validation studies will necessarily rely on self-reported measures of dietary intake, and while these should be the best measures available, interpretation of study findings should also include an understanding that biomarker-diet associations will be attenuated.

Finally, studies that increase the specificity of stable isotope measures for foods of interest are especially warranted. The use of multiple isotope ratios is an approach that has been relatively underexplored, particularly the potential for using $\delta^{34}S$ values to resolve whether elevated $\delta^{15}N$ values reflect fish or meat intake. Measuring the isotope ratios of specific molecules, such as amino acids or plasma glucose, is an additional approach that has demonstrated promise but requires additional research. One drawback to this approach is an increase in cost and reduction in throughput efficiency; therefore, it would also be worth investigating whether the isotope ratios of specific molecules can be used to calibrate *bulk* markers. Until the issue of specificity has been resolved, it is imperative that any validation study reports associations of stable isotope biomarkers with all potential confounders, and not just the food or nutrient of interest. For example, studies validating the carbon isotope biomarker of sweets should also report associations with animal protein (meat/fish) and whole corn intake, to assess the degree of confounding from other dietary sources.

14.10 CONCLUSIONS

Stable isotope ratios hold tremendous promise as dietary biomarkers, because several foods of epidemiologic interest have distinct isotope ratios (signatures) that are passed onto consumers, and because consumer isotope ratios can be easily measured in a variety of tissues or biological molecules. Although stable isotope ratios of many light elements, including carbon, nitrogen, oxygen, hydrogen and sulfur, have the potential to be informative regarding dietary intake, the vast majority of studies to date have focused on carbon and nitrogen. The carbon isotope ratio is elevated in C4 plants (including corn and sugarcane), and animal protein, and has demonstrated associations with both food groups when measured in sample types including serum, plasma, RBC, and hair (Davy et al. 2011; Nash et al. 2013; Nash et al. 2014b; Yeung et al. 2010). The nitrogen isotope ratio is elevated in animal protein sources, especially fish, and has demonstrated strong associations with both self-reported and biomarker-based measures of intake when measured in RBC and hair (Nash et al. 2009; O'Brien et al. 2009; Petzke et al. 2005b). Key strengths of stable isotope biomarkers include their ease of collection, storage and analysis, and the fidelity with which they are captured in tissues and biological molecules. Importantly, stable isotope ratios are not affected by long-term storage and can be measured in a wide variety of biological sample types, meaning that these markers could be analyzed among archived specimens from epidemiologic studies. However, despite their potential, further validation work is necessary before stable isotope biomarkers can be fully embraced as by the epidemiologic community. An approach utilizing both controlled and population-based studies will provide the most comprehensive evaluation of the questions still outstanding in this ongoing validation process.

REFERENCES

Ambrose, S. H. 1990. Preparation and characterization of bone and tooth collagen for isotopic analysis. *J Archaeol Sci* 17(4): 431–451.

Ambrose, S. H. and L. Norr. 1993. Experimental evidence for the relationship of the carbon isotope ratios of whole diet and dietary protein to those of bone collagen and carbonate. In *Prehistoric Human Bone*, pp. 1–37. Berlin, Heidelberg: Springer.

Aslibekyan, S., L. K. Vaughan, H. W. Wiener et al. 2013. Evidence for novel genetic loci associated with metabolic traits in Yup'ik people. *Am J Hum Biol* 25(5): 673–680.

Aslibekyan, S., H. W. Wiener, P. J. Havel et al. 2014. DNA (docosahexaenoic acid) methylation patterns are associated with n-3 fatty acid intake in Yup'ik people. *J Nutr* 144(4): 425–430.

Bahar, B., S. M. Harrison, A. P. Moloney, F. J. Monahan, O. Schmidt. 2014. Isotopic turnover of carbon and nitrogen in bovine blood fractions and inner organs. *Rapid Commun Mass Spectrom* 28(9): 1011–1018.

Bartelink, E. J., G. E. Berg, M. M. Beasley, L. A. Chesson. 2014. Application of stable isotope forensics for predicting region of origin of human remains from past wars and conflicts. *Ann Anthropol Pract* 38(1): 124–136.

Beaulieu-Jones, B. R., D. M. O'Brien, S. E. Hopkins, J. H. Moore, B. B. Boyer, D. Gilbert-Diamond. 2015. Sex, adiposity, and hypertension status modify the inverse effect of marine food intake on blood pressure in Alaska Native (Yup'ik) people. *J Nutr* 145(5): 931–938.

Bentzen, T. W., E. H. Follmann, S. C. Amstrup, G. York, M. Wooller, T. O'Hara. 2007. Variation in winter diet of southern Beaufort sea polar bears inferred from stable isotope analysis. *Can J Zool* 85(5): 596–608.

Bersamin, A., B. R. Luick, E. Ruppert, J. S. Stern, S. Zidenberg-Cherr. 2006. Diet quality among Yup'ik Eskimos living in rural communities is low: The center for Alaska native health research pilot study. *J Am Diet Assoc* 106(7): 1055–1063.

Bersamin, A., S. Zidenberg-Cherr, J. S. Stern, B. R. Luick. 2007. Nutrient intakes are associated with adherence to a traditional diet among Yup'ik Eskimos living in remote Alaska native communities: The CANHR Study. *Int J Circumpolar Health* 66(1): 62–70.

Bingham, S. A., R. Luben, A. Welch, N. Wareham, K. T. Khaw, N. Day. 2003. Are imprecise methods obscuring a relation between fat and breast cancer? *Lancet* 362(9379): 212–214.

Bostic J. N., S. J. Palafox, M. E. Rottmueller, A. H. Jahren. 2015. Effect of baking and fermentation on the stable carbon and nitrogen isotope ratios of grain-based food. *Rapid Commun Mass Spectrom* 29(10): 937–947.

Bowen, G. J., J. R. Ehleringer, L. A. Chesson, E. Stange, T. E. Cerling. 2007. Stable isotope ratios of tap water in the contiguous United States. *Water Resour Res* 43(3).

Braun, A., K. Auerswald, A. Vikari, H. Schnyder. 2013. Dietary protein content affects isotopic carbon and nitrogen turnover. *Rapid Commun Mass Spectrom* 27(23): 2676–2684.

Camin, F., L. Bontempo, K. Heinrich et al. 2007. Multi-element (H, C, N, S) stable isotope characteristics of lamb meat from different European regions. *Anal Bioanal Chem* 389(1): 309–320.

Carleton, S. A. and C. Martinez del Rio. 2005. The effect of cold-induced increased metabolic rate on the rate of 13C and 15N incorporation in house sparrows (Passer domesticus). *Oecologia* 144(2): 226–232.

Chesson, L. A., D. W. Podlesak, T. E. Cerling, J. R. Ehleringer. 2009. Evaluating uncertainty in the calculation of non-exchangeable hydrogen fractions within organic materials. *Rapid Commun Mass Spectrom* 23(9): 1275–1280.

Chesson, L. A., D. W. Podlesak, A. H. Thompson, T. E. Cerling, J. R. Ehleringer. 2008. Variation of hydrogen, carbon, nitrogen, and oxygen stable isotope ratios in an American diet: Fast food meals. *J Agric Food Chem* 56(11): 4084–4091.

Chi, D. L., S. Hopkins, D. O'Brien, L. Mancl, E. Orr, D. Lenaker. 2015. Association between added sugar intake and dental caries in Yup'ik children using a novel hair biomarker. *BMC Oral Health* 15(1): 121.

Choy, K., S. H. Nash, A. R. Kristal, S. Hopkins, B. B. Boyer, D. M. O'Brien. 2013. The carbon isotope ratio of alanine in red blood cells is a new candidate biomarker of sugar-sweetened beverage intake. *J Nutr* 143(6): 878–884.

Cohen, R. M., R. S. Franco, P. K. Khera et al. 2008. Red cell life span heterogeneity in hematologically normal people is sufficient to alter HbA1c. *Blood* 112(10): 4284–4291.

Cook, C. M., A. L. Alvig, Y. Q. Liu, D. A. Schoeller. 2010. The natural 13C abundance of plasma glucose is a useful biomarker of recent dietary caloric sweetener intake. *J Nutr* 140(2): 333–337.

Corr, L. T., R. Berstan, R. P. Evershed. 2007. Optimisation of derivatisation procedures for the determination of δ13C values of amino acids by gas chromatography/combustion/isotope ratio mass spectrometry. *Rapid Commun Mass Spectrom* 21(23): 3759–3771.

Davy, B., A. Jahren, V. Hedrick, W. You, J. Zoellner. 2016. Influence of an intervention targeting a reduction in sugary beverage intake on the δ13C sugar intake biomarker in a predominantly obese, health-disparate sample. *Public Health Nutr.* 2017;20(1): 25–29.

Davy, B. M., A. H. Jahren, V. E. Hedrick, D. L. Comber. 2011. Association of delta(1)(3)C in fingerstick blood with added-sugar and sugar-sweetened beverage intake. *J Am Diet Assoc* 111(6): 874–878.

del Rio, C. M., N. Wolf, S. A. Carleton, L. Z. Gannes. 2009. Isotopic ecology ten years after a call for more laboratory experiments. *Biol Rev Camb Philos Soc* 84(1): 91–111.

DeNiro, M. J. and S. Epstein. 1977. Mechanism of carbon isotope fractionation associated with lipid synthesis. *Science* 197(4300): 261–263.

Ehleringer, J. R., G. J. Bowen, J. A. Chesson, A. G. West, D. W. Podlesak, T. E. Cerling. 2008. Hydrogen and oxygen isotope ratios in human hair are related to geography. *Proc Natl Acad Sci USA* 105(8): 2788–2793.

Fakhouri, T. H., A. H. Jahren, L. J. Appel, L. Chen, R. Alavi, C. A. Anderson. 2014. Serum carbon isotope values change in adults in response to changes in sugar-sweetened beverage intake. *J Nutr* 144(6): 902–905.

Farquhar, G. D., J. R. Ehleringer, K. T. Hubick. 1989. Carbon isotope discrimination and photosynthesis. *Annu Rev Plant Biol* 40(1): 503–537.

Fohner, A. E., Z. Wang, J. Yracheta et al. 2016. Genetics, diet, and season are associated with serum 25-Hydroxycholecalciferol concentration in a Yup'ik study population from Southwestern Alaska. *J Nutr* 146(2): 318–325.

Fraser, I., W. Meier-Augenstein, R. M. Kalin. 2006. The role of stable isotopes in human identification: A longitudinal study into the variability of isotopic signals in human hair and nails. *Rapid Commun Mass Spectrom* 20(7): 1109–1116.

Fuller, B. T., J. L. Fuller, N. E. Sage, D. A. Harris, T. C. O'Connell, R. E. Hedges. 2004. Nitrogen balance and delta15N: Why you're not what you eat during pregnancy. *Rapid Commun Mass Spectrom* 18(23): 2889–2896.

Fuller, B. T., J. L. Fuller, N. E. Sage, D. A. Harris, T. C. O'Connell, R. E. Hedges. 2005. Nitrogen balance and delta15N: Why you're not what you eat during nutritional stress. *Rapid Commun Mass Spectrom* 19(18): 2497–2506.

Gannes, L. Z., C. Martinez del Rio, P. Koch. 1998. Natural abundance variations in stable isotopes and their potential uses in animal physiological ecology. *Comp Biochem Physiol A Mol Integr Physiol* 119(3): 725–737.

Haley, S. 2013. *Electronic Outlook Report from the Economic Research Service. Sugar and Sweeteners Outlook/SSS-M-293.*: U.S. Department of Agriculture.

Hatch, K. A., M. A. Crawford, A. W. Kunz et al. 2006. An objective means of diagnosing anorexia nervosa and bulimia nervosa using 15N/14N and 13C/12C ratios in hair. *Rapid Commun Mass Spectrom* 20(22): 3367–3373.

Hayes, J. M. 2001. Fractionation of carbon and hydrogen isotopes in biosynthetic processes. *Rev Mineral Geochem* 43(1): 225–277.

Hedrick, V. E., B. M. Davy, G. A. Wilburn, A. H. Jahren, J. M. Zoellner. 2016. Evaluation of a novel biomarker of added sugar intake ($\delta^{13}C$) compared with self-reported added sugar intake and the Healthy Eating Index-2010 in a community-based, rural U.S. sample. *Public Health Nutr* 19(03): 429–436.

Hedrick, V. E., J. M. Zoellner, A. H. Jahren, N. A. Woodford, J. N. Bostic, B. M. Davy. 2015. A dual-carbon-and-nitrogen stable isotope ratio model is not superior to a single-carbon stable isotope ratio model for predicting added sugar intake in southwest virginian adults. *J Nutr* 145(6): 1362–1369.

Huelsemann, F., U. Flenker, K. Koehler, W. Schaenzer. 2009. Effect of a controlled dietary change on carbon and nitrogen stable isotope ratios of human hair. *Rapid Commun Mass Spectrom* 23(16): 2448–2454.

Huelsemann, F., K. Koehler, H. Braun, W. Schaenzer, U. Flenker. 2013. Human dietary delta(15)N intake: Representative data for principle food items. *Am J Phys Anthropol* 152(1): 58–66.

Jahren, A. H., J. N. Bostic, B. M. Davy. 2014. The potential for a carbon stable isotope biomarker of dietary sugar intake. *J Anal At Spectrom* 29(5): 795–816.

Jahren, A. H. and R. A. Kraft. 2008. Carbon and nitrogen stable isotopes in fast food: Signatures of corn and confinement. *Proc Natl Acad Sci USA* 105(46): 17855–17860.

Jahren, A. H., C. Saudek, E. H. Yeung, W. H. Kao, R. A. Kraft, B. Caballero. 2006. An isotopic method for quantifying sweeteners derived from corn and sugar cane. *Am J Clin Nutr* 84(6): 1380–1384.

Jenab, M., N. Slimani, M. Bictash, P. Ferrari, S. A. Bingham. 2009. Biomarkers in nutritional epidemiology: Applications, needs and new horizons. *Hum Genet* 125(5–6): 507–525.

Johnson, R. K. and B. A. Yon. 2010. Weighing in on added sugars and health. *J Am Diet Assoc* 110(9): 1296–1299.

Katzenberg, M. A. and H. R. Krouse. 1989. Application of stable isotope variation in human tissues to problems in identification. *Can Soc Forensic Sci J* 22(1): 7–19.

Kipnis, V., A. F. Subar, D. Midthune et al. 2003. Structure of dietary measurement error: Results of the OPEN biomarker study. *Am J Epidemiol* 158(1): 14–21; discussion 22–26.

Klimentidis, Y. C., D. J. Lemas, H. H. Wiener et al. 2014. CDKAL1 and HHEX are associated with type 2 diabetes-related traits among Yup'ik people. *J Diabetes* 6(3): 251–259.

Kraeer, K., L. S. Arneson, S. E. MacAvoy. 2014. The intraspecies relationship between tissue turnover and metabolic rate in rats. *Ecol Res* 29(5): 937–947.

Kraft, R. A., A. H. Jahren, C. D. Saudek. 2008. Clinical-scale investigation of stable isotopes in human blood: Delta13C and delta15N from 406 patients at the Johns Hopkins Medical Institutions. *Rapid Commun Mass Spectrom* 22(22): 3683–3692.

Kuhnle, G. G., A. M. Joosen, C. J. Kneale, T. C. O'Connell. 2013. Carbon and nitrogen isotopic ratios of urine and faeces as novel nutritional biomarkers of meat and fish intake. *Eur J Nutr* 52(1): 389–395.

Lee, T. N., C. L. Buck, B. M. Barnes, D. M. O'Brien. 2012. A test of alternative models for increased tissue nitrogen isotope ratios during fasting in hibernating arctic ground squirrels. *J Exp Biol* 215(19): 3354–3361.

Lemas, D. J., Y. C. Klimentidis, H. H. Wiener et al. 2013. Obesity polymorphisms identified in genome-wide association studies interact with n-3 polyunsaturated fatty acid intake and modify the genetic association with adiposity phenotypes in Yup'ik people. *Genes Nutr* 8(5): 495–505.

Lemas, D. J., H. W. Wiener, D. M. O'Brien et al. 2012. Genetic polymorphisms in carnitine palmitoyltransferase 1A gene are associated with variation in body composition and fasting lipid traits in Yup'ik Eskimos. *J Lipid Res* 53(1): 175–184.

Macko, S. A., M. H. Engel, V. Andrusevich, G. Lubec, T. C. O'Connell, R. E. Hedges. 1999. Documenting the diet in ancient human populations through stable isotope analysis of hair. *Philos Trans R Soc Lond B Biol Sci* 354(1379): 65–75; discussion 75–76.

Makhoul, Z., A. R. Kristal, R. Gulati et al. 2010. Associations of very high intakes of eicosapentaenoic and docosahexaenoic acids with biomarkers of chronic disease risk among Yup'ik Eskimos. *Am J Clin Nutr* 91(3): 777–785.

Marriott, B. P., L. Olsho, L. Hadden, P. Connor. 2010. Intake of added sugars and selected nutrients in the United States, National Health and Nutrition Examination Survey (NHANES) 2003–2006. *Crit Rev Food Sci Nutr* 50(3): 228–258.

Martínez del Rio, C., N. Wolf, S. A. Carleton, L. Z. Gannes. 2009. Isotopic ecology ten years after a call for more laboratory experiments. *Biological Reviews* 84(1): 91–111.

McCutchan, J. H., W. M. Lewis, C. Kendall, C. C. McGrath. 2003. Variation in trophic shift for stable isotope ratios of carbon, nitrogen, and sulfur. *Oikos* 102(2): 378–390.

Mekota, A. M., G. Grupe, S. Ufer, U. Cuntz. 2006. Serial analysis of stable nitrogen and carbon isotopes in hair: Monitoring starvation and recovery phases of patients suffering from anorexia nervosa. *Rapid Commun Mass Spectrom* 20(10): 1604–1610.

Minagawa, M. 1992. Reconstruction of human diet from $\sigma^{13}C$ and $\sigma^{15}N$ in contemporary Japanese hair: A stochastic method for estimating multi-source contribution by double isotopic tracers. *Appl Geochem* 7(2): 145–158.

Minagawa, M. and E. Wada. 1984. Stepwise enrichment of 15N along food chains: Further evidence and the relation between δ15N and animal age. *Geochim Cosmochim Acta* 48(5): 1135–1140.

Mook, W., J. Bommerson, W. Staverman. 1974. Carbon isotope fractionation between dissolved bicarbonate and gaseous carbon dioxide. *Earth Planet Sci Lett* 22(2): 169–176.

Nakamura, K., D. A. Schoeller, F. J. Winkler, H. L. Schmidt. 1982. Geographical variations in the carbon isotope composition of the diet and hair in contemporary man. *Biomed Mass Spectrom* 9(9): 390–394.

Nardoto, G. B., R. S. S. Murrieta, L. E. G. Prates et al. 2011. Frozen chicken for wild fish: Nutritional transition in the Brazilian Amazon region determined by carbon and nitrogen stable isotope ratios in fingernails. *Am J Hum Biol* 23(5): 642–650.

Nardoto, G. B., S. Silva, C. Kendall et al. 2006. Geographical patterns of human diet derived from stable-isotope analysis of fingernails. *Am J Phys Anthropol* 131(1): 137–146.

Nash, S. H., A. Bersamin, A. R. Kristal et al. 2012. Stable nitrogen and carbon isotope ratios indicate traditional and market food intake in an indigenous circumpolar population. *J Nutr* 142(1): 84–90.

Nash, S. H., A. R. Kristal, A. Bersamin et al. 2014a. Isotopic estimates of sugar intake are related to chronic disease risk factors but not obesity in an Alaska native (Yup'ik) study population. *Eur J Clin Nutr* 68(1): 91–96.

Nash, S. H., A. R. Kristal, A. Bersamin, S. E. Hopkins, B. B. Boyer, D. M. O'Brien. 2013. Carbon and nitrogen stable isotope ratios predict intake of sweeteners in a Yup'ik study population. *J Nutr* 143(2): 161–165.

Nash, S. H., A. R. Kristal, B. B. Boyer, I. B. King, J. S. Metzgar, D. M. O'Brien. 2009. Relation between stable isotope ratios in human red blood cells and hair: Implications for using the nitrogen isotope ratio of hair as a biomarker of eicosapentaenoic acid and docosahexaenoic acid. *Am J Clin Nutr* 90(6): 1642–1647.

Nash, S. H., A. R. Kristal, S. E. Hopkins, B. B. Boyer, D. M. O'Brien. 2014b. Stable isotope models of sugar intake using hair, red blood cells, and plasma, but not fasting plasma glucose, predict sugar intake in a Yup'ik study population. *J Nutr* 144(1): 75–80.

O'Brien, D. M. 2015. Stable isotope ratios as biomarkers of diet for health research. *Ann Rev Nutr* 35: 565.

O'Brien, D. M., K. E. Thummel, L. R. Bulkow et al. 2016. Declines in traditional marine food intake and vitamin D levels from the 1960s to present in young Alaska Native women. *Public Health Nutr* 1: 1–8.

O'Brien, D. M., M. L. Fogel, C. L. Boggs. 2002. Renewable and nonrenewable resources: Amino acid turnover and allocation to reproduction in Lepidoptera. *Proc Natl Acad Sci USA* 99(7): 4413–4418.

O'Brien, D. M., A. R. Kristal, M. A, Jeannet, M. J. Wilkinson, A. Bersamin, B. Luick. 2009. Red blood cell delta15N: A novel biomarker of dietary eicosapentaenoic acid and docosahexaenoic acid intake. *Am J Clin Nutr* 89(3): 913–919.

O'Brien, D. M., A. R. Kristal, S. H. Nash et al. 2014. A stable isotope biomarker of marine food intake captures associations between n-3 fatty acid intake and chronic disease risk in a Yup'ik study population, and detects new associations with blood pressure and adiponectin. *J Nutr* 144(5): 706–713.

O'Brien, D. M. and M. J. Wooller. 2007. Tracking human travel using stable oxygen and hydrogen isotope analyses of hair and urine. *Rapid Commun Mass Spectrom* 21(15): 2422–2430.

O'Connell, T. C., R. E. Hedges, M. Healey, A. Simpson. 2001. Isotopic comparison of hair, nail and bone: Modern analyses. *J Archaeol Sci* 28(11): 1247–1255.

O'Connell, T. C., C. J. Kneale, N. Tasevska, G. G. Kuhnle. 2012. The diet-body offset in human nitrogen isotopic values: A controlled dietary study. *Am J Phys Anthropol* 149(3): 426–434.

O'Leary, M. H. 1988. Carbon isotopes in photosynthesis. *Bioscience* 38(5): 328–336.

Ocke, M. C. and R. J. Kaaks. 1997. Biochemical markers as additional measurements in dietary validity studies: Application of the method of triads with examples from the European Prospective Investigation into Cancer and Nutrition. *Am J Clin Nutr* 65(Suppl. 4): 1240S–1245S.

Oppel, S., R. N. Federer, D. M. O'Brien, A. N. Powell, T. E. Hollmén. 2010. Effects of lipid extraction on stable isotope ratios in avian egg yolk: Is arithmetic correction a reliable alternative? *The Auk* 127(1): 72–78.

Patel, P. S., A. J. Cooper, T. C. O'Connell et al. 2014. Serum carbon and nitrogen stable isotopes as potential biomarkers of dietary intake and their relation with incident type 2 diabetes: The EPIC-Norfolk study. *Am J Clin Nutr* 100(2): 708–718.

Perriello, G., R. Jorde, N. Nurjhan et al. 1995. Estimation of glucose-alanine-lactate-glutamine cycles in post-absorptive humans: Role of skeletal muscle. *Am J Physiol* 269(3 Pt 1): E443–E450.

Petzke, K. J., H. Boeing, S. Klaus, C. C. Metges. 2005a. Carbon and nitrogen stable isotopic composition of hair protein and amino acids can be used as biomarkers for animal-derived dietary protein intake in humans. *J Nutr* 135(6): 1515–1520.

Petzke, K. J., H. Boeing, C. C. Metges. 2005b. Choice of dietary protein of vegetarians and omnivores is reflected in their hair protein 13C and 15N abundance. *Rapid Commun Mass Spectrom* 19(11): 1392–1400.

Petzke, K. J., T. Feist, W. E. Fleig, C. C. Metges. 2006. Nitrogen isotopic composition in hair protein is different in liver cirrhotic patients. *Rapid Commun Mass Spectrom* 20(19): 2973–2978.

Petzke, K. J. and S. Lemke. 2009. Hair protein and amino acid 13C and 15N abundances take more than 4 weeks to clearly prove influences of animal protein intake in young women with a habitual daily protein consumption of more than 1 g per kg body weight. *Rapid Commun Mass Spectrom* 23(16): 2411–2420.

Pikosky, M. A., P. C. Gaine, W. F. Martin et al. 2006. Aerobic exercise training increases skeletal muscle protein turnover in healthy adults at rest. *J Nutr* 136(2): 379–383.

Podlesak, D. W., G. J. Bowen, S. O'Grady, T. E. Cerling, J. R. Ehleringer. 2012. Δ2H and Δ18O of human body water: A GIS model to distinguish residents from non-residents in the contiguous USA. *Isotopes Environ Health Stud* 48(2): 259–279.

Podlesak, D. W. and S. R. McWilliams. 2006. Metabolic routing of dietary nutrients in birds: Effects of diet quality and macronutrient composition revealed using stable isotopes. *Physiol Biochem Zool* 79(3): 534–549.

Polet, Y. 2015. *EU-28 Sugar Annual Report*: U.S. Department of Agriculture.

Popp, B. N., B. S. Graham, R. J. Olson et al. 2007. Insight into the trophic ecology of yellowfin tuna, Thunnus albacares, from compound-specific nitrogen isotope analysis of proteinaceous amino acids. *Terrestrial Ecology* 1: 173–190.

Post, D. M., C. A. Layman, D. A. Arrington, G. Takimoto, J. Quattrochi, C. G. Montana. 2007. Getting to the fat of the matter: Models, methods and assumptions for dealing with lipids in stable isotope analyses. *Oecologia* 152(1): 179–189.

Richards, M. P., B. T. Fuller, M. Sponheimer, T. Robinson, L. Ayliffe. 2003. Sulfur isotopes in palaeodietary studies: A review and results from a controlled feeding experiment. *Int J Osteoarchaeol* 13(1–2): 37–45.

Richards, M. P., R. Jacobi, J. Cook, P. B. Pettitt, C. B. Stringer. 2005. Isotope evidence for the intensive use of marine foods by Late Upper Palaeolithic humans. *J Hum Evol* 49(3): 390–394.

Santamaria-Fernandez, R., J. Giner Martinez-Sierra, J. M. Marchante-Gayon, J. I. Garcia-Alonso, R. Hearn. 2009. Measurement of longitudinal sulfur isotopic variations by laser ablation MC-ICP-MS in single human hair strands. *Anal Bioanal Chem* 394(1): 225–233.

Sauer, P. E., A. Schimmelmann, A. L. Sessions, K. Topalov. 2009. Simplified batch equilibration for D/H determination of non-exchangeable hydrogen in solid organic material. *Rapid Commun Mass Spectrom* 23(7): 949–956.

Schmidt, O., J. Quilter, B. Bahar et al. 2005. Inferring the origin and dietary history of beef from C, N and S stable isotope ratio analysis. *Food Chem* 91(3): 545–549.

Schoeller, D. A. 1999. Isotope fractionation: Why aren't we what we eat? *J Archaeol Sci* 26(6): 667–673.

Schoeller, D. A., P. D. Klein, J. B. Watkins, T. Heim, W. C. MacLean, Jr. 1980. 13C abundances of nutrients and the effect of variations in 13C isotopic abundances of test meals formulated for 13CO2 breath tests. *Am J Clin Nutr* 33(11): 2375–2385.

Schoeller, D. A., M. Minagawa, R. Slater, I. Kaplan. 1986. Stable isotopes of carbon, nitrogen and hydrogen in the contemporary North American human food web. *Ecol Food Nutr* 18(3): 159–170.

Schoeninger, M. J. and K. Moore. 1992. Bone stable isotope studies in archaeology. *J World Prehist* 6(2): 247–296.

Schwarcz, H. P. 1991. Some theoretical aspects of isotope paleodiet studies. *J Archaeol Sci* 18(3): 261–275.

Sharp, Z. D., V. Atudorei, H. O. Panarello, J. Fernández, C. Douthitt. 2003. Hydrogen isotope systematics of hair: Archeological and forensic applications. *J Archaeol Sci* 30(12): 1709–1716.

Silfer, J., M. Engel, S. Macko, E. Jumeau. 1991. Stable carbon isotope analysis of amino acid enantiomers by conventional isotope ratio mass spectrometry and combined gas chromatography/isotope ratio mass spectrometry. *Anal Chem* 63(4): 370–374.

Steele, K. and R. M. Daniel. 1978. Fractionation of nitrogen isotopes by animals: A further complication to the use of variations in the natural abundance of 15 N for tracer studies. *J Agric Sci* 90(01): 7–9.

Subar, A. F., V. Kipnis, R. P. Troiano et al. 2003. Using intake biomarkers to evaluate the extent of dietary misreporting in a large sample of adults: The OPEN study. *Am J Epidemiol* 158(1): 1–13.

Sweeting, C. J., N. V. Polunin, S. Jennings. 2006. Effects of chemical lipid extraction and arithmetic lipid correction on stable isotope ratios of fish tissues. *Rapid Commun Mass Spectrom* 20(4): 595–601.

Takano, Y., Y. Kashiyama, N. O. Ogawa, Y. Chikaraishi, N. Ohkouchi. 2010. Isolation and desalting with cation-exchange chromatography for compound-specific nitrogen isotope analysis of amino acids: Application to biogeochemical samples. *Rapid Commun Mass Spectrom* 24(16): 2317–2323.

Tasevska, N., D. Midthune, L. T. Tinker et al. 2014. Use of a urinary sugars biomarker to assess measurement error in self-reported sugars intake in the nutrition and physical activity assessment study (NPAAS). *Cancer Epidemiol Biomarkers Prev* 23(12): 2874–2883.

Tasevska, N., S. A. Runswick, A. McTaggart, S. A. Bingham. 2005. Urinary sucrose and fructose as biomarkers for sugar consumption. *Cancer Epidemiol Biomarkers Prev* 14(5): 1287–1294.

Thompson, A. H., L. A. Chesson, D. W. Podlesak, G. J. Bowen, T. E. Cerling, J. R. Ehleringer. 2010. Stable isotope analysis of modern human hair collected from Asia (China, India, Mongolia, and Pakistan). *Am J Phys Anthropol* 141(3): 440–451.

U.S. Environmental Protection Agency. (2014). *Estimated Fish Consumption Rates for the US Population and Selected Subpopulations (NHANES 2003–2010)*. Washington, D.C.

Vaughan, L. K., H. W. Wiener, S. Aslibekyan et al. 2015. Linkage and association analysis of obesity traits reveals novel loci and interactions with dietary n-3 fatty acids in an Alaska Native (Yup'ik) population. *Metabolism* 64(6): 689–697.

Waterhouse, C. and J. Keilson. 1978. The contribution of glucose to alanine metabolism in man. *J Lab Clin Med* 92(5): 803–812.

Whigham, L. D., D. E. Butz, L. K. Johnson et al. 2014. Breath carbon stable isotope ratios identify changes in energy balance and substrate utilization in humans. *Int J Obes (Lond)* 38(9): 1248–1250.

Wilkinson, M. J., Y. Yai, D. M. O'Brien. 2007. Age-related variation in red blood cell stable isotope ratios (delta13C and delta15N) from two Yupik villages in southwest Alaska: A pilot study. *Int J Circumpolar Health* 66(1): 31–41.

Yaemsiri, S., N. Hou, M. M. Slining, K. He. 2010. Growth rate of human fingernails and toenails in healthy American young adults. *J Eur Acad Dermatol Venereol* 24(4): 420–423.

Yeung, E. H., C. D. Saudek, A. H. Jahren et al. 2010. Evaluation of a novel isotope biomarker for dietary consumption of sweets. *Am J Epidemiol* 172(9): 1045–1052.

15 The Food Metabolome and Dietary Biomarkers

Augustin Scalbert, Joseph A. Rothwell, Pekka Keski-Rahkonen, and Vanessa Neveu

CONTENTS

15.1 INTRODUCTION

Diet is by nature extremely complex. It is composed of foods originating from a wide diversity of plant and animal sources, each having its own chemical composition. In addition, foods are often processed and cooked, resulting in the degradation of some food constituents and in the formation of others. They may also contain additives and contaminants. All food constituents may exert some effects on the human organism. Macronutrients and some micronutrients are essential for life and must be found in sufficient amounts in the diet to fulfill the key functions in the organism. A wide variety of other compounds although not essential, may also exert biological effects on the organism. These effects can be beneficial or detrimental and limit or increase the risk of various chronic diseases.

Great efforts have been made to assess the complex dietary exposure of individuals and to characterize its interactions with health and diseases in nutritional epidemiological studies. This assessment has been largely based on the use of questionnaires (food frequency questionnaires [FFQ] or dietary recalls) to describe habitual or acute intake of hundreds or thousands of food items (Chapter 1). However, a number of biases and errors, widely studied over the last 20 years and amply described in this volume, limit their accuracy (Chapters 2 and 10). In addition, food composition databases, needed to assess individual exposures to dietary compounds, are often missing or insufficient for nonnutrient compounds, food additives, or food contaminants, making the determination of intake for such compounds difficult or impossible (Chapter 19). Dietary supplements may also significantly contribute to the intake of some nutrients or food bioactives and their intake is not easily estimated as this information has not been systematically collected in cohort studies. These limitations may prevent us from detecting associations with disease outcomes.

In this chapter, we first give an overview of current knowledge on dietary biomarkers, used as indicators of dietary exposures in place or as a complement to dietary questionnaires. We then discuss critical parameters to be considered for validation of dietary biomarkers, with a particular focus on biomarker reproducibility over time. We describe new opportunities to measure dietary biomarkers in alternative biospecimens such as hair, nails, and teeth, and finally review recent advances in the application of metabolomics to identify novel dietary biomarkers.

15.2 BACKGROUND

Dietary biomarkers (also called nutritional biomarkers) are (bio)chemical indicators of dietary exposures measured in blood, urine, and other biospecimens through various analytical methods. They are most often food compounds absorbed through the gut barrier, eventually transformed into some metabolites by the gut microbiota and tissular enzymes. They can also be biomarkers of early biological effects, directly linked to the exposure of food constituents. These last biomarkers are often indicators of the status of essential nutrients (e.g., enzymes that require vitamins or minerals as cofactors) or indicators of toxicity (e.g., enzymes released in blood upon tissue damage by high exposure to alcohol).

Two main types of dietary biomarkers with different applications can be distinguished (Jenab et al. 2009): (1) *recovery biomarkers* provide an estimate of absolute intake of a nutrient. These are mainly limited to urinary nitrogen and potassium as biomarkers of protein and potassium intake, respectively. Together with doubly labeled water (Chapter 10), they have been used to validate or calibrate dietary assessment tools based on questionnaires. (2) Most known dietary biomarkers are *concentration biomarkers*. They differ from recovery biomarkers as only a (small) fraction of their food precursors is recovered in biospecimens (most often blood or urine) where measured. They have been mainly used to monitor dietary exposures and to compare and classify subjects according to dietary exposures in population-based studies. Concentration biomarkers are the focus of this chapter.

Biomarkers have increasingly been used in nutritional epidemiology over the last 20 years and their use parallels the development of molecular epidemiology (Chapters 13 and 18). In contrast to dietary questionnaires, dietary biomarkers are objective measurements of dietary exposures which do not rely on self-reporting. They may help detecting associations of dietary factors with disease risk that may have been overlooked when measuring intake of foods or food constituents with questionnaires. Some examples clearly show the potential benefits of using biomarkers to assess dietary exposures. Measurement of sucrose in urine was found to be associated with an increased risk of obesity in a cohort of obese and lean British subjects whereas no association could be observed with sugar intake estimated with a FFQ because of some underreporting of sugar intake (Bingham et al. 2007). Blood concentrations of carotenoids were more strongly associated with reduced breast cancer risk than carotenoid intake measured with FFQs (Aune et al. 2012). This is most likely explained by errors made in estimating carotenoid intake, due the variability of carotenoid contents in foods, their wide distribution in a large variety of foods, and errors made in the estimation of food intake. The estimation of intake of food contaminants can also be particularly difficult due to the large variability of their content in foods. In a study on aflatoxins and hepatocellular carcinoma risk, aflatoxin B1 contents were measured in 2,000 foods to calculate aflatoxin intake, and no association was observed between intake and disease risk in 18,244 men from Shanghai, whereas a strong association could be found with aflatoxin B1 measured in urine (Qian et al. 1994).

Over one hundred dietary biomarkers have been measured so far in various populations (Neveu et al. 2017). However, many more biomarkers remain to be identified, considering the 27,000 chemical constituents already known in various foods (http://foodb.ca/). Many of these compounds are absorbed from the gut and found in human biofluids and other biospecimens where they constitute what has been called the food metabolome (Figure 15.1) (Scalbert et al. 2014). Modern analytical

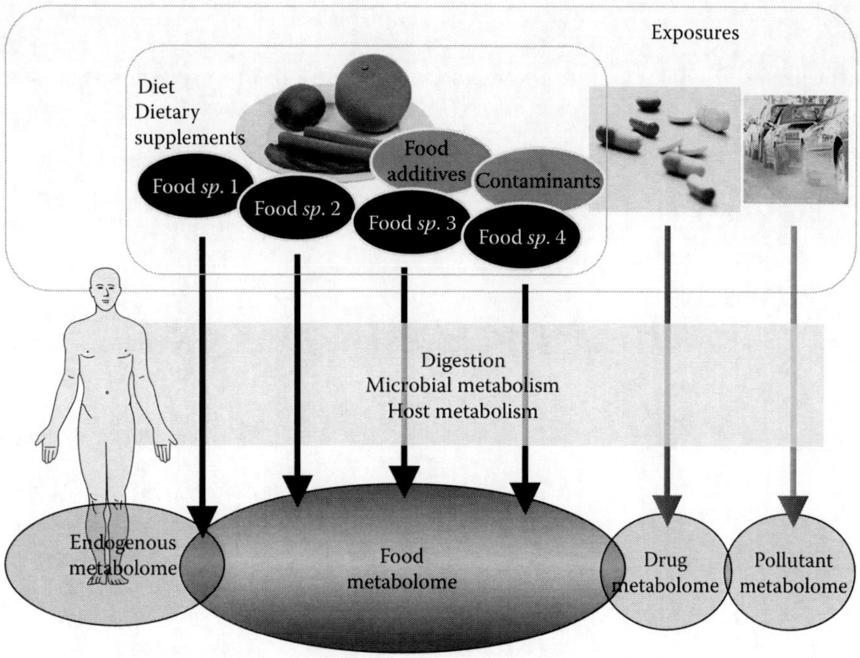

FIGURE 15.1 The food metabolome is the sum of all metabolites that are derived from the digestion of the various foods constituting the diet of an individual. Components of the food metabolome include natural food constituents, as well as food additives and contaminants. Food derivatives are often transformed in the host tissues and by the resident microbiota. (Reproduced from Scalbert, A. et al., *Am. J. Clin. Nutr.*, 99, 1286–1308, 2014.)

instruments like mass spectrometers allow today the measurement of a large fraction of the food metabolome. Hundreds of food-derived compounds can be detected in a single analytical run, and this rich information can be mined to identify dietary markers for more comprehensive and accurate characterization of dietary exposures (Chapter 16).

15.3 DIETARY BIOMARKERS: AN OVERVIEW

Biomarkers of intake for dietary macronutrients and foods have been recently discussed based on the content of a selection of 33 articles (Hedrick et al. 2012). They have been used to assess intake of sugars, lipids, proteins, and other food components as well as of foods and beverages such as olive oil, citrus fruits, cocoa, wholegrain cereals, and wine. Recently, information on dietary biomarkers measured in the general population has been collected from 220 peer-reviewed publications and compiled in Exposome-Explorer, a new database on biomarkers of exposure to environmental factors (http://exposome-explorer.iarc.fr/) (Neveu et al. 2017). This online database contains detailed information on biomarkers: type of samples where measured, analytical methods used for their measurement, populations where measured, concentrations in biospecimens, correlations with other methods of exposure assessment, and reproducibility over time. In total, data on over 160 dietary biomarkers have so far been collected.

Dietary biomarkers measured in various population-based studies belong to eight main classes: vitamins, amino acids, carotenoids, fatty acids, phytosterols, polyphenols, alkaloids, and dietary contaminants such as aflatoxins or compounds formed during cooking such as acrylamide or heterocyclic amines (Table 15.1). Most of these biomarkers have been measured in blood or urine (122 and 58 biomarkers measured in blood and urine respectively in the Exposome-Explorer database).

TABLE 15.1
The Main Classes of Dietary Biomarkers as Documented in the Exposome-Explorer Database

Class	Number of Biomarkers[a]	Biospecimens	Examples of Biomarkers
Vitamins	26	Blood and urine	25-Hydroxycholecalciferol, pyridoxal 5′-phosphate, alpha-tocopherol, ascorbic acid, biotin, folates, nicotinamide, pantothenic acid, retinol, riboflavin, thiamin
Amino acids	5	Blood and urine	1-Methylhistidine, 3-methylhistidine
Carotenoids	16	Blood and adipose tissue	α-Carotene, β-carotene, lycopene, β-cryptoxanthin, lutein, zeaxanthin
Fatty acids	56	Blood and adipose tissue	EPA, DHA, linoleic acid, palmitic acid, arachidonic acid
Phytosterols	2	Blood	β-Sistosterol, campesterol
Polyphenols	48	Blood and urine	Daidzein, genistein, enterolactone, gallic acid, naringenin, hesperetin, 5-*n*-heneicosylresorcinol, quercetin, 4-*O*-methylgallic acid
Purine alkaloids	2	Blood	Caffeine, paraxanthine
Dietary contaminants	11	Blood and urine	Acrylamide, 2-amino-1-methyl-6-phenylimidazo[4,5-b]pyridine, aflatoxins

[a] Number of biomarkers curated in the Exposome-Explorer database.

The selection of blood or urine for biomarker measurements depends on the availability of biospecimens in cohort studies, the concentration of biomarkers, and their reproducibility over time in each type of samples. Plasma/serum samples are more commonly collected in large cohort studies than urine. Respective concentrations in blood or urine vary with the nature of the biomarkers. Some biomarkers largely excreted in urine like polyphenols, show much higher concentrations (50–500-fold) in urine when compared with plasma or serum (Achaintre, Rinaldi and Scalbert, unpublished results) due to water reabsorption in the kidney. In contrast, lipophilic compounds are mainly found in blood, a matrix rich in lipids. This explains why lipophilic markers such as carotenoids, fatty acids, and tocopherols have been measured in blood only, whereas other more hydrophilic biomarkers like polyphenols have been measured in both urine and blood (Table 15.1).

A fraction of the dietary biomarkers measured in blood or urine were found to correlate with the consumption of specific foods and have eventually been used as biomarkers of food intake. These correlation values have been systematically extracted from the scientific literature and compiled in the Exposome-Explorer database (Neveu et al. 2017). High correlation values suggest that the corresponding biomarkers could potentially be used as indicators of food intake (Hedrick et al. 2012). About 8000 correlation values could be collected. Biomarkers whose measured concentrations best correlated with either habitual or acute food intake (measured respectively with FFQs or dietary recalls) are described in Table 15.2.

For meat, both 1-methylhistidine and fatty acids with odd numbers of carbons formed by the rumen microbiota have been used as biomarkers of intake. Phytanic acid formed from chlorophyll in the rumen of ruminants and the same fatty acids with odd number of carbons have also been used as biomarkers of milk and dairy food intake. For fish, *n*-3 fatty acids have mainly been used

TABLE 15.2
Blood and Urinary Biomarkers Reported to Correlate ($r > 0.3$) with either Acute or Habitual Food Intake, According to the Exposome-Explorer Database

Food	Blood	Urine
Animal Foods		
Ruminant meat	Pentadecylic acid*	
Red meat		1-Methylhistidine*
Poultry		1-Methylhistidine*
Fish	n-3-PUFA*, EPA*, DHA*, DPA*, margaric acid*, 1-methylhistidine*	
Dairy Foods		
Milk	Phytanic acid*	Iodine*
Butter, cheese	Phytanic acid*, margaric acid*, pentadecylic acid*	
Cereals		
Wholegrain cereals	Alkylresorcinols (C25:0, C23:0, C21:0, C19:0, C17:0)	3-(3,5-Dihydroxyphenyl)-1-propanoic acid*, 3,5-dihydroxybenzoic acid*
Fruits and Vegetables		
Fruits and vegetables (total)	Vitamin C*, α-carotene*, β-carotene*, β-cryptoxanthin*, lutein*, lycopene*, phytoene*, zeaxanthin*, retinol*	Apigenin*, eriodictyol*, phloretin*, hesperetin*, naringenin*, hippuric acid
Fruits (total)	Vitamin C*, α-carotene*, β-carotene*, β-cryptoxanthin*, lutein*, lycopene*, zeaxanthin*, retinol*	Phloretin*, hesperetin*, naringenin*, gallic acid, 4-*O*-methylgallic acid, isorhamnetin, kaempferol
Citrus fruits	Vitamin C*, β-carotene*, β-cryptoxanthin*, zeaxanthin	Hesperetin*, naringenin
Apple		Phloretin, *m*-coumaric acid, isorhamnetin, kaempferol
Grapefruit		Naringenin
Orange		Hesperetin, caffeic acid
Papaya	β-Cryptoxanthin*	
Melon	Vitamin C*, β-carotene*, retinol*	
Vegetables (total)	Vitamin C*, α-carotene*, β-carotene*, β-cryptoxanthin*, lycopene*, lutein*, retinol*	Enterolactone
Carrot	α-Carotene*	
Tomato	Lycopene*, lutein*	
Beans		
Soyfood	Genistein*, daidzein*	Genistein, daidzein, *O*-desmethylangolensin
Beverages		
Wine		Resveratrol metabolites*, caffeic acid, gallic acid, 4-*O*-methylgallic acid
Coffee		Chlorogenic acid, dihydroferulic acid sulfate*, atractyligenin*, cyclo(isoleucyl-prolyl)*, trigonelline*
Tea		Kaempferol*, gallic acid, 4-*O*-methylgallic acid*

Note: Asterisks indicate at least one documented correlation value with habitual food intake estimated by food frequency questionnaire.

as biomarkers of intake. Alkylresorcinols found in bran of wholegrain cereals and their metabolites (3-(3,5-dihydroxyphenyl)-1-propanoic acid, 3,5-dihydroxybenzoic acid) formed in the colon by the human microbiota have been used to assess intake of wholegrain cereal products. Several nutrients and other constituents commonly found in fruits and vegetables, mainly vitamin C and carotenoids (Van den Berg et al. 2000, Holden et al. 1999), have been used as generic biomarkers for fruit and/or vegetable intake (Baldrick et al. 2011). Some food constituents can be specific for particular fruits or vegetables. α-Carotene is particularly abundant in carrot and lycopene in tomato (Van den Berg et al. 2000) and both compounds were used as biomarkers of intake for each of these vegetables (Brantsaeter et al. 2007, Al-Delaimy et al. 2005). Hesperetin and naringenin are two flavanones abundant in orange and grapefruit, respectively and have been used as biomarkers for these two fruits. Phloretin is a dihydrochalcone abundant in apple and has been described by several authors as a possible biomarker of apple intake.

Several biomarkers of beverage intake are also known. Gallic acid and its *O*-methylated product formed in the liver were proposed as biomarkers for wine intake. However gallic acid is also present in tea (Perez-Jimenez et al. 2010b) and these two biomarkers may lack specificity in a population consuming both beverages. Gallic acid ethyl ester was proposed more recently as a biomarker of red wine intake and was found to be a better predictor of intake in the European Prospective Investigation into Cancer and Nutrition (EPIC) cohort than the nonesterified gallic acid or its *O*-methylated metabolite (Edmands et al. 2015; Zamora-Ros et al. 2016). Gallic acid ethyl ester is formed in wine by transesterification of other galloylated proanthocyanidins with ethanol (Viriot et al. 1993).

Several compounds were proposed as biomarkers for coffee intake: chlorogenic acid, dihydroferulic acid sulfate (a metabolite of chlorogenic acid), atractyligenin, cyclo(isoleucyl-prolyl), and trigonelline (Edmands et al. 2015; Zamora-Ros et al. 2016; Rothwell et al. 2014). Gallic acid and its 4-*O*-methylated metabolite, (−)-epicatechin and kaempferol have been found to be correlated with tea intake and could be used as biomarkers (Edmands et al. 2015; Zamora-Ros et al. 2016). Several catechins including (−)-epicatechin, (−)-epicatechin gallate, (−)-epigallocatechin, (−)-epigallocatechin gallate and 4′-*O*-methyl-(−)-epigallocatechin measured in urine or plasma have been used as surrogates for tea intake in several cohort studies (Iwasaki et al. 2010; Luo et al. 2010; Sasazuki et al. 2008; Yuan et al. 2007).

Biomarkers of food intake are not limited to natural food constituents. Food additives such as iodine in milk and dairy products may also serve as indicators of intake for specific foods when regulations have been in place for mandatory fortification of these foods (Brantsaeter et al. 2009).

Dietary biomarkers discussed earlier may not have been fully validated. We examine below key parameters that need to be considered for biomarker validation: specificity, dose–response relationship, and reproducibility over time for biomarkers of habitual food intake.

15.4 SPECIFICITY OF DIETARY BIOMARKERS

Correlation of biomarkers with intake of a particular food or food group does not establish their specificity for these foods. Some of these biomarkers may not be specific enough if other significant dietary sources of the same metabolite are also consumed in the population studied. For example, caffeic acid concentrations in blood or urine have been reported to correlate with citrus fruit and wine intake (Table 15.2), but this association would be confounded by intake of coffee, a major source of caffeic acid (Edmands et al. 2015).

Evaluation of the specificity of a particular biomarker requires knowledge of the chemical composition of foods and the metabolism of food compounds. The occurrence of precursors to a given biomarker can be very specific to a particular food, such as alkylresorcinols in bran and wholegrain cereal products, gallic acid ethyl ester in wine, and phloretin in apple. Their exclusive occurrence in the food or food group of interest can be established with the aid of comprehensive food composition databases such as Phenol-Explorer for polyphenols (Neveu et al. 2010). However, for many food compounds, their distribution in foods is often not known in sufficient detail, and the existence of confounders cannot be ruled out. One way to identify possible confounders is to study correlations between the biomarker and foods or food groups consumed in the population (Edmands et al. 2015; Zamora-Ros et al. 2016). Another more labor intensive approach would be to test all possible confounders in a dietary intervention study (Zhang et al. 1999).

A single biomarker may not be specific enough for a given food if it also originates from other foods or if it is also produced endogenously. In this case, measurement of a combination of biomarkers may improve specificity. For example, the ratio of two alkylresorcinols (5-*n*-heptadecylresorcinol/5-*n*-Heneicosylresorcinol) was found to vary between cereal species. When measured in plasma samples, this ratio was found to discriminate consumers of wholegrain rye and wholegrain wheat (Andersson et al. 2011). It was also observed that profiles of 34 polyphenols measured in urine were better predictors of intake of polyphenol-containing foods such as citrus fruits, apple and pear, olives, coffee, tea, and wine, than any single biomarker considered individually (Noh et al. 2016).

15.5 DOSE–RESPONSE RELATIONSHIP

The dose–response relationship is explored in controlled intervention studies where various doses of foods or dietary compounds are consumed. A systematic analysis of these relationships has been conducted for 40 dietary polyphenols as described in 162 intervention studies (Perez-Jimenez et al. 2010a). Linear relationships between the level of intake and urinary excretion were observed for most polyphenols, but the magnitude of the correlations was found to vary between polyphenols. The highest correlations were observed for hydroxytyrosol and some isoflavones and lignans. Most often no saturation of urinary recovery was observed. In contrast, other food intake markers reach a plateau with increasing intake. A classic example is ascorbic acid for which a plateau in plasma concentration is reached when the dose exceeds 200–1000 mg/day due to transporter saturation (Levine et al. 1996). Intervention studies also inform about biomarker sensitivity, and the lowest dose above which an increase of the biomarker level can be observed.

15.6 BIOMARKER REPRODUCIBILITY AND ELIMINATION HALF-LIFE

In most large cohort studies, biospecimens are collected at baseline only when a new participant is recruited. To classify individuals according to their diet, dietary biomarkers should be reproducible over time and ideally show a constant value characteristic of each subject. Reproducibility of a biomarker will mainly depend on frequency of intake and its elimination half-life. A dietary biomarker showing rapid elimination after intake (i.e., short half-life) will show large fluctuations over time in a particular individual, and the measured concentrations in biospecimens will depend on the time of sampling after intake. Concentrations in plasma or serum will be high in the postprandial phase and lower in samples collected in fasting conditions.

Half-lives of dietary biomarkers vary widely. In plasma, serum, or urine, they may be as short as 1–10 hours for polyphenols and hydroxylated polyaromatic hydrocarbons (Manach et al. 2005; Li et al. 2012) and as long as 5–10 years for some persistent chlorinated contaminants like dioxins (FleschJanys et al. 1996). The half-lives of carotenoids vary from a few days to 60 days depending on their chemical structure (Rock et al. 1992). In general, the more lipophilic the biomarker, the longer is its elimination half-life (Sarver et al. 1997). This is explained by the accumulation of lipophilic compounds in the body adipose tissues and their slow release in blood over weeks or years. Lipophilic dietary biomarkers such as carotenoids, fatty acids, and tocopherols are usually measured in blood, a matrix rich in lipids. Their concentrations in blood slowly increase over days or years following initiation of intake, reach a stable value hardly influenced by short-term variations of intake, and slowly decrease over days or years following cessation of intake (Prince and Frisoli 1993; Katan et al. 1997; Nost et al. 2013).

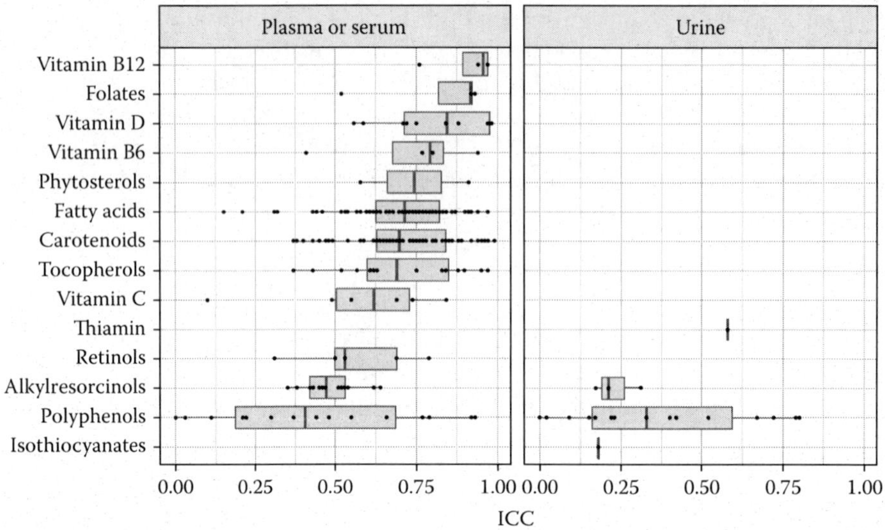

FIGURE 15.2 Intraclass correlation coefficient (ICC) for dietary biomarkers as included in the Exposome-Explorer database. Box plots include data for 303 ICC values for 40 dietary biomarkers.

Biomarker reproducibility over time is commonly measured on repeated samples collected in a selection of individuals typically at a time interval of a few weeks or months. It is expressed as the intraclass correlation coefficient (ICC), which is defined as the ratio or between-subject variance to the sum of within- and between-subject variance. ICC values range from 0 to 1, with highest value corresponding to the most reproducible biomarkers. ICC values for 40 dietary biomarkers are included in the Exposome-Explorer database (Neveu et al. 2017). The highest ICC values are observed for some vitamins such as vitamin B12 and vitamin D (Figure 15.2). High values are also observed for several classes of lipophilic biomarkers measured in plasma (fatty acids, carotenoids, tocopherols). In contrast, ICC values for the more hydrophilic polyphenols measured in plasma/serum or urine are lower. Large variations in ICC values are also observed between different polyphenols (http://exposome-explorer.iarc.fr). Enterolactone and alkylresorcinols show relatively high ICC values ranging from 0.4 to 0.8 whereas soy isoflavones showed low reproducibility when estimated in American women.

15.7 BIOMARKERS IN SPECIFIC BLOOD COMPARTMENTS

Any factor that may limit clearance rate of a diet-related compound in the kidney or liver may increase its half-life in blood and potential value as dietary biomarker. Two main mechanisms have been considered: (1) the integration of dietary fatty acids in blood cell membranes and (2) the formation of adducts between electrophilic dietary compounds and DNA and proteins.

Dietary fatty acids absorbed during digestion are integrated into different types of lipids in the liver (Hodson et al. 2008). They associate with lipoproteins and are released into the bloodstream. Phospholipids in particular contribute significantly to the composition of membranes and in particular those of erythrocytes. The supplementation of volunteers' diets with fish oil caused a doubling of mean eicosapentaenoic acid concentrations in erythrocyte membranes after three days (Katan et al. 1997). It was concluded that the fatty acid composition in erythrocytes or blood phospholipids reflects dietary intake from the preceding few weeks (Hodson et al. 2008). It was also shown in a cross-sectional study that docosahexaenoic acid (DHA) concentrations in erythrocytes was more strongly correlated with DHA intake than plasma DHA concentrations (Sun et al. 2007). This is why fatty acids are often measured in plasma phospholipids or red blood cell membranes.

Electrophilic food compounds can form adducts with blood proteins such as albumin and hemoglobin (Rappaport et al. 2011). The large molecular weight of the adducts prevents them from being excreted into urine or bile. These adducts are generally formed in blood by the reaction of an electrophilic compound with nucleophilic groups of albumin or hemoglobin. A stable covalent bond is formed between the food compound and the protein. The half-life of these adducts is directly linked to the protein half-life, 20 and 50 days for albumin and hemoglobin respectively. However the formation of hemoglobin adducts may be hampered by the inability of the electrophilic compound to cross the red blood cell membrane (van Welie et al. 1992).

Few diet-related protein adducts have so far been identified. They are contaminants or chemicals formed during food processing, main albumin adducts with aflatoxin B_1 (Wu et al. 2007), and hemoglobin adducts with acrylamide (Vesper et al. 2008, Wilson et al. 2009).

Environmental electrophiles also react in a similar way with nucleophilic groups in DNA forming a variety of DNA adducts (Balbo et al. 2014). These adducts have been widely studied for their mutagenic and carcinogenic properties and as biomarkers of exposure to carcinogenic agents. They may persist in human tissues for years. Aristolochic acid adducts were found in renal tissues 20 years after exposure had ceased (Schmeiser et al. 2014). Leukocyte DNA adducts have also been used as surrogates of tissular genotoxic adducts (Gunter et al. 2007). DNA adducts can be measured in whole blood. Rat experiments showed that their levels reach maximum within two days after exposure to genotoxic agents. They are submitted to constant DNA repair. However, for some agents, they may still be detectable in blood 20 days after exposure (Bieler et al. 2007).

DNA adducts formed with polycyclic aromatic hydrocarbons (PAHs) in white blood cells were shown to directly reflect exposure to smoking and airborne PAHs (Godschalk et al. 2003; Jedrychowski et al. 2013). A number of electrophiles present in the diet are also known to form specific adducts with DNA (Hemeryck and Vanhaecke 2016; Hossain et al. 2013; van der Woude et al. 2006). They can be food natural constituents (e.g., safrole, estragole, and methyleugenol present in several spices and herbs, polyphenols such as quercetin or gallic acid present in various foods), food contaminants (mycotoxins), and toxins formed during food heating (acrylamide, furans, PAHs and heterocyclic amines). Their value as dietary biomarkers has never been explored and warrant further research.

15.8 BIOMARKERS IN HAIR, NAILS AND TEETH

Although still rarely explored, the measurement of biomarkers in alternative biospecimens such as hair, nails, and teeth remaining in our body for a long period could clearly offers new opportunities for dietary exposure assessment. Hair is a promising biospecimen for biomarker measurement, since metabolites of both endogenous and xenobiotic origin are incorporated into it during its relatively slow growth. Hair is already used in forensics for the detection of drug abuse, and its analysis for the monitoring of exposure to organic pollutants has already been described by a number of studies (Appenzeller and Tsatsakis 2012). A limited volume of literature describes the measurement of compounds of dietary origin in hair. One such compound is 2-amino-1-methyl-6-phenylimidazo[4,5-b]pyridine (PhIP), a carcinogenic heterocyclic aromatic amine produced during the high temperature cooking of meat. PhIP, as measured in the hair of 20 subjects, was highly correlated with the intake of grilled or stir-fried meat (Kobayashi et al. 2005) over a 28-day period as assessed by weighed dietary records. This finding was later reproduced in a controlled intervention study in which cooked meat was fed to volunteers, and hair samples were analyzed for PhIP (Turesky et al. 2013). Another important food group to which exposure has been assessed using hair samples is alcoholic beverages. Studies have so far focused on the measurement of ethyl glucuronide, a metabolite of ethanol, in hair. Ethyl glucuronide was the most promising of nine potential biomarkers of alcohol intake extracted from hair samples (Pirro et al. 2011b). The compounds were measured in the hair of subjects previously categorized as nondrinkers, social drinkers, or heavy drinkers. Hair ethyl glucuronide was a specific and sensitive discriminant of heavy drinking, allowing the detection of heavy drinkers not detected by measurement of rapidly changing blood biomarkers. Hair ethyl glucuronide was also found to reflect alcohol intake in alcohol abuse patients (Appenzeller et al. 2007). Using different hair segments collected in a same subject, could be observed across several month, and indications of times of drinking cessation could be collected.

Hair can therefore be used as an indicator of long-term exposure to exogenous compounds of interest, and potentially of long-term exposures to foods or beverages such as meat and alcohol. Hair analysis does present some new technical challenges, such as the separation of metabolites accumulated from the diet with compounds derived from airborne exposure (Turesky and Le Marchand 2011), normalization of concentrations for different hair types (Turesky et al. 2013), and the lack of comparison of metabolite composition in head hear and in hair from other parts of the body (Pirro et al. 2011a). Although few dietary compounds have so far been measured in hair, recent analytical work suggests that a hair metabolome will soon be exploitable for the discovery of many exposure biomarkers (Sulek et al. 2014).

Nails, in the same way as hair, may accumulate derivatives of external exposure during slow growth. In particular exposure to trace minerals has been measured in nails (He 2011). However, exposure to organic compounds has seldom been evaluated using nails. Polyamines, whose biosynthesis and accumulation may be altered by tumorigenesis, have successfully been measured in nails (Min et al. 2011). Although at present no biomarkers of food intake have been proposed in nails, such compounds may also be diet-derived and therefore show potential as intake biomarkers.

Teeth are an even slower growing, longer term biospecimen for the potential analysis of dietary and environmental exposures during early life and childhood. Given that there is currently much interest in the effect of early life exposures upon health and disease status in later life, exposure profiles derived from teeth could be valuable in epidemiology. Tooth analysis may constitute a retrospective exposome for prenatal and early childhood environmental and dietary exposures (Andra et al. 2015). One study demonstrated that distributions of barium in teeth reflects weaning, the infant's transition from mother's milk to solid food (Austin et al. 2013). The analysis of teeth for assessment of environmental exposures is reviewed in detail by Andra et al. (2015). The main caveat of teeth as a practical biospecimen is the difficulty of sample collection.

15.9 METABOLOMIC STUDIES FOR DIETARY BIOMARKER DISCOVERY

Most dietary biomarkers discussed earlier have been identified through hypothesis-driven approaches, based on knowledge of food composition and on metabolism of food compounds. These biomarkers have usually been tested in small intervention studies and assayed in various populations. Advances in analytical technologies and metabolomics have made possible the implementation of data-driven approaches to discover novel dietary biomarkers. In such studies, statistical approaches are used to compare metabolic profiles of consumers and nonconsumers of a particular food, and signals whose intensities are elevated in consumers are identified. Workflows used in metabolomics, from study design to chemical and statistical analyses are described in detail in Chapter 16. Here, we describe the main results of metabolomic studies aiming to discover novel dietary biomarkers in the last 10 years. About 35 studies have searched for biomarkers of intake for fruits, vegetables, cereals, beverages, cocoa, nuts, and dietary fibers (Table 15.3).

The earliest studies used nuclear magnetic resonance (NMR) spectroscopy and searched for novel metabolites formed from compounds in tea and chamomile tea (Wang et al. 2005; Daykin et al. 2005; Van Dorsten et al. 2006). A main limitation of NMR spectroscopy for data acquisition is its lack of sensitivity. As a consequence, a limited number of metabolites are usually detected, and those detected often lack specificity. Most studies on the food metabolome have been carried out since 2010 using high-resolution mass spectrometry techniques (Chapter 16).

TABLE 15.3
Metabolomic Studies on Dietary Biomarker Discovery

Dietary Factor	Study Type	Dietary Assessment Tool	Biospecimen	Analytical Technique	Biomarkers	References
Fruits						
Citrus fruits	Free-living subjects	24HR	Urine	NMR	Proline betaine	Heinzmann et al. (2010)
Citrus fruits	Free-living subjects	FFQ	Urine	HRMS	Proline betaine, 4-hydroxyproline betaine	Lloyd et al. (2011a)
Citrus fruits	Single meal intervention	N/A	Urine	HRMS	Proline betaine, hydroxyproline betaine, limonene 8,9-diol glucuronide, nootkatone 13,14-diol glucuronide, hesperetin 3′-*O*-glucuronide, naringenin 7-*O*-glucuronide, *N*-methyltyramine sulfate	Pujos-Guillot et al. (2013)
Citrus fruits	Free-living subjects	24HR	Urine	HRMS	Proline betaine, hesperetin-3-glucuronide	Andersen et al. (2014)
Citrus fruits	Free-living subjects	24HR	Urine	HRMS	Naringenin glucuronide	Edmands et al. (2015)
Raspberries	Short/medium term intervention	N/A	Urine	HRMS	Caffeic acid sulfate, methylepicatechin sulfate	Lloyd, Favé et al. (2011c)
Strawberries	Free-living subjects	24HR	Urine	HRMS	2,5-Dimethyl-4-methoxy-3(2H)-furanone	Andersen et al. (2014)
Strawberries	Single meal intervention	N/A	Urine	HRMS	*p*-Coumaric acid sulfate, 4-hydroxyhippuric acid, pelargonidin monoglucuronide, furaneol glucuronide	Cuparencu et al. (2016)
Sea buckthorn	Single meal intervention	N/A	Urine	HRMS	5-Hydroxyindole acetic acid, 2,3-dihydro-2-oxo-1H-indole-3-acetic acid	Cuparencu et al. (2016)
Apple and pear	Free-living subjects	24HR	Urine	HRMS	Phloretin glucuronide	Edmands et al. (2015)
Vegetables						
Vegetables (total)	Free-living subjects	Food diary	Urine	NMR	Phenylacetylglutamine	O'Sullivan et al. (2011)
Broccoli	Short/medium term intervention	N/A	Urine	HRMS	Tetronic acid, xylonate/lyxonate, threitol/erythritol	Lloyd, Favé et al. (2011c)
Cruciferous vegetables	Short/medium term intervention	N/A	Urine	NMR	*S*-Methyl-L-cysteine sulfoxide	Edmands et al. (2011)

(*Continued*)

TABLE 15.3 (*Continued*)
Metabolomic Studies on Dietary Biomarker Discovery

Dietary Factor	Study Type	Dietary Assessment Tool	Biospecimen	Analytical Technique	Biomarkers	References
Cruciferous vegetables	Single meal intervention	N/A	Urine	HRMS	Sulforaphane *N*-acetylcysteine, *N*-acetyl-(*N*′-benzylthiocarbamoyl)cysteine, sulforaphane *N*-cysteine, *N*-acetyl-*S*-(*N*-3-methylthiopropyl)cysteine, *N*-acetyl-*S*-(*N*-allylthiocarbamoyl)cysteine, iberin *N*-acetylcysteine, 4-iminopentylisothiocyanate, erucin *N*-acetylcysteine	Andersen et al. (2013)
Red cabbage	Free-living subjects	24HR	Urine	HRMS	3-Hydroxy-3-(methylsulfinyl)propanoic acid, 3-hydroxyhippuric acid sulfate, 3-hydroxyhippuric acid, iberin *N*-acetyl-cysteine, *N*-acetyl-*S*-(*N*-3-methylthiopropyl) cysteine, *N*-acetyl-*S*-(*N*-allylthiocarbamoyl)cysteine sulforaphane *N*-acetylcysteine	Andersen et al. (2014)
Beetroot	Free-living subjects	24HR	Urine	HRMS	4-Ethyl-5-aminopyrocatechol sulfate, 4-ethyl-5-methylaminopyrocatechol sulfate, 4-methylpyridine-2-carboxylic acid glycine conjugate	Andersen et al. (2014)
Cereals						
Wholegrain rye bread	Short/medium term intervention	N/A	Urine	HRMS	3-(3,5-Dihydroxyphenyl)-1-propanoic acid sulfate and glucuronide, enterolactone glucuronide, azelaic acid, 2-aminophenol sulfate, 2–4-dihydroxy-1,4-benzoxazin-3-one sulfate, indolylacryloylglycine, ferulic acid sulfate, 3,5-dihydroxyphenylethanol sulfate, 3,5-dihydroxycinnamic acid sulfate	Bondia-Pons et al. (2013)
Wholegrain bread	Free-living subjects	FFQ	Urine	HRMS	3-(3,5-Dihydroxyphenyl)propanoic acid, 2-hydroxy-7-methoxy-2H-1,4-benzoxazin-3-one, 5-(3,5-dihydroxyphenyl)pentanoic acid sulfate, dihydroferulic acid sulfate, enterolactone glucuronide, 3-indolecarboxylic acid glucuronide	Garcia-Aloy, Llorach, Urpi-Sarda, Tulipani, Salas-Salvadó et al. (2014)

(*Continued*)

TABLE 15.3 (*Continued*)
Metabolomic Studies on Dietary Biomarker Discovery

Dietary Factor	Study Type	Dietary Assessment Tool	Biospecimen	Analytical Technique	Biomarkers	References
Wholegrain rye	Free-living subjects	24HR	Urine	HRMS	Hydroxyhydroxyphenyl acetamide sulfate, 3,5-dihydroxyphenylpropionic acid sulfate, caffeic acid sulfate, and hydroxyphenyl acetamide sulfate	Hanhineva et al. (2015)
Meat and Fish						
Red meat	Free-living subjects	Food diary	Urine	NMR	*O*-Acetylcarnitines	O'Sullivan et al. (2011)
Red meat (beef)	Single meal intervention	N/A	Plasma	HRMS	β-Alanine, 4-hydroxyproline	Ross et al. (2015)
Herring	Single meal intervention	N/A	Plasma	HRMS	Docosahexaenoic acid, cetoleic acid	Ross et al. (2015)
Salmon	Short/medium term intervention	N/A	Urine	HRMS	Anserine, methylhistidine, trimethylamine-*N*-oxide	Lloyd, Favé et al. (2011c)
Oily fish	Free-living subjects	FFQ	Urine	HRMS	Methylhistidine	Lloyd et al. (2013b)
Seafoods	Single meal intervention	N/A	Urine	HRMS	Trimethylamine-*N*-oxide	Andersen et al. (2013)
Beverages						
Chamomile tea	Short/medium term intervention	N/A	Urine	NMR	Hippuric acid	Wang et al. (2005)
Black tea	Single meal intervention	N/A	Urine	NMR	Hippuric acid, gallic acid, 1,3-dihydroxyphenyl-2-*O*-sulfate	Daykin et al. (2005)
Tea	Short/medium term intervention	N/A	Urine	NMR	Hippuric acid, 1,3-dihydrophenyl-2-*O*-sulfate	Van Dorsten et al. (2006)
Green tea	Single meal intervention	N/A	Urine	NMR	Hippuric acid	Law et al. (2008)
Black tea	Single meal intervention	N/A	Urine	NMR	Hippuric acid, 4-hydroxyhippuric acid, 1,3-dihydrophenyl-2-*O*-sulfate, gallic acid, 4-*O*-methylgallic acid	van Velzen et al. (2009)
Tea	Free-living subjects	24HR, FFQ	Urine	HRMS	4-*O*-Methylgallic acid	Edmands et al. (2015)
Coffee	Single meal intervention	N/A	Urine	HRMS	*N*-Methylpyridinium, trigonelline	Lang et al. (2011)
Coffee	Free-living subjects	FFQ	Urine	HRMS	*N*-Methylpyridinium, trigonelline	Lang et al. (2011)
Coffee	Free-living subjects	FFQ	Urine	HRMS	Dihydrocaffeic acid	Lloyd et al. (2013b)

(*Continued*)

TABLE 15.3 (*Continued*)
Metabolomic Studies on Dietary Biomarker Discovery

Dietary Factor	Study Type	Dietary Assessment Tool	Biospecimen	Analytical Technique	Biomarkers	References
Coffee	Free-living subjects	24HR, FFQ	Urine	HRMS	Atractyligenin glucuronide, trigonelline, cyclo(isoleucyl-prolyl), caffeine metabolites	Rothwell et al. (2014)
Coffee	Free-living subjects	FFQ	Serum	HRMS	Trigonelline, quinate, caffeine, caffeine metabolites	Guertin et al. (2015)
Coffee	Free-living subjects	24HR, FFQ	Urine	HRMS	Dihydroferulic acid sulfate	Edmands et al. (2015)
Coffee	Single meal intervention	N/A	Urine	NMR	2-Furoylglycine	Heinzmann et al. (2015)
Wine	Short/medium term intervention	N/A	Urine	NMR	Tartrate, 4-hydroxyphenylacetate, mannitol, ethanol	Vazquez-Fresno et al. (2012)
Wine (red)/ grape juice	Short/medium term intervention	N/A	Urine	HRMS	Hippuric acid, 3-hydroxyhippuric acid, 4-hydroxyhippuric acid, 4-hydroxybenzoic acid, 1,2,3-trihydroxybenzene, vanillic acid, isovanillic acid, syringic acid, 3-hydroxyphenylacetic acid, 4-hydroxymandelic acid, vanillylmandelic acid, ferulic acid, 3-hydroxyphenylpropionic acid, 3,4-dihydroxyphenylpropionic acid, 3-(3-hydroxyphenyl)-3-hydroxypropionic acid, catechol, pyrogallol, citrate, betaine	van Dorsten et al. (2010) and Jacobs et al. (2012)
Wine (red)	Free-living subjects	24HR, FFQ	Urine	HRMS	Gallic acid ethyl ester	Edmands et al. (2015)
Sugar-sweetened beverages	Free-living subjects	Food record	Urine	NMR	Citrulline, formate, isocitrate, taurine	Gibbons et al. (2015)
Other Foods						
Cocoa Powder	Single meal intervention	N/A	Urine	HRMS	Vanilloylglycine, 6-amino-5-[*N*-methylformylamino]-1-methyluracil, 3-methyluric acid, 7-methyluric acid, 3-methylxanthine, 7-methylxanthine, dimethyluric acid, theobromine, caffeine, trigonelline, hydroxynicotinic acid, 3,5-diethyl-2-methylpyrazine, hydroxyacetophenone, diketopiperazines, epicatechin sulfate, *O*-methylepicatechin, vanillic acid, phenylvaleric acid and phenylvalerolactone derivatives, furoylglycine, xanthurenic acid	Llorach et al. (2009)

(*Continued*)

TABLE 15.3 (*Continued*)
Metabolomic Studies on Dietary Biomarker Discovery

Dietary Factor	Study Type	Dietary Assessment Tool	Biospecimen	Analytical Technique	Biomarkers	References
Cocoa Powder	Short/medium term intervention	N/A	Urine	HRMS	Hydroxynicotinic acid, 6-amino-5-[*N*-methylformylamino]-1-methyluracil, 7- and 3-methyluric acid, 7- and 3-methylxanthine, 3,7-dimethylruric acid, cyclo(propylalanyl), 3,5-diethyl-2-methylpyrazine, theobromine, vanillic acid glucuronide and sulfoglucuronide, vanilloylglycine, phenylvaleric acid and phenylvalerolactone derivatives, (epi)catechin glucuronide and sulfoglucuronide, methyl-(epi)catechin sulfate, *N*-[4′-hydroxy-3′-methoxy-E-cinnamoyl]-L-aspartic acid, *N*-[4′-hydroxycinnamoyl]-L-aspartic acid	Llorach et al. (2013)
Chocolate	Free-living subjects	24HR	Urine	HRMS	6-Amino-5-[*N*-methylformylamino]-1-methyluracil, theobromine, 7-methyluric acid	Andersen et al. (2014)
Chocolate	Free-living subjects	24HR	Urine	HRMS	Methyl(epi)catechin sulfate	Edmands et al. (2015)
Almond skin extract	Single meal intervention	N/A	Urine	HRMS	(Epi)catechin sulfate, *O*-methyl(epi)catechin sulfate, naringenin *O*-glucuronide, phenylvaleric acid and phenylvalerolactone derivatives, 2-(dihydroxyphenyl) acetic acid glucuronide and sulfate, 2-(hydroxymethoxyphenyl)acetic acid glucuronide, 2-(hydroxyphenyl)acetic acid sulfate, 3-(hydroxyphenyl) propionic acid glucuronide, 3-(dihydroxyphenyl)propionic acid sulfate, vanillic acid glucuronide, hydroxyhippuric acid, ferulic acid glucuronide	Llorach et al. (2010)

(*Continued*)

TABLE 15.3 (*Continued*)
Metabolomic Studies on Dietary Biomarker Discovery

Dietary Factor	Study Type	Dietary Assessment Tool	Biospecimen	Analytical Technique	Biomarkers	References
Nuts	Short/medium term intervention	N/A	Urine	HRMS	10-Hydroxydecene-4,6-diynoic acid sulfate, tridecadienoic/tridecynoic acid glucuronide, dodecanedioic acid, 1,3-dihydroxyphenyl-2-O-sulfate, *p*-coumaroyl alcohol glucuronide and sulfate, *N*-acetylserotonine sulfate, 5-hydroxyindoleacetic acid, urolitin A glucuronide, sulfate and sulfoglucuronide	Tulipani et al. (2011)
Walnut	Free-living subjects	24HR	Urine	HRMS	5-Hydroxyindole-3-acetic acid	Andersen et al. (2014)
Walnut	Free-living subjects	FFQ	Urine	HRMS	Urolithin A glucuronide	Garcia-Aloy, Llorach, Urpi-Sarda, Tulipani, Estruch et al. (2014)
Nuts	Short/medium term intervention	N/A	Plasma	HRMS	Urolithin A glucuronide	Mora-Cubillos et al. (2015)
Nutrients						
Dietary fiber	Short/medium term intervention	Dietary record	Urine	NMR	Hippuric acid	Rasmussen et al. (2012)
Dietary fiber	Short/medium term intervention	N/A	Plasma	HRMS	2-Aminophenol sulfate, 2,6-dihydroxybenzoic acid, hydroxynuategenin glucuronide	Johansson-Persson et al. (2013)

Note: Precise identities may still need to be fully confirmed by comparison with an authentic standard often not commercially available (see Chapter 16 for metabolite annotation). NMR, nuclear magnetic resonance spectroscopy; HRMS, high-resolution mass spectrometry.

A wide diversity of compounds was found to be associated to the consumption of various foods in dietary intervention studies or in free-living subjects. Some of these compounds were already known (see Table 15.1). These include 1-methylhistidine, DHA, alkylresorcinols, hesperetin/naringenin, and 4-*O*-methylgallic acid respectively associated to intake of red meat, fish, wholegrain cereals, citrus fruits, and tea.

Other compounds were newly identified, and these results illustrate the value of agnostic metabolomic approaches for biomarker discovery. Examples of newly discovered biomarkers are proline betaine, limonene, and nootkatone for citrus fruit intake (Heinzmann et al. 2010, Pujos-Guillot et al. 2013), furaneol and pelargonidin for strawberry (Cuparencu et al. 2016), *S*-methyl-L-cysteine sulfoxide for cruciferous vegetables (Edmands et al. 2011), 2-aminophenol for wholegrain cereals and dietary fibers (Bondia-Pons et al. 2013; Johansson-Persson et al. 2013), *N*-methylpyridinium, trigonelline, and atractyligenin for coffee (Lang et al. 2011; Rothwell et al. 2014), gallic acid ethyl ester for red wine (Edmands et al. 2015), and vanilloylglycine for cocoa (Llorach et al. 2009).

Most biomarkers identified in metabolomic studies and listed in Table 15.3 are surrogates for acute food intake. They were most often identified in short-term intervention studies or in observational studies in which biomarker levels were compared to acute food intake as assessed with either 24HR or dietary records on the day when biospecimens were collected. Such biomarkers would not necessarily be effective in estimating habitual intake of the same foods. In a few recent studies, however, metabolites were tested against habitual food intake as measured with FFQs. Still, most biomarkers found to be associated with habitual food intake (e.g., proline betaine, 3-(3,5-dihydroxyphenyl)propanoic acid, methylhistidine, dihydrocaffeic acid or urolithin A) have elimination half-lives of 24 hours or less (Heinzmann et al. 2010; Sjolin et al. 1987; Manach et al. 2005). Significant associations between these markers and habitual (usual) food intake could therefore be explained by the frequent consumption of the corresponding foods (Edmands et al. 2015).

15.10 FUTURE RESEARCH

It makes no doubt that the highly sensitive analytical techniques and metabolomic approaches increasingly used by nutritionists will lead to the discovery of more dietary biomarkers. So far most of these studies used urine due to ease of collection and high concentrations of food-derived metabolites (Table 15.3). However, most biomarkers in urine are typically short-term biomarkers (Figure 15.2) and may only be used when repeated samples are available. They would then be of limited value in large cohort studies except for foods or food compounds most frequently consumed. More metabolomic studies should be conducted on blood samples most commonly stored in biobanks, to identify lipophilic biomarkers more reproducible over time. Formation of adducts with DNA or blood proteins also increases half-life of dietary biomarkers and more work should also be done on such adducts as biomarkers of dietary exposures. Such adducts are more reproducible over time but the number of known adducts with food compounds is still limited. Untargeted adductomic approaches, similar to metabolomics, should lead to the discovery of new DNA and protein adducts formed with dietary electrophiles (Balbo et al. 2014, Chou et al. 2010; Rappaport et al. 2011). More work should also be done on other biospecimens with longer time of residence such as hair, nails, or teeth, which accumulate food-derived compounds over time.

Many of the dietary biomarkers recently discovered through metabolomics (Table 15.3) have not yet been validated and their utility for population studies is still unclear. More efforts should be made to validate these biomarkers. Biological plausibility should be assessed for each novel biomarker. Their parentage with metabolic precursors present in the food of interest should be explored. Many of these biomarkers were discovered by comparing high and low food consumers in fully controlled studies on small number of subjects, and their robustness and ability to classify individuals in populations still need to be ascertained (Chapter 17).

Specificity of the newly proposed biomarkers is also often unclear. Distribution of a newly discovered biomarker or of its precursors in any potentially confounding food should be explored.

Their distribution should be specific to the food of interest, or this food should represent its major dietary source in the population where the biomarker is intended to be measured. A number of candidate biomarkers can be eliminated on this basis. The evaluation of biological plausibility thus requires comprehensive knowledge of the content of all main biomarker precursors in all foods consumed in a population. This information is easily available for common nutrients but usually not for many nonnutrient compounds. Some food precursors like proline betaine (de Zwart et al. 2003; Heinzmann et al. 2010) or 1- and 3-methylhistidine (Sjolin et al. 1987) have been analyzed in various foods to provide information on their dietary origin. However these data are often incomplete and more comprehensive databases such as Phenol-Explorer for polyphenols (Neveu et al. 2010) are needed. The development of FooDB, a database on all known dietary compounds and their occurrence in different foods, is a big step in this direction (University of Alberta 2016).

Measurement of candidate biomarkers in cross-sectional studies and the analysis of their correlation with food intake may allow to evaluate their sensitivity and specificity and to identify unexpected dietary confounders (Edmands et al. 2015; Zamora-Ros et al. 2016). However the response of a biomarker to the ingestion of varying amounts of food should be explored in fully controlled intervention studies, the only studies where the amount of food or food compounds is precisely known.

The different biomarkers proposed for a particular dietary exposure should be compared, and biomarkers should be ranked according to their sensitivity and specificity. A frame should be defined to make the collection and evaluation of the data on the various biomarkers easier (Scalbert et al. 2014). For each type of dietary exposure, candidate biomarkers could be ranked, and further studies could be developed to further validate biomarkers that ranked higher.

More work is also needed on the evaluation of markers made of combinations of metabolites, particularly when no single biomarker is specific enough to reliably estimate the exposure to a particular food. Combinations of biomarkers should also be useful to characterize dietary patterns and to measure compliance in dietary intervention studies (Marklund et al. 2014).

As the number of dietary and diversity of dietary biomarkers will grow, robust analytical methods will be needed for routine analyses of large panels of dietary biomarkers in nutrient-wide association studies (Tzoulaki et al. 2012; Patel et al. 2010). Analytical methods have been developed and used in epidemiological studies to analyze tens of dietary compounds from particular chemical classes such as fatty acids or polyphenols sometimes present in quite different concentrations (Achaintre et al. 2016; Wang et al. 2013; Chajes et al. 2008). Robust and fast methods able to measure sets of biomarkers as diverse as those shown in Table 15.1 and able to capture complex dietary exposures have not yet been developed and are today needed to conduct nutrient-wide association studies.

15.11 CONCLUSIONS

Dietary biomarkers may substantially improve accuracy and comprehensiveness of dietary exposure assessment. A wide diversity of food constituents have been proposed as dietary biomarkers, and a number of them such as vitamins, fatty acids, or carotenoids, have already been used to validate other methods of dietary assessment, to study associations of dietary exposures with disease or to check compliance to a dietary intervention study. However, many of the proposed biomarkers (Tables 15.2 and 15.3) have not been fully validated. Their qualification as dietary biomarker requires the evaluation of the dose–response relationship, the identification of possible confounders, the measurement of reproducibility over time, and the identification of the range of exposures and populations to which a particular biomarker is applicable.

More efforts should be made to identify longer term biomarkers able to characterize dietary exposures over the previous weeks or months. In particular, more emphasis should be put on the more lipophilic biomarkers and on DNA or protein adducts, and on setting up new studies with appropriate designs for discovery of such biomarkers of longer term intake. Most dietary biomarkers have been measured in blood or urine. More studies on other types of biospecimens like hair, nails, or teeth that may characterize dietary exposures on longer periods should be conducted and more of these samples should be collected in cohort studies.

The complexity of the food metabolome has only been realized recently with the emergence of more sensitive and reliable analytical techniques. Metabolomic approaches have already made possible the identification of new dietary biomarkers. Techniques for metabolomics are not yet fully mature but the discipline is progressing quickly (Chapter 16). Metabolomics should make possible the identification of new biomarkers for a still wider diversity of foods that may differ according to their mode of production or cooking. Dietary biomarkers may allow the characterization of dietary exposures to a level of details not possible with questionnaires.

Lastly, the limitations of biomarkers should also be considered. Different genetic backgrounds may induce variations in absorption, metabolism, and elimination of food compounds and in biomarker levels between individuals (Lampe 2009; Reszka et al. 2006; Etemadi et al. 2013), and this may result in some misclassification of subjects. Cost effectiveness of dietary biomarkers should also be compared to that of other approaches of dietary assessment. Costs will depend on the nature and speed of biomarker analyses and on the number and volume of samples needed for a particular study. Dietary biomarkers will certainly continue to be used in conjunction with other dietary assessment approaches. However, it is expected that a growing number of dietary biomarkers will be measured in future studies, and this should open new perspectives for nutritional epidemiology and should help unraveling associations with diseases hitherto unseen.

ACKNOWLEDGMENTS

Support for the writing of this review paper from the FoodBAll project funded by the BIO-NH call under the Joint Programming Initiative, "a Healthy Diet for a Healthy Life" (grant number 529051002) is acknowledged.

REFERENCES

Achaintre, D., A. Bulete, C. Cren-Olive, L. Li, S. Rinaldi, A. Scalbert. 2016. Differential isotope labeling of 38 dietary polyphenols and their quantification in urine by liquid chromatography electrospray ionization tandem mass spectrometry. *Analytical Chemistry* 88:2637–2644.

Al-Delaimy, W. K., P. Ferrari, N. Slimani et al. 2005. Plasma carotenoids as biomarkers of intake of fruits and vegetables: Individual-level correlations in the European Prospective Investigation into Cancer and Nutrition (EPIC). *European Journal of Clinical Nutrition* 59:1387–1396.

Andersen, M.-B., H. Reinbach, Å. Rinnan, T. Barri, C. Mithril, L. Dragsted. 2013. Discovery of exposure markers in urine for Brassica-containing meals served with different protein sources by UPLC-qTOF-MS untargeted metabolomics. *Metabolomics* 9:984–997.

Andersen, M. B., M. Kristensen, C. Manach et al. 2014. Discovery and validation of urinary exposure markers for different plant foods by untargeted metabolomics. *Analytical and Bioanalytical Chemistry* 406:1829–1844.

Andersson, A., M. Marklund, M. Diana, R. Landberg. 2011. Plasma alkylresorcinol concentrations correlate with whole grain wheat and rye intake and show moderate reproducibility over a 2- to 3-month period in free-living Swedish adults. *Journal of Nutrition* 141:1712–1718.

Andra, S. S., C. Austin, M. Arora. 2015. Tooth matrix analysis for biomonitoring of organic chemical exposure: Current status, challenges, and opportunities. *Environmental Research* 142:387–406.

Appenzeller, B. M. R., R. Agirman, P. Neuberg, M. Yegles, R. Wennig. 2007. Segmental determination of ethyl glucuronide in hair: A pilot study. *Forensic Science International* 173:87–92.

Appenzeller, B. M. R. and A. M. Tsatsakis. 2012. Hair analysis for biomonitoring of environmental and occupational exposure to organic pollutants: State of the art, critical review and future needs. *Toxicology Letters* 210:119–140.

Aune, D., D. S. M. Chan, A. R. Vieira et al. 2012. Dietary compared with blood concentrations of carotenoids and breast cancer risk: A systematic review and meta-analysis of prospective studies. *American Journal of Clinical Nutrition* 96:356–373.

Austin, C., T. M. Smith, A. Bradman et al. 2013. Barium distributions in teeth reveal early-life dietary transitions in primates. *Nature* 498:216–219.

Balbo, S., R. J. Turesky, P. W. Villalta. 2014. DNA adductomics. *Chemical Research in Toxicology* 27:356–366.

Baldrick, F. R., J. V. Woodside, J. S. Elborn, I. S. Young, M. C. McKinley. 2011. Biomarkers of fruit and vegetable intake in human intervention studies: A systematic review. *Critical Reviews in Food Science and Nutrition* 51:795–815.

Bieler, C. A., M. G. Cornelius, M. Stiborova et al. 2007. Formation and persistence of DNA adducts formed by the carcinogenic air pollutant 3-nitrobenzanthrone in target and non-target organs after intratracheal instillation in rats. *Carcinogenesis* 28:1117–1121.

Bingham, S., R. Luben, A. Welch, N. Tasevska, N. Wareham, K. T. Khaw. 2007. Epidemiologic assessment of sugars consumption using biomarkers: Comparisons of obese and nonobese individuals in the European Prospective Investigation of Cancer Norfolk. *Cancer Epidemiology Biomarkers & Prevention* 16:1651–1654.

Bondia-Pons, I., T. Barri, K. Hanhineva et al. 2013. UPLC-QTOF/MS metabolic profiling unveils urinary changes in humans after a whole grain rye versus refined wheat bread intervention. *Molecular Nutrition and Food Research* 57:412–422.

Brantsaeter, A. L., M. Haugen, K. Julshamn, J. Alexander, H. M. Meltzer. 2009. Evaluation of urinary iodine excretion as a biomarker for intake of milk and dairy products in pregnant women in the Norwegian Mother and Child Cohort Study (MoBa). *European Journal of Clinical Nutrition* 63:347–354.

Brantsaeter, A. L., M. Haugen, S. E. Rasmussen, J. Alexander, S. O. Samuelsen, H. M. Meltzer. 2007. Urine flavonoids and plasma carotenoids in the validation of fruit, vegetable and tea intake during pregnancy in the Norwegian Mother and Child Cohort Study (MoBa). *Public Health Nutrition* 10:838–847.

Chajes, V., A. C. M. Thiebaut, M. Rotival et al. 2008. Association between serum trans-monounsaturated fatty acids and breast cancer risk in the E3N-EPIC study. *American Journal of Epidemiology* 167:1312–1320.

Chou, P. H., S. Kageyama, S. Matsuda et al. 2010. Detection of lipid peroxidation-induced DNA adducts caused by 4-oxo-2(E)-nonenal and 4-oxo-2(E)-hexenal in human autopsy tissues. *Chemical Research Toxicology* 23:1442–1448.

Cuparencu, C. S., M. B. S. Andersen, G. Gurdeniz et al. 2016. Identification of urinary biomarkers after consumption of sea buckthorn and strawberry, by untargeted LC-MS metabolomics: A meal study in adult men. *Metabolomics* 12:31. doi:10.1007/s11306-015-0934-0.

Daykin, C. A., J. P. Van Duynhoven, A. Groenewegen, M. Dachtler, J. M. Van Amelsvoort, T. P. Mulder. 2005. Nuclear magnetic resonance spectroscopic based studies of the metabolism of black tea polyphenols in humans. *Journal of Agricultural and Food Chemistry* 53:1428–1434.

de Zwart, F. J., S. Slow, R. J. Payne et al. 2003. Glycine betaine and glycine betaine analogues in common foods. *Food Chemistry* 83:197–204.

Edmands, W., P. Ferrari, J. A. Rothwell et al. 2015. The polyphenol metabolome in human urine and its association with intake of polyphenol-rich foods across European countries. *American Journal of Clinical Nutrition* 902:905–913.

Edmands, W. M. B., O. P. Beckonert, C. Stella et al. 2011. Identification of human urinary biomarkers of cruciferous vegetable consumption by metabonomic profiling. *Journal of Proteome Research* 10:4513–4521.

Etemadi, A., F. Islami, D. H. Phillips et al. 2013. Variation in PAH-related DNA adduct levels among nonsmokers: The role of multiple genetic polymorphisms and nucleotide excision repair phenotype. *International Journal of Cancer* 132:2738–2747.

FleschJanys, D., H. Becher, P. Gurn et al. 1996. Elimination of polychlorinated dibenzo-p-dioxins and dibenzofurans in occupationally exposed persons. *Journal of Toxicology and Environmental Health* 47:363–378.

Garcia-Aloy, M., R. Llorach, M. Urpi-Sarda et al. 2014. Novel multimetabolite prediction of walnut consumption by a urinary biomarker model in a free-living population: The PREDIMED study. *Journal of Proteome Research* 13:3476–483.

Garcia-Aloy, M., R. Llorach, M. Urpi-Sarda et al. 2014. Nutrimetabolomics fingerprinting to identify biomarkers of bread exposure in a free-living population from the PREDIMED study cohort. *Metabolomics* 11:155–165.

Gibbons, H., B. A. McNulty, A. P. Nugent et al. 2015. A metabolomics approach to the identification of biomarkers of sugar-sweetened beverage intake. *American Journal of Clinical Nutrition* 101:471–477.

Godschalk, R. W. L., F. J. Van Schooten, H. Bartsch. 2003. A critical evaluation of DNA adducts as biological markers for human exposure to polycyclic aromatic compounds. *Journal of Biochemistry and Molecular Biology* 36:1–11.

Guertin, K. A., E. Loftfield, S. M. Boca et al. 2015. Serum biomarkers of habitual coffee consumption may provide insight into the mechanism underlying the association between coffee consumption and colorectal cancer. *The American Journal of Clinical Nutrition* 101:1000–1011.

Gunter, M. J., R. L. Divi, M. Kulldorff et al. 2007. Leukocyte polycyclic aromatic hydrocarbon-DNA adduct formation and colorectal adenoma. *Carcinogenesis* 28:1426–1429.

Hanhineva, K., C. Brunius, A. Andersson et al. 2015. Discovery of urinary biomarkers of whole grain rye intake in free-living subjects using non-targeted LC-MS metabolite profiling. *Molecular Nutrition & Food Research* 59:2315–2325.

He, K. 2011. Trace elements in nails as biomarkers in clinical research. *European Journal of Clinical Investigation* 41:98–102.

Hedrick, V. E., A. M. Dietrich, P. A. Estabrooks, J. Savla, E. Serrano, B. M. Davy. 2012. Dietary biomarkers: Advances, limitations and future directions. *Nutrition Journal* 11:109. doi: 10.1186/1475-2891-11-109.

Heinzmann, S. S., I. J. Brown, Q. Chan et al. 2010. Metabolic profiling strategy for discovery of nutritional biomarkers: Proline betaine as a marker of citrus consumption. *American Journal of Clinical Nutrition* 92:436–443.

Heinzmann, S. S., E. Holmes, S. Kochhar, J. K. Nicholson, P. Schmitt-Kopplin. 2015. 2-Furoylglycine as a candidate biomarker of coffee consumption. *Journal of Agriculture Food Chemistry* 63:8615–8621.

Hemeryck, L. Y. and L. Vanhaecke. 2016. Diet-related DNA adduct formation in relation to carcinogenesis. *Nutrition Reviews* 74:475–489.

Hodson, L., C. M. Skeaff, B. A. Fielding. 2008. Fatty acid composition of adipose tissue and blood in humans and its use as a biomarker of dietary intake. *Progress in Lipid Research* 47:348–380.

Holden, J. M., A. L. Eldridge, G. R. Beecher et al. 1999. Carotenoid content of U.S. foods: An update of the database. *Journal of Food Composition and Analysis* 12:169–196.

Hossain, M. Z., S. F. Gilbert, K. Patel, S. Ghosh, A. K. Bhunia, S. E. Kern. 2013. Biological clues to potent DNA-damaging activities in food and flavoring. *Food and Chemical Toxicology* 55:557–567.

Iwasaki, M., M. Inoue, S. Sasazuki et al. 2010. Plasma tea polyphenol levels and subsequent risk of breast cancer among Japanese women: A nested case-control study. *Breast Cancer Research and Treatment* 124:827–834.

Jacobs, D. M., J. C. Fuhrmann, F. A... van Dorsten et al. 2012. Impact of short-term intake of red wine and grape polyphenol extract on the human metabolome. *Journal of Agricultural and Food Chemistry* 60:3078–3085.

Jedrychowski, W. A., F. P. Perera, D. L. Tang et al. 2013. The relationship between prenatal exposure to airborne polycyclic aromatic hydrocarbons (PAHs) and PAH-DNA adducts in cord blood. *Journal of Exposure Science and Environmental Epidemiology* 23:371–377.

Jenab, M., N. Slimani, M. Bictash, P. Ferrari, S. A. Bingham. 2009. Biomarkers in nutritional epidemiology: applications, needs and new horizons. *Human Genetics* 125:507–525.

Johansson-Persson, A., T. Barri, M. Ulmius, G. Onning, L. O. Dragsted. 2013. LC-QTOF/MS metabolomic profiles in human plasma after a 5-week high dietary fiber intake. *Analytical and Bioanalytical Chemistry* 405:4799–4809.

Katan, M. B., J. P. Deslypere, A. P. van Birgelen, M. Penders, M. Zegwaard. 1997. Kinetics of the incorporation of dietary fatty acids into serum cholesteryl esters, erythrocyte membranes, and adipose tissue: An 18-month controlled study. *Journal of Lipid Research* 38:2012–2022.

Kobayashi, M., T. Hanaoka, H. Hashimoto, S. Tsugane. 2005. 2-amino-1-methyl-6-phenylimidazo 4, 5-b pyridine (PhIP) level in human hair as biomarkers for dietary grilled/stir-fried meat and fish intake. *Mutation Research-Genetic Toxicology and Environmental Mutagenesis* 588:136–142.

Lampe, J. W. 2009. Interindividual differences in response to plant-based diets: Implications for cancer risk. *American Journal of Clinical Nutrition* 89:S1553–S1557.

Lang, R., A. Wahl, T. Stark, T. Hofmann. 2011. Urinary N-methylpyridinium and trigonelline as candidate dietary biomarkers of coffee consumption. *Molecular Nutrition and Food Research* 55:1613–1623.

Law, W. S., P. Y. Huang, E. S. Ong, C. N. Ong, S. F. Li, K. K. Pasikanti, E. C. Chan. 2008. Metabonomics investigation of human urine after ingestion of green tea with gas chromatography/mass spectrometry, liquid chromatography/mass spectrometry and (1)H NMR spectroscopy. *Rapid Communications in Mass Spectrometry* 22:2436–2446.

Levine, M., C. Conry-Cantilena, Y. Wang et al. 1996. Vitamin C pharmacokinetics in healthy volunteers: Evidence for a recommended dietary allowance. *Proceedings of the National Academy Sciences USA* 93:3704–3709.

Li, Z., L. Romanoff, S. Bartell et al. 2012. Excretion profiles and half-lives of ten urinary polycyclic aromatic hydrocarbon metabolites after dietary exposure. *Chemical Research in Toxicology* 25:1452–1461.

Llorach, R., I. Garrido, M. A. Monagas et al. 2010. Metabolomics study of human urinary metabolome modifications after intake of almond (prunus dulcis (Mill.) D.A. Webb) skin polyphenols. *Journal of Proteome Research* 9:5859–5867.

Llorach, R., M. Urpi-Sarda, O. Jauregui, M. Monagas, C. Andres-Lacueva. 2009. An LC-MS-based metabolomics approach for exploring urinary metabolome modifications after cocoa consumption. *Journal of Proteome Research* 8:5060–5068.

Llorach, R., M. Urpi-Sarda, S. Tulipani, M. Garcia-Aloy, M. Monagas, C. Andres-Lacueva. 2013. Metabolomic fingerprint in patients at high risk of cardiovascular disease by cocoa intervention. *Molecular Nutrition and Food Research* 57:962–973.

Lloyd, A. J., M. Beckmann, G. Fave, J. C. Mathers, J. Draper. 2011a. Proline betaine and its biotransformation products in fasting urine samples are potential biomarkers of habitual citrus fruit consumption. *British Journal of Nutrition* 106:812–824.

Lloyd, A. J., M. Beckmann, S. Haldar, C. Seal, K. Brandt, J. Draper. 2013b. Data-driven strategy for the discovery of potential urinary biomarkers of habitual dietary exposure. *American Journal of Clinical Nutrition* 97:377–389.

Lloyd, A. J., G. Favé, M. Beckmann, W. Lin, K. Tailliart, L. Xie, J. C. Mathers, and J. Draper. 2011c. Use of mass spectrometry fingerprinting to identify urinary metabolites after consumption of specific foods. *The American Journal of Clinical Nutrition* 94:981–991.

Luo, J., Y.-T. Gao, W.-H. Chow et al. 2010. Urinary polyphenols and breast cancer risk: Results from the Shanghai Women's Health Study. *Breast Cancer Research and Treatment* 120:693–702.

Manach, C., G. Williamson, C. Morand, A. Scalbert, C. Remesy. 2005. Bioavailability and bioefficacy of polyphenols in humans. I. Review of 97 bioavailability studies. *The American Journal of Clinical Nutrition* 81:230S–242S.

Marklund, M., O. K. Magnusdottir, F. Rosqvist et al. 2014. A dietary biomarker approach captures compliance and cardiometabolic effects of a healthy nordic diet in individuals with metabolic syndrome. *Journal of Nutrition* 144:1642–1649.

Min, J. Z., H. Yano, A. Matsumoto et al. 2011. Simultaneous determination of polyamines in human nail as 4-(N, N-dimethylaminosulfonyl)-7-fluoro-2,1,3-benzoxadiazole derivatives by nano-flow chip LC coupled with quadrupole time-of-flight tandem mass spectrometry. *Clinica Chimica Acta* 412:98–106.

Mora-Cubillos, X., S. Tulipani, M. Garcia-Aloy, M. Bullo, F. J. Tinahones, C. Andres-Lacueva. 2015. Plasma metabolomic biomarkers of mixed nuts exposure inversely correlate with severity of metabolic syndrome. *Molecular Nutrition & Food Research* 59:2480–2490.

Neveu, V., A. Moussy, H. Rouaix et al. 2017. Exposome-Explorer: A manually-curated database on biomarkers of exposure to dietary and environmental factors. *Nucleic Acids Research* 45:D979-D984.

Neveu, V., J. Perez-Jiménez, F. Vos et al. 2010. Phenol-Explorer: An online comprehensive database on polyphenol contents in foods. *Database* 2010:bap024.

Noh, H., H. Freisling, N. Assi et al. 2016. Urinary polyphenol profiles as predictors of food intake in the European Prospective Investigation into Cancer and Nutrition (EPIC) study.

Nost, T. H., K. Breivik, O.-M. Fuskevag, E. Nieboer, J. O. Odland, T. M. Sandanger. 2013. Persistent organic pollutants in Norwegian men from 1979 to 2007: Intraindividual changes, age-period-cohort effects, and model predictions. *Environmental Health Perspectives* 121:1292–1298.

O'Sullivan, A., M. J. Gibney, L. Brennan. 2011. Dietary intake patterns are reflected in metabolomic profiles: Potential role in dietary assessment studies. *American Journal of Clinical Nutrition* 93:314–321.

Patel, C. J., J. Bhattacharya, A. J. Butte. 2010. An environment-wide association study (EWAS) on type 2 diabetes mellitus. *PLoS One* 5:e10746.

Perez-Jimenez, J., J. Hubert, K. Ashton et al. 2010a. Urinary metabolites as biomarkers of polyphenol intake in humans: A systematic review. *American Journal of Clinical Nutrition* 92:801–809.

Perez-Jimenez, J., V. Neveu, F. Vos, A. Scalbert. 2010b. Systematic analysis of the content of 502 polyphenols in 452 foods and beverages: An application of the phenol-explorer database. *Journal of Agricultural and Food Chemistry* 58:4959–4969.

Pirro, V., D. Di Corcia, S. Pellegrino, M. Vincenti, B. Sciutteri, A. Salomone. 2011a. A study of distribution of ethyl glucuronide in different keratin matrices. *Forensic Science International* 210:271–277.

Pirro, V., V. Valente, P. Oliveri, A. De Bernardis, A. Salomone, M. Vincenti. 2011b. Chemometric evaluation of nine alcohol biomarkers in a large population of clinically-classified subjects: Pre-eminence of ethyl glucuronide concentration in hair for confirmatory classification. *Analytical and Bioanalytical Chemistry* 401:2153–2164.

Prince, M. R. and J. K. Frisoli. 1993. Beta-carotene accumulation in serum and skin. *American Journal of Clinical Nutrition* 57:175–181.

Pujos-Guillot, E., J. Hubert, J.-F. Martin et al. 2013. Mass spectrometry-based metabolomics for the discovery of biomarkers of fruit and vegetable intake: Citrus fruit as a case study. *Journal of Proteome Research* 12:1645–1659.

Qian, G. S., R. K. Ross, M. C. Yu et al. 1994. A follow-up study of urinary markers of aflatoxin exposure and liver cancer risk in Shanghai, People's Republic of China. *Cancer Epidemiology Biomarkers & Prevention* 3:3–10.

Rappaport, S. M., H. Li, H. Grigoryan, W. E. Funk, E. R. Williams. 2011. Adductomics: Characterizing exposures to reactive electrophiles. *Toxicology Letters* 213:83–90.

Rasmussen, L. G., H. Winning, F. Savorani et al. 2012. Assessment of dietary exposure related to dietary GI and fibre intake in a nutritional metabolomic study of human urine. *Genes and Nutrition* 7:281–293.

Reszka, E., W. Wasowicz, J. Gromadzinska. 2006. Genetic polymorphism of xenobiotic metabolising enzymes, diet and cancer susceptibility. *British Journal of Nutrition* 96:609–619.

Rock, C. L., M. E. Swendseid, R. A. Jacob, R. W. McKee. 1992. Plasma carotenoid levels in human subjects fed a low carotenoid diet. *Journal of Nutrition* 122:96–100.

Ross, A. B., C. Svelander, I. Undeland, R. Pinto, A. S. Sandberg. 2015. Herring and Beef Meals Lead to Differences in Plasma 2-Aminoadipic Acid, β-Alanine, 4-Hydroxyproline, Cetoleic Acid, and Docosahexaenoic Acid Concentrations in Overweight Men. *The Journal of Nutrition* 145:2456–2463.

Rothwell, J. A., Y. Fillâtre, J.-F. Martin et al. 2014. New biomarkers of coffee consumption identified by the non-targeted metabolomic profiling of cohort study subjects. *PLoS One* 9:e93474.

Sarver, J. G., D. White, P. Erhardt, K. Bachmann. 1997. Estimating xenobiotic half-lives in humans from rat data: Influence of log P. *Environmental Health Perspectives* 105:1204–1209.

Sasazuki, S., M. Inoue, T. Miura, M. Iwasaki, S. Tsugane. 2008. Plasma tea polyphenols and gastric cancer risk: A case-control study nested in a large population-based prospective study in Japan. *Cancer Epidemiology Biomarkers & Prevention* 17:343–351.

Scalbert, A., L. Brennan, C. Manach et al. 2014. The food metabolome: A window over dietary exposure. *American Journal of Clinical Nutrition* 99:1286–1308.

Schmeiser, H. H., J. L. Nortier, R. Singh et al. 2014. Exceptionally long-term persistence of DNA adducts formed by carcinogenic aristolochic acid I in renal tissue from patients with aristolochic acid nephropathy. *International Journal of Cancer* 135:502–507.

Sjolin, J., G. Hjort, G. Friman, L. Hambraeus. 1987. Urinary excretion of 1-methylhistidine: A qualitative indicator of exogenous 3-methylhistidine and intake of meats from various sources. *Metabolism-Clinical and Experimental* 36:1175–1184.

Sulek, K., T. L. Han, S. G. Villas-Boas et al. 2014. Hair metabolomics: identification of fetal compromise provides proof of concept for biomarker discovery. *Theranostics* 4:953–359.

Sun, Q., J. Ma, H. Campos, S. E. Hankinson, F. B. Hu. 2007. Comparison between plasma and erythrocyte fatty acid content as biomarkers of fatty acid intake in US women. *American Journal of Clinical Nutrition* 86:74–81.

Tulipani, S., R. Llorach, O. Jáuregui et al. 2011. Metabolomics unveils urinary changes in subjects with metabolic syndrome following 12-week nut consumption. *Journal of Proteome Research* 10:5047–5058.

Turesky, R. J. and L. Le Marchand. 2011. Metabolism and biomarkers of heterocyclic aromatic amines in molecular epidemiology studies: Lessons learned from aromatic amines. *Chemical Research in Toxicology* 24:1169–1214.

Turesky, R. J., L. Liu, D. Gu et al. 2013. Biomonitoring the cooked meat carcinogen 2-amino-1-methyl-6-phenylimidazo [4,5-b] pyridine in hair: Impact of exposure, hair pigmentation, and cytochrome P450 1A2 phenotype. *Cancer Epidemiology Biomarkers & Prevention* 22:356–364.

Tzoulaki, I., C. J. Patel, T. Okamura et al. 2012. A nutrient-wide association study on blood pressure. *Circulation* 126:2456–2464.

University_of_Alberta. 2016. FooDB—The food component database. Accessed http://foodb.ca/.http://www.foodb.ca/.

Van den Berg, H., R. Faulks, H. F. Granado et al. 2000. The potential for the improvement of carotenoid levels in foods and the likely systemic effects. *Journal of the Science of Food and Agriculture* 80:880–912.

van der Woude, H., M. G. Boersma, G. M. Alink, J. Vervoort, I. Rietjens. 2006. Consequences of quercetin methylation for its covalent glutathione and DNA adduct formation. *Chemico-Biological Interactions* 160:193–203.

Van Dorsten, F. A., C. A. Daykin, T. P. J. Mulder, J. P. M. Van Duynhoven. 2006. Metabonomics approach to determine metabolic differences between green tea and black tea consumption. *Journal of Agricultural and Food Chemistry* 54:6929–6938.

van Dorsten, F. A., C. H. Grun, E. J. J. van Velzen, D. M. Jacobs, R. Draijer, J. P. M. van Duynhoven. 2010. The metabolic fate of red wine and grape juice polyphenols in humans assessed by metabolomics. *Molecular Nutrition and Food Research* 54:897–908.

van Velzen, E. J., J. A. Westerhuis, J. P. van Duynhoven et al. 2009. Phenotyping tea consumers by nutrikinetic analysis of polyphenolic end-metabolites. *Journal of Proteome Research* 8:3317–30.

van Welie, R. T. H., R. G. J. M. van Dijck, N. P. E. Vermeulen, N. J. Vansittert. 1992. Mercapturic acids, protein adducts, and DNA adducts as biomarkers of electrophilic chemicals. *Critical Reviews in Toxicology* 22:271–306.

Vazquez-Fresno, R., R. Llorach, F. Alcaro et al. 2012. 1H-NMR-based metabolomic analysis of the effect of moderate wine consumption on subjects with cardiovascular risk factors. *Electrophoresis* 33:2345–2354.

Vesper, H. W., N. Slimani, G. Hallmans et al. 2008. Cross-sectional study on acrylamide hemoglobin adducts in subpopulations from the European Prospective Investigation into Cancer and Nutrition (EPIC) Study. *Journal of Agriculture and Food Chemistry* 56:6046–6053.

Viriot, C., A. Scalbert, C. Lapierre, M. Moutounet. 1993. Ellagitannins and lignins in ageing of spirits in oak barrels. *Journal Agricultural and Food Chemistry* 41:1872–1879.

Wang, L. Y., K. Summerhill, C. Rodriguez-Canas et al. 2013. Development and validation of a robust automated analysis of plasma phospholipid fatty acids for metabolic phenotyping of large epidemiological studies. *Genome Medicine* 5:39. doi: 10.1186/gm443.

Wang, Y., H. Tang, J. K. Nicholson, P. J. Hylands, J. Sampson, E. Holmes. 2005. A metabonomic strategy for the detection of the metabolic effects of chamomile (Matricaria recutita L.) ingestion. *Journal of Agricultural and Food Chemistry* 53:191–196.

Wilson, K. M., H. W. Vesper, P. Tocco et al. 2009. Validation of a food frequency questionnaire measurement of dietary acrylamide intake using hemoglobin adducts of acrylamide and glycidamide. *Cancer Causes and Control* 20:269–278.

Wu, H.-C., Q. Wang, L.-W. Wang et al. 2007. Polycyclic aromatic hydrocarbon- and aflatoxin-albumin adducts, hepatitis B virus infection and hepatocellular carcinoma in Taiwan. *Cancer Letters* 252:104–114.

Yuan, J. M., Y. T. Gao, C. S. Yang, M. C. Yu. 2007. Urinary biomarkers of tea polyphenols and risk of colorectal cancer in the Shanghai Cohort Study. *International Journal of Cancer* 120:1344–1350.

Zamora-Ros, R., D. Achaintre, J. A. Rothwell et al. 2016. Urinary excretions of 34 dietary polyphenols and their associations with lifestyle factors in the EPIC cohort study. *Scientific Reports* 6:26905.

Zhang, A. Q., S. C. Mitchell, R. L. Smith. 1999. Dietary precursors of trimethylamine in man: A pilot study. *Food and Chemical Toxicology* 37:515–520.

16 Metabolomic Techniques to Discover Food Biomarkers

Pekka Keski-Rahkonen, Joseph A. Rothwell, and Augustin Scalbert

CONTENTS

16.1 INTRODUCTION

Traditionally researchers have been limited to hypothesis-driven approaches to discover biomarkers of food intake. A detailed knowledge of the composition of the food of interest is required, as well as a good understanding of the absorption and metabolism of major components. This greatly limits the potential for biomarker discovery. The recent emergence of metabolomics has allowed a more comprehensive data-driven approach. The technique takes advantage of the fact that a great number of circulating metabolites are food derived, and many of these are highly specific to a single food or food group. The principle is that every measurable signal in a biospecimen such as blood or urine sample is tested for an association with known food intake, and therefore minimal knowledge of food composition or metabolism is required. In practice, this process consists of four major stages: (1) analysis of samples with an untargeted method, (2) processing of the acquired raw data to extract all detectable metabolites, (3) statistical analysis to find discriminant metabolites, and (4) identification of the metabolites. A common workflow is illustrated in Table 16.1. The aim of this chapter is to give an overview of these stages and how untargeted metabolomics is applied to biospecimens from intervention or observational studies for biomarker discovery. As the design of the study also plays an important role in the biomarker discovery process, this will be discussed at the end of this chapter, before finishing with future perspectives with emphasis on targeted and quantitative metabolomics.

TABLE 16.1
Outline of Untargeted Metabolomics Process for Biomarker Discovery

Sample preparation	Removal of proteins, extraction of metabolites, deconjugation, derivatization, and addition of internal standards as needed
Untargeted analysis	Acquisition of untargeted data: LC–MS, GC–MS, or NMR
Raw data processing	Feature finding from the acquired data, alignment of features, Output as a data matrix: Intensity of each feature across all samples
Statistical analysis	Univariate or multivariate analysis to find discriminant metabolites for food intake
Identification of discriminants	Identification of unknown metabolites with available chemical information

16.2 BACKGROUND

The most widely used analytical technologies for metabolomics are currently based on either nuclear magnetic resonance (NMR) or mass spectrometry (MS) coupled with liquid chromatography (LC) or gas chromatography (GC). NMR and MS enable both quantitative and structural analysis of low molecular weight compounds, which together with the recent advances in the instrument performance and in the computational processing of the data generated, are the main reasons for their applicability and success in metabolomics. A detailed description of the operational principles is beyond the scope of this chapter, but before focusing on how these techniques are used to discover food biomarkers, we will begin with a brief introduction.

LC–MS and GC–MS are combinations of two analytical techniques, where the purpose of the chromatographic system, either GC or LC, is to separate the organic molecules in a sample by their physicochemical properties such as lipophilicity before detection by MS. In LC, a small volume (microliters) of sample is introduced into a chromatographic column that is typically a 50–150 mm long steel tube filled with particulate packing (stationary phase) that is flushed with a mixture of solvents. Compounds in a sample will then elute through the column in different amounts of time depending on their interactions with the stationary phase and the solvent. In order to detect them with MS, the solvent is evaporated and the molecules ionized with the help of heat and a stream of nitrogen, inside the ion source that interfaces the LC and MS systems. In GC, the sample is injected into a heated inlet that transfers the sample vapors to a hollow capillary column. The column is typically a 15–60 m long, coiled fused-silica capillary with a thin film applied on its inner surface, and the separation is achieved by flushing the capillary with a neutral gas while gradually increasing the temperature of the column. As in LC, the compounds elute from the column in the order of their degree of retention to the stationary phase, and are ionized before entering the mass spectrometer. Irrespective of the chromatographic system used, the molecules are detected as ions with the mass spectrometer, and the final output is a chromatogram showing the intensity of each ion over time.

NMR is based on fundamentally different operational principles. The analysis starts by placing the sample, most commonly as a solution in a narrow-walled glass tube, into a magnet with a very high static magnetic field. This will align the nuclear spins of the atoms in the sample molecules with the magnetic field. The sample is then irradiated with a short electromagnetic pulse, a broad spectrum of defined radio waves that move the nuclear spins out of alignment. After the pulse, the spins align back with the magnetic field, which can be detected by the NMR spectrometer. Differences in response for a given atom reflect influences of nearby nuclei and electrons, giving information about the structure of the molecule where the atoms reside. The output of the experiment is an NMR spectrum, where the resonance frequency of the nuclei is expressed as chemical shift and intensity, the latter being directly proportional to the number of atoms. Both NMR and MS-based systems can be equipped with automatic samplers to enable analyzing large series of samples without manual operation.

16.3 ANALYTICAL TECHNIQUES

NMR has traditionally been used for the characterization of pure compounds, but advances in the instrumentation and data processing have made profiling complex biospecimens feasible (Bothwell and Griffin 2011; Forseth and Schroeder 2011). Currently the total number of measurements achieved from plasma or serum by established proton-NMR platforms is around 200 (Soininen et al. 2015). The main advantage of NMR in metabolomics lies in its quantitative nature and in its ability to detect certain unique features such as blood lipoproteins. However, the relatively low number of metabolites detected in biospecimens limits its utility in finding dietary biomarkers. NMR is able to detect molecule-specific signals for a limited, relatively well-characterized group of molecules with concentrations at micromolar to millimolar levels including some amino acids, fatty acids, and other predominant metabolites like cholesterol, citrate, and hippuric acid (Scalbert et al. 2014; Soininen et al. 2015). MS is far more sensitive technique (Pan and Raftery 2007), and depending on the method and instrumentation can detect a wide variety of compounds from nanomolar to millimolar levels, with the total numbers ranging from several hundreds to few thousands per sample. In contrast to NMR, however, MS-based methods in general do not permit absolute quantification without calibration standards, and thus the analysis of untargeted data is based on relative quantification using measured intensities rather than concentrations. The data generated during sample analysis also does not enable full elucidation of chemical structures, requiring the analysis of reference standards for compound identification. However, due to its superior sensitivity, MS is now preferred to NMR for dietary biomarker research, which is reflected in the greater number of dietary biomarkers discovered so far (Scalbert et al. 2014). In this chapter, the focus will thus be mostly on MS-based metabolomics.

The main prerequisite for effective biomarker discovery with MS-based systems is the acquisition of high-resolution mass spectra with accurate mass measurement. High resolution is required for the selective detection of compounds with near identical masses, such as L-canavanin (a compound found in some leguminous plants) and serotonin, which are measured at 176.09093 and 176.09495 Da, respectively, having a mass difference of only 0.00402 Da. In complex biological samples, the resolution can determine whether a compound is observed or missed due to masking interferences. Combining high resolution with accuracy then enables the calculation of elemental compositions that match with the measured mass, facilitating the identification of unknown compounds. There are currently three types of MS analyzers capable of performance that is considered sufficient for metabolomics: (1) time-of-flight (TOF), (2) Fourier transform ion cyclotron resonance (FT-ICR), and (3) orbital ion trap (Orbitrap). FT-ICR and Orbitrap can achieve very high resolution, but have been less commonly used for dietary biomarker research than TOF (Scalbert et al. 2014). Earlier TOF instruments had relatively narrow dynamic range, which limited their suitability for metabolomics, but the more recent instruments can commonly cover a concentration range greater than four orders of magnitude, significantly increasing the number of compounds for which mass can be accurately measured. Orbitrap is the youngest of the technologies and can achieve both greater resolution and dynamic range than TOF. The resolution of some Orbitrap instruments is close to that achievable with FT-ICR, which has long been the reference in ultimate resolution.

Regardless of the type of the MS instrument, metabolomics methods can be broadly divided between direct analysis of the sample and separation-based techniques. In direct analysis, the sample is typically introduced into the mass spectrometer with the help of LC pump and autosampler, relying on the MS for selective compound detection. The main advantages of this technique are its high throughput and robustness due to the lack of chromatographic separation, which commonly takes at least 10 minutes per sample (Draper et al. 2013). However, despite the high mass resolution achieved by current instruments, chromatography has remained important in metabolomics, as separation prior to detection adds retention time as an additional dimension to the acquired data, enabling selective analysis of isobaric compounds and reduction of interferences from the sample matrix. In the case of LC–MS, the chromatographic methods are most commonly based on reversed phase columns, with a number of different column dimensions and selectivities being used (Tulipani et al. 2015). Reversed phase chromatography has an established position in LC–MS metabolomics, but typically cannot provide sufficient retention and separation for highly polar molecules such as carbohydrates, phosphates, amino acids, and other short-chain organic acids that are likely to represent a large portion of dietary metabolites. Fortunately, alternative chromatographic techniques exist, and applying two or more orthogonal methods can significantly extend the number of compounds detected (Boudah et al. 2014; Ramakrishnan et al. 2016). Hydrophilic interaction chromatography (HILIC) in particular has been found to be complementary to reversed phase in metabolomics (Kloos et al. 2013) and has enabled successful analysis of highly polar dietary metabolites such as malic acid (Pekkinen et al. 2014) and amino acid-derived betaines (Pekkinen et al. 2015).

Interfacing of the LC system to MS is made through an ion source, which is used to evaporate the sample and ionize the dissolved molecules for their detection as ions. Different ion sources are available which will cause major differences in metabolite coverage. Most published metabolomics methods have been based on electrospray ionization (ESI), which is most suitable for the analysis of compounds that are ionizable in solution. ESI is a soft ionization technique that does not induce excessive fragmentation of chemically labile compounds, but is somewhat susceptible to coeluting matrix interferences from the biological background. Matrix effects, such as ion suppression, can mask the concentration–response relationship or render a compound undetectable. Alternative ion sources such as atmospheric pressure chemical ionization (APCI) and atmospheric pressure photoionization (APPI) are available, which both enable more efficient ionization of neutral compounds and suffer less from matrix effects, although with less efficient ionization of highly polar metabolites and more in-source fragmentation. The potential benefits of APCI and APPI in metabolomics have been well recognized, but they have remained less commonly used (Ernst et al. 2014; Mirnaghi and Caudy 2014). Irrespective of the ion source type, an important parameter is the ionization mode used to create the ions. Depending on the molecular structure, it may be possible to detect a metabolite only as a cation or an anion, and it is thus advantageous to perform sample analyses with both polarities. In addition to ion sources, a promising new technology for LC–MS metabolomics is ion mobility MS. This is achieved with a hybrid instrument where the ions generated in the ion source are taken into a gas-phase separation system before mass analysis. This enables measuring the drift time for the ions, which is analogous to LC retention time, and can add another dimension to the data that can increase the number of selectively detected compounds (Dwivedi et al. 2010; May et al. 2015). One challenge of the technique lies in the processing of the acquired data, which require software that can fully exploit the additional drift time measurements.

GC–MS is less commonly used for untargeted profiling in dietary biomarker studies, but offers another mode of chromatographic separation and ionization (Pasikanti et al. 2008). A characteristic feature of GC–MS analysis is the need for chemical derivatization of compounds that are not volatile or thermally stabile (perhaps most of the metabolites), which is not necessary for LC–MS analysis. The separation efficiency however can be significantly higher for many compounds that are difficult to analyze with LC–MS. Also the most commonly employed ionization technique for GC–MS, electron ionization, creates fragmentation patterns characteristic to particular molecules regardless of the instrumentation used, being readily compatible with large spectral libraries. In addition to electron ionization, chemical ionization can be used with GC–MS, being comparable with that employed in LC–MS. Two-dimensional GC coupled to MS has also been successfully applied to metabolomics to further increase the separation efficiency and amount of information from the injected sample (Rocha et al. 2012). In this technique, compounds eluting from the first column are introduced into a second, complementary column for increased separation efficiency (Welthagen et al. 2005).

In conclusion, a range of complementary instrument techniques exists and can be used in combination, or based on the expected metabolites of interest. So far LC–MS, and LC–QTOF–MS in particular, has been the most common technique for dietary biomarker research (Scalbert et al. 2014). It is clear however that no single instrument or method enables analysis of all the anticipated metabolites, and the challenge remains to cover a reasonable fraction of the broad range of compounds in a sample, with sufficient throughput and robustness.

16.4 PREPARATION OF BIOSPECIMENS

Sample preparation methods for metabolomics are largely dictated by the requirements of the analytical technique used. In untargeted metabolomics, the aim is usually to achieve maximum coverage of low molecular weight compounds with minimal method-related variability, while achieving sufficient sample clean up or other processing needed (Chen et al. 2016). Sample preparation has a major effect on the acquired data, but its evaluation is difficult due to the lack of specific target compounds. The applicability of different existing methods to untargeted metabolomics has been a topic of several investigations (Rico et al. 2014; Sarafian et al. 2014; Schimek et al. 2016; Yanes et al. 2011; Whiley et al 2012), but because of different ways to assess a method's performance various approaches have been proposed as optimal.

Overall, the main objective of blood sample preparation for LC–MS and GC–MS analysis is removal of abundant proteins. The protein content in human blood is high, typically around 3–5 g albumin per 100 mL of serum (Rustad et al. 2004), and this will form an insoluble precipitate that must be removed before injection. This has been traditionally performed by mixing the sample with organic solvent such as different alcohols or acetonitrile, followed by removal of the precipitate by centrifugation or filtration. In the case of LC–MS, the resulting sample can then be analyzed without further processing. Since metabolomics typically aims to retain the broadest possible range of compounds in a sample, protein precipitation has become a popular method as it does not involve fractionation and removal of soluble sample components. A disadvantage is that the wide diversity of compounds present can cause some analytical problems including matrix effects (Souverain et al. 2004). Several other methods also exist and have been used for metabolomics, alone or combined, being usually based on some form of fractionation to isolate the anticipated analytes of interest from potential matrix interferences, or otherwise reduce the amount of compounds that may interfere with the analytes (Armirotti et al. 2014; Whiley et al. 2012). Fractionation is commonly achieved by liquid–liquid or solid phase extraction, and has been found to be beneficial for the untargeted profiling of groups of specific metabolites such as lipids (Sarafian et al. 2014). Splitting the sample into fractions based on lipophilicity and employing the chromatographic methods best suited to each fraction has been proposed as an approach to explore a wide polarity range from sugars ($\mathrm{Log}P < -2$) to triacylglycerols ($\mathrm{Log}P > 10$) (Armirotti et al. 2014). In recent years, methods for selective removal of blood phospholipids, which are considered a major source of ion suppression and ion source contamination, have become increasingly popular. These are based on classical protein precipitation and filtration, but employ a filter element that is designed to selectively retain phospholipids. Using such a technique, minimal number of other compounds that might be biologically relevant are removed from the sample. The technique has been successfully used for untargeted LC–MS metabolomics and found to be effective in reducing matrix effects when compared with deproteinization alone (Tulipani et al. 2013, 2015).

For the LC–MS analysis of urine samples and other biospecimens with very low protein and particle content, the aforementioned techniques may not be needed, and simply diluting the sample with suitable solvent can be sufficient. However, the variability associated with large differences in urine sample concentrations has to be controlled, irrespective of the analytical technique. There are different methods of normalization to achieve this, and these can be divided in those that are performed before or after the sample analysis. Normalization before the sample analysis has been made through measuring the specific gravity of the samples by refractometry and diluting the samples to the lowest measured gravity, or by adjusting the injection volume based on the measured creatinine levels in the samples. Post-acquisition normalization methods include adjusting the acquired data using specific gravity, urine volume, or combinations of these (Edmands et al. 2014).

For GC–MS metabolomics, sample preparation is slightly less straightforward. Derivatization is needed for the analysis of samples containing nonvolatile, polar or thermally labile molecules, regardless of the sample matrix used. After the initial extraction or protein precipitation, evaporation of the sample is typically required, followed by reconstitution of the residue in a derivatization mixture. A great number of derivatization reagents are available, with various approaches having been developed for untargeted metabolomics (Karamani et al. 2013; Khakimov et al. 2013). Commonly employed reactions are based on silylation, often following oximation (Abbiss et al. 2015). A challenge is to achieve complete, reproducible derivatization of a wide range of metabolites within a given reaction time (Kanani et al. 2008). Although derivatization adds complexity to the sample preparation, it can also be used for LC–MS metabolomics to enhance ionization or separation of compounds that are difficult to detect or separate. A common technique for labeling various nucleophilic compounds employs dansyl chloride, a fluorescent-labeling reagent traditionally used for amino acid analysis, which has also been used in an untargeted fashion for metabolomics (Guo et al. 2011; Peng et al. 2014). This can be a particularly efficient method to broaden the metabolite coverage of a method to compounds such as dietary polyphenols that are better separated on the chromatographic column and easier to ionize after derivatization (Achaintre et al. 2016). In addition to these techniques, enzymatic hydrolysis of conjugated metabolites can be performed as part of the sample preparation. This may be used to reduce the total number of different conjugates of given metabolites or to facilitate their analysis by GC–MS when performed prior to derivatization (Grün et al. 2008).

In quantitative MS-based bioanalysis, addition of internal standards is usually performed during the sample preparation. Quantification is then based on the response ratio of the analyte to the internal standard, calculated as a function of the concentration of the analyte. This is an efficient way of compensating for multiple sources of variability, and can cover the entire analytical process from sample extraction to detection. However, due to physicochemical differences between the analytes, any single compound cannot be used as a universal internal standard, and a common practice is to incorporate a stable isotope labeled analogue of each analyte whenever possible. In untargeted metabolomics, the use of individual internal standards is not possible, but it may be feasible to employ carefully selected internal standards to assist in normalizing the data for variability in injection volumes or other factors that are not compound specific.

The sample volumes required for metabolomics do not differ from common bioanalysis, being generally in the range from tens to hundreds of microliters. However, there are some limitations that are related to the untargeted nature of the analyses. Lower detection limits can generally be achieved by injecting higher molar amounts, but due to the large range of concentrations in biological samples detector saturation for some of the compounds is commonly reached even with diluted, sub-50-μL samples. Thus, increasing the injected amount further would result in increased number of both saturated and detected compounds. This is particularly common with TOF instruments that have relatively narrow dynamic range. Smaller injected molar amounts may also be an effective way of preventing ion source contamination and the resulting changes in response over time, enabling the analysis of larger batches of samples. As a result, relatively small sample volumes are commonly used for MS-based metabolomics.

Although NMR has been much less commonly used in dietary biomarker studies, it benefits from a straightforward sample preparation. For the commonly analyzed blood and urine, minimal work is needed, and although similar fractionation methods as those discussed earlier can be employed, mixing of a particulate-free sample with a buffer solution for pH control is often the only necessary step. Typically the added buffer solution also contains an internal reference such as trimethylsilyl propionate in deuterium oxide (D_2O) and a suitable bacterioside such as sodium azide (Soininen et al. 2009). Due to the nondestructive sample preparation, valuable samples also remain available for other types of analysis.

Finally, regardless of the analytical technique used, it is important to ensure the comparability of the samples by minimizing all sources of variability during sample preparation. In addition to the factors such as extraction recovery, chemical stability of the samples is of major concern.

Assessment of analyte stability is an important part of bioanalytical assay validation in regulated laboratories (Viswanathan et al. 2007), and although stability studies can be difficult to perform with untargeted methods, it is obvious that chemical degradation during uncontrolled conditions can introduce unexpected differences between samples. There are evidence of sample storage at room temperature and at 4°C, as well as freeze–thaw cycles and storage time affecting the metabolomics results (Peakman and Elliott 2008; Teahan et al. 2006), highlighting the importance of controlling the conditions during sample processing.

16.5 PROCESSING OF ACQUIRED RAW DATA

The raw data generated by untargeted analyses require processing to obtain the final intensity matrix to be submitted to statistical analysis. In the case of NMR this includes elucidating of individual molecules from the spectra that are often complex due to the great variety of molecules in a biological matrix. By employing techniques including baseline correction, linefitting, and regression analysis, quantification is achieved for the compounds detected (Chang et al. 2007; Mihaleva et al. 2014). A typical workflow for a dietary assessment study using NMR metabolomics is described by O'Sullivan et al. (2011). Processing of MS data is fundamentally different. Targeted, quantitative analysis is commonly performed by monitoring the abundance of preselected ions. These are detected as chromatographic peaks, with their intensity serving as a basis for quantification, using calibration curves of external or internal standards. In untargeted metabolomics, however, intensities of all the ions are recorded throughout the chromatographic run, with no *a priori* knowledge of which would be meaningful. After the data acquisition, algorithm-based data mining is performed, including several steps that aim at finding ions resembling organic molecules by their chromatographic behavior. Strongly correlated ions are grouped if they match with the predicted ions that are known to originate from common organic molecules (isotopes, fragments, or different ion species of the same compound). The data from individual samples are then aligned so that each feature receives the same mass and retention time across samples, despite variation in the measurements along the analytical batch. The output is typically a table of intensities for each feature and sample analyzed. Preparation for statistical analysis such as transformation, normalization, and imputation of missing values may be beneficial (Godzien et al. 2013; Veselkov et al. 2011), although the need for and the best use of such steps differs depending on the subsequent statistical analysis, and may be very different for univariate and multivariate analyses (Di Guida et al. 2016).

Processing of metabolomics data relies on algorithms. Several solutions for MS data exist, both commercial software from the instrument manufacturers and open-source approaches including XCMS (Smith et al. 2006) and its many extensions (Mahieu et al. 2016), MZmine (Pluskal et al. 2010), MetAlign (Lommen 2009) and others. Regardless of the algorithm, a common aim is to reproducibly find all the detectable compounds in every sample, with high sensitivity and minimal amount of noise. There are some differences in the way the various available solutions achieve this, and it has been found that they may not produce entirely identical sets of features from the same raw data (Coble and Fraga 2014; Koh et al. 2010; Niu et al. 2014; Rafieiand Sleno 2015). The outcome may also vary depending on individual operators' subjective evaluation of the raw data quality, and there have been attempts to decrease the operator influence on the process (Libiseller et al. 2015; Peters et al. 2009). However, despite the challenges related to the richness of the data, recent developments in processing algorithms and other bioinformatics tools have had a fundamental role in the success and advancement of metabolomics (Johnson et al. 2015).

16.6 QUALITY CONTROL

Established guidelines for bioanalytical method validation (FDA 2001; Viswanathan et al. 2007) are often used in quantitative bioanalysis for assessing assay performance and setting quality control (QC) practices. However, these are difficult to implement in metabolomics due to the lack of predefined target analytes. A common practice in metabolomics is to include QC samples in the analytical batch, such as samples from a pool that has been created by mixing small aliquots of study samples. These are then analyzed in regular intervals throughout the batch. Different approaches for using the obtained data have been proposed, often aiming at correcting or reducing analytical variation (Dunn et al. 2012; van der Kloet et al. 2009). Some traditional quality measures such as accuracy are not applicable as concentrations are not measured, and thus QC is usually limited to assessing reproducibility of detection and intensity measurement (Godzien et al. 2015; Smilde et al. 2009). The data from replicates may also be used to exclude poorly reproduced features before statistical analyses, or serve as a basis for accepting or rejecting the batch. Different solutions have been developed for correcting and improving the data using repeated samples, such as adjustments for drifts in overall response and retention times (Chen et al. 2014; Dunn et al. 2011; Edmands et al. 2015; Fernández-Albert et al. 2014).

A widely used target for reproducibility is the relative standard deviation (RSD) of repeated measurement, which is usually acceptable when less than 15% throughout the measured concentration range, except at the lower limit of quantification where it can be 20%. In untargeted metabolomics, it is possible to use data from repeated analysis of a pooled QC sample to calculate RSD for a great number of features, but due to heteroscedasticity of the data and a single concentration level analyzed, using the 15%–20% RSD as an acceptance criterion may be too strict. To overcome this, each sample can also be injected multiple times to calculate technical variability for each feature (Go et al. 2015). Irrespective of the approach, assessing reproducibility is important and reflects the level of expected nonbiological variation, and needs to be taken into account when filtering the data, and as part of the study design. Assessing the variability in untargeted data using QC samples may also be helpful in evaluating and improving the methods (Smilde et al. 2009). It may be considered as a form of method validation or used as a system suitability testing with predefined acceptance criteria (e.g., for response and retention time variability) to ensure that the method is producing acceptable results. Finally it may be necessary to acquire analyze large series of samples as several analytical batches or by employing more than one instrument, and due to potential differences in measured intensities or retention times, it may be necessary to perform adjustments or corrections of the data to enable its pooling into a single feature table. To achieve this, inclusion of common reference samples or QCs in all the batches can be an effective way to enable post-acquisition data adjustments (Dane et al. 2014).

16.7 STATISTICAL ANALYSES

To discover biomarkers of food intake, the food intake of intervention or observational study participants (the y-variable) is first modeled using the metabolomics data (x-variables) to find those features, which are associated. The food intake of a study group follows different distributions, depending on the nature of the food and the study design. In intervention studies, this is a dichotomous or discrete variable, and in the simplest case, only consumer and nonconsumer levels may be present. Measurement error is negligible since the dose is carefully controlled by the experimenter. For discovery studies, intake as calculated from dietary survey is continuous, but subject to self-reporting inaccuracy, and unlikely to follow normal distributions. For example, many low and high consumers may be observed, with fewer moderate consumers, or the overall distribution may be uniform. The distribution may also contain a nonnegligible number of high outliers. Such distributions are typically transformed, where possible, to normality prior to statistical analyses. For continuous data, nonconsumers may cause problems for subsequent statistical models.

Exploratory analyses of metabolomics datasets are systematically performed to prepare them for modeling. As a preliminary check, the raw data file size or the number of features may be examined for each sample. This aids in identifying erroneous profiles produced by technical failure and experimental samples can be compared to embedded QCs and blanks. At the next level, a principal component analysis (PCA) is usually employed to characterize the correlation structure of the data and check that sufficient biological variation has been captured by the analysis (Bro and Smilde 2014). Principal component scores are plotted to identify outliers and assess the relative magnitudes of sources of variability in the experiment. Temporal changes in experimental conditions that may cause the drift of sample profiles can also be monitored. A limitation of PCA in untargeted metabolomics is that the datasets commonly include various amounts of noise and background features from solvents and other contaminants, so that not all information is originating from the biospecimens analyzed.

Once outliers have been removed and the integrity of metabolomics data checked, either univariate or multivariate models may be used to test for associations of spectral features with food intake. Statistical analysis of data derived from intervention studies, with a discrete *y*-variable, is the simplest case. Intensities from each signal can be compared between consumer and control groups using a *t*-test, ANOVA, or nonparametric equivalents. Where subject data are paired (such as two different treatments at different times on the same subject), paired *t*-tests or equivalent may increase statistical power. The association of each feature with food intake will generate a *p*-value for significance and the resulting vector of *p*-values should then be adjusted for multiple testing to limit the possibility of false positive associations. These *p*-values are used in conjunction with fold changes between control and exposed groups and average signal intensity to identify those features to be subsequently annotated.

For continuous intake data, studies so far used to identify discriminant signals for food intake in metabolomics data have taken one of the two approaches. The first is to create discrete intake groups by splitting data into quantiles, for example low and high consumers or consumers and nonconsumers, and then model high and low groups as a dichotomous variable. This is sometimes used to circumvent the problem of skewed intake distributions. In addition, uncertainty regarding intake measurements may be reduced, as reports of high and low intake may be more reliable than reports of moderate intake. Confounding, due to imbalances in participant location, sex, or age, is difficult to control for, and high and low intake groups must be matched as closely as possible. Once discrete groups are established, statistical models can be applied as described earlier. As well as univariate models, many studies have used multivariate predictive models such as partial least squares discriminant analysis (PLS-DA), random forests or support vector machines. Instead of performing hypothesis tests on each feature, the contribution of each variable to the predictive model is used as a measure of importance, and in PLS-DA for example, features with a variable importance in projection (VIP) > 1.5 are usually retained for further investigation. The ability of discovered markers to classify subjects into low or high intake groups is frequently testing using ROC (receiver operating characteristic) analysis. Areas under ROC curves summarize the efficacy of the classifier, accounting for both specificity and sensitivity over all possible cut-points (Xia et al. 2013). When discrete intake groups are used, univariate and multivariate analyses are often performed on the same data and only features that perform well in both analyses are retained.

A second approach is the calculation of correlation coefficients. For the Pearson product-moment correlation, intake data must first be transformed to follow a sufficiently normal distribution. Otherwise, the Spearman rank correlation may be used, but at the expense of statistical power. The correlations of each signal with transformed food intake are computed and those which are statistically significant, or exceed a certain cut-point, are retained for further investigation. The use of partial correlation coefficients allows adjustment for potential confounders of food intake by the inclusion of covariates. Intake might, for example, be confounded by country or geographical location, and might also be associated with the intake of other foods or lifestyle habits that influence metabolism, such as smoking or physical activity (Guertin et al. 2015).

p-values generated by univariate models are commonly adjusted to reduce the risk of false positives, of which two are most commonly employed. The simplest, and most conservative, is the Bonferroni family-wise multiple testing adjustment. *p*-values are multiplied by the number of statistical tests before acceptance or rejection of null hypotheses. This, however, is prone to false negatives when thousands of tests are performed, and recently the false discovery rate (FDR) method (Benjamini and Hochberg 1995) has been gaining popularity. FDR involves a weighted adjustment of p-values depending on a predetermined FDR and p-value rank. Finally, features found to be significantly associated with food intake after adjustment should be checked for plausibility as potential biomarkers. The discriminant feature should be detectable in the majority of food consumers; if this is not the case, the discriminant is unlikely to pass further validation tests. Once any signal not matching these criteria have been filtered out, those remaining are passed to the subsequent annotation steps.

16.8 IDENTIFICATION OF METABOLITES

The final step after the statistical analysis is to convert the discriminant features to identified molecules. In LC–MS-based metabolomics, clusters of discriminant ions at the same retention time are often observed and may correspond to a single metabolite only. Also, where features originate from the same compound, the intensities will be highly correlated across subjects. Closely related features can therefore be grouped by pairwise calculation of correlation coefficients followed by hierarchical clustering. Using this approach, it is often possible to deduce parent masses to prioritize in subsequent annotation steps. The elemental composition of the molecule can then be calculated from the measured monoisotopic mass, which is determined from the mass-to-charge (m/z) value of the ion detected. The conversion is commonly performed by the algorithms used for feature finding or later stages of processing, and is based on predefined common ion species such as proton and sodium adducts. Manual raw data inspection and mass assignment may also be necessary to ensure correctness of the algorithm-based assignments, which may be prone to errors due to the presence of undefined ion species such as in-source fragments. When the accurate mass is known, it is then possible to compute a combination of elements encountered in common metabolites (such as C, H, O, N, P, S) that results in the same mass. This is a simple process that relies on the high mass accuracy, but it works well only for very low molecular weights: the number of different elemental compositions with almost identical mass grows exponentially with the mass, resulting in too many matching compounds. It has been demonstrated that it is not feasible to assign a unique formula for common metabolites based on mass alone, even with very high mass accuracy (<ppm) (Kind and Fiehn 2006). Fortunately the mass spectra also contain isotope peaks of each element of a compound, and when grouped together during the processing, an isotope pattern can be obtained. Comparing both the mass and the isotope pattern of an unknown metabolite to those generated from a candidate elemental composition enables significantly higher confidence in the compound identification. Additional structural information for the compounds can be obtained by fragmentation, commonly termed MS/MS. This is performed by breaking down the chemical structures in the mass spectrometer into fragments that are detected as ions in a similar fashion as intact molecules. However, deciphering the identity of an unknown compound from its mass, isotopes, and fragments is difficult, and to facilitate the process, databases have been developed that enable searches by mass, elemental composition and MS/MS spectra. The use of spectral databases for compound identification in GC–MS has a long history due to the compound-specific fragment ions generated in electron ionization. Large databases have been routinely used for decades, with the number of compounds included reaching well over 100,000 already by the late 1980s (McLafferty et al. 1991). For LC–MS, such extensive databases have been more difficult to create due to the differences in spectra depending on the instrumentation used for their creation (Rathahao-Paris et al. 2016). However, current metabolite repositories such as Metlin, HMDB, MassBank, and mzCloud include a growing number of MS/MS spectra specific to LC–MS. Dedicated databases for food-derived compounds, such as FoodDB, has also been developed (Scalbert et al. 2011). A comprehensive review of the databases for MS-based metabolomics has been made by Vinaixa et al. (2016).

With the help of existing databases, an unknown compound with a known mass can be placed in three categories: (1) known metabolites that can be found in databases that are searchable by

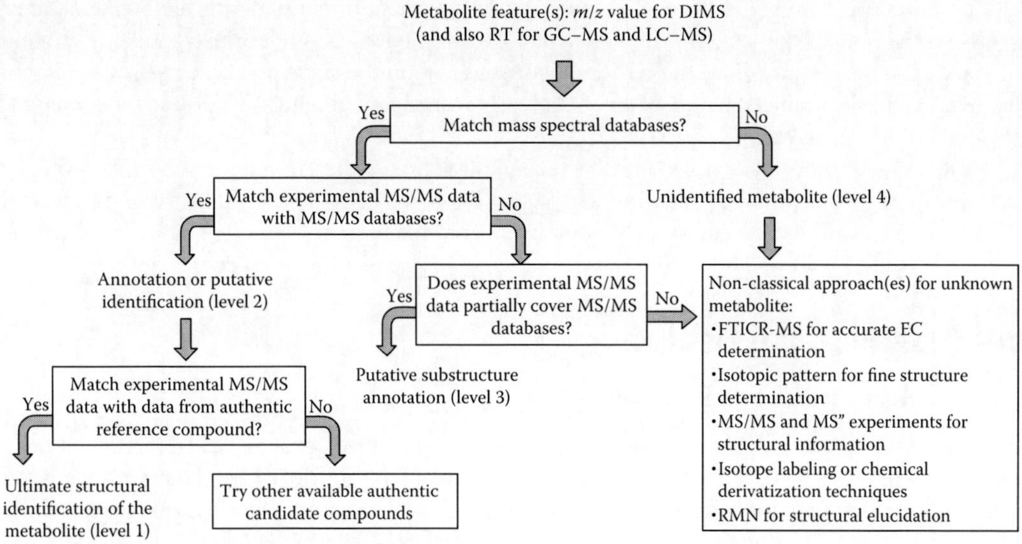

FIGURE 16.1 A strategy for annotation of unknown metabolites. (From Rathatao-Paris et al. *Metabolomics*, 12, 10, 2016.)

mass, (2) known metabolites that are not included in such databases, and (3) metabolites that are not known to exist. The identification strategy for the first category is based on confirmation of the database matches, which ultimately relies on comparing the data obtained from sample analysis against that of a pure chemical standard under identical conditions. For the second and third category, identification is based on structural elucidation using accurate mass and isotope pattern and additional information including MS/MS spectra. An example of such strategy is presented in Figure 16.1. Although MS/MS spectra are commonly acquired using targeted analysis, for example, isolation of the ions of interest for subsequent fragmentation, it can be performed in an untargeted fashion already during sample analysis with data-dependent or data-independent methods. The first works by cycling between the full scan and MS/MS modes and selecting some abundant ions from full scan for fragmentation. The second usually fragments a broader range of ions, creating a complex spectrum that is then deconvoluted to pair the precursor and productions. These and other available techniques employed for the structural elucidation of unknowns has been recently reviewed by Rathahao-Paris et al. (2016).

Definite identification of an unknown metabolite is not always achieved, and to harmonize the reporting standards for the identification, The Chemical Analysis Working Group of the Metabolomics Standards Initiative (MSI) has set four levels of identification (Figure 16.1) (Goodacre et al. 2007). Level 1, *Identified compounds* requires at least two independent and orthogonal data relative to an authentic compound analyzed under identical experimental conditions; Level 2, *Putatively annotated compounds* are identified without reference standards; Level 3, *Putatively characterized compound classes* are as above but related to chemical classes, and Level 4 is a category *unknown compounds*. Since reaching a definite identification is far faster through the confirmation of a database match than by chemistry-based inference, the rapid increase in number and quality of compound databases has been a notable advancement in metabolomics. In addition, the development of software tools such as CSI:FingerID, xMSannotator, and MetFusion that employ various metabolite databases and the chemical information acquired (isotope pattern, MS/MS fragmentation) can significantly speed up annotation.

Identification of unknowns from NMR spectra enables a comparable untargeted approach. NMR has traditionally been used for the unambiguous identification of molecules, both known and novel, and is commonly the technique required by scientific journals for the confirmation of synthesized organic compounds. Although a biological sample is a complex matrix, identification of the unknown metabolites from the NMR spectra can rely on the same principles as is used in confirming synthesis products, but also benefits greatly from spectral databases of known compounds similarly to MS (Larive et al. 2015).

Regardless of other evidence collected for annotation of a discriminant feature, its biological plausibility should always be taken into account. This is particularly relevant for the discovery of food biomarkers, whose precursor food components are likely to be well documented in published journal articles, books and open source or commercial databases (Scalbert et al. 2011).

16.9 EXPERIMENTAL DESIGN

16.9.1 Requirements for Biomarker Discovery

The basis of all study designs is a group of participants who have consumed different quantities of the food of interest. Intakes are either set by the experimenter or retrieved from dietary records and other information on temporality or repeated doses may also be used. Over the full set of subjects in the study, these intake data will form the response variable (y-variable) for statistical modeling, a topic covered in more detail in Chapters 7 and 13 of this book. Biospecimens from all participants are then required for metabolite profiling. The resulting dataset consists of n observations of k variables and will be used to model food intake using either univariate or multivariate techniques. The biomarkers identified will partly depend on the biospecimen analyzed, although some overlap between biomarkers from different biospecimens is often found.

16.9.2 Intervention and Observational Study Designs

Biomarkers of food intake are discovered either by intervention studies, where the experimenter controls the behavior of study participants prior to biospecimen collection, or by observational studies, where stored biospecimens are analyzed and associated with previously collected food intake data. Intervention designs carefully control the food intake of subjects, standardizing the dose, background diet and the time period between food intake and biospecimen collection. A wash-out, where the subjects avoid the food of interest and other possible interferences for a set time period prior to food ingestion, is included in the design. Duration of the wash-out period is determined in order to warrant the return of the biomarker level to baseline before the test food is ingested. Since the time from food administration to sample collection is known, and samples can be collected at multiple time points, interventions are useful for additionally investigating the metabolism and kinetics of potential biomarkers, or establishing a dose response (Scalbert et al. 2014). A cross-over design may be used to control for changing conditions over the course of the intervention. Although a high degree of control and customization by the researcher is permitted, markers found may only reflect intake over a known time period and therefore may not be applicable to subjects consuming their habitual diets. In addition, interventions have high labor costs due to the need for subject recruitment and sample collection.

With the arrival of more sensitive metabolite profiling techniques, it has become possible to use observational cross-sectional or cohort studies for the discovery of food biomarkers (Lloyd et al. 2013; Pujos-Guillot et al. 2013). Individual food intake data are obtained from dietary surveys such as 24-hour dietary recalls (24HRs; assessing acute food intake) or from food frequency questionnaires (FFQ; assessing habitual food intake), in which subjects report their consumption of a range of foods or food groups (Penn et al. 2010). Biospecimens may be collected on the same day as 24HR and metabolomics data used to identify biomarkers of acute dietary intake as most often done in intervention studies. The same metabolomics data combined with habitual (usual) dietary data collected with either multiple dietary recalls or FFQ can also be exploited to identify biomarkers

TABLE 16.2
The Advantages and Limitations of Intervention and Observational Study Designs for Discovery of Food Intake Biomarkers

	Controlled Intervention Study	Observational Study (Cohort or Cross-Sectional)
Food intake data	Precise dose and carefully controlled diet	Data obtained retrospectively from dietary surveys
Self-reported bias	Eliminated by nature of study design	Substantial, depending on the food
Biospecimens	Can be selected	Available retrospectively
Signal to noise ratio	High	Low, as diet is not controlled
Temporality of markers found	Limited to timescale of the study	Depends on dietary data available
Specificity of markers found	High as interferences removed	Not guaranteed
Utility of markers found	Not guaranteed as origin from other sources cannot be excluded	Should be robust to other interferences
Costs	High due to recruitment of subjects and sample collection	Low as samples obtained retrospectively

of habitual food intake. One drawback here is the possibility of self-reporting bias, where certain foods are under- or over-reported (Fave et al. 2009). Food intake may also be reported differentially in multicentre studies, and this should be controlled for when building statistical models. Although discriminant signal to noise ratio is lower, labor costs are lower than those of intervention studies since biospecimens are obtained retrospectively. Furthermore, the larger study groups and detailed food intake data allow the testing of multiple hypotheses from a single dataset. Care must be taken to check that intake of the food of interest is not correlated with that of other foods, or identified discriminants of intake may be spurious. Markers found in observational studies must be carefully checked for specificity for the food of interest. This should be the only or predominant source of the putative marker. The advantages and limitations of intervention and observational studies are summarized in Table 16.2.

16.10 FUTURE RESEARCH: TARGETED ANALYSES

Although metabolomics as a technique is usually associated with untargeted analysis, targeted quantitative analysis of a number of compounds can also be performed and can be considered as targeted metabolomics. This is traditionally based on multiplex or multianalyte methods, including commercially available kits. The common challenge of many multianalyte methods in quantitative analysis is the difficulty to fulfill quantification criteria described in the validation guidelines for each analyte. For instance, using established quantification practices including an individual isotope labeled internal standards and calibration curves for hundreds or even thousands of compounds is rarely possible, necessitating the use of common standards for compounds within chemical class or other groups of analytes. Regardless of the need for accurate quantification, the targeted approach also limits the potential intake markers to those included in the method. This may however be acceptable if there are anticipated metabolites of interest that are within certain groups of metabolites that are best analyzed with a dedicated method. This can be particularly important for metabolites that require specific sample preparation or instrument conditions that cannot be used in a method intended for untargeted analysis such as compounds with extremely high or low polarity. An example of a larger group of metabolites for which a targeted approach has been successfully applied is dietary polyphenols (Achaintre et al. 2016; Zamora-Ros et al. 2016).

However, it may not always be necessary to reach the requirements typically set for quantitative bioanalysis, and targeted metabolomics may also rely on estimating the concentrations with semiquantitative methodology, or relative quantification. In this case, the difference with untargeted approach is that the metabolites in targeted analysis would be detected with proven specificity and concentration–response relationship. There has recently been increased interest in merging the untargeted and targeted methods to enable quantification. One approach is termed chemical isotope labeling or isotope coded derivatization, where the sample is derivatized with a reagent such as dansyl chloride, and then spiked with a reference solution that is separately derivatized with isotopically labeled reagent, an approach reviewed by Bruheim et al. (2013). Another promising technique, reference standardization, is based on calibrating the response for various detected metabolites to those obtained from the analysis of a reference sample such as NIST's reference human plasma SRM1950, and using this as a one-point calibrator (Go et al. 2015). It is also possible to employ an untargeted method while including traditional calibrators in the method for selected compounds, or embedding time segments or other compound-specific analytical condition in the otherwise untargeted method. The potential for such techniques is partly related to the capabilities of MS instrumentation used, as the targeted analysis often requires compound-specific MS/MS transitions or other operational parameters that limit the applicability of the method for simultaneous untargeted analysis. However, the increased performance of MS instruments has made this increasingly feasible. This topic has been recently reviewed and discussed by Cajka and Fiehn (2016).

Other major future research interests that are aimed at facilitating the transformation of untargeted profiling results into structures and absolute concentrations include development of different bioinformatics and other computational tools to overcome the challenges in annotation of the unknown metabolites (Cho et al. 2014). For food biomarker studies, developing methods for the analysis of alternative biospecimens to blood and urine such as hair is also of interest (Chapter 15).

16.11 CONCLUSION

Metabolomics is inherently suited to food biomarker discovery. The food metabolome is a rich source of potential biomarkers, and the data-driven approach permits the discovery of metabolites reflecting food intake that are not yet documented in published literature. In particular, untargeted LC–MS is effective due to its high sensitivity and wide coverage of metabolites. Instrumentation is evolving, and anticipating improving computational capacity and software algorithms to increase the efficiency of the data analysis metabolomics will play a prominent role in food biomarker discovery in the years to come.

REFERENCES

Abbiss, H., C. Rawlinson, G. L. Maker et al. 2015. Assessment of automated trimethylsilyl derivatization protocols for GC-MS-based untargeted metabolomic analysis of urine. *Metabolomics* 11:1908–1921.

Achaintre, D., A. Buleté, C. Cren-Olivé et al. 2016. Differential isotope labeling of 38 dietary polyphenols and their quantification in urine by liquid chromatography electrospray ionization tandem mass spectrometry. *Anal. Chem.* 88:2637–2644.

Armirotti, A., A. Basit, N. Realini et al. 2014. Sample preparation and orthogonal chromatography for broad polarity range plasma metabolomics: Application to human subjects with neurodegenerative dementia. *Anal. Biochem.* 455:48–54.

Benjamini, Y. and Y. Hochberg. 1995. Controlling the false discovery rate: A practical and powerful approach to multiple testing. *J. R. Statist. Soc. B* 57:289–300.

Bothwell, J. H. and J. L. Griffin. 2011. An introduction to biological nuclear magnetic resonance spectroscopy. *Biol. Rev. Camb. Philos. Soc.* 86:493–510.

Boudah, S., M. F. Olivier, S. Aros-Calt et al. 2014. Annotation of the human serum metabolome by coupling three liquid chromatography methods to high-resolution mass spectrometry. *J. Chromatogr. B* 966:34–47.

Bro, R. and A. K. Smilde. 2014. Principal component analysis. *Anal. Methods* 6:2812–2831.

Bruheim, P., H. F. N. Kvitvang, S. G. Villas-Boas. 2013. Stable isotope coded derivatizing reagents as internal standards in metabolite profiling. *J. Chromatogr. A* 1296:196–203.

Cajka, T. and O. Fiehn. 2016. Toward merging untargeted and targeted methods in mass spectrometry-based metabolomics and lipidomics. *Anal. Chem.* 88:524–545.

Chang, D., C. D. Banack, S. L. Shah. 2007. Robust baseline correction algorithm for signal dense NMR spectra. *J. Magn. Reson.* 187:288–292.

Chen, M., R. S. Rao, Y. Zhang et al. 2014. A modified data normalization method for GC-MS-based metabolomics to minimize batch variation. *Springerplus* 3:439. doi:10.1186/2193-1801-3-439.

Chen, Y., J. Xu, R. Zhang et al. 2016. Methods used to increase the comprehensive coverage of urinary and plasma metabolomes by MS. *Bioanalysis* 8:981–997.

Cho, K., N. G. Mahieu, S. L. Johnson, G. J. Patti. 2014. After the feature presentation: Technologies bridging untargeted metabolomics and biology. *Curr. Opin. Biotechnol.* 28:143–148.

Coble, J. B. and C. G. Fraga. 2014. Comparative evaluation of preprocessing freeware on chromatography/mass spectrometry data for signature discovery. *J. Chromatogr. A* 1358:155–164.

Dane, A. D., M. M. Hendriks, T. H. Reijmers et al. 2014. Integrating metabolomics profiling measurements across multiple biobanks. *Anal. Chem.* 86:4110–4114.

Di Guida, R., J. Engel, J. W. Allwood et al. 2016. Non-targeted UHPLC-MS metabolomic data processing methods: A comparative investigation of normalisation, missing value imputation, transformation and scaling. *Metabolomics* 12:93.

Draper, J., A. J. Lloyd, R. Goodacre et al. 2013. Flow infusion electrospray ionisation mass spectrometry for high throughput, non-targeted metabolite fingerprinting: A review. *Metabolomics* 9:S4–S29.

Dunn, W. B., D. Broadhurst, P. Begley et al. 2011. Procedures for large-scale metabolic profiling of serum and plasma using gas chromatography and liquid chromatography coupled to mass spectrometry. *Nat. Protoc.* 6:1060–1083.

Dunn, W. B., I. D. Wilson, A. W. Nicholls et al. 2012. The importance of experimental design and QC samples in large-scale and MS-driven untargeted metabolomic studies of humans. *Bioanalysis* 4:2249–2264.

Dwivedi, P., A. J. Schultz, H. H. Hill. 2010. Metabolic profiling of human blood by high resolution ion mobility mass spectrometry (IM-MS). *Int. J. Mass Spectrom.* 298:78–90.

Edmands, W. M., D. K. Barupal, A. Scalbert. 2015. MetMSLine: An automated and fully integrated pipeline for rapid processing of high-resolution LC-MS metabolomic datasets. *Bioinformatics* 31:788–790.

Edmands, W. M., P. Ferrari, A. Scalbert. 2014. Normalization to specific gravity prior to analysis improves information recovery from high resolution mass spectrometry metabolomic profiles of human urine. *Anal. Chem.* 86:10925–10931.

Ernst, M., D. B. Silva, R. R. Silva et al. 2014. Mass spectrometry in plant metabolomics strategies: From analytical platforms to data acquisition and processing. *Nat. Prod. Rep.* 31:784–806.

Fave, G., M. E. Beckmann, J. H. Draper et al. 2009. Measurement of dietary exposure: a challenging problem which may be overcome thanks to metabolomics? *Genes Nutr.* 4:135–141.

Food and Drug Administration (FDA). 2001. *Guidance for Industry, Bioanalytical Method Validation.* Rockville, MD: US Department of Health and Human Services, FDA, CDER.

Fernández-Albert, F., R. Llorach, M. Garcia-Aloy et al. 2014. Intensity drift removal in LC/MS metabolomics by common variance compensation. *Bioinformatics* 30:2899–2905.

Forseth, R. R. and F. C. Schroeder. 2011. NMR-spectroscopic analysis of mixtures: From structure to function. *Curr. Opin. Chem. Biol.* 15:38–47.

Go, Y. M., D. I. Walker, Y. Liang et al. 2015. Reference standardization for mass spectrometry and high-resolution metabolomics applications to exposome research. *Toxicol. Sci.* 148:531–543.

Godzien, J., V. Alonso-Herranz, C. Barbas et al. 2015. Controlling the quality of metabolomics data: new strategies to get the best out of the QC sample. *Metabolomics* 11:518–528.

Godzien, J., M. Ciborowski, S. Angulo et al. 2013. From numbers to a biological sense: How the strategy chosen for metabolomics data treatment may affect final results. A practical example based on urine fingerprints obtained by LC-MS. *Electrophoresis* 34:2812–2826.

Goodacre, R., D. Broadhurst, A. K. Smilde et al. 2007. Proposed minimum reporting standards for data analysis in metabolomics. *Metabolomics* 3:231–241.

Grün, C. H., F. A. van Dorsten, D. M. Jacobs et al. 2008. GC-MS methods for metabolic profiling of microbial fermentation products of dietary polyphenols in human and in vitro intervention studies. *J. Chromatogr. B* 871:212–219.

Guertin, K. A., E. Loftfield, S. M. Boca et al. 2015. Serum biomarkers of habitual coffee consumption may provide insight into the mechanism underlying the association between coffee consumption and colorectal cancer. *Am. J. Clin. Nutr.* 101:1000–1011.

Guo, K., F. Bamforth, L. Li. 2011. Qualitative metabolome analysis of human cerebrospinal fluid by 13C-/12C-isotope dansylation labeling combined with liquid chromatography Fourier transform ion cyclotron resonance mass spectrometry. *J. Am. Soc. Mass Spectrom.* 22:339–347.

Johnson, C. H., J. Ivanisevic, H. P. Benton et al. 2015. Bioinformatics: The next frontier of metabolomics. *Anal. Chem.* 87:147–156.

Kanani, H., P. K. Chrysanthopoulos, M. I. Klapa. 2008. Standardizing GC-MS metabolomics. *J. Chromatogr. B* 871:191–201.

Karamani, A. A., Y. C. Fiamegos, G. Vartholomatos et al. 2013. Fluoroacetylation/fluoroethylesterification as a derivatization approach for gas chromatography-mass spectrometry in metabolomics: Preliminary study of lymphohyperplastic diseases. *J. Chromatogr. A* 1302:125–132.

Khakimov, B., M. S. Motawia, S. Bak et al. 2013. The use of trimethylsilyl cyanide derivatization for robust and broad-spectrum high-throughput gas chromatography-mass spectrometry based metabolomics. *Anal. Bioanal. Chem.* 405:9193–9205.

Kind, T. and O. Fiehn. 2006. Metabolomic database annotations via query of elemental compositions: Mass accuracy is insufficient even at less than 1 ppm. *BMC Bioinformatics* 7:234.

Kloos, D. P., H. Lingeman, W. M. Niessen et al. 2013. Evaluation of different column chemistries for fast urinary metabolic profiling. *J. Chromatogr. B* 927:90–96.

Koh, Y., K. K. Pasikanti, C. K. Yap et al. 2010. Comparative evaluation of software for retention time alignment of gas chromatography/time-of-flight mass spectrometry-based metabonomic data. *J. Chromatogr. A* 1217:8308–8316.

Larive, C. K., G. A. Barding, Jr., M. M. Dinges. 2015. NMR spectroscopy for metabolomics and metabolic profiling. *Anal. Chem.* 87:133–146.

Libiseller, G., M. Dvorzak, U. Kleb et al. 2015. IPO: A tool for automated optimization of XCMS parameters. *BMC Bioinformatics* 16:118.

Lloyd, A. J., M. Beckmann, S. Haldar et al. 2013. Data-driven strategy for the discovery of potential urinary biomarkers of habitual dietary exposure. *Am. J. Clin. Nutr.* 97:377–389.

Lommen, A. 2009. MetAlign: interface-driven, versatile metabolomics tool for hyphenated full-scan mass spectrometry data preprocessing. *Anal. Chem.* 81:3079–3086.

Mahieu, N. G., J. L. Genenbacher, G. J. Patti. 2016. A roadmap for the XCMS family of software solutions in metabolomics. *Curr. Opin. Chem. Biol.* 30:87–93.

May, J. C., C. R. Goodwin, J. A. McLean. 2015. Ion mobility-mass spectrometry strategies for untargeted systems, synthetic, and chemical biology. *Curr. Opin. Biotechnol.* 31:117–121.

McLafferty, F. W., D. B. Stauffer, S. Y. Loh. 1991. Comparative evaluations of mass spectral data bases. *J. Am. Soc. Mass Spectrom.* 2:438–440.

Mihaleva, V. V., S. P. Korhonen, J. van Duynhoven et al. 2014. Automated quantum mechanical total line shape fitting model for quantitative NMR-based profiling of human serum metabolites. *Anal. Bioanal. Chem.* 406:3091–3102.

Mirnaghi, F. S. and A. A. Caudy. 2014. Challenges of analyzing different classes of metabolites by a single analytical method. *Bioanalysis* 6:3393–3416.

Niu, W., E. Knight, Q. Xia et al. 2014. Comparative evaluation of eight software programs for alignment of gas chromatography-mass spectrometry chromatograms in metabolomics experiments. *J. Chromatogr. A* 1374:199–206.

O'Sullivan, A., M. J. Gibney, L. Brennan. 2011. Dietary intake patterns are reflected in metabolomic profiles: Potential role in dietary assessment studies. *Am. J. Clin. Nutr.* 93:314–321.

Pan, Z. and D. Raftery. 2007. Comparing and combining NMR spectroscopy and mass spectrometry in metabolomics. *Anal. Bioanal. Chem.* 387:525–527.

Pasikanti, K. K., P. C. Ho, E. C. Chan. 2008. Gas chromatography/mass spectrometry in metabolic profiling of biological fluids. *J. Chromatogr. B* 871:202–211.

Peakman, T. C. and P. Elliott. 2008. The UK Biobank sample handling and storage validation studies. *Int J Epidemiol.* 37(Suppl 1):i2–i6

Pekkinen, J., N. N. Rosa, O. I. Savolainen et al. 2014. Disintegration of wheat aleurone structure has an impact on the bioavailability of phenolic compounds and other phytochemicals as evidenced by altered urinary metabolite profile of diet-induced obese mice. *Nutr. Metab.* 11:1. doi:10.1186/1743-7075-11-1.

Pekkinen, J., N. Rosa-Sibakov, V. Micard et al. 2015. Amino acid-derived betaines dominate as urinary markers for rye bran intake in mice fed high-fat diet-A nontargeted metabolomics study. *Mol. Nutr. Food Res.* 59:1550–1562.

Peng, J., Y. T. Chen, C. L. Chen et al. 2014. Development of a universal metabolome-standard method for long-term LC-MS metabolome profiling and its application for bladder cancer urine-metabolite-biomarker discovery. *Anal. Chem.* 86, 6540–6547.

Penn, L., H. Boeing, C. J. Boushey et al. 2010. Assessment of dietary intake: NuGO symposium report. *Genes Nutr.* 5:205–213.

Peters, S., E. van Velzen, H. G. Janssen, 2009. Parameter selection for peak alignment in chromatographic sample profiling: Objective quality indicators and use of control samples. *Anal. Bioanal. Chem.* 394:1273–1281.

Pluskal, T., S. Castillo, A. Villar-Briones et al. 2010. MZmine 2: Modular framework for processing, visualizing, and analyzing mass spectrometry-based molecular profile data. *BMC Bioinformatics* 11:395.

Pujos-Guillot, E., J. Hubert, J. F. Martin et al. 2013. Mass spectrometry-based metabolomics for the discovery of biomarkers of fruit and vegetable intake: Citrus fruit as a case study. *J. Proteome Res.* 12:1645–1659.

Rafiei, A. and L. Sleno. 2015. Comparison of peak-picking workflows for untargeted liquid chromatography/high-resolution mass spectrometry metabolomics data analysis. *Rapid Commun. Mass Spectrom.* 29:119–127.

Ramakrishnan, P., S. Nair, K. Rangiah. 2016. A method for comparative metabolomics in urine using high resolution mass spectrometry. *J. Chromatogr. A* 1443:83–92.

Rathahao-Paris, E., S. Alves, C. Junot et al. 2016. High resolution mass spectrometry for structural identification of metabolites in metabolomics. *Metabolomics* 12:10.

Rico, E., O. González, M. E. Blanco et al. 2014. Evaluation of human plasma sample preparation protocols for untargeted metabolic profiles analyzed by UHPLC-ESI-TOF-MS. *Anal. Bioanal. Chem.* N406:7641–7652.

Rocha, S. M., M. Caldeira, J. Carrola et al. 2012. Exploring the human urine metabolomic potentialities by comprehensive two-dimensional gas chromatography coupled to time of flight mass spectrometry. *J. Chromatogr. A* 1252:155–163.

Rustad, P., P. Felding, L. Franzson et al. 2004. The Nordic Reference Interval Project 2000: recommended reference intervals for 25 common biochemical properties. *Scand. J. Clin. Lab. Invest.* 64:271–284.

Sarafian, M. H., M. Gaudin, M. R. Lewis et al. 2014. Objective set of criteria for optimization of sample preparation procedures for ultra-high throughput untargeted blood plasma lipid profiling by ultra performance liquid chromatography-mass spectrometry. *Anal. Chem.* 86:5766–5774.

Scalbert A., C. Andres-Lacueva, M. Arita et al. 2011. Databases on food phytochemicals and their health-promoting effects. *J. Agric. Food Chem.* 59:4331–4348.

Scalbert, A., L. Brennan, C. Manach et al. 2014. The food metabolome: A window over dietary exposure. *Am. J. Clin. Nutr.* 99:1286–1308.

Schimek, D., K. A. Francesconi, A. Mautner et al. 2016. Matrix removal in state of the art sample preparation methods for serum by charged aerosol detection and metabolomics-based LC-MS. *Anal. Chim. Acta* 915:56–63.

Smilde, A. K., M. J. van der Werf, J. P. Schaller et al. 2009. Characterizing the precision of mass-spectrometry-based metabolic profiling platforms. *Analyst* 134:2281–2285.

Smith, C. A., E. J. Want, G. O'Maille et al. 2006. XCMS: Processing mass spectrometry data for metabolite profiling using nonlinear peak alignment, matching, and identification. *Anal. Chem.* 78:779–787.

Soininen, P., A. J. Kangas, P. Wuertz et al. 2015. Quantitative serum nuclear magnetic resonance metabolomics in cardiovascular epidemiology and genetics. *Circ. Cardiovasc. Genet.* 8:192–206.

Soininen, P., A. J. Kangas, P. Würtz et al. 2009. High-throughput serum NMR metabonomics for cost-effective holistic studies on systemic metabolism. *Analyst* 134:1781–1785.

Souverain, S., S. Rudaz, J. L. Veuthey. 2004. Matrix effect in LC-ESI-MS and LC-APCI-MS with off-line and on-line extraction procedures. *J. Chromatogr. A* 1058:61–66.

Teahan, O., S. Gamble, E. Holmes, J. Waxman, J. K. Nicholson, C. Bevan, H. C. Keun. 2006. Impact of analytical bias in metabonomic studies of human blood serum and plasma. *Anal Chem.* 78:4307–18.

Tulipani, S., R. Llorach, M. Urpi-Sarda et al. 2013. Comparative analysis of sample preparation methods to handle the complexity of the blood fluid metabolome: When less is more. *Anal. Chem.* 85:341–348.

Tulipani, S., X. Mora-Cubillos, O. Jáuregui et al. 2015. New and vintage solutions to enhance the plasma metabolome coverage by LC-ESI-MS untargeted metabolomics: The not-so-simple process of method performance evaluation. *Anal. Chem.* 87:2639–2647.

van der Kloet, F. M., I. Bobeldijk, E. R. Verheij et al. 2009. Analytical error reduction using single point calibration for accurate and precise metabolomic phenotyping. *J. Proteome Res.* 8:5132–5141.

Veselkov, K. A., L. K. Vingara, P. Masson et al. 2011. Optimized preprocessing of ultra-performance liquid chromatography/mass spectrometry urinary metabolic profiles for improved information recovery. *Anal. Chem.* 83:5864–5872.

Vinaixa, M., E. L. Schymanski, S. Neumann et al. 2016. Mass spectral databases for LC/MS- and GC/MS-based metabolomics: State of the field and future prospects. *Trends Anal. Chem.* 78:23–25.

Viswanathan, C. T., S. Bansal, B. Booth et al. 2007. Quantitative bioanalytical methods validation and implementation: Best practices for chromatographic and ligand binding assays. *Pharm. Res.* 24:1962–1973.

Welthagen, W., R. A. Shellie, J. Spranger et al. 2005. Comprehensive two-dimensional gas chromatography-time–of-flight mass spectrometry (GC × GC-TOF) for high resolution metabolomics: biomarker discovery on spleen tissue extracts of obese NZO compared to lean C57BL/6 mice. *Metabolomics* 1:65–73.

Whiley, L., J. Godzien, F. J. Ruperez et al. 2012. In-vial dual extraction for direct LC-MS analysis of plasma for comprehensive and highly reproducible metabolic fingerprinting. *Anal. Chem.* 84:5992–5999.

Xia, J. G., D. I. Broadhurst, M. Wilson et al. 2013. Translational biomarker discovery in clinical metabolomics: An introductory tutorial. *Metabolomics* 9:280–299.

Yanes, O., R. Tautenhahn, G. J. Patti et al. 2011. Expanding coverage of the metabolome for global metabolite profiling. *Anal. Chem.* 83:2152–2161.

Zamora-Ros, R., D. Achaintre, J. A. Rothwell, et al. 2016. Urinary excretions of 34 dietary polyphenols and their associations with lifestyle factors in the EPIC cohort study. *Sci. Rep.* 6: 26905.

17 The Validation of Dietary Biomarkers

Pietro Ferrari

CONTENTS

17.1 INTRODUCTION

A dietary biomarker can be defined as a biochemical indicator of short- or long-term dietary intake, or an index of nutritional status with respect to the metabolism of dietary constituents, and can be interpreted as markers of the biological consequences of dietary intake (Potischman and Freudenheim 2003). One of the convincing reasons to use dietary biomarkers is that they are objective measures, in the sense that they do not rely on individuals' ability to recall past dietary exposure. In statistical terms, this corresponds to the hypothesis that the errors in biomarker measurements are independent from errors in (self-reported) assessments (Kaaks et al. 1997, Day and Ferrari 2002). The unknown true dietary exposure is the result of different factors, whose mode of action is indeed difficult to account with a satisfactory degree of accuracy. Among others, they include the extent of cooking and the combination of foods eaten together, which determine nutrients absorption, and the availability of accurate food composition tables, which, in turn, may well depend on processing procedures, including cooking (Potischman 2003).

17.2 BACKGROUND

Dietary biomarkers have become increasingly popular because they can provide information about nutrients' bioavailability, defined as the available effective internal dose after absorption and pre-circulatory metabolism.

17.2.1 Are Dietary Biomarkers the Perfect Measurements?

They provide very useful indications of nutritional status and nutrient contents, but it is worthwhile to remember that existing biomarkers are not ideal. Although they are functional and have found wide spread applicability in modern nutritional epidemiology, they also have limitations mainly related to their physiology, that is, to the way they are absorbed, the tissue turnover, and their excretion (Kaaks et al. 1997, Potischman 2003). Many of these metabolic factors are not well understood, particularly on the way they determine individuals' differences in concentration levels.

17.2.2 Biomarker Subtypes

From a methodological point of view, it is important to distinguish two major classes of dietary biomarkers. The first class are markers based on the metabolic balance between intake and excretion of specific chemical components, that is, the percent recovery of the compound or its metabolites in excretion products, mostly in urinary samples, over a fixed period of time (Kaaks et al. 1997). Biomarkers in this class are defined recovery biomarkers, and they provide estimates of absolute intakes. Unfortunately, the list of recovery biomarkers is very limited, that is, the doubly labeled water collected in urines to estimate absolute levels of total energy intake (Schoeller 1988), as extensively described in Chapter 10, urinary nitrogen and urinary potassium for dietary protein (Bingham and Cummings 1985) and potassium (Williams and Bingham 1986), respectively. Since the quantitative relationship between recovery-based markers and dietary intake is known, and the recovery is known to be a fixed proportion of intake (Kaaks et al. 2002), recovery biomarkers, M, are related to unknown true intake, T, according to the following relationship

$$M_i = T_i + \varepsilon_{Mi} \tag{17.1}$$

For subject i, $i = 1, \ldots, N$, where the terms ε_{Mi} expresses random measurement error in marker M, and it is assumed that $E(\varepsilon_{Mi} | T) = 0$, and $\mathrm{Var}(\varepsilon_{Mi}) = \sigma^2_{eM}$. Equation 17.1 indicates that the variation that may occur over time in the marker is assumed to be random, and not correlated with true intake, and that the biomarkers offer a common reference scale (Plummer and Clayton 1993, Kaaks et al. 2002, Kipnis et al. 2003, Ferrari et al. 2009). Model (17.1) is known as the classical measurement error model.

The second class of biomarkers are measured as concentration of specific compounds in biological fluids, for example, serum, blood, urine, in specific tissues or cells, in cellular membranes, and in DNA (Kaaks et al. 1997). They are known as concentration markers, and they generally do not have the same quantitative relationship with dietary intake levels for every individual in a study population, and they cannot be translated into absolute levels of intake. The vast majority of dietary biomarkers are concentration biomarkers, like, for example, the concentration of vitamin E in lipoproteins, specific carotenoids, and the relative fatty acid composition of circulating phospholipids. The relationship between concentration markers and T is expressed in the following measurement error model, as

$$M_i = \alpha_M + \beta_M T_i + \varepsilon_{Mi} \tag{17.2}$$

where the terms α_M and β_M refer to as constant and proportional scaling bias, respectively. The classical measurement error model in Equation 17.1 is a special case of Equation 17.2, if $\alpha_M = 0$ and $\beta_M = 1$. In essence, concentration markers correlate with intakes of corresponding foods or nutrients and can provide relative ranking by intake level of specific food or nutrients (Kaaks et al. 1997). Fully comprehending the characteristics of each biomarker is, therefore, a necessary step for their use in epidemiological investigations.

A third type of biomarkers was recently introduced, the predictive biomarkers (Tasevska et al. 2005). These markers can be translated into absolute amounts of dietary intake, however, unlike recovery biomarkers, they may contain biases, that is, person-specific, intake-related, and covariate–related. In this respect, predictive markers relate to unknown intake according to model (17.2), where terms α_M, β_M, and $\sigma^2_{\varepsilon M}$ express these biases. If auxiliary information is available, ideally from well-conducted feeding studies, the biases can be quantified. Under the assumption that their magnitude do not explain a large portion of variability in the biomarker, and that these biases are stable between individuals and possibly across populations, the biomarker can be calibrated, and the calibrated measurements can be used as reference measurements (Tasevska et al. 2011) Sucrose and fructose measured in 24-hour urine, a biomarker of total sugars intake, is, to date, the only biomarker in this category (Tasevska 2015).

17.3 BIOMARKERS REPRODUCIBILITY

Biomarkers have traditionally been used as additional objective measurements in validation studies to assess the accuracy of dietary questionnaires (Kipnis et al. 2003, Potischman 2003, Ferrari et al. 2009), particularly since it emerged that the comparison of questionnaires with reference dietary measurements, such as 24-hour dietary recalls or food records, was confounded by error correlation in self-reported assessments (Day and Ferrari 2002, Kaaks et al. 2002). It has been argued that measuring *habitual* diet through self-reported assessment instrument is a challenging task, not compatible with realistic possibilities of existing methodology and logistic constraints (Kristal et al. 2005). Other than further refining the methodology and the strategy of dietary assessment in large-scale epidemiological investigations, the use of biomarkers has been advocated as the way to complement the investigation of the etiology of chronic diseases in relation to dietary and nutritional factors (Prentice et al. 2004, Freedman et al. 2010). To achieve accurate estimation of measures of risk associated to level of biomarker measurements, it is essential to carry out thorough evaluations of the spectrum of features and the performance of biomarker measurements in *ad hoc* designed studies, including its validity and reliability (White et al. 2008).

17.3.1 Reliability

The term reliability refers to the reproducibility of a measure, that is, how consistently a measurement can be repeated on the same subject. The estimation of reproducibility can involve the quantification of different aspects.

The first one is applied during the very first step of an assay development, and it was treated in Chapter 16 on Metabolomics techniques, and refers to the consistency of a specific laboratory procedure, technique, or even a laboratory instrument to measure a given compound in biological specimens. Reliability is assessed by comparison of two or more laboratory outcomes of the same specimens from a set of study subjects taken in one single time point. Most often, this assessment turns out to be extremely demanding and eventually impractical; instead, the same sample from one or few study subjects with heterogeneous concentration levels is aliquoted repeatedly, up to 5 or 10 times. In this setting, the coefficient of variation provides an estimate of the reliability of the technique, as CV = s/m, where s and m are the sample standard deviation and arithmetic means, respectively, over the replicates. A CV value close to zero would indicate good internal consistency across successive measurements. The internal consistency can be defined as repeatability (Vineis and Garte 2008).

The second notion of reliability refers to the assessment of the consistency of the measurements evaluated across samples taken from the same study subjects in two or more occasions. This is the reliability mostly relevant for association studies, when biomarker measurements are related to disease outcomes. White et al. refer to this quantity as the intramethod reliability (White et al. 2008). In extension to Equation 17.1, a simple statistical model to estimate reliability reads as

$$M_{ij} = T_i + \varepsilon_{Mij} \tag{17.3}$$

for subject $i = 1, \ldots, N$, and replicate $j = 1, \ldots, J$, with M_{ij} indicating the concentration level on subject i on occasion j. Concentration levels $M_1, M_2, \ldots, M_J$ are available for each subject. It is assumed that $T_i \sim (\mu_T, \sigma_T^2)$ and $\varepsilon_{Mij} \sim (0, \sigma_{\varepsilon M}^2)$, where the error terms reflect within-subject variability among replicates. Model (17.3) further assumes that $\mathrm{Cov}(T_i, \varepsilon_{Mij}) = 0$ and $\mathrm{Cov}(\varepsilon_{Mij}, \varepsilon_{Mik}) = 0$ for any $j \varepsilon k$. These assumptions state that random errors are independent from true level for any given subject and replicate, and that the errors in the different replicates are of similar magnitude across replicates, with one single variance term, $\sigma_{\varepsilon M}^2$, capturing variability, and that they are mutually independent. These assumptions, which are consistent with the description in Chapter 13, are known as the model of the parallel test (White et al. 2008), and they are often not clearly spelled out in reliability studies. An estimate of the reliability coefficient is provided by the intraclass correlation coefficient (ICC), which is estimated as

$$\hat{\mathrm{ICC}} = \frac{\hat{\sigma}_T^2}{\hat{\sigma}_T^2 + \hat{\sigma}_{\varepsilon M}^2} \tag{17.4}$$

Equation 17.4 indicates that the ICC expresses the amount of true variability between person variability over the total variability observed in the biomarker measurements, corresponding to the sum of the true variability plus the variability due to errors. Estimates of σ_T^2 and $\sigma_{\varepsilon M}^2$ can be obtained as variance components in a mixed model, which can be implemented in any popular statistical software. Alternatively, estimating equations could be derived from the sum of squares decomposition of the analysis of variance (ANOVA) (Dunn 1989). The between-subject and within-subjects variances can be, respectively, defined as

$$s_B^2 = \frac{1}{n-1}\sum_{i=1}^{n}\left(\bar{M}_i - \bar{M}\right)^2$$

$$s_W^2 = \frac{1}{J-1}\sum_{i=1}^{n}\sum_{j=1}^{J}\left(M_{ij} - \bar{M}_i\right)^2$$

It can be shown (Dunn 1989) that

$$E(s_B^2) = J\sigma_T^2 + \sigma_{\varepsilon M}^2 \tag{17.5}$$

$$E(s_W^2) = \sigma_{\varepsilon M}^2 \tag{17.6}$$

Substituting Equation 17.6 into Equation 17.5 gives

$$\hat{\sigma}_T^2 = \frac{s_B^2 - s_W^2}{J} \tag{17.7}$$

Therefore, replacing terms of Equations 17.6 and 17.7 into Equation 17.4 gives

$$\hat{\mathrm{ICC}} = \frac{\hat{\sigma}_T^2}{\hat{\sigma}_T^2 + \hat{\sigma}_{\varepsilon M}^2} = \frac{s_B^2 - s_W^2}{s_B^2 + (J-1)s_B^2}$$

The estimation is greatly simplified if only two measurements are available, that is, $J = 2$. Let us define the sample correlation coefficients (r) as

$$r = \frac{\text{Cov}(M_{1i}, M_{2i})}{\sqrt{\text{Var}(M_{1i})\,\text{var}(M_{2i})}} = \frac{\text{Cov}(T_i + \varepsilon_{M1i}, T_i + \varepsilon_{M2i})}{\sqrt{\text{Var}(T_i + \varepsilon_{M1i})\,\text{Var}(T_i + \varepsilon_{M2i})}} = \frac{\text{Var}(T_i)}{\text{Var}(T_i) + \text{Var}(\varepsilon_{Mi})}$$

relying on the assumptions of mutual independence between errors, and with T. It follows that

$$r = \frac{\hat{\sigma}_T^2}{\hat{\sigma}_T^2 + \hat{\sigma}_{\varepsilon M}^2} = \hat{\text{ICC}} \tag{17.8}$$

If the model of the parallel test applies, the ICC can be estimated by the sample correlation coefficient for two biomarker replicates (Dunn 1989, White et al. 2008).

17.3.2 Accounting for Different Sources of Variability

Model (17.3) offers a working model when the interest is in the evaluation of biomarker measurements over a predefined time window, that is, a few weeks, 6 months or 10 years. After assessing the analytical repeatability of measurements, usually the focus lies in the evaluation of biomarker measurements for their use in association studies. Typically two replicates, say, six months apart would be collected. Estimates of the ICC would be informative on the use of biomarkers as baseline measurements of exposure, assuming that a certain time lag would elapse between exposure and disease onset. This is the usual assumption in prospective studies where disease-free individuals are recruited.

A very fundamental concept in biomarker reliability is that of its variability, which is the result of different sources, some of it is being to measurement error, but a large proportion of human biomarker assay variability is due to variation between- and within-subjects (Vineis and Garte 2008). In extension to models previously introduced (Taioli et al. 1994), if one could conceive the perfect design with unlimited financial and logistical resources, a comprehensive, yet demanding, statistical model to evaluate reliability could reads as

$$M_{ijtk} = \mu + a_i + b_j + \varepsilon_{Mijk} \tag{17.9}$$

for subject $i = 1, \ldots, N$, at time $j = 1, \ldots, J$, and replicate $k = 1, \ldots, K$. The term μ is the population mean, and a_i capture the deviations from the unknown subject-specific true values T_i, with $T_i = \mu + a_i$, in analogy with models (17.1) and (17.3). It is assumed that $a_i \sim (0, \sigma_a^2)$, $b_j \sim (0, \sigma_b^2)$ and $\varepsilon_{ijk} \sim (0, \hat{\sigma}_{\varepsilon M}^2)$, and that these terms are mutually independent. Model (17.9) could well describe a study where biomarker measurements are acquired on N subjects, six-month apart ($J = 2$), with samples from subject i aliquoted twice ($K = 2$), at time t_1 and t_2. The term σ_a^2 would estimate between-subject variability, and corresponds to the term σ_T^2, introduced in model (17.3). This variability reflects biological diversity between individuals, which is the result of factors like sex, gender, age, but also dietary differences and differences in metabolism efficiency, in turn dependent on lifestyle (e.g., physical activity) or genetic characteristics. The term σ_b^2 refers to within-subject variability, and may reflect differences in unknown subject-specific biological variation, and also changes in dietary exposure and/or metabolic efficiency across consecutive replicates. The term $\sigma_{\varepsilon M}^2$ reflects analytical or laboratory variation, that could be attributable to variation between and within batches, and if the laboratory design is complex, it could also refer to variation originating from plates within batches, as well as random variation within the measurement error itself (Taioli et al. 1994).

It is noteworthy to stress that the way factors related to terms a and b in model (17.9) influence between- and within-subject variability is largely unknown. There is a tendency to believe that these factors can be controlled during the research process, while everything that is unknown is in the residual error terms. In practice, all variance components contain (varying) elements of uncertainty. An estimate of ICC is obtained as

$$\hat{ICC} = \frac{\hat{\sigma}_a^2}{\hat{\sigma}_a^2 + \hat{\sigma}_b^2 + \hat{\sigma}_{\varepsilon M}^2} \tag{17.10}$$

Estimates of ICC in Equation 17.10 based on model (17.9) should be the target of reliability studies, thus accounting for between-, within-person, and laboratory variability. The within-person variability in model (17.3) does not account for analytical variation, if replicate measurements were not aliquoted at each sample occasion. As a consequence, ICC values in Equation 17.10 can be substantially lower than the estimates in Equation 17.4, if the magnitude of analytical variation is sizeable, and may simply not tell the whole story. Sampson and colleagues provided adapted definitions of the ICC (Sampson et al. 2013). Specifically, they first provided a definition of ICC as in Equation 17.10 (π_T^B in the text), but also defined the quantity $\sigma_a^2 + \sigma_b^2$ as the biologic variation, that is, attributed to biological diversity. In addition, the ICC estimates in Equation 17.4 was defined as the proportion of biological variability due variation across individuals (π_{BW}^B in the text). Interestingly, a definition of technical ICC was also provided as

$$\hat{ICC}_{\text{technical}} = \frac{\hat{\sigma}_a^2 + \hat{\sigma}_b^2}{\hat{\sigma}_a^2 + \hat{\sigma}_b^2 + \hat{\sigma}_{\varepsilon M}^2}$$

which can be viewed as a measure of laboratory accuracy (Sampson et al. 2013).

17.3.3 A Working Model

It should be noted that the reliability of biomarker levels is based on model (17.3) or on its extension (17.9), in analogy with the classical measurement error model (17.1) for recovery biomarkers. In practice, the vast majority of biomarker measurements are concentration markers, for which model (17.2), with components of systematic errors, applies. In addition, models (17.3) and (17.9) assume that any systematic errors are randomly distributed across subjects, and replicates. This is an opportunistic decision, which makes the model identifiable, that is, simplifying assumptions are introduced in the statistical methodology to render the model parameters estimable. In practice, biomarker measurements are characterized by sources of systematic errors not necessarily randomly distributed across subjects and by correlation between errors. The between variation in concentration markers is generally determined not only by dietary intake of a given compound, but also by variations in absorption, distribution over body compartments, endogenous synthesis and metabolism, and excretion (Kaaks et al. 1997). For example, plasma levels of β-carotene depend on intake levels, but also on factors affecting absorption, cooking method, internal metabolism and nonenzymatic internal breakdown of β-carotene because of smoking and other factors that may increase oxidative stress. Generally, these nondietary determinants are very likely to vary systematically between individuals, so that part of the variation in the marker not determined by diet would tend to be correlated over time (Kaaks et al. 2002). These elements are often overlooked in the assessment of biomarkers. Indeed, they lead to violate the assumptions of the model of the parallel test, thus introducing dependency between terms a and b in Equation 17.9. As a result, ICC values in Equation 17.10 are spuriously overestimated.

Estimating the reproducibility is the first move of the assessment of a biomarker, a necessary step before further evaluating other quantities related to its performance, validity first of all, but also specificity and sensitivity (Vineis and Garte 2008). Validity and reliability may provide divergent evidence. A measurement may be perfectly reliable (repeatable if successively aliquoted and reproducible over time) but consistently wrong. Interest lies in both validity and reliability, however, since validity is hard to assess, reliability is sometimes used as a surrogate.

17.4 BIOMARKERS VALIDITY

The term refers to the capacity of a set of measurements to provide estimates of what they are meant to measure (Willett 1989). For biomarkers, this could be an ambiguous concept, as measurements would often refer to specific dietary constituents where the research interest lies, not necessarily to the true value of a laboratory assay. Validity can be assessed through comparison with a standard, which in turn is a reference measurement that represents the truth. Truth is simply not measurable in many behavioral sciences, including nutritional epidemiology. A thorough way to assess the validity of dietary biomarkers is to conduct highly controlled feeding studies, where replicate measurements of the biomarker taken over multiple days are compared to the truth, that is, known dietary intakes. For example, to investigate the use of 24-hour urinary nitrogen to validate habitual protein intake on an individual level, four men and four women were given their usual varying diet over a 28-d period while they lived in a metabolic suite (Bingham and Cummings 1985). Duplicates of individual daily diets, and 24-hour urine and fecal collections made daily during this period, were then measured for their nitrogen contents. Mean 28-d nitrogen excretion was highly correlated with nitrogen intake ($r = 0.99$). After accounting for other nitrogen losses (e.g., fecal, blood, and skin), the mean ratio of urinary to dietary nitrogen was found to be 0.81 (Bingham and Cummings 1985), an estimate later confirmed by a meta-analysis of five feeding studies (Kipnis et al. 2001).

The absence of systematic errors in recovery biomarkers that could determine error correlations among replicates from the same subjects has been the object of research (Schoeller et al. 1986, Black and Cole 2000, Bingham 2003). Willet challenged this assumption in his commentary to the Observing Protein & Energy Nutrition (OPEN) study, a large validation study for total energy intake and dietary protein that used recovery biomarkers (Kipnis et al. 2003), suggesting that the within-person variation in doubly labeled water measurements increases over time (Black and Cole 2000), thus indicating that errors in doubly labeled water replicates would be correlated, if measurements are taken close in time, like in OPEN (Kipnis et al. 2003). Kipnis and colleagues presented evidence from an analysis of urinary nitrogen data with repeated measurements over a period of one to nine months indicating absence of statistically significant relation between within-person variability and time, consistently in men and women (Kipnis et al. 2003). Interestingly, these observations stimulated further research to accommodate potential changes of true intake over time (Freedman et al. 2015).

The evaluation of the validity is as challenging for concentration and predictive biomarkers. The predictive biomarker of sugars intake was validated in a 30-day feeding study with 13 subjects consuming their usual varied diet and collecting 24-hour urine samples daily, under highly controlled conditions (Tasevska et al. 2005, 2009). In this study, 30-day mean total sugars intake was highly correlated with 30-day mean urinary sugars ($r = 0.84$), explaining 72% of the variability in sugars excretion. A recently developed measurement error model for predictive biomarkers was then fitted to the sugars biomarker data from the feeding study (Tasevska et al. 2005), in order to estimate the random and systematic components of the measurement error parameters of the biomarker as shown in model (17.2) in Section 17.2, and generate a biomarker calibration equation for this biomarker (Tasevska et al. 2011).

Feeding studies usually involve a relatively small number of individuals, but the characterization of the biomarker and the dietary assessments are very accurate. Although they may be logistically and financially very demanding, they constitute the reference for proper validation purposes.

17.5 FURTHER OPPORTUNITIES FOR BIOMARKERS EVALUATION

Ongoing prospective epidemiological investigations that aim at evaluating the role of dietary, lifestyle, and metabolic factors on the onset of chronic conditions assessed diet and collected biological samples at baseline in disease-free participants. Despite far from the ideal-controlled conditions of feeding studies, these settings create the opportunity to quantify the proximity of concentration biomarker measurements with unknown truth, through the estimated validity coefficient, which estimates the correlation coefficient between observed measurements and T, as ρ_{MT}.

In validation studies of dietary questionnaires (Q), it became apparent that the comparison with self-reported reference measurements, such as 24-hour dietary recall data or dietary record (R), was biased by unwanted correlation between errors in Q and R measurements. Errors in Q can be correlated to error in R because both assessment methods depend on many of the same cognitive processes, including memory and perception of serving sizes (Willett 1989, Kaaks and Riboli 1997, Kipnis et al. 1999). Biomarkers have been originally suggested as useful instrumental variables to validate self-reported assessments as they provided objective measurements, whose errors could be assumed to be independent from Q and R (Kaaks et al. 1994, Bingham et al. 1997, Freedman et al. 2010).

A measurement error model by means of structural equation models to relate observed variables to unknown true intake can be formulated (Bentler and Weeks 1989, Skrondal and Rabe-Hesketh 2008). In structural equation models, the validity of all involved assessment quantities in the system is investigated, primarily questionnaires, but also biomarkers and ultimately even the reference measurements. The performance of a biomarker as an accurate assessment of a given dietary constituent can be evaluated through the comparison with self-reported dietary measurements, relying on three principles, (1) that all available informative components should be used during the evaluation process; (2) that all observed quantities (Q, R, and M) contain measurements errors, with errors in Q and R likely being correlated; and (3) that the errors structure should be estimated and comprehended.

In a relatively simple form, a structural equation model involving two distinct dietary variables, with two latent factors T_1 and T_2, reads as

$$\begin{aligned}
M_{1i} &= \alpha_{M1} + \beta_{M1} T_{1i} + \varepsilon_{M1i} \\
M_{2i} &= \alpha_{M2} + \beta_{M2} T_{2i} + \varepsilon_{M2i} \\
Q_{1i} &= \alpha_{Q1} + \beta_{Q1} T_{1i} + \varepsilon_{Q1i} \\
Q_{2i} &= \alpha_{Q2} + \beta_{Q2} T_{2i} + \varepsilon_{Q2i} \\
R_{1i} &= T_{1i} + \varepsilon_{R1i} \\
R_{2i} &= T_{2i} + \varepsilon_{R2i}
\end{aligned} \tag{17.11}$$

with $i = 1,\ldots,N$ and $T_{1i} \sim (\mu_{T1}, \sigma^2_{T1})$, $T_{1i} \sim (\mu_{T2}, \sigma^2_{T2})$ and $\mathrm{Cov}(T_{1i}, T_{2i}) = \sigma^2_{T1T2}$. In addition, for $s = 1, 2$, $\varepsilon_{Msi} \sim (0, \sigma^2_{\varepsilon M})$, $\varepsilon_{Qsi} \sim (0, \sigma^2_{\varepsilon Q})$, $\varepsilon_{Rsi} \sim (0, \sigma^2_{\varepsilon R})$. It is also assumed that $\mathrm{Cov}(\varepsilon_{Msi}, \varepsilon_{Qsi}) = (\varepsilon_{Msi}, \varepsilon_{Rsi}) = \mathrm{Cov}(\varepsilon_{M1i}, \varepsilon_{M2i}) = 0$, while $\mathrm{Cov}(\varepsilon_{Qi}, \varepsilon_{Ri})$ is not forced to be null. In other words, it is assumed that errors in Q and R share some level of dependency, whereas errors in M measurements are mutually independent and independent from errors in Q and R. The terms α_Q and β_Q are the constant and proportional scaling bias for questionnaire measurements, in analogy with model (17.2). To estimate the model parameters, maximum likelihood method is used, which requires that observed variables be normally distributed (Skrondal and Rabe-Hesketh 2004, Ferrari et al. 2007).

The assumption that $\mathrm{Cov}(\varepsilon_{M1i}, \varepsilon_{M2i}) = 0$ relies on the hypothesis that, when two different markers are involved, the error dependencies due to factors related to marker absorption and metabolism can possibly be assumed to be marginal. Once maximum likelihood estimation is carried out, the validity coefficient for the first biomarker is obtained as a function of parameter estimates, as

$$\hat{\rho}_{M_1T_1} = \frac{\mathrm{Cov}(M_1, T_1)}{\sqrt{\mathrm{Var}(M_1)\mathrm{Var}(T_1)}} = \frac{\hat{\beta}_{M1}\sigma^2_T}{\sqrt{\left(\hat{\beta}^2_{M1}\hat{\sigma}^2_T + \hat{\sigma}^2_{\varepsilon M1}\right)\sigma^2_T}} = \frac{1}{\sqrt{1+\left(\hat{\sigma}^2_{\varepsilon M1}/\hat{\beta}^2_{M1}\hat{\sigma}^2_T\right)}} \tag{17.12}$$

Estimates of the validity coefficients for the biomarker of the second dietary variable, as well as for the questionnaires and the reference measurements, can be obtained by replacing the corresponding estimates in Equation 17.13. Sensitivity analyses can be performed to evaluate the extent of nonzero error correlations either between different biomarkers.

17.6 INVESTIGATING THE VARIABILITY OF METABOLOMICS DATA

Another biological acquisition of dietary exposure can be observed using recently developed *omics* technologies, such as metabolomics, the study of metabolic responses defined as the comprehensive analysis of all measurable metabolite concentrations under a given set of conditions (Nicholson and Lindon 2008). As a result, characteristic patterns of metabolites that reflect the metabolic phenotype of the organism were generated, which in turn can be used to explore molecular mechanisms of chronic disease etiology (Jenab et al. 2009). The measurement of hundreds or thousands of metabolites in metabolomic experiments now allows the characterization of individual phenotypes, in terms of dietary biomarkers (Scalbert et al. 2014). Given the large amount of information generated in metabolomics studies, data-driven research not requiring prior knowledge of the metabolites has been advocated as a viable approach to discover novel biomarkers for a number of foods, nutrients, or diets, as extensively described in Chapter 16. The acquisition of metabolic profiles generated by the spectrometric analysis of biological samples involve several steps, which include the full characterization of this data, from data preprocessing and data alignment to data normalization and signal correction, followed by statistical analysis (Issaq et al. 2009). From a statistical point of view, the characterization of metabolomics data involves the evaluation of analytical and temporal variability, but also variability determined by specific factors.

17.6.1 Reliability of Metabolomics

The evaluation of the reliability of metabolomics has been the focus of recent research. The between- and within-person variation of the concentrations of 163 serum metabolites acquired with targeted technology was determined to estimate the metabolite ICC values (Floegel et al. 2011). Samples were taken from 100 healthy individuals from the European Prospective Investigation into Cancer and Nutrition (EPIC)-Potsdam study who had provided two fasting blood samples four months apart. The median ICC of the 163 metabolites was 0.57, with high values determined for hexose (ICC = 0.76), sphingolipids (median ICC = 0.66, range: 0.24–0.85), amino acids (median ICC = 0.58, range: 0.41–0.72), and glycerophospholipids (median ICC = 0.58, range: 0.03–0.81). The authors concluded that, for most of the metabolites, a single measurement may be sufficient for risk assessment in epidemiologic studies (Floegel et al. 2011).

Within the Shanghai Physical Activity Study, 385 metabolites were measured in 60 women at baseline and year one, and the observed patterns were confirmed in the Prostate, Lung, Colorectal, and Ovarian Cancer Screening study (Sampson et al. 2013). Metabolites were acquired using liquid chromatography–mass spectrometry (LC–MS) and gas chromatography–mass spectroscopy (GC–MS) platforms. It was observed that reliability over time within an individual was low, although high technical reliability values were determined, with a median ICC equal to 0.8. In light of these findings, the study also provides interesting power calculations to determine minimal relative risks detectable with a given sample size. Large sample sizes are needed to observe weak associations typical of epidemiological investigations, thus indicating that future studies should plan for, but not be discouraged by, the potentially high intraindividual variability in metabolites (Nicholson et al. 2011, Sampson et al. 2013).

Similarly, the temporal stability of targeted metabolite levels was investigated to anticipate potential limitations for their use in epidemiological investigations. The reliability over a two-year period of 158 metabolites acquired through mass spectrometry in fasting and nonfasting serum samples from two EPIC centers was evaluated (Carayol et al. 2015). Overall, the reproducibility was higher in fasting sample, with a median ICC of 0.70, than nonfasting samples, with a median of 0.54. The authors concluded that a single measurement per individual may be sufficient for the study of 73% and 52% of the metabolites that showed ICC values greater than 0.50, in fasting and nonfasting samples, respectively (Carayol et al. 2015).

17.6.2 Systematic Variation in Metabolomics

Dealing with metabolomics can be cumbersome as the output of acquisition technologies can be large (hundreds of metabolites), at times very large (thousands of signals). Analytical solutions to deal with such large dimension have been conceived, but relatively simple questions like having an idea of what drives overall variability in metabolomics data can be challenging. Metabolomics profiles of human samples display variability due to dietary and lifestyle habits, as well as to intrinsic physiologic characteristics, that is, genetic drift, hormonal levels/status, medication use, as well as variability due to study protocols and sample processing. It is therefore essential to explore potential sources of variation in metabolomics data, possibly using analytical tools that allow the different sources to be disentangled. The PC-PR2 method was introduced as a descriptive tool to provide a quantitative overview of the different sources of variation in metabolomics data, integrating two popular statistical instruments: (1) principal component analysis (PCA) and (2) multivariable linear regression (Fages et al. 2014). The PC-PR2 involves the investigation of a large set of metabolomic profiles, whose variability is explained by a list of explanatory variables, in turn technical factor related to the sampling and other laboratory processing features (batch, fasting status, time between collection and final storage), as well as individual characteristics of the sample (age, gender, country of origin, smoking status). A summarizing value of the amount of variability in the set of metabolites that each explanatory variable contribute to explain, conditional on other covariates in the model, is provided by the R^2_{partial} statistics. The method provides an interesting snapshot of factors that mostly determine variation in a large set of metabolites/signals and can be useful to identify patterns of unwanted systematic variability in the data. PC-PR2 is a promising versatile method to explore variability in large sets of data, in terms of a set of intercorrelated explanatory variables (Fages et al. 2014).

17.6.3 Statistical Challenges in Metabolomics

The statistical analysis for -omics data for the identification of potential dietary biomarkers can be overwhelming. Multivariate existing methods for explorative and pattern recognition can be unsupervised, such as PCA or partial least squares (PLS) (Assi et al. 2015), or supervised, such as PLS-discriminative analysis (PLS-DA) (Gromski et al. 2015) and other more complex versions (Tzoulaki et al. 2014). PCA and PLS techniques aim at identifying the best linear combinations of variable that maximize, in turn, the total variability of a set of variable PCA or the covariance between two related sets of variables. These techniques have already been successfully applied in the context of epidemiological investigations, either in the discovery of relevant signals or for etiological purposes within the remit of the meeting-in-the-middle principle (Vineis and Perera 2007, Chadeau-Hyam et al. 2011, Assi et al. 2015).

It has been argued that dimension reduction techniques may facilitate the treatment of very complex sets of data but may encompass challenges in the interpretation of findings, if the various linear combinations, the components, do not make biological sense. To overcome this issue, several methods have been proposed to screen out the list of variables to deal with any regression model relating an outcome of interest to potential predictors (Chadeau-Hyam et al. 2011).

In the era of the exposome concept, there is an abundance of methodological choices to analyze metabolomics data in nutritional and lifestyle epidemiology, in order to tackle marginal and jointly complex exposure and -omics datasets (Chadeau-Hyam et al. 2013). This research calls for a sensitive use of available methodology, possibly making sure that adapted and appropriate analytical frameworks, rather than black-box type of statistical analyses, are developed and applied to overwhelming, yet vastly unexplored, large amount of data.

17.7 CONCLUDING REMARKS

The assessment validity of concentration biomarkers is a necessary yet challenging step, which involves the use of statistical methodology and complex study design to combine knowledge on behavioral, metabolic, laboratory, physiological, and nutritional aspects of the measured quantities. The definition of a measurement error model for biomarkers is a useful way to resume these features into a relationship with unknown true dietary intake, thus making explicit the different sources of bias. Recovery, concentration, and predictive markers are characterized by distinct error models, as shown in models (17.1) and (17.2).

The concepts of reliability and validity were described and discussed in this chapter. Reliability refers about the consistency of replicate measurements of the same biomarker over time. It has a technical, a subject-specific, and a between-person component. Reliability has a time dimension, from replicates of the same samples to samples taken, say, days, months, and even years apart. In other words, to fully give sense to estimates of reliability, other than capturing the basic background statistical notions, it is essential to acquire full understanding of the experimental design that generated the measurements. Reliability offers invaluable evidence on the laboratory repeatability of the acquirement process, and on the nature of a given marker in terms of the specific time window, it refers to, that is, whether it reflects dietary exposure on a short-, medium- or long-term exposure that ultimately depends on the type of biological samples used, urine vs. serum or blood, and/or the compounds measured. Despite different measurement error models characterize the type of dietary biomarkers, the estimation of reliability coefficient always relies on the assumptions that errors sum up in an additive way, that is, according to classical measurement error model. As the terms expressing constant and proportional scaling bias in (17.2) are set to 0 and 1 respectively, no components of systematic bias characterize the relationship, and errors in replicate measurements are assumed to be independent and of similar magnitude, as discussed around model (17.9). Although these are necessary assumptions to make the model identifiable and the various parameters estimable, it is noteworthy that potential limitations of reliability quantities are rarely addressed and discussed.

Reliability can be considered the first necessary step of a biomarker evaluation process, before moving into the assessment of its validity, which provides information on whether the measurements truly reflect what they are supposed to, through the quantification of the strength of association of the measurements with the true dietary variable of interest. The validation of dietary biomarkers is ideally assessed in studies that are carried out in very controlled conditions to determine a very accurate comparison of biological measurements to study participants' dietary intake over multiple days. This was the case for all existing recovery biomarkers, that is, urinary doubly labelled water (DLW), nitrogen and potassium for energy intake (Schoeller 1988), dietary protein (Bingham and Cummings 1985) and potassium (Williams and Bingham 1986), respectively, as well as for the predictive biomarker of urinary sugars for dietary sugars (Tasevska et al. 2005).

Measurements of concentration biomarkers have often been compared to dietary assessments in correlation studies in study design with varying degrees of complexity, in terms of sample size, type of dietary instruments, and number of replicates. Although it can be assumed that errors in dietary and biomarker measurements are independent, these correlation studies do not account for the errors affecting both dietary and biomarker measurements.

Beaton et al. (1997) stated that: "There will always be error in dietary assessments. The challenge is to understand, estimate, and make use of the error structure during analysis." This is the spirit of the evaluation of biomarker measurements through structural equation modeling, where the comparison with self-reported dietary assessments, that is, dietary questionnaires and short-term reference measurements, acknowledges the fact that all these quantities contain component of random and systematic errors, and that the errors in self-reported estimates are correlated.

Metabolomics is increasingly used for the identification of biomarkers with application in large-scale epidemiological investigations, thanks to technological spectral acquisition progresses and to developments in automation of sample preparation, sample handling, and data processing, together with the application of multivariate statistical methods for data analysis. Metabolomics is coming across as a laboratory technique with great opportunities for dietary biomarker identification, with a number of challenges related to the possibility to explore an overwhelming mass of information. In this setting, it will be of paramount importance that the variability of this large amount of data could be exhaustively investigated and comprehended, in terms of variation over time and due to specific dietary, lifestyle, and laboratory factors. The identification of a common language among people with different backgrounds, expertise, and talents will likely be a necessary step for metabolomics to display its full potential.

ACKNOWLEDGMENTS

I would like to thank Natasha Tasevska (Arizona State University, United States), Sabina Rinaldi, Augustin Scalbert (International Agency for Research on Cancer [IARC], France), Rudolf Kaaks (Deutsches Krebsforschungszentrum [DKFZ], Germany), and Victor Kipnis (NCI, United States) for invaluable scientific exchange, as well as Dale Schoeller (University of Wisconsin, United States) and Margriet Westerterp (University of Maastricht, The Netherlands) for their continuous incitement and guidance.

REFERENCES

Assi, N., A. Fages, P. Vineis et al. 2015. A statistical framework to model the meeting-in-the-middle principle using metabolomic data: Application to hepatocellular carcinoma in the EPIC study. *Mutagenesis* 30 (6):743–53. doi:10.1093/mutage/gev045.

Beaton, G.H., J. Burema, and C. Ritenbaugh. 1997. Errors in the interpretation of dietary assessments. *Am. J. Clin. Nutr* 65 (Suppl 4):1100S–1107S.

Bentler, P.M. and D.G. Weeks. 1989. Linear structural equations with latent variables. *Psychometrika* 45:289–308.

Bingham, S.A., C. Gill, A. Welch et al. 1997. Validation of dietary assessment methods in the UK arm of EPIC using weighed records, and 24-hour urinary nitrogen and potassium and serum vitamin C and carotenoids as biomarkers. *Int J Epidemiol* 26 (Suppl 1):S137–S151.

Bingham, S.A. 2003. Urine nitrogen as a biomarker for the validation of dietary protein intake. *J. Nutr* 133 (Suppl 3):921S–924S.

Bingham, S.A. and J.H. Cummings. 1985. Urine nitrogen as an independent validatory measure of dietary intake: A study of nitrogen balance in individuals consuming their normal diet. *Am. J. Clin. Nutr* 42 (6):1276–1289.

Black, A.E. and T.J. Cole. 2000. Within- and between-subject variation in energy expenditure measured by the doubly-labelled water technique: Implications for validating reported dietary energy intake. *Eur J Clin Nutr* 54 (5):386–394.

Carayol, M., I. Licaj, D. Achaintre et al. 2015. Reliability of serum metabolites over a two-year period: A targeted metabolomic approach in fasting and non-fasting samples from EPIC. *PLoS One* 10 (8):e0135437. doi:10.1371/journal.pone.0135437.

Chadeau-Hyam, M., T.J. Athersuch, H.C. Keun et al. 2011. Meeting-in-the-middle using metabolic profiling–A strategy for the identification of intermediate biomarkers in cohort studies. *Biomarkers* 16 (1):83–88.

Chadeau-Hyam, M., G. Campanella, T. Jombart et al. 2013. Deciphering the complex: Methodological overview of statistical models to derive OMICS-based biomarkers. *Environ Mol Mutagen* 54 (7):542–57. doi:10.1002/em.21797.

Day, N.E. and P. Ferrari. 2002. Some methodological issues in nutritional epidemiology. *IARC Sci. Publ* 156:5–10.

Dunn, G. 1989. *Design and Analysis of Reliability Studies*. New York: Oxford University Press. Reprint, Not in File.

Fages, A., P. Ferrari, S. Monni et al. 2014. Investigating sources of variability in metabolomic data in the EPIC study: The Principal Component Partial R-square (PC-PR2) method. *Metabolomics* 10 (6):1074–1083. doi:10.1007/s11306-014-0647-9.

Ferrari, P., C. Friedenreich, and C.E. Matthews. 2007. The role of measurement error in estimating levels of physical activity. *Am. J. Epidemiol* 166 (7):832–840.

Ferrari, P., A. Roddam, M.T. Fahey et al. 2009. A bivariate measurement error model for nitrogen and potassium intakes to evaluate the performance of regression calibration in the European Prospective Investigation into Cancer and Nutrition study. *Eur. J. Clin. Nutr* 63 (Suppl 4):S179–S187.

Floegel, A., D. Drogan, R. Wang-Sattler et al. 2011. Reliability of serum metabolite concentrations over a 4-month period using a targeted metabolomic approach. *PLoS One* 6 (6):e21103.

Freedman, L.S., D. Midthune, K.W. Dodd, R.J. Carroll, and V. Kipnis. 2015. A statistical model for measurement error that incorporates variation over time in the target measure, with application to nutritional epidemiology. *Stat Med* 34 (27):3590–3605. doi:10.1002/sim.6577.

Freedman, L.S., V. Kipnis, A. Schatzkin, N. Tasevska, and N. Potischman. 2010. Can we use biomarkers in combination with self-reports to strengthen the analysis of nutritional epidemiologic studies? *Epidemiol. Perspect. Innov* 7 (1):2.

Gromski, P.S., H. Muhamadali, D.I. Ellis et al. 2015. A tutorial review: Metabolomics and partial least squares-discriminant analysis–A marriage of convenience or a shotgun wedding. *Anal Chim Acta* 879:10–23. doi:10.1016/j.aca.2015.02.012.

Issaq, H.J., Q.N. Van, T.J. Waybright, G.M. Muschik, and T.D. Veenstra. 2009. Analytical and statistical approaches to metabolomics research. *J Sep Sci* 32 (13):2183–2199. doi:10.1002/jssc.200900152.

Jenab, M., N. Slimani, M. Bictash, P. Ferrari, and S.A. Bingham. 2009. Biomarkers in nutritional epidemiology: Applications, needs and new horizons. *Hum. Genet* 125 (5–6):507–525.

Kaaks, R., P. Ferrari, A. Ciampi, M. Plummer, and E. Riboli. 2002. Uses and limitations of statistical accounting for random error correlations, in the validation of dietary questionnaire assessments. *Public Health Nutr* 5 (6A):969–976.

Kaaks, R. and E. Riboli. 1997. Validation and calibration of dietary intake measurements in the EPIC project: Methodological considerations. European Prospective Investigation into Cancer and Nutrition. *Int. J. Epidemiol* 26 (Suppl 1):S15–S25.

Kaaks, R., E. Riboli, J. Esteve, A.L. van Kappel, and W.A. Van Staveren. 1994. Estimating the accuracy of dietary questionnaire assessments: Validation in terms of structural equation models. *Stat. Med* 13 (2):127–142.

Kaaks, R., E. Riboli, and R. Sinha. 1997. Biochemical markers of dietary intake. *IARC Sci. Publ* (142):103–126.

Kipnis, V., R.J. Carroll, L.S. Freedman, and L. Li. 1999. Implications of a new dietary measurement error model for estimation of relative risk: Application to four calibration studies. *Am J Epidemiol* 150 (6):642–651.

Kipnis, V., D. Midthune, L.S. Freedman et al. 2001. Empirical evidence of correlated biases in dietary assessment instruments and its implications. *Am J Epidemiol* 153 (4):394–403.

Kipnis, V., A.F. Subar, D. Midthune et al. 2003. Structure of dietary measurement error: Results of the OPEN biomarker study. *Am. J. Epidemiol* 158 (1):14–21.

Kristal, A.R., U. Peters, and J.D. Potter. 2005. Is it time to abandon the food frequency questionnaire? *Cancer Epidemiol Biomarkers Prev* 14 (12):2826–2828. doi:10.1158/1055-9965.epi-12-ed1.

Nicholson, G., M. Rantalainen, A.D. Maher et al. 2011. Human metabolic profiles are stably controlled by genetic and environmental variation. *Mol. Syst. Biol* 7:525. doi:10.1038/msb.2011.57.:525.

Nicholson, J.K. and J.C. Lindon. 2008. Systems biology: Metabonomics. *Nature* 455 (7216):1054–1056. doi:10.1038/4551054a.

Plummer, M. and D. Clayton. 1993. Measurement error in dietary assessment: An investigation using covariance structure models. Part I. *Stat. Med* 12 (10):925–935.

Potischman, N. 2003. Biologic and methodologic issues for nutritional biomarkers. *J. Nutr* 133 (Suppl 3): 875S–880S.

Potischman, N. and J.L. Freudenheim. 2003. Biomarkers of nutritional exposure and nutritional status: An overview. *J. Nutr* 133 (Suppl 3):873S–874S.

Prentice, R.L., W.C. Willett, P. Greenwald et al. 2004. Nutrition and physical activity and chronic disease prevention: Research strategies and recommendations. *J Natl Cancer Inst* 96 (17):1276–87. doi:10.1093/jnci/djh240.

Sampson, J.N., S.M. Boca, X.O. Shu et al. 2013. Metabolomics in epidemiology: Sources of variability in metabolite measurements and implications. *Cancer Epidemiol. Biomarkers Prev* 22 (4):631–640.

Scalbert, A., L. Brennan, C. Manach et al. 2014. The food metabolome: A window over dietary exposure. *Am J Clin Nutr* 99 (6):1286–308. doi:10.3945/ajcn.113.076133.

Schoeller, D.A., E. Ravussin, Y. Schutz, K.J. Acheson, P. Baertschi, and E. Jequier. 1986. Energy expenditure by doubly labeled water: Validation in humans and proposed calculation. *Am J Physiol* 250 (5):R823–R830.

Schoeller, D.A. 1988. Measurement of energy expenditure in free-living humans by using doubly labeled water. *J. Nutr* 118 (11):1278–1289.

Skrondal, A. and S. Rabe-Hesketh. 2004. *Generalized Latent Variable Modeling: Multilevel, Longitudinal and Structural Equation Models*. Boca Raton, FL: Chapman & Hall/CRC. Reprint.

Skrondal, A. and S. Rabe-Hesketh. 2008. Latent variable modelling. *Stat. Methods Med. Res* 17 (1):3–4. doi: 17/1/3[pii];10.1177/0962280207081235.

Taioli, E., P. Kinney, A. Zhitkovich et al. 1994. Application of reliability models to studies of biomarker validation. *Environ Health Perspect* 102 (3):306–309.

Tasevska, N. 2015. Urinary sugars—A biomarker of total sugars intake. *Nutrients* 7 (7):5816–5833. doi:10.3390/nu7075255.

Tasevska, N., D. Midthune, N. Potischman et al. 2011. Use of the predictive sugars biomarker to evaluate self-reported total sugars intake in the Observing Protein and Energy Nutrition (OPEN) study. *Cancer Epidemiol Biomarkers Prev* 20 (3):490–500. doi:10.1158/1055-9965.EPI-10-0820.

Tasevska, N., S.A. Runswick, A. McTaggart, and S.A. Bingham. 2005. Urinary sucrose and fructose as biomarkers for sugar consumption. *Cancer Epidemiol Biomarkers Prev* 14 (5):1287–94. doi:10.1158/1055-9965.epi-04-0827.

Tasevska, N., S.A. Runswick, A.A. Welch, A. McTaggart, and S.A. Bingham. 2009. Urinary sugars biomarker relates better to extrinsic than to intrinsic sugars intake in a metabolic study with volunteers consuming their normal diet. *Eur J Clin Nutr* 63 (5):653–659. doi:10.1038/ejcn.2008.21.

Tzoulaki, I., T.M. Ebbels, A. Valdes, P. Elliott, and J.P. Ioannidis. 2014. Design and analysis of metabolomics studies in epidemiologic research: A primer on -omic technologies. *Am J Epidemiol* 180 (2):129–139. doi:10.1093/aje/kwu143.

Vineis, P. and S. Garte. 2008. Biomarker validation. In *Molecular Epidemiology of Chronic Diseases*, pp. 71–81. John Wiley & Sons.

Vineis, P. and F. Perera. 2007. Molecular epidemiology and biomarkers in etiologic cancer research: the new in light of the old. *Cancer Epidemiol. Biomarkers Prev* 16 (10):1954–1965.

White, E., B.K. Armstrong, and R. Saracci. 2008. *Principles of Exposure Measurement in Epidemiology: Collecting, Evaluating, and Improving Measures of Disease Risk Factors*. Oxford: Oxford University Press.

Willett, W. 1989. An overview of issues related to the correction of non-differential exposure measurement error in epidemiologic studies. *Stat. Med* 8 (9):1031–1040.

Williams, D.R. and S.A. Bingham. 1986. Sodium and potassium intakes in a representative population sample: estimation from 24 h urine collections known to be complete in a Cambridgeshire village. *Br. J. Nutr* 55 (1):13–22.

18 Targeted and Untargeted Metabolomics for Specific Food Intake Assessment

Whole Grains as an Example

Carl Brunius, Huaxing Wu, and Rikard Landberg

CONTENTS

18.1 INTRODUCTION

Cereal foods constitute a major food group, which is one of the main contributors to energy and dietary fiber intake in the Western diet. Ever since the advent of agriculture about 10,000 years ago and during the majority of that time, cereals have been consumed as whole grain (Spiller 2002). It is only within the last hundred years that most people have consumed refined grains and this appears detrimental for health (Slavin 2005). The positive health effects of whole grain were recognized early and during the past two hundred years physicians and scientists have recommended whole grain to prevent constipation. In the early 1970s, the *fiber hypothesis* was proposed by Burkitt and colleagues (1974). It was suggested that nonrefined foods, such as whole grains, fruits, and vegetables, which provide dietary fiber along with other constituents, have a protective effect against *Western* diseases such as coronary heart disease and colon cancer, and consistent inverse associations have indeed been found for whole grain intake and the risk of developing CVD, type 2 diabetes, and certain cancers, whereas the risk was shown to be unaffected or even increased if refined grains are consumed (Sun et al. 2010; Aune et al. 2011, 2013; Ye et al. 2012). As a result, the official authorities in many countries, advise the population to increase consumption of whole grain foods, at the expense of refined grains as a way of reducing the risk of developing chronic conditions (Frølich et al. 2013). However, the role of separate whole grains in disease prevention has not yet been disentangled despite the fact that cereals have a large difference in total amount and quality of dietary fiber and bioactive compounds. This is mainly due to the problems to accurately estimate their separate intake by means of food frequency questionnaires. Moreover, it is yet unclear to what extent the individual components or their combinations contribute to the protective associations found in observational studies. Dietary biomarkers that reflect the intake of specific whole grains may be used to overcome such problems. In this chapter, we provide an overview of the current available knowledge about alternative approaches to address food intakes by using dietary biomarkers. Biomarkers of whole grain intake identified by targeted and untargeted metabolomics approaches will be used as illustrative examples.

18.2 BACKGROUND

The human diet constitutes a large diversity of foods from different plant, animal, and fungal origins and is composed of thousands of different molecules. Environmental and genetic factors cause large variation in the content and composition of compounds in foods. This is further affected by food processing and cooking, which induce even larger variation due to formation and breakdown of compounds. The diet, along with other life style factors, plays a very important role in health and disease and is therefore an attractive target for prevention and treatment of chronic diseases and their risk factors. Great efforts have been made to investigate the impact of specific dietary patterns, foods, and their constituents on human health using observational and interventional study designs.

Cereal foods constitute a major source of energy intake and dietary fiber in the Western diet and are therefore an interesting food group in relation to human health. Cereals are mostly consumed as refined grains, that is, the nutrient-rich bran and germ have been removed, but whole grain based foods, that is, where all parts of the grain kernel is present in cracked, intact, or milled form is reaching wider acceptance among consumers and is advocated by governmental authorities in many countries due to beneficial health effects (van der Kamp et al. 2014). Whole grain food intake has been consistently associated with lower risk of T2D in different populations (Fung et al. 2002; Montonen et al. 2003; Aune et al. 2013). Whole grains are rich in dietary fiber, vitamins, minerals, unsaturated fatty acids, and phytochemicals, which all may contribute to protective effects (Okarter and Liu 2010). Different cereals differ in the content and composition of these components, but this has typically not been accounted for in observational or randomized controlled studies (Fung et al. 2002; Andersson et al. 2007; Brownlee et al. 2010; Tighe et al. 2010) partly due to difficulties in assessing the intake from different grains through self-reporting methods.

Dietary intake assessment of the habitual diet through self-reporting techniques such as food frequency questionnaires, dietary recalls, or food records is a great challenge in nutritional epidemiology. The general utility and limitations of such techniques have been discussed in Chapters 1 and 4 of this book. On top of more general challenges with self-reported intakes, specific problems for accurate assessment of whole grain intake may include difficulties for consumers in distinguishing whole grain products from other products, or poor precision in intake estimation due to large differences in whole grain content in whole grain products (Kantor et al. 2002). No uniform definition of whole grain products or serving size has been used across studies (Lang and Jebb 2003; Frølich and Åman 2010). This may lead to misclassification, which is likely to attenuate the association between whole grain and disease toward null and preventing existing associations with disease outcomes to be revealed or cause underestimation of associations that may be stronger than observed (Jensen et al. 2004).

Using a specific biomarker for a food or a nutrient/compound as an alternative or as a complement is one approach, which has been used in attempts to tackle the problem of large measurement errors in dietary assessment methods. One of the key features of a biomarker is that it is independent of subjects' recognition capacity, memory, and motivation and in that sense could be regarded as *objective* (Freudenheim and Marshall 1988). This feature has made some biomarkers useful for validation of dietary assessment methods (Kaaks et al. 1997; Livingstone and Black 2003). This is because random errors in the biomarker (any variation that is uncorrelated to the individual's *true* habitual intake) can be assumed to be statistically independent of the subject's response in traditional dietary assessment (Kaaks et al. 1997).

The biomarker may be used as a ranking tool, replacing or complementing traditional survey-based dietary assessment as a predictor of disease risk, or it may be used in combination with self-reporting to strengthen the analysis, as recently suggested (Freedman et al. 2010). Biomarkers that meet certain criteria may also be used for studying the measurement errors in traditional dietary assessment methods and for calibration (Bingham 2002). Various aspects of dietary biomarker discovery and their validation through OMICs techniques have been excellently covered in Chapters 13 and 15 through 17.

In this chapter, we address discovery, validation, and implementation of biomarkers of specific foods using targeted, and with particular emphasis, untargeted metabolomics approaches with whole grain foods as an example of a food category important to human health.

18.3 TARGETED VERSUS UNTARGETED APPROACHES FOR BIOMARKER DISCOVERY AND DEVELOPMENT

Dietary biomarkers can be classified as *recovery* and *concentration* biomarkers depending on their characteristic. Recovery biomarkers reflect the balance between intake and excretion of a specific chemical component on an absolute scale over a specific time period, whereas concentration biomarkers are correlated with intake (Kaaks et al. 1997). Recovery biomarkers represent the highest standard and can be used to calibrate other dietary instruments. Sometimes prediction biomarkers are mentioned as a third category, falling in between the recovery biomarkers and concentrations biomarkers (Tasevska 2015). Prediction biomarkers are less affected by nondietary factors, variation in bioavailability, and metabolism compared with concentration biomarkers, and thus they are generally reflecting the intake more accurately (Tasevska 2015). However, prediction biomarkers are not quantitatively recovered to the same extent as recovery biomarkers (Tasevska 2015). Following recent advances in the last years, metabolomics has become a fundamental tool to study phenotypic changes caused by the impact of environmental agents, such as diet, on several disease or other outcomes. Using chemometric, multivariate tools, new possibilities are consequently being provided for the discovery of biomarker panels, which may improve prediction compared with single concentration biomarkers (Chapters 13, 16, and 17). Metabolomics methodologies can be divided into targeted and untargeted approaches (Patti et al. 2012): In targeted metabolomics, a defined set of well-characterized and annotated metabolites are analyzed. Untargeted approaches, on the other hand, aim at maximizing the metabolite coverage in a set of biological samples, even though the vast majority of measured metabolic features are unidentified. Metabolite features of special interest are then annotated/identified at a later stage in the analytical pipeline. Inherent to the wide coverage, untargeted approaches are well suited for exploratory biomarker studies, whereas targeted approaches are more suited for validation of hypotheses and for practical use in epidemiological investigations (Brunius et al. 2015). Targeted methods may be developed for any exposure or disease biomarker or biomarker panel identified through untargeted metabolomics to reach cost effective, wide application for disease prediction, diagnostics, or prognosis in clinical practice (Klein and Shearer 2016).

Based on knowledge of the unique presence of alkylresorcinols, a group of phenolic lipids, in the outer parts of wheat and rye grains, targeted metabolomics has been successfully employed for the validation of them as specific concentration biomarkers of whole grain wheat and rye intake (Landberg et al. 2014). Currently, no specific biomarkers of whole grain oats, barley, rice, sorghum, or millet have been reported.

18.4 ALKYLRESORCINOLS AS BIOMARKERS OF WHOLE GRAIN WHEAT AND RYE INTAKE

Alkylresorcinols (AR) are phenolic lipids mainly found in the bran of wheat and rye grains and not in other commonly consumed foods. The hydrocarbon chain of AR consists of odd number of carbon atom from 17 to 25 (C17:0–C25:0). The C17:0/C21:0 ratio is unique in different cereals:

around 0.01 in durum wheat, 0.1 in common wheat, and 1 in rye and can therefore be used to differentiate the source of whole grains (Chen et al. 2004). AR are resistant to food processing (Chen et al. 2004) and about 50%–60% of AR homologues are absorbed at the small intestine (Ross et al. 2003). Although the detailed mechanism of absorption is still unknown, AR are most likely absorbed by passive diffusion, incorporated into chylomicrons and collected into the lymphatic system, similarly to vitamin E (Rigotti 2007). In blood, AR are mainly assembled into lipoprotein and erythrocyte membranes (Linko and Adlercreutz 2005; Linko et al. 2005; Linko-Parvinen et al. 2007; Landberg et al. 2014). The apparent half-lives of AR homologues have been estimated to five hours in blood (Landberg et al. 2006). However, due to an apparent absorption half-life of about six to eight hours, fluctuations in plasma are relatively small upon regular intake (Landberg et al. 2009; Landberg et al. 2013). AR are metabolized in the liver through ω-oxidization, followed by β-oxidation similarly to tocopherols with two resulting main metabolites: 3,5-dihydroxybenzoic acid (DHBA) and 3-(3,5-dihydroxyphenyl)-1-propanoic acid (DHPPA), which are excreted in urine (Ross et al. 2004; Söderholm et al. 2011; Marklund et al. 2013b). Recent studies have also identified other metabolites, but they only account for a small proportion of the total metabolites and their utility as biomarkers have not yet been investigated in detail (Wierzbicka et al. 2015).

AR in plasma and their main metabolites in plasma and urine have been validated as biomarkers in both controlled feeding studies and in free-living populations. In controlled feeding studies, their pharmacokinetics, dose-response, and reproducibility under controlled and frequent intakes have been assessed (Landberg et al. 2006; Linko-Parvinen et al. 2007; Landberg et al. 2009a; Landberg et al. 2009b). Three studies have investigated the plasma AR concentrations after controlled whole grain intakes at different doses in humans (Linko et al. 2005; Landberg et al. 2009a; Landberg et al. 2009b). In all studies, plasma AR concentration increased linearly with increasing intake under controlled conditions. The correlation between calculated whole grain intake and plasma AR concentration was in the range 0.30–0.58 (Ross et al. 2012). A high C17:0/C21:0 ratio (>0.6) in plasma indicated whole grain/bran rye intake, whereas a lower ratio (0.2–0.3) was observed after mixed whole grain consumption in which whole grain wheat dominated (Landberg et al. 2009a). Studies under free-living conditions in different populations have been conducted to evaluate the long-term reproducibility and potential as a biomarker (Aubertin-Leheudre et al. 2008, 2010; Andersson et al. 2011; Landberg et al. 2013).

The plasma concentrations of intact AR and of their metabolites in plasma and urine mainly reflect whole grain intake of wheat and rye intake over a relatively short period. This makes them suitable for measuring compliance in whole grain intervention studies. Despite a short half-life, plasma AR concentrations remain stable in populations where the consumption of whole grain wheat and rye is high and frequent. The reproducibility is comparable to that of other biomarkers typically used in epidemiological research. The correlations of plasma AR versus self-reported whole grain wheat and rye intakes range in the order of $r \sim 0.2$–0.5, depending on population and dietary instrument used, showing a potential to reflect whole grain of wheat and rye intake. Reproducibility of AR in plasma and their metabolites in plasma and urine is comparable and typically in the range ICC = 0.3–0.5, depending on matrix and population (Montonen et al. 2010; Landberg et al. 2012, 2013; Marklund et al. 2013a).

The relatively short half-lives of AR in blood, approximately six hours (Landberg et al. 2006; Landberg et al. 2009a), limit their use as biomarkers of long-term intake of whole grain wheat, and rye, particularly in population with irregular whole grain consumption (van Dam and Hu 2008). Lipophilic compounds typically have slower turnover rate in adipose tissue than in blood (Beynen et al. 1980; Katan et al. 1997) and therefore AR concentrations in adipose tissue biopsy samples (10–50 mg) were measured and evaluated as long-term biomarkers of whole grain wheat and rye intake (Jansson et al. 2010; Wu et al. 2015). Results showed that adipose tissue concentrations responded to differences in intake under controlled intervention conditions and were associated with estimated long-term whole grain product intake from food frequency questionnaires. Studies are currently being undertaken to assess nondietary determinants and compare performance with plasma AR in free-living populations.

Inherent to the encouraging findings from biomarker validation studies, we and others have recently conducted several experiments where AR in plasma samples or their main metabolites in urine samples have been used as surrogates of whole grain wheat and rye intake in epidemiological investigations as well as measurement of compliance in whole grain intervention studies (Knudsen et al. 2014; Kyrø et al. 2014; Magnusdottir et al. 2014a; Biskup et al. 2016; McKeown et al. 2016). Based on findings from these studies, we believe that incorporation of well-characterized single dietary biomarkers in parallel or in combination with traditional dietary assessment provide additional important information for improved evaluation of the role of specific whole grains in relation to human health (Biskup et al. 2016). Further efforts are needed to discover and validate biomarkers of other commonly consumed grains such as oats and rice. For that purpose, untargeted metabolomics applied to samples from controlled feeding studies or sufficiently large cross-sectional studies in free-living populations could be used.

18.5 DISCOVERY AND VALIDATION OF WHOLE GRAIN INTAKE BIOMARKERS USING UNTARGETED METABOLOMICS

Although metabolomics is rapidly becoming a standard methodology for untargeted discovery of dietary biomarkers, there are still a limited number of studies applying this methodology to assess whole grain and refined grain consumption. In fact, a present day literature search* on Scopus yielded 36 scientific papers of which only 6 dealt with metabolic profiles in humans. These papers were all from the most recent years and an increase in such scientific work and publications could be expected with an increasing maturity in the field of metabolomics in general and nutritional metabolomics specifically. In the European Union, a large project (FOODBALL: The Food Biomarker Alliance 2014–2017) is currently ongoing with the overall aim to carry out a systematic exploration and validation of biomarkers of specific foods widely consumed including whole grains, in different population groups within Europe. This will be achieved by applying metabolomics to discover biomarkers, exploring use of easier sampling techniques and body fluids, revising the current dietary biomarker classification and developing a validation scoring system, applying these on selected new biomarkers, and exploring biological effects using biomarkers of intake (See the project webpage for more detail: http://www.healthydietforhealthylife.eu/).

The remainder of this chapter will outline a suggested approach to discover urinary biomarkers of specific food consumption using whole grain rye consumption in a healthy, free-living population as an example (Hanhineva et al. 2015). This will be presented in the form of a case study, where the experimental and data analysis pipeline and rationale for some of the choices that researchers are facing within this type of experiments will be discussed.

18.6 EXPERIMENTAL DESIGN

The choice of experimental design for biomarker discovery will have a large impact on the choice of statistical techniques, what can be found in the data, and also on the workload of study participants and researchers. Although there are several possible options for study design, the main distinction is between observational studies under free-living conditions and randomized controlled trials (RCT) and their subtypes (Chapters 16 and 17). RCT are typically used in an early biomarker discovery phase to ensure sufficiently large contrast between exposed and unexposed subjects and to assess between-subject variability (Heinzmann et al. 2010). RCTs can be subdivided into different designs: In the cross-over design, the study participants undergo all treatments and thus serve as their own controls. This design is complex and requires statistical techniques adapted to the structure of dependence between samples, such as multilevel analysis (Westerhuis et al. 2010). With this study design, response variance can be subdivided into within- and between-subject variability and smaller, systematic changes between treatments can thus more easily be observed.

This is advantageous for mechanistic investigations into, for example, biological effects of exposure. Despite the advantage of cross-over design under certain conditions, many investigators use suboptimal statistical analysis approaches that do not take the advantage of the study design into account. Parallel design is more commonly used but it requires more subjects to overcome noise inherent to between-subject differences, when evaluating treatment effects, but the design may be preferred for biomarker discovery studies, since the samples are distributed more independently and thus have more power for this purpose. In RCT, lack of compliance is a major concern, which will cause misclassification in biomarker exposure, which may hamper discovery. However, such misclassification will most likely be higher when depending on self-reported intake assessment as in free-living studies.

Different observational study designs may also be used for biomarker discovery, validation, and implementation. Among the different designs, cross-sectional studies are frequently used to investigate associations between exposure and response variables, whereas prospective cohort studies can be used to investigate the association between exposure and time to disease or risk factor event. For practical and cost-related reasons, the most common observational study designs where exposure biomarkers are discovered, validated, or implemented are cross-sectional, nested case-control, or case-cohort designs. With repeated sampling in a subfraction or in the entire cohort, biomarker reproducibility can be investigated, which is crucial for epidemiological investigations.

For biomarker discovery in RCTs, the advised/prescribed treatments constitute stratification in exposures and discriminant analysis is therefore a suitable approach for statistical analysis. In observational studies, dietary exposures will typically exhibit larger variability and follow unknown distributions and under such conditions, regression techniques will be better suited. Sometimes stratification is performed in observational studies to make use of classification strategies (Holmes et al. 2008). This procedure should, however, be done with care, since it will reduce the power in the data and impose stronger similarity between subjects than is necessarily true. Moreover, if quantitative biomarkers are desired, regression methods will be better suited to investigate and optimize for linearity.

The choice of experimental design will also be reflected in which results can be obtained effectively from the data: In a controlled intervention, only such queries can be addressed that were included in the study design (i.e., corresponding to the difference in intervention diets), whereas in a free-living setting, the data can easily be queried for multiple purposes as long as there is sufficient variability in the exposure variables among study participants. The multitude of uncontrolled exposures in free-living conditions will however also impose background noise in the data and increase the risk of confounding. However, biomarkers that have been discovered in a noisy dataset may be more robust and thus have a stronger potential as biomarker candidates.

With the aim to make an exploratory, multipurpose investigation of dietary biomarkers, we decided to recruit a free-living population for an observational study (Andersson et al. 2011). In brief, healthy volunteers were instructed to adhere to their normal dietary patterns and a three-day weighed food record (3DWFR) and urine samples were collected on two separate occasions two to three months apart. Samples and data from 59 subjects were available for the present study.

Untargeted LC-qTOF-MS was selected as analytical technique (see Chapter 16 for a description of metabolomics techniques). Samples were analyzed on HILIC and C18 columns in positive and negative ionization mode in order to obtain as wide compound coverage as possible. Full description of instrument conditions are reported elsewhere (Hanhineva et al. 2015).

18.7 DATA ANALYSIS

Before raw instrumental data can be analyzed, it needs to be preprocessed into a format suitable for statistical analysis. Such preprocessing consists of peak picking and alignment (Bijlsma et al. 2006; Wei et al. 2012; Vettukattil 2015), strategies to deal with missing values (Bijlsma et al. 2006; Hrydziuszko and Viant 2012; Armitage et al. 2015; Di Guida et al. 2016), and quality control (Bijlsma et al. 2006;

Sumner et al. 2007; Dunn et al. 2011; Godzien et al. 2015). Although preprocessing parameters for untargeted metabolomics and their impact on statistical analysis has been extensively described in the literature, there is still no consensus or applied standard within the field. In our study aimed to discover biomarkers of whole grain rye consumption, we utilized a proprietary software for peak picking and alignment (Hanhineva et al. 2015), but many researchers, including ourselves, have adopted freeware and open-source solutions to gain more control over data acquisition and quality (Smith et al. 2006; Lommen 2009; Pluskal et al. 2010; Coble and Fraga 2014). There are recent initiatives to develop free and/or open-source solutions for the entire data analysis workflow (Xia et al. 2012; Giacomoni et al. 2015; Davidson et al. 2016). There is also the exciting *Optimus* software operating under Knime https://github.com/alexandrovteam/Optimus.

In untargeted LC–MS metabolomics experiments, multivariate data are generated. Typically, the number of variables (metabolic features) counts in the thousands to tens of thousands and thus far outweighs the number of samples. It is therefore often considered appropriate to analyze such data using multivariate modeling (Trygg et al. 2007). Among multivariate methods, a distinction is made between unsupervised and supervised methods. Unsupervised methods, such as principal component analysis (PCA) work by giving a representation of the measured, independent data (an X matrix; metabolomics data in this case) without the use of information from a dependent variable (normally a Y vector; such as dietary exposure in this case, although multiple responses are also possible). In supervised methods on the other hand, information from the Y variable(s) is used to guide the analysis of the X data, effectively resulting in a model relating them to each other (i.e., $Y = f(X)$). In the field of metabolomics, the partial least squares (PLS) family of methods, that is, regression and discriminant analysis using either PLS, OPLS, or sparse PLS, has become the *de facto* standard for supervised analysis (Trygg et al. 2007; Gromski et al. 2015). PLS and OPLS share the same analytical solutions and thus produce identical predictions, but in OPLS, the principal components are rotated to facilitate interpretation (Trygg et al. 2007). In sparse PLS modeling, on the other hand, a soft thresholding is employed to reduce the number of variables in the model, which often increases model performance (Lê Cao et al. 2008; Chung and Keles 2010). The overwhelming popularity of PLS methods is presumably related to the advantages they bring in terms of interpretation of the data. In fact, PCA and PLS share similarities in that they work by dimensionality reduction of the original data into sample *scores*, based on maintained variance in X for PCA and covariance between X and Y for PLS (Figure 18.1). The dimensionality-reduced scores can then easily be superimposed with variable *loadings*, and the scores and loadings together constitute the model *components*, where each component successively adds more explanatory power to the model. The combined *biplot* of scores and loadings from PCA and PLS analyses offers the possibility of simultaneous interpretation: The scores will reveal the structure or pattern in the data, whereas the loadings will give information as to which are the variables accounting for or *driving* the observed structure. In metabolomics, a combination of these strategies is often employed: Unsupervised PCA is performed initially to get a graphical overview of sample variability and identify potential outliers. Supervised PLS analysis is then performed to identify patterns in the metabolome corresponding to the research question at hand (Xia et al. 2012).

From the PCA plot, it can easily be seen, using our data, that the main variability in the metabolomics data is unrelated to the whole grain rye consumption (Figure 18.1). This should come as no surprise, since even in a Nordic population with generally high whole grain rye consumption, the main determinants of the metabolome are likely to be other lifestyle and potential disease/health status factors. Consequently, the superpositioning of variable loadings in the PCA biplot is of limited value in relation to the research question, but can be informative if examining main causes for interpersonal variability. PCA biplots may of course be more informative in studies where the research question is more likely to reflect main determinants of interpersonal variability. In the supervised PLS analysis (Figure 18.1), the loadings provide more interesting information, since the model manages to reveal a clear inherent structure in the data and the variables acting most strongly as drivers of the responses in Y can be visualized in relation to that. It should however be noted that at this stage, the metabolic features are still not annotated or identified, wherefore biochemical interpretation cannot yet be performed.

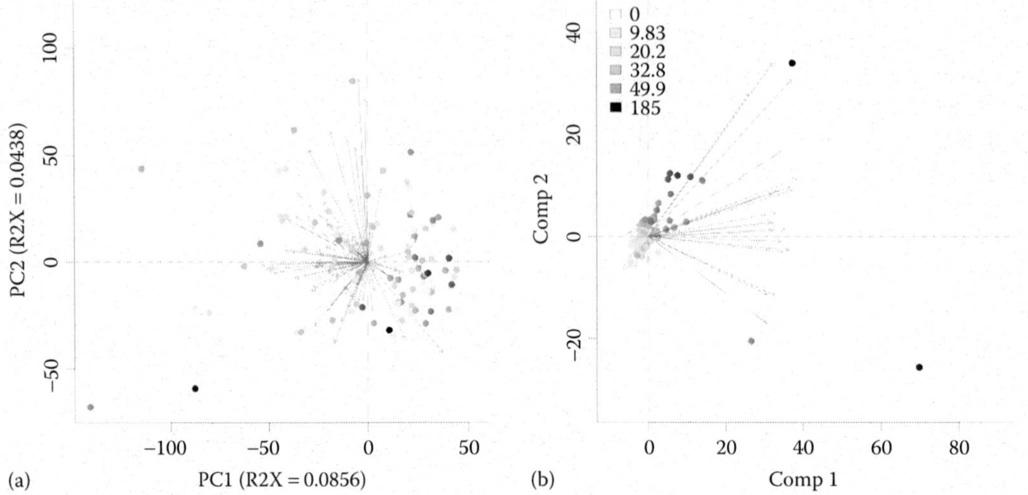

FIGURE 18.1 (See color insert.) Biplots from (a) PCA and (b) PLS analyses. Observations are grey-scaled according to whole grain rye exposure (see legend) and variable loadings are shown as red arrows. The PCA loadings (100 randomly selected variables visualized in the figure) correspond to the main variability in the metabolomics data, which however is seemingly unrelated to the response variable as seen from the observation scores. The 20 top ranked variables in the PLS analysis, shortlisted in the original paper (Hanhineva et al. 2015), show a pattern in observation scores clearly associated with the response variable.

18.7.1 Statistical Validation

As mentioned, the number of variables in untargeted metabolomics typically far outweighs the number of samples, which during modeling leads to mathematically underdetermined systems and the so-called *curse of dimensionality*, that is, these methods are prone to overfitting and false-positive findings in general (Hastie et al. 2009). This is especially true for the PLS family of methods (Westerhuis et al. 2008). To overcome the problem, statistical validation is essential (Figure 18.2) and in this sense refers to the practice of validation of the statistical modeling and should not be confused with the requirement for validation as comparison with true intake during a controlled feeding study (i.e., biomarker validation). Statistical validation is however one of the most overlooked areas in multivariate modeling and lack of appropriate validation may in worst case lead to reporting of findings that are simply not correct. In fact, in a recent survey among early stage researchers in the metabolomics field prior to a lecture on data analysis in the Training School on "Use of metabolomics in nutrition research", organized within the POSITIVe COST Action network, which aims to quantitate interindividual variation in response to consumption of plant food bioactives and determinants thereof, only 42% of respondents (n = 36) had ever used cross-validation, of which only 11% were frequent users (Brunius, personal information).

The gold standard for statistical modeling validation is to use external validation, i.e. that separate datasets are used for model construction and validation. The constructed model is used for prediction of independent (testing) data never used for model construction. External validation is then achieved by comparing predicted model outcome to known values (such as accurate whole grain rye consumption) to confirm that the underlying explanatory variables predict the desired outcome under independent conditions. Although the most unbiased validation statistics are provided using this approach, external validation requires large sample sets and most often represents an expensive and inefficient use of samples in a nutritional metabolomics setting. The most frequently used validation procedure is instead n-fold cross-validation, where the data

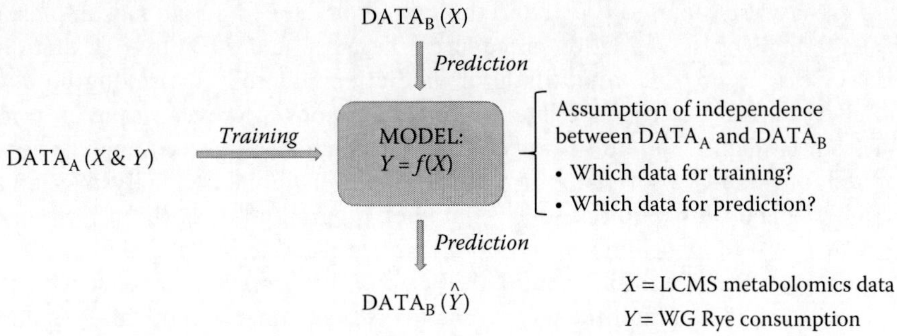

Type of validation	A and B Constraints	Comments
External	Different data sets	Best practice Requires many samples (expensive)
Cross-validation (*n*-fold)	Subsampled from complete data into *n* segments	Effective use of data Risk of overfitting (all samples used for estimation of model parameters)
Double cross-validation	Nested subsampling	Effective use of data Reduced risk of overfitting Computationally more intensive
Repeated double cross-validation	Nested subsampling, repeated many times	As for double cross-validation. Repetitions provide increased accuracy and confidence intervals of predictions Computationally even more intensive

FIGURE 18.2 Multivariate model validation. A model is constructed (trained) using $DATA_A$ and model predictions are performed on (independent) $DATA_B$. Predicted model outcomes ($\hat{Y}_B$) are then compared to the true measures (Y_B) to generate validation statistics.

are separated into segments, which are then used for model training and prediction in different folds (Figure 18.3). Holdout sets in each cross-validation fold are used to generate outcome predictions for estimation of performance as described earlier. It should however be noted, that employing an *n*-fold cross-validation does not safeguard against overfitting, since all samples are used for estimation of model parameters, even if samples are held out from training (Westerhuis et al. 2008). Other, more complex validation schemes have been suggested, such as the double cross-validation (Westerhuis et al. 2008) and repeated double cross-validation (Filzmoser et al. 2009) (Figure 18.2), which effectively reduce the risk of overfitting by

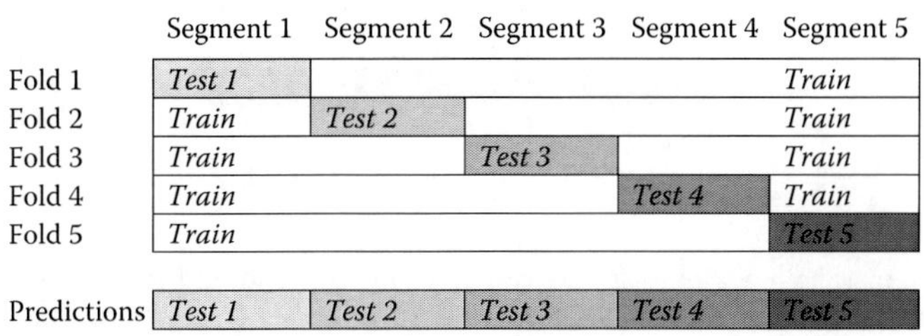

FIGURE 18.3 *n*-fold cross-validation. The complete data are split into segments. For each fold, one test segment is held out from model training and later predicted for validation performance evaluation. However, all folds and thus all segments are used for tuning of final model parameters, introducing risk of overfitting.

nesting of cross-validation loops to avoid tuning the model using holdout samples. Moreover, the repeated double cross-validation procedure reiterates the nested cross-validation in a resampling procedure to provide a multitude of predictions, thereby improving prediction accuracy and also gaining a confidence interval for predictions, however, at the cost of increased computational burden. It should be noted, however, that these more complex schemes are not incorporated into standard software for multivariate or chemometric analysis, such as SIMCA or Unscrambler and thus requires some computer programming skills in languages such as MATLAB or R where algorithms are available, to incorporate into metabolomics workflows. Another important consideration in validation is to ensure sample independence between data segments. If, for example, multiple measurements are taken per individual, then all such measures are dependent and therefore should be kept together in the same validation segment to reduce the likelihood of overfitting and false-positive findings. Again, these procedures are not incorporated in designated, proprietary software.

In the example study, there were a limited number of samples with a continuous response variable, repeated on two occasions, which made us choose a PLS regression analysis, employing a repeated double cross-validation type of validation and ensuring that both samples per individual were always cosampled during cross-validation segmentation. Although multivariate methods, such as PLS, are more robust to collinearities between variables compared to classical statistical methods, such as stepwise regression, they are not insensitive to noisy variables. Instead, predictive performance can in general be vastly improved by removing from the data such variables that do not contribute relevant information to the model. In fact, the process of variable selection for PLS has received considerable attention in the last years (Lê Cao et al. 2008; Fernández Pierna et al. 2009; Mehmood et al. 2012; Varmuza et al. 2013; Rinnan et al. 2014), and later approaches have also included random forest (RF) modeling for variable selection purposes (e.g., Genuer et al. 2015). However, most variable selection processes work on the entire set of observations before statistical analysis, thereby increasing the risk of overfitting. To avoid this, we developed an approach, where variable selection occurs within the repeated double cross-validation framework, ensuring that holdout samples were never included in the variable selection process (Figure 18.4).

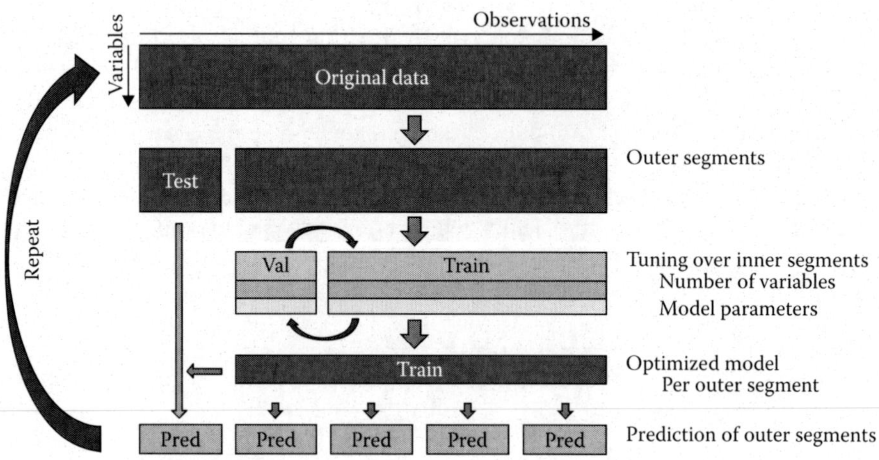

FIGURE 18.4 Multivariate modeling with unbiased variable selection. This procedure build upon the concept of repeated double cross-validation (Westerhuis et al. 2008; Filzmoser, Liebmann and Varmuza, 2009), but the model also performs an unbiased variable selection, represented by the shrinking variable space during tuning. This procedure is performed without inclusion of holdout (test) samples and increases the information density and predictive performance with minimized risk of overfitting.

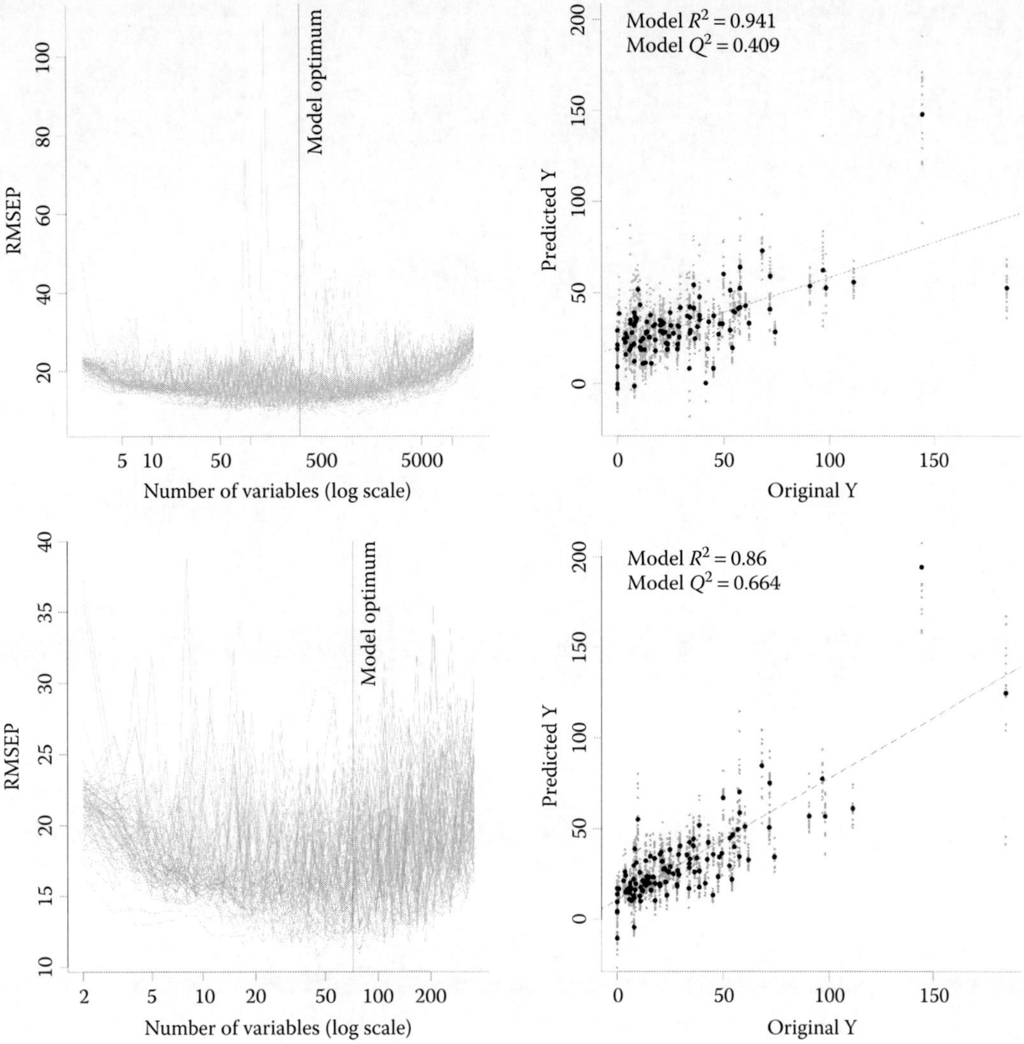

FIGURE 18.5 Validation (left) and prediction (right) performance of PLS modeling with unbiased variable selection using original, full data with 16,392 variables (upper) and univariate variable selection prefiltering (lower). Prediction error, represented by root-mean-squared-error-of-prediction (RMSEP), decreases initially as the number of variables decrease through the removal of uninformative, noisy variables. At a certain point, the prediction error starts to increase, corresponding to the removal of truly informative features. The optimal model returns for each observation a number of predictions (gray dots), which help to improve prediction accuracy (black dots) and also provides a confidence interval per observation prediction (not shown). The large difference between R^2 and Q^2 validation statistics for the full data indicates overfitting to the variables even with the incorporated repeated double cross-validation procedure. Initial prefiltering provides increased parsimony, less overfitting, and better prediction performance.

When examining the modeling performance in detail, it is apparent that information density and modeling performance increase with the removal of noisy, irrelevant variables (Figure 18.5). However, as variables are further removed from the model, prediction performance will decrease, corresponding to removing such variables that contribute with relevant information. The optimal model, corresponding to the highest model fitness is then returned. Improved prediction performance could easily be assessed from validation characteristics (Q^2) for PLS modeling with and without unbiased variable selection (Table 18.1). In the present case of PLS regression, fitness

TABLE 18.1
Validation Characteristics Using Different Variable Selection Methods as Filters Prior to Multivariate Modeling and Validation

Modeling Technique								
Selection method	**Original Variables**	**Final Variables**	***n*Comp[e]**	R^2	Q^2	**Selection (min)**	**Validation (min)**	$p_{Permutation}$[f]
PLS; without unbiased variable selection	16,392	16,392	3	0.91	0.13	—	9.24	0.04
PLS								
Full data	16,392	309	4	0.94	0.41	—	107	0.04
Univariate[a]	354	71	3	0.86	0.66	0.7	3.36	0.04
sPLS[b]	268	104	4	0.87	0.69	0.63	2.69	0.09
RF-Boruta[c]	14	7	2	0.59	0.41	17	0.46	Not calculated
RF-VSURF-I[d]	41	15	2	0.65	0.34	259	0.91	Not calculated
RF-VSURF-P[d]	16	9	2	0.63	0.48	259	0.51	Not calculated
Random forest								
Full data	16,392	34	N/A	0.57	0.38	—	94	0.04

[a] Linear correlation p-value < 0.05 after false-discovery-rate-correction (Benjamini-Hochberg).
[b] Sparse PLS (fivefold cross-validation) (Chung and Keles 2010).
[c] Random forest application (Kursa 2010).
[d] Random forest application (Genuer et al. 2015).
[e] Model optimum of PLS components.
[f] Approximated as the cumulative probability of actual model validation Q^2 in t-distribution of permuted Q^2 population using rank data.

was estimated by prediction error, but several fitness functions are available that correspond to different types of analyses (Szymańska et al. 2012).

18.7.2 Variable Selection Filters

The algorithm for repeated double cross-validation PLS with unbiased variable selection is more computationally demanding compared to single cross-validation or even double cross-validation. It is also apparent from the validation plot that the optimum number of variables for final modeling lies in the order of tens to hundreds of variables, as opposed to the 16,392 metabolomics features originally included in the data. To decrease computational cost, variable selection techniques can also be used as filters prior to statistical analysis (Mehmood et al. 2012). It was previously stated that variable selection using all data can cause overfitting. However, when hyphenating variable selection prefilters with the unbiased selection in our algorithm, those variables which represent false-positive findings can be filtered out, resulting in a combination of improved computational efficiency with increased information density and predictive performance (unpublished). This use of complementarity may even help to reduce overfitting, since the internal variable selection mechanism will have less opportunity to overfit in the diminished dataset (unpublished). In Table 18.1, modeling characteristics is reported using a variety of variable selection techniques as prefilters. It can be seen that the reduction in input variables decreases direct model (over)fit (R^2), but improves validation performance (Q^2).

18.7.3 Permutation Tests

Although within-model statistical validation will decrease the likelihood of overfitting, an assessment of the magnitude of overfitting cannot be performed using these techniques. For such assessment, permutation analysis, in which the actual model performance is compared to a null hypothesis of not performing better than modeling random results, is used (Bijlsma et al. 2006; Westerhuis et al. 2008; Szymańska et al. 2012; Xia et al. 2012; Triba et al. 2015). In permutation analysis, the response variable is repeatedly randomly permuted and temporary models constructed to generate a population (null distribution) of the validation statistics to compare with the actual model performance. However, with decreasing sample size, the likelihood of having a large proportion of random response variables highly correlated to the original response increases, resulting in a population distribution which effectively does not correspond to a true null distribution (Lindgren et al. 1996). There are different ways to compensate for this phenomenon: One option, which we utilized in our exploration of whole grain rye biomarkers, was to compare the permuted distribution not to one single model statistical validation metric but rather to a distribution of multiple actual model results (Hanhineva et al. 2015). The assumption was that sampling randomness between several such models would generate a population of statistical metrics. Whether the actual models performed better than random was then assessed using a *t*-test between the two populations of actual versus randomly permuted models. It was found that the population of actual models performed systematically better than random models (Hanhineva et al. 2015). Another, and perhaps better, option would be to interpret the statistical validation metrics of the permuted population in relation to the correlations between permuted and actual response variables (Lindgren et al. 1996). Using either approach, care should however be taken in the event that the data analytical pipeline employs variable selection filters prior to statistical analysis. If the response variable permutation is not performed before variable selection, then the statistical test of the permuted variable is performed on a subset of variables not adapted to the permuted response variable and subsequently, artificially suppressed *p*-values are obtained (Lindgren et al. 1996; Kuligowski et al. 2013).

18.7.4 Modeling Techniques

In addition to PLS, there are also other powerful supervised multivariate modeling techniques, most notably RF and support vector machines (SVM), which have been used in metabolomics with impressive predictive performance (Hochrein et al. 2012; Scott et al. 2013; Gromski et al. 2014, 2015). These methods originate from the field of machine learning, as opposed to the statistical techniques PCA and PLS, and are therefore based on other principles for prediction. These methods tend to generate nonidentical results compared to PLS, both in terms of variable ranking/selection and predictive performance. Moreover, especially RF is a very robust methodology and insensitive to such PLS weaknesses as discontinuous or near-zero variance variables. The machine-learning background of these methods, combined with the lack of easy interpretation compared to PLS, has however resulted in so far very limited use in metabolomics. In the same survey previously mentioned, the proportion of respondents that was not at all familiar with multivariate techniques ranged from 3% for PCA, via 17% for PLS and 44% for RF all the way up to 61% for SVM (Brunius, personal information).

The repeated double cross-validation algorithm with unbiased variable selection is not intrinsically bound to PLS modeling, but can be extended to incorporate any multivariate methodology. So far we have utilized both PLS (Hanhineva et al. 2015) and RF (Buck et al. 2016) core modeling and the incorporation of more methods is in the pipeline. For comparison, results from RF modeling of whole grain rye exposure are included in Table 18.1. As expected, RF resulted in a more parsimonious modeling with similar Q^2 compared to PLS modeling of full data, albeit without the large degree of overfitting measured in R^2.

18.8 METABOLITE IDENTIFICATION AND INTERPRETATION

Metabolite features (i.e., measured variables) in LC–MS metabolomics number, as mentioned, in the tens of thousands and the vast majority are never annotated or identified in a typical metabolomics experiment. The process of metabolite identification is instead normally performed on a selected subset of variables. This is performed either before statistical analysis, whereby the inherently untargeted LC–MS methodology is converted to a targeted approach; or afterwards, in which case the statistical analysis serves to single out those features that most strongly drive the biological research question being investigated. Even after subset selection, metabolite identification in LC–MS metabolomics is a highly laborious process, and in fact, one of the main bottlenecks in the metabolomics workflow, highly dependent on database queries using either manual approaches, data-mining algorithms, or combinations thereof (Scalbert et al. 2009; Xiao et al. 2012; Dunn et al. 2013). Since liquid chromatographic conditions differ significantly between systems, most researchers rely most strongly on MS and MS–MS fragmentation data for annotation/identification, at least on the lower levels of metabolite identification (Sumner et al. 2007; Salek et al. 2013). For accurate identification at the highest level, however, retention times as well as exact mass and fragmentation pattern must always correspond to those of authentic standards (Sumner et al. 2007).

It is apparent from practical experience in metabolomics that the choice of techniques for variable selection, variable pretreatment or modeling technique (see Sections 18.7.2 and 18.7.4) will have a large impact on which metabolites are selected during modeling as biomarker candidates, either alone or as main drivers in a multivariate predictive marker panel. However, systematic evaluation of how different techniques and settings affects candidate biomarker discovery is severely under-investigated in the metabolomics literature. In PLS modeling, the importance of individual variables is frequently assessed using the variable importance in projection (VIP), which aggregates loadings across all components, weighted by the variance accounted for by each component (Trygg et al. 2007). A VIP value >1, or some other arbitrarily chosen number, is frequently used as a criterion for final variable selection. This is however a doubtful procedure (Mehmood et al. 2012). Our own procedure described earlier provides an unbiased selection procedure optimized for prediction purposes, which however may be impractical for final decisions on how many and which metabolite features undergo identification, especially if selected variables count in the hundreds (Table 18.1). Considering the large efforts involved in metabolite identification, researchers will often find themselves limited in resources and will simply work from the top of the list and downward until a sufficient number of metabolites have been covered, making this represent yet another area within the field of metabolomics in which standards are desirable but sorely missed.

In the original investigation of whole grain rye biomarkers, the 20 highest ranked metabolite features were chosen for identification. These could be divided into five distinct categories (Hanhineva et al. 2015): (1) Low-abundant metabolites, where annotation and identification was not possible; (2) metabolites/derivatives of phenolic compounds normally found in the rye bran, related to DHPPA (Ross et al. 2004; Marklund et al. 2013a) and caffeic acid sulfate; (3) phenylacetamides possibly related to benzoxazinoid metabolism (Zikmundova et al. 2002; Beckmann et al. 2013; Hanhineva et al. 2014); (4) dicarboxylic acids (i.e., pimelic acid) (Bondia-Pons et al. 2013; Pekkinen et al. 2014); and (5) carnitines of unknown structure. Whereas a majority of the annotated compounds have been either previously suggested as potential biomarkers of rye or belonging to such a class, the presence of carnitines represented a novel and highly interesting finding in relation to whole grain rye exposure. Carnitines have previously been linked to, for example, metabolism of betaine (Pekkinen et al. 2013, 2015) and are frequently associated with cardiometabolic disturbances (Brunius et al. 2015). This implies that these compounds may be markers not only of dietary exposure but also of metabolic effects of dietary exposure (Hanhineva et al. 2015).

Modeling parameters, such as the use of variable selection filters or core modeling method, will have an impact not only on the predictive performance (Table 18.1) but also on the variable ranking and final selection for identification (Figure 18.6). Of the original top 20 selected features, 19 were also unbiasedly selected in the PLS models employing univariate or sPLS prefiltering. RF modeling provided a somewhat smaller overlap of 17 variables, presumably owing to the higher degree of parsimony in RF modeling, whereas Boruta-filtered PLS consisted of only 7 variables in the final model, accounting for the substantially lower overlap of only 5 variables. There was even smaller

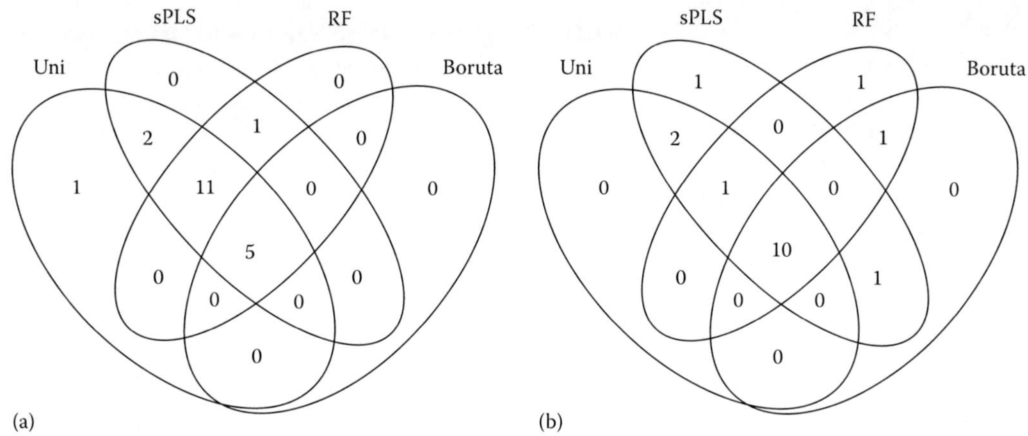

FIGURE 18.6 Overlap of top-ranking metabolite features from different multivariate modeling conditions. (a) Shows the distribution of the 20 top ranking features from the original modeling (Hanhineva et al., 2015) within the unbiasedly selected variables using different variable selection filters prior to PLS modeling (Univariate (Uni), sPLS or Random Forest (Boruta) filters) or Random Forest (RF) modeling using the full dataset (Table 18.1). Uni, sPLS and RF show considerable overlap in selected variables with the original 20 features. The Boruta-filtered model selected only seven variables, accounting for the smaller overlap. (b) Shows a similar distribution, but this time among the top 20 ranked variables per model. Maximum overlap is seen for the sPLS-filtered model followed by Uni and RF. The Boruta-filtered model contained only 14 variables, accounting for the smaller overlap.

overlap when looking only at the 20 highest ranking features per model variant (except Boruta-filter, which was limited to only 14 variables) instead of the full list of unbiasedly selected variables (Figure 18.6). Unsurprisingly, sPLS-filtering showed the greatest overlap (15 variables) with full model results, since both prefiltering and core modeling employed PLS variants. Again, the RF methods provided smaller overlap, presumably for the same reason as previously mentioned.

The observational design explores associations between dependent and independent variables and was thus well suited for exploratory biomarker discovery but poorly adapted for mechanistic investigations. Biochemical interpretations, such as pathway analysis, were therefore not undertaken but otherwise constitute the final activities in the metabolomics workflow (Hendriks et al. 2011; Xia et al. 2012; Chagoyen and Pazos 2013; Giacomoni et al. 2015). Care should however be taken in such analyses to limit pathway investigations only to significantly affected, properly identified metabolites (Scalbert et al. 2009; Hendriks et al. 2011). Moreover, adequate experimental designs and techniques, such as cross-over, repeated measures, multilevel analysis and metabolic flux analysis, should be employed to ensure that results are stringent and indicative of causality rather than mere associations (van Velzen et al. 2009; Zamboni et al. 2009; Nieman et al. 2013).

So far, no study has used putatively identified specific biomarkers of whole grain rye intake in relation to any disease end point. However, several studies have used the alkylresorcinol C17:0/C21:0 homologue ratio in plasma, which reflects the whole grain rye to whole grain wheat + rye intake ratio, in relation to cardiometabolic risk factors and type 2 diabetes incidence. These studies have in general found inverse associations between C17:0/C21:0 ratio and cardiometabolic risk factors and type 2 diabetes risk (Magnusdottir et al. 2014a; Magnusdottir et al. 2014b; Biskup et al. 2016). Further studies using specific biomarkers of whole grains from different cereals identified by untargeted metabolomics in relation to disease end points are warranted.

18.9 CONCLUSIONS

Targeted and untargeted metabolomics provide complementary approaches to discover and validate dietary biomarker. Through a large number of studies under controlled interventions and free-living conditions, targeted biomarkers of whole grain wheat and rye, alkylresorcinols and their metabolites, have been evaluated and used successfully in epidemiological studies as well as for addressing compliance in intervention studies. However, these biomarkers may be of limited use in populations with irregular intake, and the search for more long-term biomarkers is needed. Moreover, biomarkers of other sources of whole grain than wheat and rye are needed, and untargeted LC–MS based metabolomics has proven a useful approach to discover such biomarkers.

We set up an explorative study using untargeted metabolomics and developed an algorithm to discover whole grain rye intake biomarkers as an example. Using that approach, previously known molecules specifically reflecting rye consumption were identified as tentative biomarkers as well as other, not previously associated with rye, showing the great potential of untargeted metabolomics for biomarker discovery as well as other biological applications. With main focus on biostatistics, we have illustrated that studies incorporating metabolomics need to consider and accurately address the entire workflow from experimental design to identification and interpretation. To avoid overfitting and false-positive findings, both prudent within-model statistical validation and permutation analysis are required. These issues were exemplified in detail by the effects of variable selection filters and modeling method on statistical validation performance, variable ranking and final selection for identification, which is an extremely complicated and time consuming task and, in fact, one of the main bottlenecks in metabolomics. Biological interpretation through pathway analysis was not performed in our case, and should otherwise be used with caution, only if allowed through experimental design and methodological choices.

The original work on whole grain exposure biomarkers constituted an exploratory study. Consequently, even though interesting biomarker candidates may be discovered and to some extent identified, these results need to be externally validated in independent studies. Furthermore, further

RCT studies will be needed for investigations on kinetics, partitioning, and dose–response relationships of promising biomarker candidates to develop truly efficient univariate biomarkers or multivariate biomarker panels accurately reflecting intake.

REFERENCES

Andersson, A., M. Marklund, M. Diana, R. Landberg. 2011. Plasma alkylresorcinol concentrations correlate with whole grain wheat and rye intake and show moderate reproducibility over a 2- to 3-month period in free-living Swedish adults. *The Journal of Nutrition* 141(9):1712–1718. doi:10.3945/jn.111.139238.

Andersson, A., S. Tengblad, B. Karlström et al. 2007. Whole-grain foods do not affect insulin sensitivity or markers of lipid peroxidation and inflammation in healthy, moderately overweight subjects. *The Journal of Nutrition* 137(6): 1401–1407.

Armitage, E. G., J. Godzien, V. Alonso-Herranz, Á. López-Gonzálvez, C. Barbas. 2015. Missing value imputation strategies for metabolomics data. *Electrophoresis* 36(24):3050–3060. doi:10.1002/elps.201500352.

Aubertin-Leheudre, M., A. Koskela, A. Marjamaa, H, Adlercreutz. 2008. Plasma alkylresorcinols and urinary alkylresorcinol metabolites as biomarkers of cereal fiber intake in Finnish women. *Cancer epidemiology, biomarkers & prevention: a publication of the American Association for Cancer Research, cosponsored by the American Society of Preventive Oncology* 17(9):2244–2248. doi:10.1158/1055-9965.EPI-08-0215.

Aubertin-Leheudre, M., A. Koskela, A. Samaletdin, H. Adlercreutz. 2010. Plasma alkylresorcinol metabolites as potential biomarkers of whole-grain wheat and rye cereal fibre intakes in women. *The British Journal of Nutrition* 103(3):339–43. doi:10.1017/S0007114509992315.

Aune, D., D. S. M. Chan, R. Lau et al. 2011. Dietary fibre, whole grains, and risk of colorectal cancer: Systematic review and dose-response meta-analysis of prospective studies. *BMJ (Clinical Research ed.)* 343:d6617. doi:10.1136/bmj.d6617.

Aune, D., T. Norat, P. Romundstad, L. J. Vatten. 2013. Whole grain and refined grain consumption and the risk of type 2 diabetes: A systematic review and dose-response meta-analysis of cohort studies. *European Journal of Epidemiology* 28(11):845–858. doi:10.1007/s10654-013-9852-5.

Beckmann, M., A. J. Lloyd, S. Haldar, C. Seal, K. Brandt, J. Draper. 2013. Hydroxylated phenylacetamides derived from bioactive benzoxazinoids are bioavailable in humans after habitual consumption of whole grain sourdough rye bread. *Molecular Nutrition & Food Research* 57(10). doi:10.1002/mnfr.201200777.

Beynen, A. C., R. J. Hermus, J. G. Hautvast. 1980. A mathematical relationship between the fatty acid composition of the diet and that of the adipose tissue in man. *The American Journal of Clinical Nutrition* 33(1):81–85.

Bijlsma, S., I. Bobeldijk, E. R. Verheij et al. 2006. Large-scale human metabolomics studies: A strategy for data (pre-) processing and validation. *Analytical Chemistry* 78(2):567–574. doi:10.1021/ac051495j.

Bingham, S. A. 2002. Biomarkers in nutritional epidemiology. *Public Health Nutrition* 5(6a):821–827. doi:10.1079/PHN2002368.

Biskup, I., C. Kyrø, M. Marklund et al. 2016. Plasma alkylresorcinols, biomarkers of whole-grain wheat and rye intake, and risk of type 2 diabetes in Scandinavian men and women. *American Journal of Clinical Nutrition* 104(1):88–96. doi:10.3945/ajcn.116.133496.

Biskup, I., C. Kyrø, M. Marklund et al. 2016. Reply to A. Abassi. *The American Journal of Clinical Nutrition* 104(6):1725–1726. doi: 10.3945/ajcn.116.140756.

Bondia-Pons, I., T. Barri, K. Hanhineva et al. 2013. UPLC-QTOF/MS metabolic profiling unveils urinary changes in humans after a whole grain rye versus refined wheat bread intervention. *Molecular Nutrition & Food Research* 57(3):412–422. doi:10.1002/mnfr.201200571.

Brownlee, I. A., C. Moore, M. Chatfield et al. 2010. Markers of cardiovascular risk are not changed by increased whole-grain intake: The WHOLEheart study, a randomised, controlled dietary intervention. *The British Journal of Nutrition* 104(1):125–134. doi:10.1017/S0007114510000644.

Brunius, C., L. Shi, R. Landberg. 2015. Metabolomics for improved understanding and prediction of cardiometabolic diseases—Recent findings from human studies. *Current Nutrition Reports* 4(4):348–364. doi:10.1007/s13668-015-0144-4.

Buck, M., L. K. J. Nilsson, C. Brunius, R. K. Dabiré, R. Hopkins, O. Terenius. 2016. Bacterial associations reveal spatial population dynamics in Anopheles gambiae mosquitoes. *Scientific Reports* 6:22806. doi:10.1038/srep22806.

Burkitt, D. P., A. R. P. Walker, N. S. Painter et al. 1974. Dietary fiber and disease. *The Journal of the American Medical Association* 229(8):1068. doi:10.1001/jama.1974.03230460018013.

Chagoyen, M. and F. Pazos. 2013. Tools for the functional interpretation of metabolomic experiments. *Briefings in Bioinformatics*. 14(6):737–744. doi:10.1093/bib/bbs055.

Chen, Y., A. B. Ross, P. Åman, A. Kamal-Eldin. 2004. Alkylresorcinols as markers of whole grain wheat and rye in cereal products. *Journal of Agricultural and Food Chemistry*. American Chemical Society, 52(26): 8242–8246. doi:10.1021/jf049726v.

Chung, D. and S. Keles. 2010. Sparse partial least squares classification for high dimensional data. *Statistical Applications in Genetics and Molecular Biology* 9(1):17. doi:10.2202/1544-6115.1492.

Coble, J. B. and C. G. Fraga. 2014. Comparative evaluation of preprocessing freeware on chromatography/mass spectrometry data for signature discovery. *Journal of Chromatography A* 1358:155–164. doi:10.1016/j.chroma.2014.06.100.

van Dam, R. M. and F. B. Hu. 2008. Are alkylresorcinols accurate biomarkers for whole grain intake? *The American Journal of Clinical Nutrition* 87(4):797–798.

Davidson, R. L., R. J. M. Weber, H. Liu et al. 2016. Galaxy-M: A Galaxy workflow for processing and analyzing direct infusion and liquid chromatography mass spectrometry-based metabolomics data. *GigaScience* 5(1):10. doi:10.1186/s13742-016-0115-8.

Dunn, W. B., D. Broadhurst, P. Begley et al. 2011. Procedures for large-scale metabolic profiling of serum and plasma using gas chromatography and liquid chromatography coupled to mass spectrometry. *Nature Protocols* 6(7):1060–1083. doi:10.1038/nprot.2011.335.

Dunn, W. B., A. Erban, R. J. M. Weber et al. 2013. Mass appeal: Metabolite identification in mass spectrometry-focused untargeted metabolomics. *Metabolomics* 9(S1):44–66. doi:10.1007/s11306-012-0434-4.

Fernández Pierna, J. A., O. Abbas, V. Baeten, P. Dardenne. 2009. A Backward Variable Selection method for PLS regression (BVSPLS). *Analytica Chimica Acta* 642(1–2):89–93. doi:10.1016/j.aca.2008.12.002.

Filzmoser, P., B. Liebmann, K. Varmuza. 2009. Repeated double cross validation. *Journal of Chemometrics* 23(4):160–171. doi:10.1002/cem.1225.

Freedman, L. S., V. Kipnis, A. Schatzkin, N. Tasevska, N. Potischman. 2010. Can we use biomarkers in combination with self-reports to strengthen the analysis of nutritional epidemiologic studies? *Epidemiologic Perspectives & Innovations* 7(1):2. doi:10.1186/1742-5573-7-2.

Freudenheim, J. L. and J. R. Marshall. 1988. The problem of profound mismeasurement and the power of epidemiological studies of diet and cancer. *Nutrition and Cancer* 11(4):243–250. doi:10.1080/01635588809513994.

Frølich, W. and P. Aman. 2010. Whole grain for whom and why? *Food & Nutrition Research* 54: 5056. doi:10.3402/fnr.v54i0.5056.

Frølich, W., P. Åman, I. Tetens. 2013. Whole grain foods and health–A Scandinavian perspective. *Food & Nutrition Research* 57:18503. doi:10.3402/fnr.v57i0.18503.

Fung, T. T., F. B. Hu, M. A. Pereir et al. 2002. Whole-grain intake and the risk of type 2 diabetes: a prospective study in men. *The American Journal of Clinical Nutrition* 76(3):535–540.

Genuer, R., J. Poggi, C. Tuleau-malot. 2015. VSURF: An R Package for Variable Selection Using Random Forests. *The R Journal* 7(December):19–33.

Giacomoni, F., G. Le Corguillé, M. Monsoor et al. 2015. Workflow4Metabolomics: A collaborative research infrastructure for computational metabolomics. *Bioinformatics* (*Oxford, England*) 31(9):1493–1495. doi:10.1093/bioinformatics/btu813.

Godzien, J., V. Alonso-Herranz, C. Barbas, E. G. Armitage. 2015. Controlling the quality of metabolomics data: New strategies to get the best out of the QC sample. *Metabolomics*. 11(3):518–528. doi:10.1007/s11306-014-0712-4.

Gromski, P. S., H. Muhamadali, D. I. Ellis et al. 2015. A tutorial review: Metabolomics and partial least squares-discriminant analysis–A marriage of convenience or a shotgun wedding. *Analytica Chimica Acta* 879:10–23. doi: 10.1016/j.aca.2015.02.012.

Gromski, P. S., Y. Xu, E. Correa, D. I. Ellis, M. L. Turner, R. Goodacre. 2014. A comparative investigation of modern feature selection and classification approaches for the analysis of mass spectrometry data. *Analytica Chimica Acta* 829:1–8. doi:10.1016/j.aca.2014.03.039.

Di Guida, R., J. Engel, J. W. Allwood et al. 2016. Non-targeted UHPLC-MS metabolomic data processing methods: A comparative investigation of normalisation, missing value imputation, transformation and scaling. *Metabolomics* 12(5):93. doi:10.1007/s11306-016-1030-9.

Hanhineva, K., C. Brunius, A. Andersson et al. 2015. Discovery of urinary biomarkers of whole grain rye intake in free-living subjects using nontargeted LC-MS metabolite profiling, *Molecular Nutrition & Food Research* 59(11), 2315–2325. doi:10.1002/mnfr.201500423.

Hanhineva, K., P. Keski-Rahkonen, J. Lappi et al. 2014. The postprandial plasma rye fingerprint includes benzoxazinoid-derived phenylacetamide sulfates. *Journal of Nutrition*. American Society for Nutrition, 144(7):1016–1022. doi:10.3945/jn.113.187237.

Hastie, T., R. Tibshirani, J. H. Friedman, J. H. 2009. *The Elements of Statistical Learning: Data Mining, Inference, and Prediction*. Springer-Verlag: New York.

Heinzmann, S. S., I. J. Brown, Q. Chan et al. 2010. Metabolic profiling strategy for discovery of nutritional biomarkers: Proline betaine as a marker of citrus consumption. *The American Journal of Clinical Nutrition* 92(2):436–443. doi:10.3945/ajcn.2010.29672.

Hendriks, M. M. W. B., F. A. van Eeuwijk, R. H. Jellema et al. 2011. Data-processing strategies for metabolomics studies. *TrAC Trends in Analytical Chemistry* 30(10):1685–1698. doi:10.1016/j.trac.2011.04.019.

Hochrein, J., M. S., Klein, H. U. Zacharias et al. 2012. Performance evaluation of algorithms for the classification of metabolic 1H NMR fingerprints. *Journal of Proteome Research* 11(12):6242–6251. doi:10.1021/pr3009034.

Holmes, E., R. L. Loo, J. Stamler et al. 2008. Human metabolic phenotype diversity and its association with diet and blood pressure. *Nature* 453(7193):396–400. doi:10.1038/nature06882.

Hrydziuszko, O. and M. R. Viant. 2012. Missing values in mass spectrometry based metabolomics: An undervalued step in the data processing pipeline. *Metabolomics* 8(S1):161–174. doi:10.1007/s11306-011-0366-4.

Jansson, E., R. Landberg, A. Kamal-Eldin, A. Wolk, B. Vessby, P. Aman. 2010. Presence of alkylresorcinols, potential whole grain biomarkers, in human adipose tissue. *The British Journal of Nutrition* 104(5):633–636. doi:10.1017/S0007114510001169.

Jensen, M. K., P. Koh-Banerjee, F. B. Hu et al. 2004. Intakes of whole grains, bran, and germ and the risk of coronary heart disease in men. *The American Journal of Clinical Nutrition* 80(6):1492–1499.

Kaaks, R., E. Riboli, R. Sinha. 1997. Biochemical markers of dietary intake. *IARC Scientific Publications* (142):103–126.

van der Kamp, J. W., K. Poutanen, C. J. Seal, D. P. Richardson. 2014. The HEALTHGRAIN definition of "whole grain." *Food & Nutrition Research* 58. doi:10.3402/fnr.v58.22100.

Kantor, L., J. Variyam, J. Allshouse et al. 2002. *Whole-Grain Foods in Health and Disease*. St Paul, MN: Eagan Press, pp. 301–325.

Katan, M. B., J. P. Deslypere, A. P. van Birgelen, M. Penders, M. Zegwaard. 1997. Kinetics of the incorporation of dietary fatty acids into serum cholesteryl esters, erythrocyte membranes, and adipose tissue: An 18-month controlled study. *Journal of Lipid Research* 38(10):2012–2022.

Klein, M. S. and J. Shearer. 2016. Metabolomics and type 2 diabetes: Translating basic research into clinical application. *Journal of Diabetes Research* 1–10. doi:10.1155/2016/3898502.

Knudsen, M. D., C. Kyrø, A. Olsen et al. 2014. Self-reported whole-grain intake and plasma alkylresorcinol concentrations in combination in relation to the incidence of colorectal cancer. *American Journal of Epidemiology* 179(10):188–196. doi:10.1093/aje/kwu031.

Kuligowski, J., D. Pérez-Guaita, J. Escobar et al. 2013. Evaluation of the effect of chance correlations on variable selection using partial least squares-discriminant analysis. *Talanta*, 116:835–840. doi:10.1016/j.talanta.2013.07.048.

Kursa, M. B. and W. R. Rudnicki 2010. Feature selection with the Boruta package. *Journal of Statistical Software* 36(11). doi: 10.18637/jss.v036.i11.

Kyrø, C., A. Olsen, R. Landberg et al. 2014. Plasma alkylresorcinols, biomarkers of whole-grain wheat and rye intake, and incidence of colorectal cancer. *Journal of the National Cancer Institute* 106(1):352. doi:10.1093/jnci/djt352.

Landberg, R., P. Aman, L. E. Friberg, B. Vessby, H. Adlercreutz, A. Kamal-Eldin. 2009a. Dose response of whole-grain biomarkers: Alkylresorcinols in human plasma and their metabolites in urine in relation to intake. *The American Journal of Clinical Nutrition* 89(1):290–296. doi:10.3945/ajcn.2008.26709.

Landberg, R., P. Aman, G. Hallmans, I. Johansson. 2013. Long-term reproducibility of plasma alkylresorcinols as biomarkers of whole-grain wheat and rye intake within Northern Sweden Health and Disease Study Cohort. *European Journal of Clinical Nutrition* 67(3):259–263. doi:10.1038/ejcn.2013.10.

Landberg, R., A. Kamal-Eldin, S.-O. Andersson et al. 2009b. Reproducibility of plasma alkylresorcinols during a 6-week rye intervention study in men with prostate cancer. *Journal of Nutrition* 139(5): 975–980. doi:10.3945/jn.108.099952.

Landberg, R., A.-M. Linko, A. Kamal-Eldin, B. Vessby, H. Adlercreutz, P. Aman. 2006. Human plasma kinetics and relative bioavailability of alkylresorcinols after intake of rye bran. *The Journal of Nutrition* 136(11):2760–2765.

Landberg, R., M. Marklund, A. Kamal-Eldin, P. Åman. 2014. An update on alkylresorcinols–Occurrence, bioavailability, bioactivity and utility as biomarkers. *Journal of Functional Foods* 7: 77–89. doi:10.1016/j.jff.2013.09.004.

Landberg, R., M. K. Townsend, N. Neelakantan et al. 2012. Alkylresorcinol metabolite concentrations in spot urine samples correlated with whole grain and cereal fiber intake but showed low to modest reproducibility over one to three years in U.S. women. *The Journal of Nutrition* 142(5):872–877. doi:10.3945/jn.111.156398.

Lang, R. and S. A. Jebb. 2003. Who consumes whole grains, and how much? *Proceedings of the Nutrition Society* 62(1):123–127. doi:10.1079/PNS2002219.

Lê Cao, K.-A., D. Rossouw, C. Robert-Granié, P. Besse. 2008. A sparse PLS for variable selection when integrating omics data. *Statistical Applications in Genetics and Molecular Biology* 7(1). doi:10.2202/1544-6115.1390.

Lindgren, F., B. Hansen, W. Karcher, M. Sjöström, L. Eriksson. 1996. Model validation by permutation tests: Applications to variable selection. *Journal of Chemometrics.* 10(5–6):521–532. doi:10.1002/(SICI)1099-128X(199609)10:5/6<521::AID-CEM448>3.0.CO;2-J.

Linko, A.-M. and H. Adlercreutz. 2005. Whole-grain rye and wheat alkylresorcinols are incorporated into human erythrocyte membranes. *The British Journal of Nutrition* 93(1):11–13.

Linko, A.-M., K. S. Juntunen, H. M. Mykkänen, H. Adlercreutz. 2005. Whole-grain rye bread consumption by women correlates with plasma alkylresorcinols and increases their concentration compared with low-fiber wheat bread. *The Journal of Nutrition* 135(3):580–583.

Linko-Parvinen, A.-M., R. Landberg, M. J. Tikkanen, H. Adlercreutz, J. L. Peñalvo. 2007. Alkylresorcinols from whole-grain wheat and rye are transported in human plasma lipoproteins. *The Journal of Nutrition* 137(5):1137–1142.

Livingstone, M. and A. Black. 2003. Biomarkers of nutritional exposure and nutritional status: an overview. *The Journal of Nutrition* 133 Suppl(3):873S–874S.

Lommen, A. 2009. MetAlign: Interface-driven, versatile metabolomics tool for hyphenated full-scan mass spectrometry data preprocessing. *Analytical Chemistry* 81(8):3079–3086. doi:10.1021/ac900036d.

Magnusdottir, O. K., R. Landberg, I. Gunnarsdottir et al. 2014a. Plasma alkylresorcinols C17:0/C21:0 ratio, a biomarker of relative whole-grain rye intake, is associated to insulin sensitivity: A randomized study. *European Journal of Clinical Nutrition*, 68(4):453–458. doi:10.1038/ejcn.2014.12.

Magnusdottir, O. K., R. Landberg, I. Gunnarsdottir et al. 2014b. Whole grain rye intake, reflected by a biomarker, is associated with favorable blood lipid outcomes in subjects with the metabolic syndrome–A randomized study. *PLoS One*, 9(10):e110827. doi:10.1371/journal.pone.0110827.

Marklund, M., R. Landberg, A. Andersson, P. Åman, A. Kamal-Eldin. 2013a. Alkylresorcinol metabolites in urine correlate with the intake of whole grains and cereal fibre in free-living Swedish adults. *The British Journal of Nutrition* 109(1):129–136. doi:10.1017/S0007114512000621.

Marklund, M., N. M. McKeown, J. B. Blumberg, C.-Y. O Chen. 2013b. Hepatic biotransformation of alkylresorcinols is mediated via cytochrome P450 and β-oxidation: A proof of concept study. *Food Chemistry* 139(1):925–930. doi:10.1016/j.foodchem.2013.01.122.

McKeown, N. M., A. Hruby, R. Landberg, D. M. Herrington, A. H. Lichtenstein. 2016. Plasma alkylresorcinols, biomarkers of whole-grain intake, are not associated with progression of coronary artery atherosclerosis in postmenopausal women with coronary artery disease. *Public Health Nutrition* 19(2):326–331. doi:10.1017/S1368980015001123.

Mehmood, T., K. H. Liland, L. Snipen, S. Sæbø. 2012. A review of variable selection methods in partial least squares regression. *Chemometrics and Intelligent Laboratory Systems* 118:62–69. doi:10.1016/j.chemolab.2012.07.010.

Montonen, J., P. Knekt, R. Järvinen, A. Aromaa, A. Reunanen. 2003. Whole-grain and fiber intake and the incidence of type 2 diabetes. *The American Journal of Clinical Nutrition* 77(3):622–629.

Montonen, J., R. Landberg, A. Kamal-Eldin et al. 2010. Reliability of fasting plasma alkylresorcinol concentrations measured 4 months apart. *European Journal of Clinical Nutrition* 64(7):698–703. doi:10.1038/ejcn.2010.71.

Nieman, D. C., R. A. Shanely, N. D. Gillitt, K. L. Pappan, M. A. Lila. 2013. Serum metabolic signatures induced by a three-day intensified exercise period persist after 14 h of recovery in runners. *Journal of Proteome Research* 12(10):4577–4584. doi:10.1021/pr400717j.

Okarter, N. and R. H. Liu. 2010. Health benefits of whole grain phytochemicals. *Critical Reviews in Food Science and Nutrition* 50(3):193–208. doi:10.1080/10408390802248734.

Patti, G. J., O. Yanes, G. Siuzdak. 2012. Metabolomics: The apogee of the omics trilogy. *Nature Reviews Molecular Cell Biology* 13(4):263–269. doi:10.1038/nrm3314.

Pekkinen, J., K. Olli, A. Huotari et al. 2013. Betaine supplementation causes increase in carnitine metabolites in the muscle and liver of mice fed a high-fat diet as studied by nontargeted LC-MS metabolomics approach. *Molecular Nutrition & Food Research* 57(11):1959–1968. doi:10.1002/mnfr.201300142.

Pekkinen, J., N. N. Rosa, O.-I. Savolainen et al. 2014. Disintegration of wheat aleurone structure has an impact on the bioavailability of phenolic compounds and other phytochemicals as evidenced by altered urinary metabolite profile of diet-induced obese mice. *Nutrition & Metabolism* 11(1):1. doi:10.1186/1743-7075-11-1.

Pekkinen, J., N. Rosa-Sibakov, V. Micard et al. 2015. Amino acid-derived betaines dominate as urinary markers for rye bran intake in mice fed high-fat diet-A nontargeted metabolomics study. *Molecular Nutrition & Food Research* 59(8):1550–1562. doi:10.1002/mnfr.201500066.

Pluskal, T., S. Castillo, A. Villar-Briones et al. 2010. MZmine 2: Modular framework for processing, visualizing, and analyzing mass spectrometry-based molecular profile data. *BMC Bioinformatics* 11(1):395. doi:10.1186/1471-2105-11-395.

Rigotti, A. 2007. Absorption, transport, and tissue delivery of vitamin E. *Molecular Aspects of Medicine* 28(5):423–436. doi:10.1016/j.mam.2007.01.002.

Rinnan, Å., M. Andersson, C. Ridder, S. B. Engelsen. 2014. Recursive weighted partial least squares (rPLS): An efficient variable selection method using PLS. *Journal of Chemometrics* 28(5):439–447. doi:10.1002/cem.2582.

Ross, A. B., A. Bourgeois, H. N. Macharia et al. 2012. Plasma alkylresorcinols as a biomarker of whole-grain food consumption in a large population: Results from the WHOLEheart Intervention Study. *American Journal of Clinical Nutrition* 95(1):204–211. doi:10.3945/ajcn.110.008508.

Ross, A. B., A. Kamal-Eldin, E. A. Lundin, J.-X. Zhang, G. Hallmans, P. Aman. 2003. Cereal alkylresorcinols are absorbed by humans. *The Journal of Nutrition* 133(7):2222–2224.

Ross, A. B., P. Åman, A. Kamal-Eldin. 2004. Identification of cereal alkylresorcinol metabolites in human urine—Potential biomarkers of wholegrain wheat and rye intake. *Journal of Chromatography B* 809(1):125–130. doi:10.1016/j.jchromb.2004.06.015.

Salek, R. M., C. Steinbeck, M. R. Viant, R. Goodacre, W. B. Dunn. 2013. The role of reporting standards for metabolite annotation and identification in metabolomic studies. *GigaScience* 2(1):13. doi:10.1186/2047-217X-2-13.

Scalbert, A., L. Brennan, O. Fiehn et al. 2009. Mass-spectrometry-based metabolomics: Limitations and recommendations for future progress with particular focus on nutrition research. *Metabolomics* 5(4):435–458. doi:10.1007/s11306-009-0168-0.

Scott, I. M., W. Lin, M. Liakata et al. 2013. Merits of random forests emerge in evaluation of chemometric classifiers by external validation. *Analytica Chimica Acta*, 801:22–33. doi:10.1016/j.aca.2013.09.027.

Slavin, J. 2005. Whole grains and cardiovascular disease. *Whole Grains and Health*. Iowa, USA: Blackwell Publishing.

Smith, C. A., E. J. Want, G. O'Maille, R. Abagyan, G. Siuzdak. 2006. XCMS: Processing mass spectrometry data for metabolite profiling using nonlinear peak alignment, matching, and identification. *Analytical Chemistry* 78(3):779–787. doi:10.1021/ac051437y.

Spiller, G. 2002. Whole grains, whole wheat and white flours in history. *Whole-Grain Foods in Health and Disease*. St. Paul, Minnesota, USA: American Association of Cereal Chemists.

Sumner, L. W., A. Amberg, D. Barrett et al. 2007. Proposed minimum reporting standards for chemical analysis. *Metabolomics* 3(3):211–221. doi:10.1007/s11306-007-0082-2.

Sun, Q., D. Spiegelman, R. M. van Dam et al. 2010. White rice, brown rice, and risk of type 2 diabetes in US men and women. *Archives of Internal Medicine* 170(11):961. doi:10.1001/archinternmed.2010.109.

Szymańska, E., E. Saccenti, A. K. Smilde, J. A. Westerhuis. 2012. Double-check: Validation of diagnostic statistics for PLS-DA models in metabolomics studies. *Metabolomics: Official Journal of the Metabolomic Society* 8(Suppl 1):3–16. doi:10.1007/s11306-011-0330-3.

Söderholm, P. P., J. E. Lundin, A. H. Koskela, M. J. Tikkanen, H. C. Adlercreutz. 2011. Pharmacokinetics of alkylresorcinol metabolites in human urine. *The British Journal of Nutrition* 106(7):1040–1044. doi:10.1017/S0007114511001383.

Tasevska, N. 2015. Urinary sugars—A biomarker of total sugars intake. *Nutrients* 7(7):5816–5833. doi:10.3390/nu7075255.

Tighe, P., G. Duthie, N. Vaughan et al. 2010. Effect of increased consumption of whole-grain foods on blood pressure and other cardiovascular risk markers in healthy middle-aged persons: A randomized controlled trial. *The American Journal of Clinical Nutrition* 92(4):733–740. doi:10.3945/ajcn.2010.29417.

Triba, M. N., L. Le Moyec, R. Amathieu et al. 2015. PLS/OPLS models in metabolomics: The impact of permutation of dataset rows on the K-fold cross-validation quality parameters. *Molecular BioSystems* 11(1):13–19. doi:10.1039/C4MB00414K.

Trygg, J., E. Holmes, T. Lundstedt. 2007. Chemometrics in metabonomics. *Journal of Proteome Research* 6(2):469–479. doi:10.1021/pr060594q.

Varmuza, K., P. Filzmoser, M. Dehmer. 2013. Multivariate linear QSPR/QSAR models: Rigorous evaluation of variable selection for PLS. *Computational and Structural Biotechnology Journal* 5(6):e201302007. doi:10.5936/csbj.201302007.

Wei, X., X. Shi, S. Kim et al. 2012. Data preprocessing method for liquid chromatography–mass spectrometry based metabolomics. *Analytical Chemistry* 84(18):7963–7971. doi:10.1021/ac3016856.

van Velzen, E. J. J., J. A. Westerhuis, J. P. M. van Duynhoven et al. 2009. Phenotyping tea consumers by nutrikinetic analysis of polyphenolic end-metabolites. *Journal of Proteome Research* 8(7):3317–3330. doi:10.1021/pr801071p.

Westerhuis, J. A., H. C. J. Hoefsloot, S. Smit et al. 2008. Assessment of PLSDA cross validation. *Metabolomics* 4(1):81–89. doi:10.1007/s11306-007-0099-6.

Westerhuis, J. A., E. J. J. van Velzen, H. C. J. Hoefsloot, A. K. Smilde. 2010. Multivariate paired data analysis: Multilevel PLSDA versus OPLSDA. *Metabolomics* 6(1):119–128. doi:10.1007/s11306-009-0185-z.

Vettukattil, R. 2015. Preprocessing of raw metabonomic data. *Metabonomics: Methods and Protocols* 1277:123–136. doi:10.1007/978-1-4939-2377-9_10.

Wierzbicka, R., H. Wu, M. Franek, A. Kamal-Eldin, R. Landberg. 2015. Determination of alkylresorcinols and their metabolites in biological samples by gas chromatography–Mass spectrometry. *Journal of Chromatography B* 1000:120–129. doi:10.1016/j.jchromb.2015.07.009.

Wu, H., M. Kolehmainen, H. Mykkänen et al. 2015. Alkylresorcinols in adipose tissue biopsies as biomarkers of whole-grain intake: An exploratory study of responsiveness to advised intake over 12 weeks. *European Journal of Clinical Nutrition* 69(11):1244–1248. doi:10.1038/ejcn.2015.138.

Xia, J., R. Mandal, I. V. Sinelnikov, D. Broadhurst, D. S. Wishart. 2012. MetaboAnalyst 2.0–A comprehensive server for metabolomic data analysis. *Nucleic Acids Research* 40:W127–W133. doi:10.1093/nar/gks374.

Xiao, J. F., B. Zhou, H. W. Ressom. 2012. Metabolite identification and quantitation in LC-MS/MS-based metabolomics. *TrAC Trends in Analytical Chemistry* 32:1–14. doi:10.1016/j.trac.2011.08.009.

Ye, E. Q., S. A. Chacko, E. L. Chou, M. Kugizaki, S. Liu. 2012. Greater whole-grain intake is associated with lower risk of type 2 diabetes, cardiovascular disease, and weight gain. *The Journal of Nutrition* 142(7):1304–1313. doi:10.3945/jn.111.155325.

Zamboni, N., S.-M. Fendt, M. Rühl, U. Sauer. 2009. 13C-based metabolic flux analysis. *Nature Protocols* 4(6):878–892. doi:10.1038/nprot.2009.58.

Zikmundova, M., K. Drandarov, L. Bigler, M. Hesse, C. Werner. 2002. Biotransformation of 2-benzoxazolinone and 2-hydroxy-1,4-benzoxazin-3-one by endophytic fungi isolated from Aphelandra tetragona. *Applied and Environmental Microbiology* 68(10):4863–4870. doi:10.1128/AEM.68.10.4863-4870.2002.

19 Strengths and Limitations of Food Composition Databases

Phyllis Stumbo and Gary Beecher

CONTENTS

> The nutritional and dietetic treatment of disease, as well as research into problems of human nutrition, demand an exact knowledge of the chemical composition of food.
>
> **Robert McCance and Elsie Widdowson, 1940**

19.1 INTRODUCTION

Food composition tables and databases provide the foundation on which food and nutrition research, policy, and practice are based (Ahuja et al. 2013). As a result, data in these tools must be accurate and representative of foods consumed. They are usually compiled for a specific purpose such as characterizing the components in a person's diet, supporting a national survey of food intake, providing supporting data for dietary guidance, or facilitating new product development. The foundation for most databases is compilations by national governments that characterize the foods available within the country or region. Commercial applications often require addition of ingredients unique to the specific operation such as stabilizers used in processed foods or flavorings and additives used in commercial kitchens.

19.2 BACKGROUND

Some aspects of food composition are evident by merely seeing, tasting, or preparing the food. Sugar and salt are evident by taste, carotene by its orange color, and fat content by rendering. Other essential factors are impossible to identify without chemical analysis, and all require chemical analysis for an accurate estimate of amount present. Although the quantity of any component, whether obvious or hidden, may be determined by chemical analysis, it is more commonly estimated from data in food composition tables. Thus, as indicated in the quote above, the ideal food composition database should be both accurate and complete. Most of the tools discussed in this book utilize food table values to describe their results. Even verification of intake by biomarker relies on a food table to identify the foods carrying the marker in question. In addition, food recalls, records, and questionnaires all use food composition tables to interpret their results.

Maintaining a database to represent all foods commonly available is a tremendous task, owing to the frequent addition of new products in the market, changes in product formulation, updates to analytical methods, and addition of newly recognized components. Within the past several years, changes in fat level in meat available in the market, improvements in analytical methods such as when cholesterol assays were improved, reducing reported cholesterol values nearly in half-required reanalysis of many of the foods listed on the database. Couple this with the over 85,000 uniquely formulated foods available today in the market place that are all subject to change; maintaining an up-to-date database is a major effort.

19.3 PRIMARY RESOURCES FOR FOOD COMPOSITION DATA

The food label is probably the most accessible source of food data for the consumer; however, the student and researcher require a more comprehensive source of data. The United States Department of Agriculture's Agricultural Research Service (USDA-ARS) issues two food composition databases that are comprehensive for the United States, and are freely available on the Internet. The first is the Standard Reference listing over 8,000 foods and up to 230 components including, in addition to familiar components assigned to all or nearly all foods there are several uncommon fatty acid isomers and other components with values available for only a few foods (USDA National Nutrient Database for Standard Reference 2016). Analyzed values are given for most foods. Additional values are calculated for recipes (stews and other mixed dishes) and for commercial products that have both a nutritional label and list of ingredients in descending order that facilitate accurate calculations; to maximize completeness of the data set. For example, breakfast cereals in the United States are often enriched and/or fortified with vitamins and minerals. Thus, breakfast cereals are often listed by brand to account for their unique enrichment and fortification.

The second database is the Food and Nutrient Database for Dietary Studies that is customized to support the National Health and Nutrition Examination Survey (NHANES) conducted by the National Center of Health Statistics of the Centers for Disease Control and Prevention (CDC-NCHS). This database is developed by the USDA-ARS Survey Research Group and in 2016 contained 64 nutrient components for 7600 foods. The nutrient components are selected for their public health importance and because a value is either available for each component reported or can be imputed for every food on the database (USDA Food and Nutrient Database for Dietary Studies 2016).

Recent additions to government food composition databases are not strictly of food, but of dietary supplements. Historically national food consumption surveys were limited to components from food. This limitation was troublesome because a supplement could provide as much of a vitamin and/or mineral as the entire day's food intake. In 1995 the Office of Dietary Supplements was created within the National Institutes of Health to help correct this obvious omission in databases. Their stated objectives were to explore the potential role of dietary supplements in health and to conduct and coordinate research relating to dietary supplements. Two databases were created by leadership

TABLE 19.1
Food Data Sets from the United States Department of Agriculture (USDA) Agricultural Research Service (ARS)

Database	Source/Availability	Primary Objective	Unique Features
SR28-Standard Reference	USDA/ARS Nutrient Data Laboratory	Primary data set for use in software and wherever basic information about nutrients in food is required	8789 foods 150 components
FNDDS 2011–2012 Food and Nutrient Database for Dietary Studies	USDA/ARS Food Surveys Research Group	Data set based on the SR with added terms to facilitate interpretation of food intake reported in the vernacular	7600 foods 64 components
NIH/USDA Dietary Supplement Ingredient Database (DSID)	NIH/USDA	Data for a selected set of dietary supplements; distributed as analytical values as label values to probable content	Analytical data compared to label data for child and adult multivitamin and mineral supplements
NIH/NLM Dietary Supplement Label Database (DSLD)	NIH/National Library of Medicine	Data for a sample of dietary supplements marketed in the United States. Data taken from label information. Applicable for population studies of intake. Updated annually.	Label data from approximately 55,000 products

from this office that are of interest to users of food composition databases, namely the Dietary Supplement Label Database (DSLD) and the Dietary Supplement Ingredient Database (DSID). The ingredient database verifies composition of the most popular dietary supplements through chemical analysis and the label database provides composition of supplements as reported on the product label and this latter database is formatted to facilitate incorporation into existing food composition databases. The goal of the DSLD is to list all 80,000 supplements sold in the United States and by 2016 it had tabulated data for approximately 55,000 supplement products. The DSLD was first used in the United States National Health and Nutrition Examination Survey (NHANES) in 2010–2011. Table 19.1 summarizes the main characteristics of these national databases.

In addition to these government sponsored databases several commercial and academic enterprises have customized the government resources for use in software offered for sale to the public. Examples of these products are listed in the International Nutrient Databank Directory (http://www.nutrientdataconf.org/indd/).

Many countries have food databases representing foods consumed in their country; the Food and Agriculture Organization (FAO) of the United Nations maintains a list of these food tables on their website (http://www.fao.org/infoods/infoods/tables-and-databases/en/). A complete list of available food composition databases is not included here, but a few are listed to familiarize the reader with the breadth of information available. The Canadian database illustrates how information about foods in adjacent countries are similar. Many foods and eating practices in Canada mirror those in the United States; therefore, much of the Canadian data are derived from the U.S. Standard Reference Database. However enrichment practices differ between the two countries, so unique listings for processed foods are required. Also regional eating practices differ, especially when considering the native populations in the two countries, necessitating many unique entries. Numerous national databases are available on the Internet organized in a fashion to be searched online or downloaded

TABLE 19.2
Online Access to Food Composition Databases

Country, Database Name, Number of Foods, and Components	Language	Internet address
Canada: Canadian Nutrient File (CNF) 5690 Foods 152 components	English, French	https://food-nutritilso on.canada.ca/cnf-fce/ iavailable as spreadsheet or cvs file ndex-eng.jsp (search feature). Also available as spreadsheet or cvs file
Germany: Souci Fachmann Kraut	German	http://www.dfal.de/2/home/food-composition-tables.html
McCance and Widdowson	English	(https://www.gov.uk/government/publications/composition-of-foods-integrated-dataset-cofid). (spreadsheet)
The Netherlands: NEVO 2194 foods, 144 components	Dutch, plus food name and instructions in English	Nevo-online.rivm.nl (online search)
Norwegian Food Composition Table 1600 foods, 41 components	Norwegian, English	www.matvaretabellen.no on-line search, or download spreadsheet
United States: Standard Reference (SR) 8789 foods, 150 components	English	https://ndb.nal.usda.gov/ndb/search/list
West African Food Composition Table 1624 foods, 17 components	English, French	www.fao.org/docrep/003/x6877e/X6877E05.htm

as a spreadsheet. Table 19.2 lists Internet addresses for selected reference databases in the United States and Europe.

The FAO website lists other databases that may be of interest to users. Their list of European databases is available at http://www.fao.org/infoods/infoods/tables-and-databases/europe/en/.

19.4 A HISTORICAL LOOK AT FOOD COMPOSITION DATABASES

Food composition tables in the United States had their beginning in the late 1800s (though some tables were published before then) (Colombani 2011). The format of the U.S. databases underwent a dramatic evolution in content and formed during the 1900s from their initial appearance in 1894 as printed tables to large electronic databases by the end of the twentieth century.

Developing the first U.S. tables in 1894 represented both scientific and political accomplishments. Scientific, because all the tools and resources had to be built from scratch using technology learned primarily in France and Germany; and political because funding had to be won from a government agency with competing goals and before the value of food composition data was widely recognized.

The success of Wilbur Atwater, the first agriculture experiment station scientist devoted to human nutrition, in compiling these data is legendary. The team and collaborators he assembled developed calorimeters to measure energy and other specialized equipment to process and analyze food.

The 1894 tables featured the five proximates: protein (nitrogen-containing substances), fat (ether extractives), moisture, minerals (ash), and carbohydrate by difference. His untimely death in 1907 robbed him of the opportunity to experience the growth of food composition tables. In the 1920s and 1930s there was an explosion of new nutrient data as one-by-one the essentiality of vitamins were discovered and measured in foods. In addition, as the essential nature of minerals was recognized and these minerals also measured in food, these values were added to food composition databases paving the way for the more complex tables we have today.

Gradually comprehensive tables were developed in the United States and other parts of the world. Notable were Bowes and Church in the United States (Pennington 2009), McCance and Widdowson in England (Finglas et al. 2014) during the 1930s, and Souci in Germany (Souci et al. 2008) in the 1960s. McCance and Widdowson tables were released in spreadsheet format in 2008 and are freely available on the Internet (see Table 19.2).

A classic publication by USDA in 1963 was titled Agricultural Handbook No. 8, Composition of foods, raw, processed, and prepared (Watt and Merrill 1963). Entries in this publication numbered 17 for 2483 foods in the main table plus three fatty acid classes for selected foods that contained fat. In 2016 the same database is in electronic form and known as Standard Reference with 8,789 entries (Table 19.1). Documentation for the 2016 version of the database is found at http://www.ars.usda.gov/sp2UserFiles/Place/80400525/Data/SR/SR28/sr28_doc.pdf.

The 1963 *Handbook 8* data was tedious to use by clinicians because it was in 100 g portions rather than typical household measures or serving sizes. This limitation made the Bowes and Church publication (Pennington 2009) a mainstay for clinicians. Practical applications of the extensive food composition tables that were designed specifically for evaluating diets were two short databases utilizing food groups of similar composition rather than individual foods. Two notable applications were the Exchange Lists developed jointly by the American Diabetes and American Dietetic Association (Franz et al. 1987) and a more extensive short list developed for general dietary calculations (Leichsenring and Wilson 1951). In 1975 the Handbook 8 data was published in a more user-friendly format as Handbook 456 [out of print] with data given in household units. An abbreviated form of the data is published in printed form as Home and Garden Bulletin 72 with more than 1200 food items in household units and with 19 nutrient values. An updated version of this booklet is available in PDF format on the Internet at http://www.ars.usda.gov/Services/docs.htm?docid=6282.

A large database that describes foods by brand as well as type is desirable to facilitate identification of foods as reported by consumers. The pitfall with this feature is that data listed by brand is typically limited to the six or seven values found on the food label, and the remaining components are missing and counted as zero even though the product might be a rich source. The incompleteness of data is hard to spot when the data is electronic and no printed database is available. This problem was acutely felt by users of the earliest software that depended on the government data composed largely of commodity foods as their source of information, whereas the information desired was for foods identified by brand.

As data completeness was so critical in the valid use of any database, Dr. Loretta Hoover led the compilation of a database directory that gave this information for food composition databases. The directory was a mimeographed resource available to those few scientists who attended the National Nutrient Databank.

Between about 1970 when the mainframe was becoming accessible and in 1990 as desktop computers became more widespread, the nature of available programs changed from mainframe to desktop. The desktop revolution put computer in the hands of the nutritionist. This did not solve the problem of sparse data sets, but it did make them more apparent to the user. This progression is mentioned because it suggests ways modern users should evaluate their nutrition assessment tools.

During this time two large studies sponsored by the National Heart, Lung, and Blood Institute (NHLBI) of the National Institutes of Health (NIH) were studying the effect of nutrition, especially fat intake, on the occurrence and progression of heart disease. The available databases were sufficiently complete in terms of many nutrients, but were lacking in detailed fatty acid composition especially for the many brand name products reported by research subjects. As fat and the fatty acids played a primary role in their hypotheses an effort was launched to create a database with more complete nutrient values needed for these studies. The work was conducted at the University of Minnesota and their efforts resulted in the *NHLBI* database that was subsequently marketed along with software that assisted the user to obtain detailed dietary intake data. The software and database are marketed as the Nutrition Data System for Research (Buzzard).

Development of this database and software are described in Dennis et al. (1980). The NHLBI database used the USDA standard reference database plus extensive use of industry data for their branded products. Imputed fatty acid values for brand name foods so often reported by research subjects was vital to studying the effect of nutrition on heart disease and one of the characteristics of this database that appealed to researchers. The NHLBI database was initiated to support three large studies on nutrition and heart disease that were conducted in different geographic locations but using the same methodology. As the studies were conducted in different locations it was paramount that the nutrition coding was consistent among locations. This was accomplished by not only using the same database, but also by standardizing the interviewing of subjects, the detailed query of each food, and how the resultant food list was coded for computer processing. Subjective cues were standardized to better identify which fat was used if the subject was uncertain and how to reconcile subjective decisions. This required the interviewers to be trained as a group for consistency among the multicentered studies; that food descriptions were uniform and unknowns were solved consistently. Then for uniformity the food intake record was coded at a central location by trained coders who continually added new products and updated existing data to the system. All this standardization was not without a cost, so over a period of years these processes were standardized sufficiently to be incorporated into an automated system for the desktop computer enabling the nutritionist (after training and armed with a 500 page coding manual) to match each food on the intake record with an appropriate food on the database using the same techniques as her fellow nutritionists at other centers. This system is widely believed to improve the consistency of nutrition intake data among studies using food records and recalls.

The high cost of using and maintaining this software resulted in its use primarily for research. The University of Minnesota assumed responsibility for the database. Over several years they developed user software that automated the decision making during coding making the database more accessible. This software is called NDSR (Nutrition Data System for Research) and is used by a wide research audience (Nutrition Coordinating Center 2016).

On account of its flexibility this software was utilized as a centralized coding and calculation system to support studies in 9 countries in Latin America. A harmonization procedure between each country and the NCC staff in Minnesota matched foods from each local database to an appropriate entry in the NDS system (Fisberg et al. 2015). Numerous other software serve a broad audience throughout the world and many are listed in the international database directory on the U.S. National Nutrient Databank website. (International Nutrient Databank Directory 2016).

19.5 ADVANCES IN THE ASSESSMENT OF HEALTH RELATED COMPONENTS OF FOODS

So far this discussion has centered around the availability and completeness of food composition databases, however the greater question is validity of the data to be compiled. Many procedures have been developed for analysis of nutrients and other health related components of foods. They are published in peer-reviewed journals, that is, Journal of Agricultural and Food Chemistry, Journal AOAC International, Journal of Chromatography, Journal of Food Composition and Analysis, and so on.

Many procedures have been approved for *official* status by committees of the American Association of Cereal Chemists International or the Association of Official Analytical Chemists International (AACC 2012; AOAC 2016). As information for *newer* food components and additional forms of existing nutrients is required, advances in technology and instrumentation have been integrated into analytical procedures, yet *older* methods may still play a role. Thus analysts can match requirements for details of food composition data with available instrumentation, budgetary resources, and personnel expertise.

It is beyond the scope of this treatise to critically evaluate all of the analytical procedures for food composition analyses. Lists of those procedures used in support of USDA's National Food and Nutrient Analysis Program (NFNAP) have been published (Phillips et al. 2014; USDA 2015).

Over the last quarter century a series of procedures have been developed that have standardized and greatly improved the quality of information in food composition tables and databases. Although these processes have been applied primarily to data in the USDA National Nutrient Databank and databases it supports, they are applicable to all food composition databases and are detailed below.

19.5.1 Evaluation of Food Composition Data

Analytical data may come from many sources including scientific literature, government sources, food industry laboratories, or from contractual studies. As a result, data from such diverse sources are often of uneven quality and lacking in detailed supporting documentation. Thus it is important to evaluate the quality of analytical data for its reliability and representativeness, so that only acceptable information is combined for nutrients and other health related components of foods.

Scientists at the U.S. Department of Agriculture (USDA), who maintain and disseminate the large USDA National Nutrient Database for Standard Reference, developed a system in the early 1980s to examine data for the iron content of foods (Holden et al. 2002). This system was modified and expanded as it was applied to data for other minerals (Cu, Se) and organic components (carotenoids) of foods. With this background and with comments from the food composition scientific community, software of the USDA Nutrient Databank system was modified to include a module for the evaluation of analytical data prior to incorporation into the databank (Holden et al. 2002).

The module includes five categories of evaluation, each considered important in data quality: (1) sampling plan, (2) number of analytical samples, (3) sampling handling, (4) analytical method, and (5) analytical quality control. *Sampling plan* relates to the representativeness of a particular food. Samples obtained from a local market or experimental plot receive the lowest score, whereas those selected as part of a statistically-based regional or nationwide plan are given high scores. An important component of this category is documentation—type of food, for example, retail, wholesale, and home grown; its form, for example, fresh, frozen, and cooked; sizes; cultivars and other critical details. Also information relative to basis of sampling is important, for example, market share, sales volume, and consumption patterns. *Number of samples* is critical to the estimation of the mean as well as to the magnitude of variability for a component of a food. Thus analysis of a food or food composites sampled from various retail outlets and geographic regions receive a higher score than repeated analysis of the same sample or composite. *Sample handling* impacts the validity of the nutrient or component in the sample from the time it is purchased or harvested to the time it is analyzed and thus may alter the representativeness of the food *as consumed*. Documentation of such items as storage temperature, processing, for example, dissecting, cooking, makeup of composite, homogenization and subsampling for analysis, and control of moisture content are critical information for a high score. *Analytical method* is evaluated in two parts: (1) the method itself for specificity, accuracy, and precision; and (2) the laboratory conducting the analyses for demonstrated ability to employ the method in a manner to obtain accurate results. On account of the complexity of various procedures expert panels have been assembled to identify critical steps for each method of each food component, which are subsequently used as the basis for evaluation of new procedures. A list of analytical methods for food components used by USDA contract laboratories is provided in the documentation section of the 2015 release of Nutrient Database for Standard Reference (USDA 2015). For performance of the laboratory, results of a specific procedure are examined for accuracy, precision, and day-to-day variability. Highest scores for accuracy are given when a matrix-matched certified reference material (CRM) is available and incorporated as part of routine laboratory procedures. *Analytical quality control* refers to accuracy and precision in the day-to-day performance of an analytical procedure. Values receive high scores when a quality control (QC) material is part of each batch of analytical samples. Again reference values for the QC material receive high scores when analyzed against a CRM, but in its absence values may be determined by an independent laboratory using the same method, or by a second method. Table 19.3 summarizes these factors.

TABLE 19.3
Criteria for Evaluation of Food Composition Data to Be Added to Published Database

	Sampling Plan	Number of Samples	Sample Handling	Analytical Method	Quality Control
High Score	Statistically-based regional or national plan	Food samples chosen from multiple market places and geographical regions	Document Storage temperature and cooking method, control moisture content	Validate methods for specificity, accuracy, and precision or use proven laboratory	Test reference material with each batch of unknown sample or compare results from two laboratories for same sample
Low Score	Sample from single store or experimental plot	Repeated analysis of same sample	Maintain adequate storage facilities	Untested methods, unknown laboratory	No verifications with CRM or second laboratory
Criteria	Use objective method to choose sample such as market share, sales volume, or consumption pattern	Documents number/type and source of samples analyzed	Maintain adequate storage facilities, keeps detailed record of sample handling and storage temperature	Maintains adequate record of analytical results to demonstrate reproducibility	Maintain adequate record of analytical results for CRM, reference material, duplicate samples, and/or second laboratory
Score for each factor	20	20	20	20	20

After each food component is rated for all five categories, ratings are summed to yield a quality index (QI), which is an indication of the acceptance of the value (Holden et al. 2002). The highest possible QI is 100 as each category has a maximum rating of 20. As data for a given food component–food combination are combined, QIs are also weighted and combined to give a Confidence Code—a data quality indicator that is disseminated with the nutrient value. Confidence Codes are reported as letters ranging from A to D, with *A* indicating data of highest quality.

This system has been an integral part of the National Food and Nutrient Analysis Program (NFNAP) of the USDA since the late 1990s, an important activity sponsored by several agencies of the U.S. federal government to improve the quantity and quality of data in food composition databases (Haytowitz et al. 2008). Results of evaluations of existing food composition data have been used to set priories for ongoing food sampling and analysis as part of this program. In addition, the evaluation system continues to be used to critique new data prior to incorporation into the U.S. National Nutrient Databank. Similar systems have been incorporated into several food composition databases in Europe and into harmonization procedures for databases of European Union Countries (EuroFIR) (Gry et al. 2007; Oseredczuk et al. 2009; Westenbrink et al. 2016).

19.5.2 Key Foods and Nutrients

Effective food composition databases must be routinely updated to keep pace with many developing phenomena. These include introduction of new foods, offshore imports of commodities, reformulation of existing foods, and changes in population demographics and food habits. A matrix of the huge number of food items is available (8,000+ tabulated in USDA National Nutrient Database for Standard Reference, SR28) and the large number of food components being tracked (~150 in above database) would require enormous resources for routine sampling and analyses. As a result, priories must be established relative to foods that are sampled and food components that are analyzed. The concept of a *Key Foods* list has been used by USDA scientists since the late 1980s to update databases for which they are responsible (Hepburn 1987). Key Foods are defined as those foods, which in aggregate contribute ~75% of the intake of selected nutrients or food components of public health importance from the diet. This list of foods, which is periodically updated, is based on the most recent U.S. food consumption and food composition data as well as the current nutrients or food components of health concern (Haytowitz et al. 1996; Haytowitz et al. 2002).

The most recent Key Foods list of nearly 600 foods was derived from food consumption data from What We Eat in America–NHANES Survey 2011–2012 (WWEIA–NHANES, 2011–2012), food composition data from USDA National Nutrient Database for Standard Reference issued 2009 (SR26), and for food components of public health significance as identified in the 2010 *Dietary Guidelines for Americans* (Haytowitz 2015). Targeted food components included total fat, food energy, total sugar, total dietary fiber, calcium, iron, potassium, sodium, β-carotene, retinol, vitamin B_{12}, folates, cholesterol, and saturated fatty acids. For each food reported as eaten, the nutrient content was multiplied by the amount consumed, ranked, and divided into quartiles. Eleven foods were in the First quartile (top contributors of nutrients in the diet); milk comprised nearly half of this group due to its high consumption and multinutrient contribution, including retinol. Other foods in this quartile included carrots, cheese, eggs, hamburger rolls, and tortillas as well as ice cream. There were 40 foods in the second quartile, about 90 in the third, and the remainder in the fourth quartile. Table 19.4 lists the other selected foods that collectively represent about 75% of selected nutrients.

More and more foods are consumed as *mixed dishes* away from home, as *take-out meals*, or as frozen dishes that only need to be heated before being eaten. In the past, component parts of such foods were added to their respective commodities and reported as part of Key Foods lists. Recently, a preliminary Key Foods list was developed for mixed dishes based on data reported in WWEIA–NHANES, 2011–2012 (Haytowitz 2015). Sixteen foods were listed in the first quartile of this list, of which various forms of pizza occurred five times. Other Italian-type foods (spaghetti and lasagna) and several Mexican or Tex-Mex foods (chili con carne, soft tacos, quesadillas, and burritos) were

TABLE 19.4
Key Foods for Chemical Analysis, Foods that Together Provide 75% of Intake of Selected Nutrients

Quartile for Intake From NHANES	Number of Individual Foods	Examples of Foods in Each Quartile
1	11	Milk, carrots, cheese, eggs, hamburger rolls, and tortillas
2	40	Thin crust cheese pizza, processed cheese, white bread, carbonated soda, potato chips, fries, spinach, and bananas
3	90	Thick crust cheese pizza, egg, orange juice, cookies, canned tuna, peanuts, brewed tea, ground beef, cantaloupe, and chocolate chip
4	~455	Shrimp, cabbage, ready-to-eat cereal, salsa, canned tomatoes, cheese puffs, tortilla chips, milk chocolate, chicken, celery, potatoes, frankfurter, fast food double hamburger, fast food sausage biscuit, beef, pork, and rice

also prominent, highly consumed foods. Key Foods' lists, along with other information, are used to select and prioritize foods and their components for sampling and analyses and thereby provide current, representative data to update USDA's databases, including the next release of SR.

Development of Key Food lists requires food intake data and information about the nutrient content of those foods, which are readily available for the U.S. population in general, as outlined above. However, for subpopulations and small ethnic groups, national surveys and databases may inadequately represent dietary habits and nutrient levels of respective foods. A Key Foods list developed for African Americans, who at the time represented ~12% of the U.S. population, had similar foods in the first quartile as the list for the U.S. population, but with minor ranking changes (Haytowitz 2000). However, when a Key Foods list was prepared for American Indians, using national databases, traditional foods commonly consumed by this group were not identified, thus suggesting that other methods were required to adequately track dietary habits and foods of small minority groups (Haytowitz 2000). Dietitians and other professionals familiar with dietary habits of two minority groups in the United States, Alaska Natives, and American Indians, were recruited to identify important traditional foods that may impact on the health of individuals in these subpopulations. Employing these unique techniques, nearly two hundred foods have been sampled, analyzed, and the resulting data incorporated into the current USDA National Nutrient Database for Standard Reference (SR28) (Amy and Pehrsson 2003; Pehrsson et al. 2005; Phillips et al. 2014). These observations indicate that when surveying small, minority populations it is essential to have intimate knowledge of dietary habits, so that appropriate foods can be assessed for nutrient levels in support of diet–health research and education.

19.5.3 Statistical Sampling of Foods for Analysis

Scientists at USDA's Nutrient Data Laboratory maintain food composition databases representative of nutritive values for foods consumed in the United States. A fundamental issue is how to reliably sample a large variety of foods, offered in many, many stores, across a large, diverse country. An ideal plan would involve selecting many geographically dispersed areas across the United States, selecting retail outlets in each of those areas, followed by selecting food products from an extensive list of all foods sold in those outlets (Nusser and Carriquiry 1998). Although statistically ideal, such a plan requires extensive resources of staff, time, and money. As part of the National Food and Nutrient Analysis Program instituted in the late 1990s (Haytowitz et al. 2008), an economically feasible, multistage, and self-weighting nationally representative food sampling approach was designed (Pehrsson et al. 2000). Briefly this plan consisted of (1) dividing the contiguous states into four regions with nearly equal populations; (2) further dividing each region into three strata of high, medium, or low population density; (3) identifying Generalized Consolidated Metropolitan Statistical Areas (gGCMAs) within each stratum; (4) selecting two counties proportional to urbanicity (1 rural, 1 urban) from each gGCMA; (5) selecting a grocery store outlet within each county proportional to sales volume (>$2 million/year); and (6) choosing products proportional to market share and package size (Pehrsson et al. 2000). Recently this general approach was modified to accommodate demographic changes in population, congressional redistricting, and increased popularity of warehouse-type grocery outlets (Pehrsson et al. 2013).

Although the above sampling design is appropriate for the majority subpopulations of the contiguous states, minority subpopulations, and unique food components require their own sampling plans. Thus individual plans have been designed for sampling of Alaska Native subsistence foods, for selecting traditional Northern Plains American Indian foods, and for sampling fluoride in drinking water and in selected beverages (Pehrsson et al. 2005; Pehrsson et al. 2006; Phillips et al. 2014).

Based on the above designs, composite samples of each food product were prepared at a central laboratory, and aliquots were withdrawn and distributed for analyses. As of the release of SR28 (2015), nearly 2000 foods have been sampled nationwide, analyzed for nutrients and other health related components, and the resulting data incorporated into the National Nutrient Databank, which has improved the accuracy and maintained the currency of information for critical foods (USDA 2015).

19.5.4 Accuracy in Food Component Analyses

Analysts have been measuring health-related components of foods and feeds following prescribed procedures for more than one and one-half centuries; one of the first was the *Weedan Procedure* for proximate analysis developed in 1860 (McCullom 1957). The Association of Official Agricultural Chemists (AOAC), founded in 1884, and its successor organizations, formalized and provided strict guidelines for many analytical methods that were granted the title *official* (AOAC 2013). However, all of these procedures lacked a component that addressed extent of extraction or modification of the nutrient during preparation of the sample for analysis, especially if a complex matrix was involved, for example, foods or feeds.

The National Bureau of Standards (NBS) (now National Institute of Standards and Technology [NIST]) was organized by the United States Congress in 1901 to maintain standards for commerce and the economy (Rasberry 2002). Its first *project* was the provision of standard materials (certified reference materials [CRMs]) for analysis of new cast iron alloys that had been developed for passenger rail car wheels whose fracture heretofore had been the cause of derailments and passenger deaths. Certified reference materials are standards that have a matrix similar to materials of interest and whose analytical values have metrological traceability. Extensive analytical procedures are employed to establish values of CRMs, which then can be used to establish analytical accuracy in routine laboratories. Scientists at NBS/NIST have continued to conduct research and develop standard materials as commerce, technology, and war efforts progressed.

In the 1970s a small group of scientists representing USDA and Agriculture and Agri-Food Canada coalesced at NBS in an effort to develop CRMs for the many nutrients and other important components of foods and feeds (Ihnat 2001). At the same time there was a groundswell of similar activity at the International Atomic Energy Agency and at the Commission of the European Community as the European Union was being formed. As a result of this collaboration and inclusion of representatives of metrology groups of many countries, a large number of CRMs have been developed applicable to the analysis of foods (Ihnat 2001; Phillips et al. 2006). A critical aspect of this program has been the educational effort on how these materials can be effectively and efficiently used. Thus a periodic scientific meeting was organized (International Symposium on Biological and Environmental Reference Materials [BERM]), directions were developed on the incorporation of CRMs into *official methods*, and books and national and international standards organizations have provided guidelines on the application and use of CRMs (Wolf 1993; Stoeppler et al. 2008; AOAC 2012; ISO/REMCO 2016).

CRMs have been incorporated into food analyses as part of updating the USDA Nutrient Databank and as part of the National Food and Nutrient Analysis Program. Nearly 10 years later sufficient data from CRM analyses had been generated to evaluate analytical performance of contractual laboratories (Phillips et al. 2007). In general these laboratories best performed in the analyses of selected minerals (magnesium, manganese, phosphorus, and selenium) and vitamin B_{12}. For those nutrients and food components that have a complex analytical scheme (carotenoids and fatty acids), have an undefined molecular structure (total dietary fiber), or for which a nonspecific method of detection was employed (microbiological analyses of vitamins), results were highly variable and often *out of range*. Acceptability was based on a *Z Score* calculated from results of CRM analyses and uncertainty of CRM assigned value for each nutrient (Phillips et al. 2007). For the first time these results provided a metric of the accuracy in the analysis of health related components of foods. Even though the laboratories who conducted these analyses were *prequalified* contingent on award of contract, these results highlight the extreme difficulty in routine analyses of nutrients and other components in complex matrices common in foods.

19.5.5 Advances in Instrumentation for Food Analysis

A wide variety of analytic techniques are currently in routine use to measure nutrients and health related components of foods (Phillips 2014; USDA 2015). Standard methods, including those for many of the above, have been formally approved and published (AACC 2012; AOAC 2016). These range from chemically nonspecific microbiological assays to sophisticated instrumental techniques that provide extensive atomic or molecular detail. All have benefitted from the many advances in technology and materials that have taken place over the past 50 years; a few are herewith described.

19.5.5.1 Solid-State Electronics

The invention of the transistor in 1947 and subsequent development of solid-state electronics have led to compact detectors and other stable components of instruments, greatly expanding the capability of these machines. In addition, the marriage of computers with analytical instruments has resulted in continuous control, so that instruments now run *24/7* unattended by a human, increasing the productivity of laboratories. At the same time the instrument's computer processes and stores the large amount of data these machines generate.

19.5.5.2 Advances in Materials

Advances in the formulation of plastics and metal alloys have greatly enhanced the development of analytical instrumentation and changed *the scene* of a typical laboratory from one of glassware to that of *disposable* pipette tips and other *one-time-use* labware. Hidden in many laboratory instruments is a labyrinth of stainless steel and Teflon tubing as well as many parts of polyvinyl chloride (PVC), acrylonitrile butadiene styrene (ABS), and other plastics. The formulation of special alloys of stainless steel, such as Hastelloy, that are resistant to strong acids and chlorides has permitted the application of high-performance liquid chromatography (HPLC) systems to analyze such sensitive food components as carotenoids, flavonoids, and several vitamins. In addition, the development of unique alloys has also been the impetus for development of automated instrumentation based on the Dumas method for nitrogen analysis, often called *combustion method* (Dumas 1831), thereby replacing the dangerous, time consuming, and hazardous waste-generating Kjeldahl procedure (Kjeldahl 1883).

Although laboratory instrumentation is often *improved* with the latest application of computer software or modification of its *package*, the basics of each instrument remain the same. However, two advances in laboratory instrumentation that has increased the capability and extent of food analysis over the past quarter century should be highlighted.

19.5.5.3 Inductively Coupled Plasma–Optical Emission Spectrometry (ICP–OES)

Introduction of the inductively coupled plasma torch as an excitation source, which has very high temperatures, overcame many sensitivity and interference issues of earlier atomic absorbance and emission instruments (Reed 1961; Greenfield et al. 1964). These instruments are based on the original concepts of Kirchhoff and Bunsen for detection of intensity of emission line(s) of elements (Kirchhoff and Bunsen 1860). They have undergone several generations of modifications and advancements, in terms of monochromators, detectors, and electronics, to where they have become *work-horse* instruments for mineral and other elemental analysis in many laboratories (Boss and Fredeen 2004).

19.5.5.4 Mass Spectrometers

The maturation and integration of mass spectrometers with chromatography and ICP instruments has greatly increased the capability of these machines and quality of data they generate. First developed in England in the early twentieth century and after many technological and instrument interface advances, mass spectrometers are now common detectors for many laboratory instruments involved in separation and quantification of elements and molecules (Borman 1998). Examples of nutrients that rely on mass spectrometry for detection and quantification include choline, selenium, and the various forms of vitamin D. Table 19.4 summarizes the criteria used for evaluating food composition data before it is added to national food composition databases.

19.6 STRENGTHS AND LIMITATIONS OF FOOD COMPOSITION DATABASES

A major source of nutrient data is derived from food composition assays conducted or contracted by the USDA food composition laboratory in Beltsville, MD, the United States. Other sources include studies conducted in research facilities around the world and in private laboratories operated by food manufacturers to monitor their own products. The Food and Drug Administration conduct studies of food products in their regulatory role to monitor the safety of food and validity of food labels and publish data on a limited set of nutrients including calcium, iodine, sodium, iron, molybdenum, iron, copper, magnesium, manganese, phosphorus, zinc, potassium, and sodium (Total Diet Study Analytical Results 2016). A major responsibility of database compilers is to evaluate the published data to select valid results for publication and to continually update their database to include new products and new data sources. Table 19.5 summarizes the strengths and weaknesses of food composition tables.

TABLE 19.5
Strengths and Limitations of Food Composition Databases

Parameter	Strengths	Limitations
Availability of data	Databases in the United States and many other countries are readily available on the Internet, often free of cost	Databases are difficult to use without appropriate software that may be difficult to identify and evaluate, and also costly to own
Use of data	Food composition data support dietary guidance, nutrition research, food product formulation, food labeling, and surveillance of nutritional adequacy	Continuous monitoring and revision of existing databases are because the food and dietary supplement supply are constantly changing
Analytical Procedures	Methods available for nearly all important food components. Advances in instrumentation makes possible rapid analysis of atomic and molecular species of nutrients and other health related food components	Many methods lack specificity and accuracy. Instrumentation is costly, requires highly trained technical staff, thus cost of analysis is increased
Evaluation of food composition data	Evaluates many critical aspects of food samples to assure representativeness of food as consumed and that data are valid	Trained technical staff required; increased labor burden
Key foods and critical nutrients	Resources applied to analysis of those foods that supply the majority of food components of public health importance	Key foods of important subpopulations may be overlooked, based on a nationwide food intake surveys
Statistical sample of foods for analysis	Economically feasible, multistage, and self-weighting nationally representative food sampling strategy yields reasonable estimates of food composition values for the country	Strategy must be updated to accommodate changes in food retailing and population shifts. Foods of critical subpopulations may not be sampled
Accuracy in food analysis	CRMs available for most food matrices. Procedures developed for preparation and application of laboratory quality control materials, based on CRMs	Extensive education program required. Additional staff required to evaluate results of quality control materials. Cost of analysis increased

19.7 CONCLUSION

To summarize, food composition databases are essential to meet many nutritional goals. Food composition data supports nutrition monitoring of the quality of the food supply, guides the development of dietary advice, aids in evaluation of an individual's eating habits, supports the labeling of food products, evaluation of feeding programs, and supports nutrition research and development. The availability of free, reliable food data is one of our greatest strengths; however, constant changes in our food supply challenges the timeliness and accuracy of published data. Discovery of previously unknown factors challenges scientists to develop new assays and to apply them to an ever changing food supply. Reliable food data supports the formulation of rations for extreme conditions such as space exploration, war, and famine. U.S. dietary advice is informed by the U.S. Dietary Reference Intakes (Dietary Reference Intakes 2006) and the Dietary Guidelines for Americans (DGA) (Dietary Guidelines for Americans 2015–2020). The DRI document lays out the scientific underpinning for the recommendations, and the DGA translates this information into recommended food patterns. The availability of food composition data is essential for both activities.

> There are two schools of thought about food tables. One tends to regard the figures in them as having the accuracy of atomic weight determinations; the other dismisses them as valueless on the ground that a foodstuff may be so modified by the soil, the season, or its rate of growth that no figure can be a reliable guide to its composition. The truth, of course, lies somewhere between these two points of view.
>
> **Widdowson and McCance, 1943**

REFERENCES

AACC International. 2012. *Approved Methods of Analysis.* 11th edition. St Paul: AACC International. http://mthods.aaccnet.org/cite.aspx (accessed April 13, 2016).

Ahuja, J. K. C., A. J. Moshfegh, J. M. Holden, E. Harris. 2013. USDA food and nutrition databases provide infrastructure for food and nutrition research, policy, and practice. *J Nutr* 143:241S–249S.

Amy, L. and P. R. Pehrsson. 2003. Development of the American Indian and Alaska Native foods database: Status report. *J Amer Dietet Assoc* 103(Suppl 9):55–56.

AOAC International. 2012. Guidelines for standard method performance requirements. http://www.eoma.aoac.org/app_f.pdf (accessed March 4, 2016).

AOAC International. 2016. *Official Methods of Analysis of AOAC International.* 20th edition. Gaithersburg (MD): AOAC International.

AOAC International Records. 2013. http://www.add.lib.iastate.edu/spcl/manuscripts/MS477.html (accessed April 19, 2016).

Borman, S. 1998. A brief history of mass spectrometry instrumentation. Scripps Center for Metabolomics and Mass Spectrometry. https://masspec.scripps.edu/mshistory/perspectives/sborman.php (accessed February 24, 2016).

Boss, C. C. and K. L. Fredeen. 2004. *Concepts, Instrumentation and Techniques in Inductively Coupled Plasma Optical Emission Spectrometry.* 3rd edition. Shelton CT: Perkin Elmer.

Buzzard, I. M. Microcomputer-based Interactive methods to collect and code dietary data. http://www.nutrient dataconf.org/pastconf/NDBC08/3-1_Buzzard.pdf (accessed February 24, 2016).

Colombani, P. C. 2011. On the origins of food composition tables. *J Food Comp Anal* 24:732–737.

Dennis, B., N. Ernst, M. Hjortland et al. 1980. The NHLBI nutrition data system. *J Am Dietet Assoc* 77:641–647.

Dietary Guidelines for Americans 2015–2020 8th Edition. http://www.cnpp.usda.gov/dietary-guidelines (accessed June 23, 2016).

Dietary Reference Intakes. 2006. http.www.nap.edu/read/11537/chapter/1 (accessed June 23, 2016).

Dumas, J. B. A. 1831. Procedes de l'analyse organique. *Ann Chim Phys* 247:198–213.

Finglas P. M., M. A. Roe, H. M. Pinchen, R. Berry et al. 2014. *McCance and Widdowson's The Composition of Foods*, 7th summary edition. Cambridge MA: Royal Society of Chemistry.

Fisberg M., I. Kovalskys, G. Gómez et al. 2015. Latin American Study of Nutrition and Health (ELANS): Rationale and study design. *BMC Public Health* 16:93. doi:10.1186/s12889-016-2765-y.

Franz M. J., P. Barr, H. Holler et al. 1987. Exchange lists: Revised 1986. *J Am Dietet Assoc* 87:28–34.

Greenfield, S., I. L. Jones, and C. T. Berry. 1964. High-pressure plasmas as spectroscopic emission sources. *Analyst* 89:713–720.

Gry, J., L. Black, F. D. Ericksen et al. 2007. EuroFir-BASIS–A combined composition and biological activity database for bioactive compounds in plant-based foods. *Trends Food Sci and Technol* 18:434–444.

Haytowitz, D. 2015. Updating USDA's key foods list for what we eat in America, NHANES 2011–2012. *Proc Food Sci* 4:71–78.

Haytowitz, D., P. R. Pehrsson, J. M. Holden. 2000. Adapting methods for determining priorities for the analysis of foods in diverse populations. *J Food Comp Anal* 13:425–433.

Haytowitz, D., P. R. Pehrsson, J. M. Holden. 2002. The identification of key foods for food composition research. *J Food Comp Anal* 15:183–194.

Haytowitz, D., P. R. Pehrsson, J. M. Holden. 2008. The National Food and Nutrient Analysis Program: A decade of progress. *J Food Comp Anal* 21:S94–S102.

Haytowitz, D., P. R. Pehrsson, J. Smith, S. E. Gebhardt, R. H. Matthews, B. Anderson. 1996. Key foods: Setting priorities for nutrient analyses. *J Food Comp Anal* 9:331–364.

Hepburn, F. N. 1987. Food consumption/food composition interrelationships. Admin. Report No. 382. Hyattsville (MD): US Department of Agriculture, Human Nutrition Information Service, 68–74.

Holden, J. M., S. A. Bhagwat, K. Y. Patterson. 2002. Development of a multi-nutrient data quality evaluation system. *J Food Comp Anal* 15:339–348.

Ihnat, M. 2001. Twenty five years of reference material activity at agriculture and agri-food Canada. *Fres J Anal Chem* 370:279–285.

International Nutrient Databank Directory. http://www.nutrientdataconf.org/indd/, 2016 (accessed June 30, 2016).

ISO/REMCO. 2016. Committee on reference materials. http://www.iso.org/iso/iso_catalogue/catalogue_tc/catalogue_tc_browse.htm?commid=55002 (accessed April 8, 2016).

Kirchhoff, G. and R. Bunsen. 1860. Chemical analysis by observation spectra. *Annalen der Physik und der Chemie* (Poggendorff) 110:161–189. http://www.chemteam.info/Chem-History/Kirchhoff-Bunson-1860.html (accessed February 18, 2016).

Kjeldahl, J. 1883. Neue methods Bestimmung des Stickstoffs in organishchen Körpern. *Fresenius J Anal Chem* 22:366–382.

Leichsenring, J. M. and E. D. Wilson. 1951. Food composition table for short method of dietary analysis (2nd Revision), *J Am Dietet Assoc* 27:386–389.

McCance, R. A. and E. M. Widdowson. 1940. The Chemical Composition of Foods. London: His Majesty's Stationery office.

McCollum, E. V. 1957. *A History of Nutrition. The Sequence of Ideas in Nutrition Investigations.* Boston MA: Houghton Mifflin Company. pp. 147–150.

Nusser, S. M. and A. L. Carriquiry. 1998. Sampling approaches for nutrient data bases. Contract report prepared for USDA, ARS, Nutrient Data Laboratory. Ames (IA): Iowa State University.

Oseredczuk, M., S. Salvini, M. Roe, and A. Møller. 2009. Guidelines for quality index attributuin to original data from scientific literature or reports for EuroFIR data interchange (revised edition). Technical Report D1.3.21. EuroFIR AISBL. Brussels. http://www.eurofir.org (accessed March 21, 2016).

Pehrsson, P. R., C. Perry, and M. Daniel. 2013. ARS, USDA updates food sampling strategies to keep pace with demographic shifts. *Proc Food Sci* 2:52–29.

Pehrsson, P. R., D. B. Haytowitz, J. M. Holden, C. R. Perry, and D. G. Beckler. 2000. USDA's National Food and Nutrient Analysis Program: Food sampling. *J Food Comp Anal* 13:379–389.

Pehrsson, P. R., J. Johnson, E. D. Nobmann, L. Amy, D. B. Haytowitz, and J. Holden. 2005. Sampling and analysis of Alaska Native subsistence foods. *FASEB J* 19:A1023.

Pehrsson, P. R., C. R. Perry, R. C. Cutrufelli et al. 2006. Sampling and initial findings for a study of fluoride in drinking water in the United States. *J Food Comp Anal* 19:S45–S52.

Pennington, J. A. T. 2009. History of Bowes and Church's food values of Portions commonly used. *Nutr Today* 44:250–259.

Phillips, K. M., K. Y. Patterson, A. S. Rasor et al. 2006. Quality-control materials in the USDA National Food and Nutrient Analysis Program (NFNAP). *Anal Bioanal Chem* 384:1341–1355.

Phillips, K. M., P. R. Pehrsson, W. W. Agnew et al. 2014. Nutrient composition of selected traditional United States Northern Plains Native American plant foods. *J Food Comp Anal* 34:136–152.

Phillips, K. M., W. R. Wolf, K. Y. Paterson, K. E. Sharpless, and J. M. Holden. 2007. Reference materials to evaluate measurement systems for the nutrient composition of foods: Results from USDA's National Food and Nutrient Analysis Program (NFNAP). *Anal Bioanal Chem* 389:219–229.

Rasberry, S. D. 2002. Standard reference materials—The first century. NIST Special Publication 260–150. Washington, DC: US Department of Commerce; 20p. http://www.nist.gov/srm/upload/SP260-150.pdf (accessed April 7, 2016).

Reed, T. B. 1961. Induction-coupled plasma torch. *J. Appl Phys* 32:821–824.

Souci, S. W., W. Fachmann, H. Kraut. 2008. *Food Composition Tables*, 7th edition, Stuttgart, Germany: Med Pharm Scientific Publishers.

Stoeppler, M., W. R. Wolf, and P. J. Jenks, ed. 2008. *Reference Materials for Chemical Analysis: Certification, Availability and Proper Usage.* Weinheim, Germany: Wiley-VCH.

Total Diet Study Analytical Results, Summary of Multi-Year Results, Toxic and Nutritional Elements. http://www.fda.gov/Food/FoodScienceResearch/TotalDietStudy/ucm184293.htm (accessed June 23, 2016).

Watt, B. K. and A. L. Merrill. USDA Handbook 8, 1963. PDF format available at: http://naldc.nal.usda.gov/naldc/download.xhtml?id=CAT10527663&content=PDF

US Department of Agriculture, Agricultural Research Service, Nutrient Data Laboratory. 2015. USDA National Nutrient Database for Standard Reference, Release 28, current version Sep 2015. http://www.ars.usda.gov/nea/bhnrc/ndl (accessed March 3, 2016).

USDA National Nutrient Database for Standard Reference. http://www.ars.usda.gov/services/docs.htm?docid=8964) (accessed June 29, 2016).

USDA Food and Nutrition Database for Dietary Studies. http://www.ars.usda.gov/News/docs.htm?docid=12068 (accessed June 29, 2016).

Westenbrink, S., M. Roe, M. Roe, M. Oseredczuk, I. Castanheira, and P. Finglas. 2016. EuroFIR quality approach for managing food composition data; where are we in 2014? *Food Chem* 193:69–74.

Widdowson, E. M. and R. A. McCance. 1943. Food Tables: Their scope and limitations, Lancet 241:230–232.

Wolf, W. R. 1993. Reference materials. In *Methods of Analysis for Nutritional Labeling,* D. Sullivan and D. Carpenter, ed. Gaithersburg, MD: AOAC International, pp. 115–120.

Index

Note: Page numbers followed by f and t refer to figures and tables, respectively.

E

W

Z